A MANUAL OF PALAEONTOLOGY

With an introduction on the
Principles of Palaeontology

Henry Alleyne Nicholson

2nd Edn.

1995
DAYA PUBLISHING HOUSE
Delhi-110035

First Published	1872
Second Edition	1879
Reprinted	1995

ISBN 81-7035-140-5

Published by
DAYA PUBLISHING HOUSE
1123/74, Deva Ram Park
Tri Nagar, Delhi-110035
(INDIA)
Phone 7220593

Printed at :
Raj Offset Works, Delhi

PREFACE TO THE SECOND EDITION.

The present edition of this work has not only been entirely revised and largely re-written, but it has been so largely augmented by the addition of new matter, that it may be considered as to all intents and purposes a new book. In the former edition, the final section of the work was devoted to Historical or Stratigraphical Palæontology; but this subject has been entirely omitted on the present occasion, as it is most suitably dealt with separately, and it has been treated of in a general manner in the Author's 'Ancient Life-History of the Earth.'

As in the former edition, considerably more space has been allotted to the Invertebrata than to the Vertebrata, for reasons which are obvious, and especially upon the ground that palæontological students are, as a rule, much more largely concerned with the former than the latter. An attempt has also been made to give, as far as possible, brief and general definitions of the more important and widely distributed families, or even *genera*, of the Invertebrata, as well as, to a more limited extent, of the Vertebrata. In carrying out this attempt, however, it is clear that it was necessary to make a rigid selection of material, based upon what might appear to be the relative importance of different types. All conclusions upon this subject must, however, be matters of personal opinion, and it is therefore quite likely that some of the forms which have not been alluded to may be thought to be as important and as deserving of notice as those which are actually selected for mention. So far as this point is concerned, the Author can only say that he has acted to the best of his ability, and that he by no means supposes the selection actually made to be ideally perfect.

It is hoped that most of the more important additions to our knowledge of the great and rapidly-growing science of Palæontology will be found to be incorporated, in however condensed a form, in the present edition. In this connection, however, the Author would wish to mention that the greater part of this edition was written in the early part of the year 1878, and that the whole of it was in the hands of the printers before the commencement of the present year. Owing to this fact, though considerable and unexpected delay has occurred in the actual publication of the work, it has arisen that the Author has been unable to avail himself fully, or at all, of some important palæontological works and memoirs which were published towards the close of 1878, and in the early part of the present year. Among these may be more especially mentioned the Second Part of the 'Handbuch der Palæontologie,' by Professors Zittel and Schimper; the memoirs of Munier-Chalmas and Toula upon the Dactyloporidæ; the valuable memoirs by Mr Henry B. Brady upon the Reticularian Rhizopoda of the Challenger and Arctic Expeditions; the concluding portions of the 'Beiträge zur Systematik der fossilen Spongien,' by Professor Zittel; the treatise of Professor Möbius upon the structure of Eozoön Canadense; the important Monograph by Angelin upon the Cystideans; the researches of Mr P. Herbert Carpenter upon the Oral and Apical Systems of the Echinoderms; the Croonian Lecture by Mr Moseley upon the Structure of the Stylasteridæ; and the valuable papers upon the Dwarf Crocodiles of the Jurassic, by Professor Owen, and upon the Oolitic Mammals and the Deinosaurian and Sauranodont Reptiles of North America, by Professor Marsh. The Author may add that he would have considerably modified the section dealing with the so-called "Tabulate Corals," had his investigations into the structure and relationships of these fossils been completed in time.

The number of illustrations in the present edition has been largely increased. Most of the new engravings have been drawn by the Author, and have been transferred to the wood by Mr Charles Berjeau. Lastly, the Author has to express his thanks to his friends, Mr Henry B. Brady, F.R.S., Dr Ramsay H. Traquair, F.R.S.E., and Mr R. Etheridge, jun., F.G.S., for much valuable and friendly assistance.

UNITED COLLEGE, ST ANDREWS,
September 10, 1879.

PREFACE TO THE FIRST EDITION.

THE object of the present work is to furnish the student of Geology and the general reader with a compendious account of the leading principles and facts of the vast and ever-increasing science of Palæontology. In carrying out this object, all superfluous details have been rigidly excluded, and the Author has endeavoured to restrict himself entirely to those facts which are absolutely necessary to any one who would study Palæontology as a department of science, sufficiently distinct to stand alone, and yet most closely connected with the sciences of Zoology and Botany on the one hand, and with Geology on the other hand.

In the First Part of the work is given a general account of the principles upon which the palæontological observer proceeds.

In the Second Part of the work, Palæozoology, or the past history of the Animal Kingdom, is treated of; and here much more space has been devoted to the Invertebrate than to the Vertebrate groups—upon the ground that it is chiefly, or almost exclusively, with the former that the ordinary palæontological student has to deal.

The Third Part of the work gives a brief and very general view of Palæobotany, or the past history of the Vegetable Kingdom. This department of the subject has not been treated at any length, partly because the remains of plants are comparatively rare in the stratified series, and partly because nothing less than a special treatise would suffice to handle satisfactorily this obscure and difficult branch of the subject.

The fourth and concluding portion of the work treats of Historical, or, as it might be called, Stratigraphical, Palæontology, — namely, of the application of Palæontology to the elucidation of the succession of the stratified deposits of the earth's crust. This department of the subject has also been very briefly disposed of, not because its intrinsic importance does not warrant a more extended treatment, but because it is the Author's intention, as his leisure will permit, to devote a separate treatise to the consideration of this wide and comparatively independent section of the science.

In conclusion, the Author would beg his readers to remember that there is no science which is growing so rapidly, and which is as yet so comparatively in its infancy, as Palæontology; and that there is none in which the conclusions of to-day are more liable to be vitiated by the discoveries of the morrow. Even whilst these sheets have been going through the press, facts have been brought to light which ought to have found their place in a Manual of this kind, but which have been of necessity altogether passed over, or, at best, have been merely alluded to. For all deficiencies, therefore, arising from this cause, the Author has to beg the kind indulgence of his readers.

With regard to the Illustrations, the Author has gratefully to acknowledge the kindness of Alfred R. C. Selwyn, Esq., Director of the Geological Survey of Canada, who placed at the Author's disposal a number of engravings of Silurian and Devonian fossils, from the publications of the Survey. The Author has likewise to acknowledge a similar obligation to Principal Dawson, of M'Gill University, Montreal, who kindly permitted the use of several of the illustrations of his 'Acadian Geology.' A considerable proportion of the engravings, however, are taken from D'Orbigny's beautifully illustrated 'Cours Elémentaire de Paléontologie,' by an arrangement with the publishers of that work.

University College, Toronto,
October 16, 1872.

CONTENTS

PART I.—GENERAL INTRODUCTION.

CHAPTER I.

CHAPTER II.

CHAPTER III.

CHAPTER IV.

CHAPTER V.

CHAPTER VI.

PART II.—PALÆOZOOLOGY.

CHAPTER VII.

CHAPTER VIII.

CHAPTER IX.

CHAPTER X.

CHAPTER XI.

CHAPTER XII.

CHAPTER XIII.

CHAPTER XIX.

CHAPTER XX.

CHAPTER XXI.

CHAPTER XXII.

CHAPTER XXIII.

PART I.

GENERAL INTRODUCTION

PALÆONTOLOGY.

CHAPTER I.

INTRODUCTION.

Definition of Palæontology.

Palæontology (Gr. *palaios*, ancient; *onta*, beings; *logos*, discourse) is the science which treats of the living beings, whether animal or vegetable, which have inhabited this globe at past periods in its history. It is the ancient life-history of the earth, and if its record could ever be completed, it would furnish us with an account of the structure, habits, and distribution of all the animals and plants which have at any time flourished upon the land-surfaces of the globe or inhabited its waters. From causes, however, which will be subsequently discussed, the palæontological record is most imperfect, and our knowledge is interrupted by gaps which not only bear a large proportion to our solid information, but which in many cases are of such a nature that we can never hope to have them filled.

As Zoology, then, treats of the animals now inhabiting the earth, and as Botany treats of the now existing plants, Palæontology may be considered as the Zoology and Botany of the past. Regarding it from this, the only true point of view, some knowledge of Zoology and Botany is essential to a prosecution of the study of Palæontology, and such details of these sciences as may be deemed requisite will be introduced in the proper place. The materials, again, which fall to be studied by the palæontologist, are drawn entirely from the proper province of the geologist. *Fossils* are derived from *rocks.* It will therefore be necessary to trespass to some extent upon the peculiar domain of the geologist, and to obtain some knowledge of the origin, composition, and mode of occurrence of the rocks from which Palæontology obtains its materials. Lastly, Palæontology, apart from its own importance as an independent science, is employed by the geologist to assist him in his determination of the chronological succession of the materials which compose the crust of the earth. Palæontology, therefore, in one of its aspects, is a branch of geological science, and requires separate study in its relation to historical Geology.

Definition of Fossils.

All the natural objects which come to be studied by the palæontologist are termed "fossils" (Lat. *fossus,* dug up). In most cases, fossils, or, as they are often termed, "petrifactions," are actual portions of animal or vegetable organisms, such as the shells of Molluscs, the skeletons of Corals, the bones of Vertebrate animals, the wood, bark, or leaves of plants, &c.; and these may be preserved very much in their original condition, or may have been very much altered by changes subsequent to their burial. Strictly speaking, however, by the term "fossil" is understood "any body, *or the traces of the existence of any body,* whether animal or vegetable, which has been buried in the earth by natural causes" (Lyell). We shall find, therefore, that we must include under the head of fossils objects which at no time themselves formed parts of any animal or vegetable, but which, nevertheless, point to the former existence of such organisms, and enable us to reason as to their nature. Under this head come such fossils as the moulds or "casts" of shells and the footprints left by various animals upon sand or mud.

In the great majority of cases fossils are the remains of

animals or plants which are now *extinct*—that is to say, which no longer are in existence, but have entirely disappeared from the earth's surface. In some cases, however, fossils are the remains of *recent* animals—that is, of animals which are still found in a living condition upon the globe. The term "sub-fossil," sometimes applied to these, has been more appropriately applied in another sense, and is best discarded in this connection. In any case, the fact that a given specimen belongs to an extinct species of animal or plant, or that it is referable to some existing form, does not enter in any way whatever into the determination of the question as to whether or not it is truly a *fossil*. If such a specimen is found in those portions of the earth's crust which we can show by other evidence to have been formed prior to the establishment of the existing terrestrial order, then it is a fossil; while any remains, even though belonging to the same animal, which are found in deposits which have been formed during the historical period, would properly fall to be studied by the zoologist or the botanist, and would not rightly be termed "fossils." It must be admitted, however, that in approaching the "Recent" period of the earth's history, it becomes a matter of difficulty—indeed, a matter of impossibility—to draw any precise line between fossil and recent specimens.

The terms "fauna" and "flora" are employed in Palæontology much as they are by the naturalist, to mean the entire assemblage of the animals or of the plants respectively belonging to a particular region or a particular time. Thus we may speak of the "fauna" of the Carboniferous Period, or the "flora" of the Tertiary Epoch, or the fauna of the Chalk, or of any other set of beds.

FOSSILISATION.

The term "fossilisation" may be applied in a general sense to all the processes through which an organic body passes in order to become a fossil. Here we need only consider the three leading forms in which fossils present themselves. In the first instance, the fossil is to all intents and purposes an actual organic remain, being itself a fragment of an animal or plant. Thus we may meet with fossil bones, shells, or wood, which may have undergone certain changes, such as would be produced by pressure, by the deprivation of organic matter originally present, or by more or less complete infiltration with mineral matter, but which, nevertheless, are practically the real bodies they represent. As a matter of course, it is in the more modern formations that we find fossils least changed from their primitive condition, but all formations almost contain some fossils in which the original structure is more or less completely retained.

In the second place, we very frequently meet with fossils in the state of "casts" or moulds of the original organic body. What occurs in this case will be readily understood, if we imagine any common bivalve shell, as an Oyster, or Mussel, or Cockle, embedded in clay or mud. If the clay were sufficiently soft and fluid, the first thing would be that it would gain access to the interior of the shell and would completely fill up the space between the valves. The pressure, also, of the surrounding matter would insure that the clay would everywhere adhere closely to the exterior of the shell. If now we suppose the clay to be in any way hardened so as to be converted into stone, and if we were to break up the stone, we should obviously have the following state of parts. The clay which filled the shell would form an accurate cast of the *interior* of the shell, and the clay outside would give us an exact impression or cast of the *exterior* of the shell (fig. 1). We should have, then, two casts, an interior and an exterior, and the two would be very different to one another, since the inside of a shell is very unlike the outside. In the case, in fact, of many univalve shells, the interior cast is so unlike the exterior or unlike the shell itself, that it may be difficult to determine the true origin of the former.

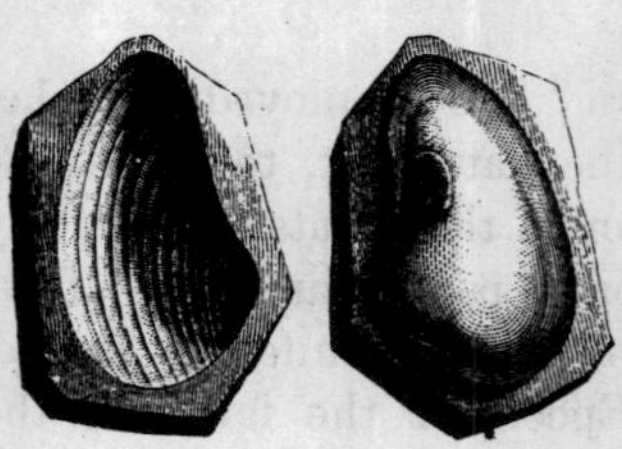

Fig. 1.—*Trigonia longa*, showing casts of the exterior and interior of the shell. Neocomian.

It only remains to add that there is sometimes a further complication. If the rock be very porous and permeable by water, it may happen that the original shell is entirely dissolved away, leaving the interior cast loose, like the kernel of a nut, within the case formed by the exterior cast. Or it may happen that subsequent to the attainment of this state of things, the space thus left vacant between the interior and exterior cast—the space, that is, formerly occupied by the shell itself—may be filled up by some foreign mineral deposited there by the infiltration of water. In this last case the splitting open of the rock would reveal an interior cast, an exterior cast, and finally a body which would have the exact form of the original shell, but which would be really a much later formation, and which would not exhibit under the microscope the minute structure of shell.

In the third class of cases we have fossils which present with the greatest accuracy the external form, and even sometimes the internal minute structure, of the original organic body, but which, nevertheless, are not themselves truly organic, but have been formed by a "replacement" of the particles of the primitive organism by some mineral substance. The most elegant example of this is afforded by fossil wood which has been "silicified" or converted into flint. In this case we have a piece of fossil wood, which presents the rings of growth and fibrous structure of wood, and which under the microscope exhibits even the minutest vessels which characterise ligneous tissue. The whole, however, instead of being composed of the original carbonaceous matter of the wood, is now converted into pure flint. The only explanation which can be given of this by no means very rare phenomenon, is that the wood must have undergone a slow process of decay in water holding silica or flint in solution. As each particle of the wood was removed by decay, its place was taken by a particle of flint deposited from the surrounding water, till ultimately the entire wood was silicified. The replacing substance is by no means necessarily flint, but may be iron-pyrites, oxide of iron, sulphur, malachite, magnesite, talc, &c.; and it is not uncommon to find many other fossils besides wood preserved in this way, such as shells, corals, or sponges.

The *replacement* of the original substance of a fossil by some foreign body is thus a matter of common occurrence, but it is by no means always easy to determine whether or not such replacement has taken place. By far the commonest mode of replacement is that whereby an originally calcareous skeleton is replaced by silica. This process of "silicification"—of the replacement of *lime* by *silica*—is not only an extremely common one, but it is also a readily intelligible one; since carbonate of lime is an easily and flint a hardly soluble substance. It is thus easy to understand that originally calcareous fossils, such as the shells of Mollusca, or the skeletons of Corals, should have in many cases suffered this change, their carbonate of lime being dissolved away, particle by particle, and replaced by precipitated silica, as they were subjected to percolation by heated or alkaline waters holding silica in solution.

When we meet with fossils, such as those alluded to above, which we *know* to have been originally calcareous, but which we now find, unchanged in form, but converted into flint, then we cannot doubt that we have to deal with cases of "silicification," and that the primitive skeleton of lime has in these cases been slowly, and more or less perfectly, replaced by silica. We cannot, however, speak in such a positive manner as to fossils which we now find to be composed of flint, but as to the original constitution of which we cannot be certain. We find, namely, some fossils which are of uncertain affinities, and which are sometimes found in a siliceous and sometimes in a calcareous state. If we are not positive as to the zoological position of these fossils, or if they belong to a group of animals in which we find the living forms to possess sometimes a calcareous and sometimes a siliceous skeleton, then it is obviously a matter of extreme difficulty to determine whether the extinct forms were really composed of lime or of flint. In such cases, we must be guided principally by the condition of preservation of the fossils which occur associated with such obscure forms in the same beds; the fact that the associated remains are converted into flint pointing to the probability that the problematical forms were originally calcareous, and *vice versa*. In the case, also, of all fossils which present themselves sometimes in a siliceous and sometimes in a calcareous form, there is always the *presumption* that the skeleton was originally composed of *lime*, this presumption being based upon the fact that the conversion of the calcareous skeletons of animals into silica by a process of replacement is an unquestionable, an extremely common, and a readily intelligible occurrence.

Until recently, indeed, naturalists never allowed themselves to contemplate the alternative possibility of an originally *siliceous* skeleton being replaced by *lime;* but we have now high authorities in favour of the view that this anomalous mode of substitution really does occur occasionally. The only case in which anything like definite proof of this abnormal mode of replacement can be said to have been brought forward is that of some of the fossil sponges found in the chalk. Certain of these sponges agree in their general form and in their microscopic structure with the living Hexactinellid Sponges (such as the Venus's Flower-basket), the skeleton of which is composed of flint, and it is therefore difficult to doubt the possession by the fossil types of a primitively siliceous skeleton. We *might*, indeed, suppose that the fossil sponges in question represented an entirely new type of sponge-structure with the *form* of the living Hexactinellids, but with a *calcareous* skeleton, and that this skeleton had in the process of fossilisation been often replaced by flint; but this supposition would be a violent one, and would not be warranted by the known facts. It has, however, been pointed out by Zittel that these sponges are sometimes found in the rocks with their skeleton in part calcareous and in part siliceous; and the same distinguished observer has further shown that in such specimens the siliceous portions of the skeleton retain their minute microscopic character, whereas the calcareous portions have entirely lost their minute structure, and are composed simply of *crystalline* calcite. Had the skeleton been originally calcareous, the lime composing it would have been in a granular and not a crystalline form, and it is therefore very difficult to account for the state of preservation of these specimens, unless we admit that the skeleton was primitively siliceous, and that we have here a case of the substitution of the hardly soluble silica by the easily soluble carbonate of lime.

In any case we must carefully distinguish between *replacement*, whether by flint or any other mineral, and *infiltration*, the latter being merely the process whereby the cavities and natural vacuities of a fossil are liable to become filled by some mineral substance, subsequent to its entombment in sediment. When such a fossil as a shell or a coral, for example, becomes buried in the sandy, calcareous, or argillaceous mud at the bottom of the sea, the surrounding sediment often does not penetrate into the deeper parts of the fossil, and there are thus left in its interior certain empty spaces, into which the surrounding water makes its way by percolation. Any mineral substances, such as carbonate of lime or silica, which may be contained in solution in the water, are then liable to undergo precipitation, and to be deposited in a solid form within the fossil. All the natural cavities of a fossil, even down to the minutest microscopic pores or tubes, may in this way become filled with some such infiltrated material, the two commonest agents in this process being lime and flint. If the skeleton of the fossil be a calcareous one, while the infiltrating material has been some less soluble substance, such as silica or some silicate, then the skeleton may be artificially or naturally dissolved away, leaving a *cast* of the internal cavities of the fossil formed of the infiltrated matter. Thus the minute shells of *Foraminifera* are often infiltrated with the silicate glauconite, and exquisitely perfect casts of their interior cavities are subsequently formed by dissolution of the shell itself. In this way, as we shall see hereafter, deposits of green sand have been sometimes produced.

Definition of Rock.

The crust of the earth consists of various different materials, produced at different successive periods, occupying certain definite spaces, and not confusedly mixed together, but, on the contrary, exhibiting a definite and discoverable order of arrangement. All these materials, however different in appearance, texture, or mineral composition, are called "rocks" by the geologist. The term "rock," then, is to be understood as applying to *all* the materials which compose the crust of the earth. In the language of geology, the finest mud, the loosest sand, and the most incoherent gravel, are just as much *rocks* as are the hardest and most compact granites or limestones.

Classification of Rocks.

For the purposes of the palæontologist all the rocks which enter into the composition of the solid exterior of the earth may be divided into two great classes: 1. The Igneous Rocks, which are formed within the body of the earth itself, and which owe their structure and origin to the action of heat; and 2. the Aqueous or Sedimentary Rocks, which are formed at the surface of the earth, and which owe their structure to the mechanical action of water. The Igneous Rocks are principally formed below the surface of the earth, are as a general rule destitute of organic remains or fossils, and are mostly in the form of *unstratified* masses. The Aqueous and Sedimentary Rocks are formed at the surface by the disintegration and reconstruction of previously existing rocks, or by the vital chemistry of animals or plants, are mostly fossiliferous, and are *stratified*—*i.e.*, are arranged in distinct layers or "strata." The Sedimentary Rocks, as containing fossils, are the only rocks which it is essential for the palæontologist to be acquainted with, and we shall very briefly consider their leading physical characters, their chief varieties, their mode of origin, and their historical succession.

CHAPTER II.

THE FOSSILIFEROUS ROCKS.

THE Sedimentary or Fossiliferous Rocks form the greater portion of that part of the earth's crust which is open to our examination, and are distinguished by the fact that they are regularly "stratified," or arranged in distinct and definite layers or "strata." These layers may consist of a single material, as in a block of sandstone, or they may consist of different materials. When examined on a large scale, they are always found to consist of alternations of layers of different mineral composition. We may examine any given area, and find in it nothing but one kind of rock —sandstone, perhaps, or limestone. In all cases, however, if we extend our examination sufficiently far, we shall ultimately come upon different rocks; and, as a general rule, the thickness of any particular set of beds is comparatively small, so that different kinds of rock alternate with one another in comparatively small spaces.

As regards the origin of the Sedimentary Rocks, they are for the most part "derivative" rocks, being derived from the wear and tear of pre-existent rock. Sometimes, however, they owe their origin to chemical or vital action, when they would more properly be spoken of simply as Aqueous Rocks. As to their mode of deposition, we are enabled to infer that the materials which compose them have formerly been spread out by the action of water, from what we see going on every day at the mouths of our great rivers, and on a smaller scale wherever there is running water. Every stream, where it runs into a lake or into the sea, carries with it a burden of mud, sand, and rounded pebbles, derived from the waste of the rocks which form its bed and banks. When these materials cease to be impelled by the force of the moving water they sink to the bottom, the heaviest pebbles, of course, sinking first, the smaller pebbles and sand next, and the finest mud last. Ultimately, therefore, as might have been inferred upon theoretical grounds, and as is proved by practical experience, every lake becomes a receptacle for a series of stratified rocks produced by the streams flowing into it. These deposits may vary in different parts of the lake, according as one stream brought down one kind of material and another stream contributed another material; but in all cases the materials will bear ample evidence that they were produced, sorted, and deposited by running water. The finer beds of clay or sand will all be arranged in thicker or thinner layers or laminæ; and if there are any beds of pebbles these will all be rounded or smooth, just like the water-worn pebbles of any brook-course. In all probability, also, we should find in some of the beds the remains of fresh-water shells or plants or other organisms which inhabited the lake at the time these beds were being deposited.

In the same way large rivers—such as the Ganges or Mississippi—deposit much of the material which they bring down at their mouths, forming in this way their "deltas." Whenever such a delta is cut through, either by man or by some channel of the river altering its course, we find that it is composed of a succession of horizontal layers or strata of sand or mud, varying in mineral composition, in structure, or in grain, according to the nature of the materials brought down by the river at different periods. Such deltas, also, will contain the remains of animals which inhabit the river, with fragments of the plants which grew on its banks, or bones of the animals which lived in its basin.

Lastly, the sea itself — irrespective of the materials delivered into it by rivers — is constantly preparing fresh stratified deposits by its own action. Upon every coast-line the sea is constantly eating back into the land and reducing its component rocks to form the shingle and sand which we see upon every shore. The materials thus produced are not, however, lost, but are ultimately deposited elsewhere in the form of new stratified accumulations, in which are buried the remains of animals inhabiting the sea at the time.

Whenever, then, we find anywhere in the interior of the land any series of beds having these characters—composed, that is, of distinct layers, the particles of which, both large and small, show distinct traces of the wearing action of water—whenever and wherever we find such rocks, we are justified in assuming that they have been deposited by water in the manner above mentioned. Either they were laid down in some former lake by the combined action of the streams which flowed into it; or they were deposited at the mouth of some ancient river, forming its delta; or they were laid down at the bottom of the ocean. In the first two cases, any fossils which the beds might contain would be the remains of fresh-water or terrestrial organisms. In the last case, the majority, at any rate, of the fossils would be the remains of marine animals.

The term "formation" is employed by geologists to express "any group of rocks which have some character in common, whether of origin, age, or composition" (Lyell); so that we may speak of stratified and unstratified formations, aqueous or igneous formations, fresh-water or marine formations, and so on.

Chief Divisions of the Aqueous Rocks.

The Aqueous Rocks may be divided into two great sections, the Mechanically-formed and the Chemically-formed, including under the last head all rocks which owe their origin to vital action, as well as those produced by ordinary chemical agencies.

A. Mechanically-formed Rocks.—These are all those Aqueous Rocks of which we can obtain proofs that their particles have been mechanically transported to their present site. Thus, if we examine a piece of conglomerate or pudding-stone, we find it to be composed of a number of rounded pebbles embedded in an enveloping paste or matrix. The pebbles are worn and rounded, and thus show that they have been subjected to much mechanical attrition, whilst they have been mechanically transported for a greater or less distance from the rock of which they originally formed part. In the case of an ordinary sandstone, the component grains of sand are equally the result of mechanical attrition, and have been equally transported from a distance. In the case of still finer rocks, such as shale, the particles have been so much water-worn that their source cannot be recognised, though a microscopical examination would reveal that their edges were all worn and rounded. It follows from this that the mechanically-formed Aqueous Rocks are such as can be proved to have been *derived* from the abrasion of other pre-existent rock: hence they are often spoken of as "Derivative Rocks." Every bed, therefore, of any mechanically-formed rock, is an exact equivalent of a corresponding amount of destruction of some older rock.

The mechanically-formed Rocks may be divided into the two groups of the Arenaceous or Siliceous Rocks, and the Argillaceous or Aluminous Rocks. In the Arenaceous group are those Aqueous Rocks which are mainly composed of smaller or larger grains of flint or silica. The chief varieties are the various kinds of sand and sandstone, grits, and most conglomerates and breccias. In the Argillaceous group are those Aqueous Rocks which contain a certain amount of clay or hydrated silicate of alumina. Under this head come clays, shales, marls, clay-slate, and most flags or flag-stones.

B. Chemically-formed Rocks.—In this section are comprised all those Aqueous Rocks which have been formed by chemical agencies. As many of these chemical agencies, however, are exerted through the medium of living beings, whether animals or plants, we get into this section a number of what may be called "organically-formed" rocks. The most important of the Chemically-formed Rocks are the so-called Calcareous Rocks, comprising all those which contain a large proportion of carbonate of lime, or are wholly made up of this substance; but there are other rocks, of different composition, which are formed by chemical or organic agency, and which may be briefly noticed.

As to the origin of the so-called *Calcareous Rocks* (Lat. *calx,* lime), carbonate of lime is soluble in water holding a certain amount of carbonic acid gas in solution and it is therefore found in larger or smaller quantity dissolved in all natural waters, both fresh and salt, since these waters are always to some extent charged with the above-mentioned solvent gas. A great number of aquatic animals, however, together with some aquatic plants, are endowed with the power of separating the lime thus held in solution in the water, and of reducing it again to its solid condition. In this way shell-fish, crustaceans, sea-urchins, corals, and an immense number of other animals, are enabled to construct their skeletons; whilst some plants form hard structures within their tissues in a precisely similar manner. We do meet with some calcareous deposits, such as the "stalactites" and "stalagmites" of caves, the "calcareous tufa" and "travertine" of some hot springs, and the spongy calcareous deposits of so-called "petrifying springs," which are purely chemical in their origin, and owe nothing to the operation of living beings. Such deposits are formed simply by the precipitation of carbonate of lime from water, in consequence of the evaporation from the water of the carbonic acid gas which formerly held the lime in solution; but, though sometimes forming masses of considerable thickness, and of geological importance, they do not concern us here. Almost all the limestones which occur in the series of the stratified rocks are, primarily at any rate, of *organic* origin, and have been, directly or indirectly, produced by the action of certain lime-making animals or plants, or both combined. The presumption as to all the calcareous rocks, which cannot be clearly shown to have been otherwise produced, is that they are thus organically formed; and in many cases this presumption can be readily reduced to a certainty. There are many varieties of the calcareous rocks, but the following are those which are of the greatest importance:—

Chalk is a calcareous rock of a generally soft and pulverulent texture, and with an earthy fracture. It varies in its purity, being sometimes almost wholly composed of carbonate of lime, and at other times more or less intermixed with foreign matter. Though usually soft and readily reducible to powder, chalk is occasionally, as in the north of Ireland, tolerably hard and compact; but it never assumes the crystalline aspect and stony density of limestone, except it be in immediate contact with some mass of igneous rock. By means of the microscope, the true nature and mode of formation of chalk can be determined with the greatest ease. In the case of the harder varieties, the examination can be conducted by means of slices ground down to a thinness sufficient to render them transparent; but in the softer kinds the rock must be disintegrated under water, and the *débris* examined microscopically. When investigated by either of these methods, chalk is found to be a genuine organic rock, being composed of the shells or hard parts of innumerable marine animals of different kinds, some entire, some fragmentary, cemented together by a matrix of very finely granular carbonate of lime. Foremost amongst the animal remains which so largely compose chalk are the shells of the minute creatures which will be subsequently spoken of under the name of *Foraminifera* (fig. 2), and which, in spite of their microscopic dimensions, play a more important part in the process of lime-making than perhaps any other of the larger inhabitants of the ocean.

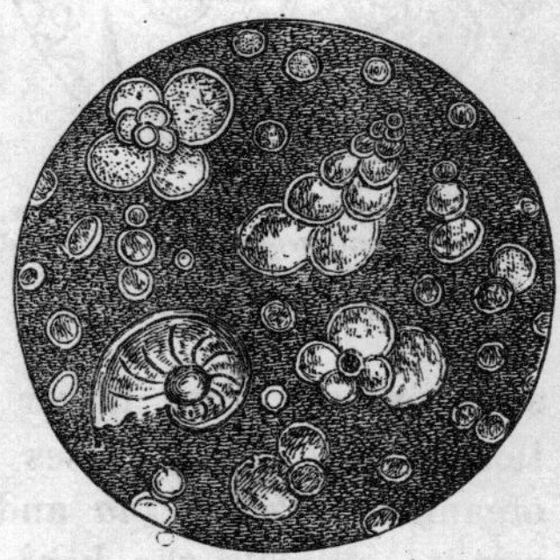

Fig. 2.—Section of Gravesend Chalk, examined by transmitted light and highly magnified. Besides the entire shells of *Globigerina*, *Rotalia*, and *Textularia*, numerous detached chambers of *Globigerina* are seen. (Original.)

As chalk is found in beds of hundreds of feet in thickness, and of great purity, there was long felt much difficulty in satisfactorily accounting for its mode of formation and origin. By the researches of Carpenter, Wyville Thomson, Huxley, Wallich, and others, it has, however, been shown that there is now forming, in the profound depths of our great oceans, a deposit which is in all essential respects identical with chalk, and which is generally known as the "Atlantic ooze," from its having been first discovered in that sea. This ooze is found at great depths (5000 to over 15,000 feet) in both the Atlantic and Pacific, covering enormously large areas of the sea-bottom, and it presents itself as a whitish-brown, sticky, impalpable mud, very like greyish chalk when dried. Chemical examination shows that the ooze is composed almost wholly of carbonate of lime, and microscopical examination proves it to be of organic origin, and to be made up of the remains of living beings. The principal forms of these belong to the *Foraminifera*, and the commonest of these are the irregularly-chambered shells of *Globigerina*, absolutely undistinguishable from the *Globigerinæ* which are so largely present in the chalk (fig. 3). Along with these occur fragments of the skeletons of other larger creatures, and a certain proportion of the flinty cases of minute animal and vegetable organisms (*Polycystina* and *Diatoms*). Though many of the minute animals, the hard parts of which form the ooze, undoubtedly live at or near the surface of the sea, others, probably, really live near the bottom; and the ooze itself forms a congenial home for numerous sponges, sea-lilies, and other marine animals which flourish at great depths in the sea. There is thus established an intimate and most interesting parallelism between the chalk and the ooze of modern oceans. Both are formed essentially in the same way, and the latter only requires consolidation to become actually converted into chalk. Both are fundamentally organic deposits, apparently requiring a considerable depth of water for their accumulation, and mainly composed of the remains of *Foraminifera*, together with the entire or broken skeletons of other marine animals of greater dimensions. It is to be remembered, however, that the ooze, though strictly representative of the chalk, cannot be said in any proper sense to be actually *identical* with the formation so called by geologists. A great lapse of time separates the two, and though composed of the remains of

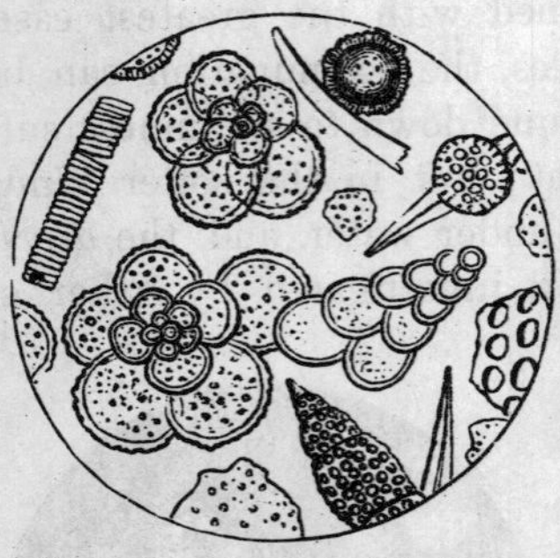

Fig. 3.—Organisms in the Atlantic Ooze, chiefly *Foraminifera* (*Globigerina* and *Textularia*), with *Polycystina* and sponge-spicules; highly magnified. (Original.)

representative classes or groups of animals, it is only in the case of the lowly-organised *Globigerinæ*, and of some other organisms of little higher grade, that we find absolutely the same kinds or *species* of animals in both.

Limestone, like chalk, is composed of carbonate of lime, sometimes almost pure, but more commonly with a greater or less intermixture of some foreign material, such as alumina or silica. The varieties of limestone are almost innumerable, but the great majority can be clearly proved to agree with chalk in being essentially of organic origin, and in being more or less largely composed of the remains of living beings. In many instances the organic remains which compose limestone are so large as to be readily visible to the naked eye, and the rock is at once seen to be nothing more than an agglomeration of the skeletons, generally fragmentary, of certain marine animals, cemented together by a matrix of carbonate of lime. This is the case, for example, with the so-called "Crinoidal Limestones," and "Encrinital Marbles" with which the geologist is so familiar, especially as occurring in great beds amongst the older formations of the earth's crust. These are seen, on weathered or broken surfaces, or still better in polished slabs, to be composed more or less exclusively of the broken stems and detached plates of sea-lilies (*Crinoids*). Similarly, other limestones are composed almost entirely of the skeletons of corals; and such old coralline limestones can readily be paralleled by formations which we can find in actual course of production at the present day. We only need to transport ourselves to the islands of the Pacific, to the West Indies, or to the Indian Ocean, to find great masses of lime formed similarly by living corals, and well known to every one under the name of "coral reefs." Such reefs are often of vast extent, both superficially and in vertical thickness, and they fully equal in this respect any of the coralline limestones of bygone ages. Again, we find other limestones—such as the celebrated "Nummulitic Limestone," which sometimes attains a thickness of some thousands of feet—to be almost entirely made up of the shells of *Foraminifera*. In the case of the "Nummulitic Limestone" just mentioned, these shells are of large size, varying from the size of a split pea up to that of a florin. Very many limestones, however, are made up of the calcareous cases of much smaller forms of *Foraminifera,* which are so minute as hardly to be visible to the naked eye. In other cases, again, we find limestones to be composed so largely of the shells of various of the true *Mollusca,* that we may regard them as essentially made up of the skeletons of this class of animals.

At the present day, then, limestone is in process of formation by the agency of various animals, amongst which the Corals, the *Foraminifera,* and the *Mollusca* are the most important. The same animals have also been the principal agents in building up the great masses of limestone which we now discover in the crust of the earth; but in the case of the older calcareous rocks we must add to the above the *Crinoids,* as having formerly contributed on an immense scale to the formation of limestone. Nor are we only to ascribe an organic origin to such limestones as are composed of fossils large enough to be visible to the unassisted eye. On the contrary, most other limestones which at first sight appear compact, more or less crystalline, and nearly devoid of traces of life, are found, when properly examined, to be also composed of the remains of various organisms. All the commoner limestones, in fact, from the Lower Silurian period onwards, can be easily proved to be thus *organic* rocks, if we investigate weathered or polished surfaces with a lens, or, still better, if we cut thin slices of the rock and grind these down till they are transparent. When thus examined, the rock is usually found to be composed of innumerable entire or fragmentary fossils, cemented together by a granular or crystalline matrix of carbonate of lime (figs. 4 and 5). When the matrix is granular, the rock is precisely similar to chalk, except that it is harder and less earthy in texture, whilst the fossils are only occasionally referable to the *Foraminifera.* In other cases, the matrix is more or less crystalline, and when this crystallisation has been carried to a great extent, the original organic nature of the rock may be greatly or completely obscured thereby. Thus, in limestones which have been greatly altered or metamorphosed" by the combined action of heat and pressure, all traces of organic remains become annihilated, and the rock becomes completely crystalline throughout. This,

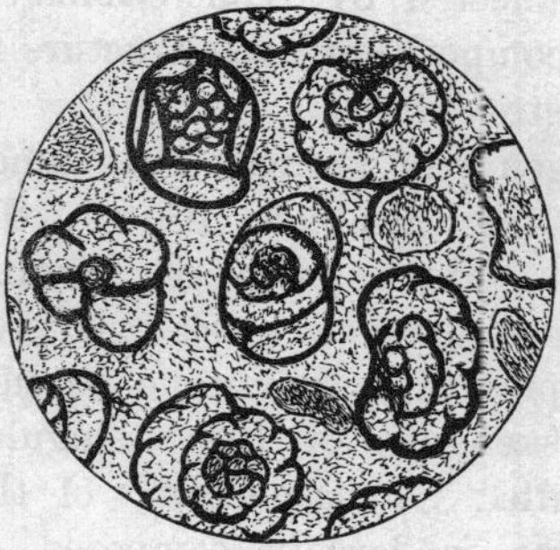

Fig. 4.—Section of Carboniferous Limestone from Spergen Hill, Indiana, U.S., showing numerous large-sized *Foraminifera* (*Endothyra*) and a few oolitic grains magnified. (Original.)

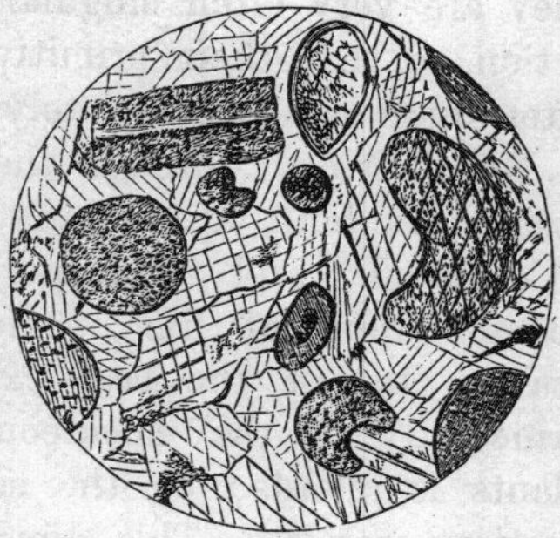

Fig. 5. — Section of Coniston Limestone (Lower Silurian) from Keisley, Westmorland; magnified. The matrix is very coarsely crystalline, and the included organic remains are chiefly stems of Crinoids. (Original.)

for example, is the case with the ordinary white "statuary marble," slices of which exhibit under the microscope nothing but an aggregate of beautifully transparent crystals of carbonate of lime, without the smallest traces of fossils. There are also other cases, where the limestone is not necessarily highly crystalline, and where no metamorphic action in the strict sense has taken place, in which, nevertheless, the microscope fails to reveal any evidence that the rock is organic. Such cases are somewhat obscure, and doubtless depend on different causes in different instances; but they do not affect the important generalisation that limestones are fundamentally the product of the operation of living beings. This fact remains certain; and when we consider the vast superficial extent occupied by calcareous deposits, and the enormous collective thickness of these, the mind cannot fail to be impressed with the immensity of the period demanded for the formation of these by the agency of such humble and often microscopic creatures as Corals, Crinoids, Foraminifers, and Mollusca.

As is the case with the ordinary limestones and marbles, so also the various kinds of *magnesian limestone* and *dolomite* are essentially organic in their origin, and are largely made up of the remains of marine animals. Magnesian limestones are, however, very often more or less highly *crystalline*, and they are very often singularly affected by "concretionary" action, so that their primitive composition and structure is often more or less completely destroyed.

Nor is it only through the agency of animals that limestones are built up. Many of the calcareous Algæ — the "Corallines" and "Nullipores" — are capable of forming accumulations of lime, sometimes upon a most extensive scale. One of the best examples of a limestone formed principally of the calcareous skeletons of these singular plants is afforded by the so-called "Leitha-Kalk" of the Tertiary series. This limestone is largely composed of nodulated masses, which exhibit no definite structure to the eye, and which were originally set down as "concretions" (fig. 6). Microscopic examination of these apparently inorganic masses shows, however, that we have to deal here with the

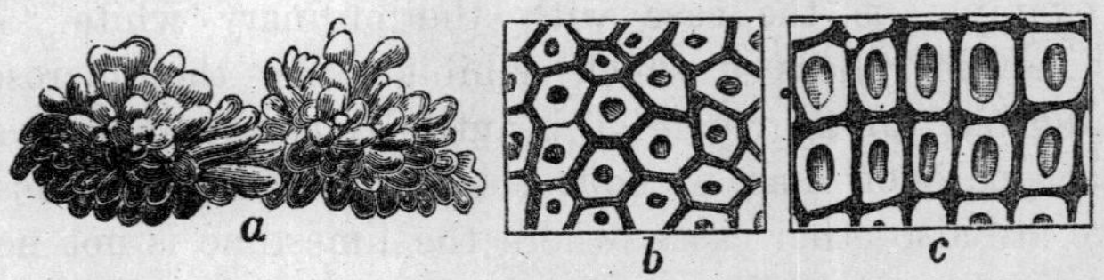

Fig. 6.—*Lithothamnium ramosissimum*, a calcareous Alga, from the Leitha-Kalk of the Vienna Basin. *a*, Portion of a mass, of the natural size; *b* and *c*, Transverse and vertical sections of the same magnified 320 diameters. After Gümbel.

calcareous skeletons of a kind of Nullipore (*Lithothamnium*). The Leitha-Kalk is not only extensively developed in the Austro-Hungarian empire, but can be traced through Bosnia into Turkey, and appears to be continued through Asia Minor into Armenia and Persia. Similar calcareous Algæ are found in many of the Secondary limestones; and the Palæozoic limestones will also, doubtless, be found in time to contain the skeletons of these plants to some extent.

Phosphate of lime is another lime-salt, which is of interest to the palæontologist. It does not occur largely in the stratified series, but it is found in considerable beds[1] in the

[1] Apart from the occurrence of phosphate of lime in actual beds in the stratified rocks, as in the Laurentian and Silurian series, this salt may also occur

Laurentian formation, and less abundantly in some later rock-groups, whilst it occurs abundantly in the form of nodules in the parts of the Cretaceous (Upper Greensand) and Tertiary deposits. Phosphate of lime forms the larger proportion of the earthy matters of the bones of Vertebrate animals, and also occurs in less amount in the skeletons of certain of the Invertebrates (*e.g.*, *Lingula*, among the Brachiopods; *Conularia* and *Theca*, among the Pteropods; and the *Crustacea* generally). It is, indeed, perhaps more distinctively than carbonate of lime, an organic compound; and though the formation of many known deposits of phosphate of lime cannot be positively shown to be connected with the previous operation of living beings, there is room for doubt whether this salt is not in reality always primarily a product of vital action. The phosphatic nodules of the Upper Greensand are erroneously called "coprolites," from the belief originally entertained that they were the fossilised excrements of extinct animals; and though this is not the case, there can be little doubt but that the phosphate of lime which they contain is in this instance of organic origin.[1] The true "coprolites"—that is, the petrified excreta of fishes, reptiles, and mammals—are also largely composed of phosphate of lime.

The last lime-salt which need be mentioned is *gypsum*, or *sulphate of lime*. This substance, apart from other modes of occurrence, is not uncommonly found interstratified with the ordinary sedimentary rocks, in the form of more or less irregular beds; and in these cases it has a palæontological

disseminated through the rock, when it can only be detected by chemical analysis. It is interesting to note that Dr Hicks has recently proved the occurrence of phosphate of lime in this disseminated form in rocks as old as the Cambrian, and that in quantity quite equal to what is generally found to be present in the later fossiliferous rocks. This affords a chemical proof that animal life flourished abundantly in the Cambrian seas.

[1] It has been maintained, indeed, that the phosphatic nodules so largely worked for agricultural purposes, are in themselves actual organic bodies or true fossils. In a few cases this admits of demonstration, as it can be shown that the nodule is simply an organism (such as a sponge) infiltrated with phosphate of lime (Sollas); but there are many other examples in which no actual structure has yet been shown to exist, and as to the true origin of which it would be hazardous to offer a positive opinion.

importance, as occasionally yielding well-preserved fossils. Whilst its exact mode of origin is uncertain, it cannot be regarded as in itself an organic rock, though clearly the product of chemical action. To look at, it is usually a whitish or yellowish-white rock, as coarsely crystalline as loaf-sugar, or more so; and the microscope shows it to be composed entirely of crystals of sulphate of lime.

We have seen that the *calcareous* or lime-containing rocks are the most important of the group of organic deposits; whilst the *siliceous* or flint-containing rocks may be regarded as the most important, most typical, and most generally distributed of the mechanically-formed rocks. We have, however, now briefly to consider certain deposits which are more or less completely formed of flint; but which, nevertheless, are essentially organic in their origin.

Flint or silex, hard and intractable as it is, is nevertheless capable of solution in water to a certain extent, and even of assuming, under certain circumstances, a gelatinous or viscous condition. Hence, some hot-springs are impregnated with silica to a considerable extent; it is present in small quantity in sea-water; and there is reason to believe that a minute proportion must very generally be present in all bodies of fresh water as well. It is from this silica dissolved in the water that many animals and some plants are enabled to construct for themselves flinty skeletons; and we find that these animals and plants are and have been sufficiently numerous to give rise to very considerable deposits of siliceous matter by the mere accumulation of their skeletons. Amongst the animals which require special mention in this connection are the microscopic *Polycystina.* These little creatures are of an extremely low grade of organisation, very closely related to the *Foraminifera,* but differing in the fact that they secrete a shell or skeleton composed of flint instead of lime. The *Polycystina* occur abundantly in our present seas; and their shells are present in some numbers in the foraminiferal ooze which is found at great depths in the Atlantic and Pacific oceans, being easily recognised by their exquisite shape, their glassy transparency, the general presence of longer or shorter spines, and the sieve-like perforations in the walls. In many places, in fact, especially in the colder portions of the great oceans, or at very great depths, the "Globigerina ooze" disappears, and its place is taken by an ooze composed almost wholly of the shells of *Polycystina*. Similar deposits, made up of the flinty skeletons of these Radiolarians, have been formed at previous periods of the earth's history, and now form part of the earth's crust. The two most famous of these deposits occur in Barbadoes and in the Nicobar Islands, the former being well known to workers with the microscope as the "Barbadoes earth" (fig. 7).

In addition to flint-producing animals, we have also the great group of fresh-water and marine microscopic plants

Fig. 7. — Shells of *Polycystina* from "Barbadoes earth;" greatly magnified. (Original.)

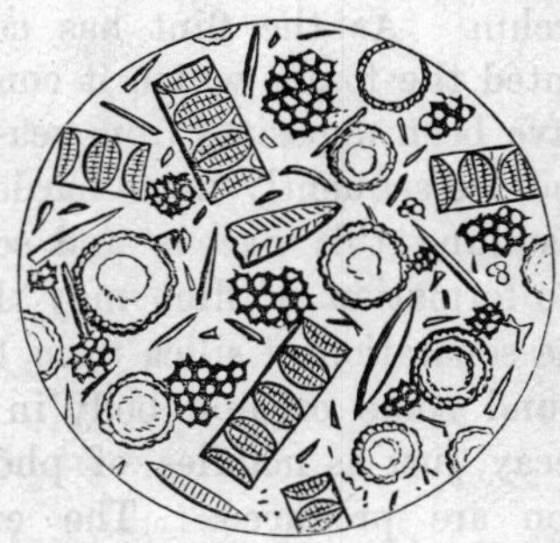

Fig. 8.—Cases of Diatoms in the Richmond "Infusorial earth;" highly magnified. (Original.)

known as *Diatoms*, which likewise secrete a siliceous skeleton, often of great beauty. The skeletons of Diatoms are found abundantly at the present day in lake-deposits, guano, the silt of estuaries, and in the mud which covers many parts of the sea-bottom; they have been detected in strata of great age; and in spite of their microscopic dimensions, they have not uncommonly accumulated to form deposits of great thickness, and of considerable superficial extent. Thus the celebrated deposit of "tripoli" ("Polir-schiefer") of Bohemia, largely worked as polishing-powder, is composed wholly, or almost wholly, of the flinty cases of Diatoms, of which it is calculated that no less that forty-one thousand millions go to make up a single cubic inch of stone. Another celebrated deposit is the so-called "Infusorial earth" of Richmond in Virginia (fig. 8), where there is a stratum, in places thirty feet thick, composed almost entirely of the microscopic shells of Diatoms.

Nodules or layers of *flint*, or the impure variety of flint known as *chert*, are found in limestones of almost all ages from the Silurian upwards; but they are especially abundant in the Chalk. When these flints are examined in thin and transparent slices under the microscope, or in polished sections, they are found to contain an abundance of minute organic bodies—such as *Foraminifera*, sponge-spicules, &c. —embedded in a siliceous basis. In many instances the flint contains larger organisms—such as a Sponge or a Sea-urchin. As the flint has completely surrounded and infiltrated the fossils which it contains, it is obvious that it must have been deposited from sea-water in a gelatinous condition, and subsequently have hardened. That silica is capable of assuming this viscous and soluble condition is known; and the formation of flint may therefore be regarded as due to the separation of silica from the sea-water and its deposition round some organic body in a state of chemical change or decay, just as nodules of phosphate of lime or carbonate of iron are produced. The existence of numerous organic bodies in flint has long been known; but it should be added that a recent observer (Mr Hawkins Johnson) asserts that the existence of an organic structure can be demonstrated by suitable methods of treatment, even in the actual matrix or basis of the flint.[1]

In addition to deposits formed of flint itself, there are other siliceous deposits formed by certain *silicates*, and also

[1] It has been asserted that the flints of the chalk are merely fossil sponges. No explanation of the origin of flint, however, can be satisfactory, unless it embraces the origin of chert in almost all great limestones from the Silurian upwards, as well as the common phenomenon of the silicification of organic bodies (such as corals and shells) which are known with certainty to have been originally calcareous. It should also be mentioned that *some* of the flints of the chalk are certainly only *secondarily* of organic origin, if even that. This is the case with the tabular masses of flint filling cracks and joints in the chalk. These masses were not produced contemporaneously with the chalk, but have been formed at a later period by the percolation into fissures of the rock of water holding silica in solution.

of organic origin. It has been shown, namely—by observations carried out in our present seas—that the shells of *Foraminifera* are liable to become completely infiltrated by silicates (such as "glauconite," or silicate of iron and potash). Should the actual calcareous shell become dissolved away subsequent to this infiltration—as is also liable to occur—then, in place of the shells of the *Foraminifera,* we get a corresponding number of green sandy grains of glauconite, each grain being the *cast* of a single shell. It has thus been shown by Dr Carpenter that the green sand found covering the sea-bottom in certain localities (as found by the Challenger expedition along the line of the Agulhas current) is really organic, and is composed of casts of the shells of *Foraminifera.* Long before these observations had been made, it had been shown by Professor Ehrenberg that the green sands of various geological formations are often composed in part of the internal casts of the shells of *Foraminifera;* and we have thus another and a very interesting example how rock-deposits of considerable extent and of geological importance can be built up by the operation of the minutest living beings.

As regards *argillaceous* deposits, containing *alumina* or *clay* as their essential ingredient, it cannot be said that any of these have been actually shown to be of organic origin. A recent observation by Sir Wyville Thomson would, however, render it not improbable that some of the great argillaceous accumulations of past geological periods may be really organic. This distinguished observer, during the cruise of the Challenger, showed that the calcareous ooze which has been already spoken of as covering large areas of the floor of the Atlantic and Pacific at great depths, and which consists almost wholly of the shells of *Foraminifera,* gave place at still greater depths to a red ooze consisting of impalpable clayey mud, coloured by oxide of iron, and devoid of traces of organic bodies. As the existence of this widely diffused red ooze, in mid-ocean, and at such great depths, cannot be explained on the supposition that it is a sediment brought down into the sea by rivers, Sir Wyville

Thomson came to the conclusion that it was probably formed by the action of the sea-water upon the shells of *Foraminifera*. These shells, though mainly consisting of lime, also contain a certain proportion of alumina, the former being soluble in the carbonic acid dissolved in the sea-water, whilst the latter is insoluble. There would further appear to be grounds for believing that the solvent power of the sea-water over lime is considerably increased at great depths. If, therefore, we suppose the shells of *Foraminifera* to be in course of deposition over the floor of the Pacific, at certain depths they would remain unchanged, and would accumulate to form a calcareous ooze; but at greater depths they would be acted upon by the water, their lime would be dissolved out, their form would disappear, and we should simply have left the small amount of alumina which they previously contained. In process of time this alumina would accumulate to form a bed of clay; and as this clay had been directly derived from the decomposition of the shells of animals, it would be fairly entitled to be considered an organic deposit. Though not finally established, the hypothesis of Sir Wyville Thomson on this subject is of the greatest interest to the palæontologist, as possibly serving to explain the occurrence, especially in the older formations, of great deposits of argillaceous matter which are entirely destitute of traces of life.[1]

It only remains, in this connection, to shortly consider the rock-deposits in which *carbon* is found to be present in greater or less quantity. In the great majority of cases where rocks are found to contain carbon or carbonaceous matter, it can be stated with certainty that this substance is of organic origin, though it is not necessarily derived from vegetables. Carbon derived from the decomposition of animal bodies is not uncommon; though it never occurs in such quantity from this source as it may do when it is derived from plants. Thus, many limestones are more or less highly bituminous; the celebrated siliceous flags or so-

[1] Further investigations have thrown doubt upon the above theory, and seem to favour the view that the red ooze is produced by the decomposition of volcanic matter.

called "bituminous schists" of Caithness are impregnated with oily matter apparently derived from the decomposition of the numerous fishes embedded in them; Silurian shales containing Graptolites, but destitute of plants, are not uncommonly "anthracitic," and contain a small percentage of carbon derived from the decay of these zoophytes; whilst the petroleum so largely worked in North America has not improbably an animal origin. That the fatty compounds present in animal bodies should more or less extensively impregnate fossiliferous rock-masses, is only what might be expected; but the great bulk of the carbon which exists stored up in the earth's crust is derived from plants; and the form in which it principally presents itself is that of *coal*. We shall have to speak again, and at greater length, of coal, and it is sufficient to say here that all the true coals, anthracites, and lignites, are of organic origin, and consist principally of the remains of plants in a more or less altered condition. The bituminous shales which are found so commonly associated with beds of coal also derive their carbon primarily from plants; and the same is certainly, or probably, the case with similar shales which are known to occur in formations younger than the Carboniferous. Lastly, carbon may occur as a conspicuous constituent of rock-masses in the form of *graphite* or *black-lead*. In this form it occurs in the shape of detached scales, or of veins or strings, or sometimes of regular layers;[1] and there can be little doubt that in many instances it has an organic origin, though this is not capable of direct proof. When present, at any rate, in quantity, and in the form of layers associated with stratified rocks, as is sometimes the case in the Laurentian formation, there can be little hesitation in regarding it as of vegetable origin, and as an altered coal.

[1] In the Huronian formation at Steel River, on the north shore of Lake Superior, there exists a bed of carbonaceous matter which is regularly interstratified with the surrounding rocks, and has a thickness of from 30 to 40 feet. This bed is shown by chemical analysis to contain about 50 per cent of carbon, partly in the form of graphite, partly in the form of anthracite; and there can be little doubt but that it is really a stratum of "metamorphic" coal.

CHAPTER III.

SUCCESSION OF FORMATIONS—CONTEMPORANEITY OF STRATA—GEOLOGICAL CONTINUITY.

DIFFERENT AGES OF THE AQUEOUS ROCKS.

THE two principal tests by which the age of any particular bed, or group of beds, may be determined, are superposition and organic remains—a third test sometimes being afforded by mineral characters. The first and most obvious test of the age of any aqueous rock is its relative position to other rocks. Any bed or set of beds of sedimentary origin is obviously and necessarily older than all the strata which surmount it, and younger than all those upon which it rests. It is to be remembered, however, that superposition can at best give us but the *relative* age of a bed as compared with other beds of the same region. It cannot give us the *absolute* age of any bed; and if we are ignorant of the age of any of the beds with which we may be dealing, we have to appeal to other tests to learn more than the mere order of succession in the particular region under examination.

The second, and in the long-run more available, test of the ages of the different sedimentary beds, is that afforded by their organic remains. Still, this test is also by no means universally applicable, nor in all cases absolutely conclusive. Many aqueous rocks are unfossiliferous through a thickness of hundreds, or even thousands, of feet of little altered sediments; and even amongst beds which do contain fossils, we often meet with strata of a few feet or yards in thickness, which are wholly destitute of any traces of life. Many fossils, again, range vertically through many groups of strata, and in some cases even through several formations. Such fossils, therefore, if occurring by themselves, or considered apart from other associated organisms, are not conclusive as to the age of any particular set of beds. As the result, however, of combined palæontological and geological researches, it is now possible for us to divide the entire series of stratified deposits into a number of definite rock-groups or formations, each of which is characterised by possessing an assemblage of organic remains which do not occur in association in any other formation. Such an assemblage of fossils, characteristic of any given formation, represents the *life* of the particular period in which the formation was deposited. It follows from this, that whenever we can get a group or collection of fossils from any particular bed or set of beds, there is rarely any difficulty in determining the precise geological horizon of the beds in which the fossils occur.

With certain limitations, however, we may go much further than this. Not only are the great formations characterised by special and characteristic assemblages of animals and plants; but, in a general way, each subdivision of each formation has its own peculiar fossils, by which it may be recognised by a skilled worker in palæontology. Whenever, for instance, we meet in Britain with the fossils known as Graptolites, we may be sure that we are dealing with Silurian Rocks. We may, however, go much further than this. If the Graptolites belong to certain genera, we may be sure that we are dealing with *Lower* Silurian Rocks. Furthermore, if certain special forms are present, we may be even able to say to what exact part or subdivision of the Lower Silurian series they belong.

All these conclusions, however, would have to be accompanied by a tacit but well-understood reservation. No Graptolites have ever been found in Britain out of rocks known upon other grounds to be Silurian; but there is no reason why they might not at any time be found in younger deposits. In the same way, the species and genera which we now regard as characteristic of the Lower Silurians, might at any time be found to have survived into the Upper Silurian period. We should never forget, therefore, in determining the age of a rock by palæontological evidence alone, that we are always reasoning upon generalisations which are the result of experience alone, and which may at any time be overthrown by fresh discoveries.

Chronological Succession of the Aqueous Rocks.

As the result of observations made upon the superposition of rocks in different localities, from their mineral characters, and from their included fossils, geologists have been able to divide the entire stratified series into a number of different divisions or formations, each characterised by a *general* uniformity of mineral composition, and by a special and peculiar *assemblage* of organic forms. Each of these primary groups is in turn divided into a series of smaller divisions, characterised and distinguished in the same way. It is not pretended for a moment that all these primary rock-groups can anywhere be seen surmounting one another regularly. There is no region upon the earth where all the stratified formations can be seen together; and, even when most of them occur in the same country, they can nowhere be seen all succeeding each other in their regular and uninterrupted succession. The reason of this is obvious. There are many places—to take a single example—where one may see the Silurian Rocks, the Old Red Sandstone, and the Carboniferous Rocks succeeding one another regularly, and in their proper order. This is because the particular region where this occurs was always submerged beneath the sea while these formations were being deposited. There are, however, many more localities in which one would find the Carboniferous Rocks resting unconformably upon the Silurians without the intervention of any strata which could be referred to the Old Red Sandstone. This might arise from one of two causes: 1. The Silurians might have been elevated above the sea immediately after their deposition, so as to form dry land during the whole of the Old Red period, in which case, of course, no strata of the age of the Old Red Sandstone could possibly be deposited. 2. The Old Red Sandstone might have been deposited upon the Silurian, and then the whole might have been elevated above the sea, and subjected to an amount of denudation sufficient to remove the Old Red Sandstone entirely. In this case, when the land was again submerged, the Carboniferous Rocks, or any younger formation, might be deposited directly upon Silurian strata. From one or other of these causes, then, or from subsequent disturbances and denudations, it happens that we can rarely find many of the primary formations following one another consecutively and in their regular order.

In no case, however, do we ever find the Old Red Sandstone resting upon the Carboniferous, or the Silurian Rocks reposing on the Old Red. We have therefore, by a comparison of many different areas, an established order of succession of the stratified formations, as shown in the subjoined ideal section of the crust of the earth (fig. 9).

The main subdivisions of the Stratified Rocks are known by the following names:—

1. Laurentian.
2. Cambrian (with Huronian ?).
3. Silurian.
4. Devonian or Old Red Sandstone.
5. Carboniferous.
6. Permian } New Red Sandstone.
7. Triassic }
8. Jurassic or Oolitic.
9. Cretaceous.
10. Eocene.
11. Miocene.
12. Pliocene.
13. Post-tertiary.

IDEAL SECTION OF THE CRUST OF THE EARTH.

Fig. 9.

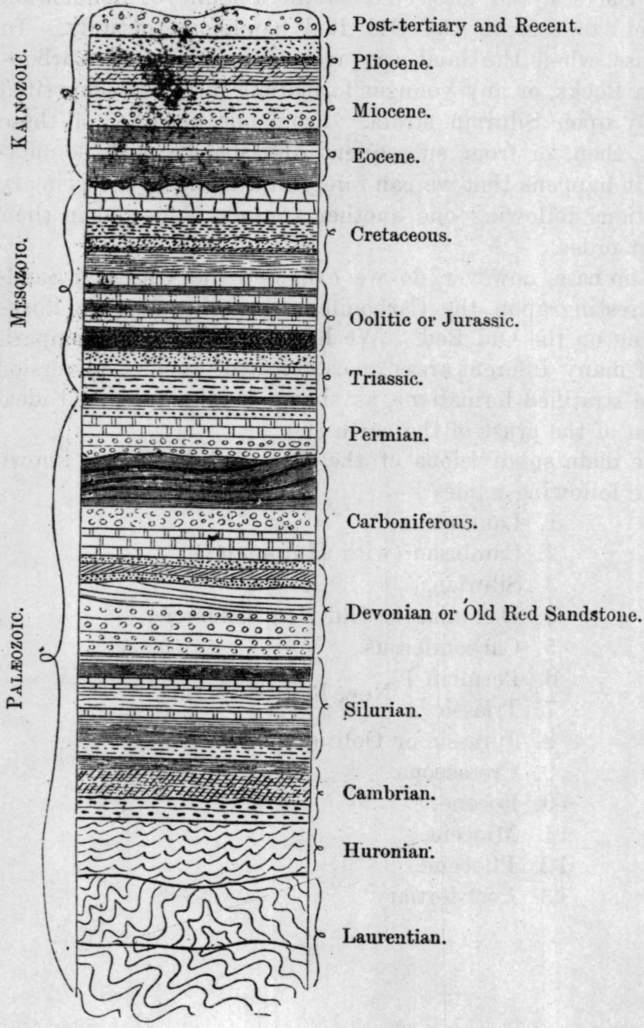

Of these primary groups, the Laurentian, Cambrian, Silurian, Devonian, Carboniferous, and Permian are collectively grouped together under the name of *Primary* or *Palæozoic* Rocks (Gr. *palaios,* ancient; *zoe,* life), because of the entire divergence of their animals and plants from any now existing upon the globe. The Triassic, Jurassic, and Cretaceous systems are grouped together as the *Secondary* or *Mesozoic* formations (Gr. *mesos,* intermediate; *zoe,* life), because their organic remains are intermediate between those of the Palæozoic period, and those of more modern strata. The Eocene, Miocene, Pliocene, and Post-tertiary Rocks are grouped together under the head of *Tertiary* or *Kainozoic* Rocks (Gr. *kainos,* new; *zoe,* life), because their organic remains approximate in character to those now existing upon the globe.

As regards the division of the entire series of stratified deposits into the above enumerated primary "formations," the value of palæontological evidence has never been disputed. In any given country, it would be possible, undoubtedly, to determine the order and relative succession of the great formations, to some extent at any rate, by a mere appeal to the mineral character and order of superposition of the rocks themselves: but it is perfectly clear that this method of procedure would necessarily break down totally the moment we came to try and determine what were the corresponding formations in some far-distant region. By the stratigraphical evidence alone we could determine the relative position and age, for example, of the Silurian, Devonian, and Carboniferous formations in Britain, but it would be an entire impossibility to identify these same formations, say in North America, except by means of the fossils which they contain. So far, then, as this goes, no question has ever been raised as to the value and powers of Palæontology; but when we come to consider the minor rock-groups included in these formations, we find much difference of opinion as to the extent to which the evidence of the fossils is available in determining stratigraphical horizons. Part of this difference of opinion is due to imperfect acquaintance on the part of stratigraphical geologists with the methods of palæontological inquiry, and needs no discussion here; but part is well founded, and either arises from actual defects in the modes of research employed by palæontologists, or is due to the fact that the laws of the distribution of fossil organisms are not always the same in different formations, and that they are liable to vary under conditions which are only partially or not at all understood. To both these points our attention may be directed for a few moments.

As regards imperfections in the methods of palæontological research, by far the most important arises from the fact that far too much weight has been attached by observers, especially in the earlier periods of the science, to the *age* of the rocks in which any given fossil occurred. So long as the opinion was current that fossils occurring in different formations *must* be different, it followed of necessity that the smallest and most trivial varietal or even individual peculiarities of form or structure were considered as sufficient to establish specific distinction. At present, however, palæontologists are tolerably agreed that the mere fact of a difference of physical position, and consequently of age, ought never to be taken into account at all in considering the true affinities and systematic position of a fossil. At the same time it is, for many reasons, most important that palæontologists should have a general personal acquaintance with the rocks in which occur any fossils that they may have to examine and describe; and many errors have arisen from the neglect of this sound rule.

Again, palæontologists are not agreed as to the relative value of different *classes* of fossils in determining the age and stratigraphical position of the rocks in which they occur. If all the fossils point towards the same conclusion, there is, of course, no difficulty in the matter; but it sometimes happens that the vegetable fossils of a given formation would lead one to conclude that it was of a given age, whereas the Invertebrate or Vertebrate fossils would induce us to place it at some different horizon in the stratified series. There has thus arisen a controversy as to the relative value of plants and animals as tests of the age of a given series of rocks; and in at least two instances this controversy has affected questions of considerable general importance. In one of the cases referred to, we find plants which are admitted to be the same as those of the Coal-measures (Carboniferous) coexisting with fishes and other animal fossils which are equally admitted to be characteristic of the later formation of the New Red Sandstone (Permian). This conjunction of an ancient flora with a more modern fauna occurs in the gas-coals of Bohemia, and much difference of opinion has been expressed as to the proper interpretation to be placed upon the facts. If we regard the plants alone, we must place the beds in question in the Carboniferous formation, whereas if we look to the animal remains alone, we should with an equal absence of hesitation refer the strata to the Permian. A still more important case of an essentially similar nature occurs in North America, and concerns the boundary-line between the Cretaceous formation and the Tertiary, two groups of rocks which in the Old World are separated by an extraordinarily abrupt and conspicuous line of demarcation. In North America we find a series of rocks which contain unquestioned Cretaceous fossils, and another great group of deposits which are charged with an equally unequivocal series of Tertiary fossils; but between these there is an immense series of beds—some four thousand feet in thickness—which contains the remains of undoubted *Cretaceous* Invertebrate and Vertebrate animals mixed with a vast number of regular and unquestionable *Tertiary* plants. If we look to the animals, we must place this series (known as the "Lignitic Series," from the presence in it of beds of lignite) in the Cretaceous; whereas from the evidence of the plants alone we should have to consider it as the base of the Tertiary. Upon the whole, however, and without entering into any detailed discussion of the question, it would appear that in all such cases plants have a much smaller value as tests of the geological position and age of the beds in which they occur than may be justifiably attached to the remains of the Marine Invertebrates, while these, again, are inferior in this respect to the remains of Vertebrates. Judging by this canon, in which most authorities are now agreed, the Bohemian gas-coals must be considered as Permian, and the great Lignitic series of North America must be considered as forming the summit of the Cretaceous series.

Lastly, there are cases in which the distribution of fossil organisms in different formations differs so much, or presents such peculiarities, that we may reasonably suppose it to have been conditioned by the special circumstances, perhaps now undiscoverable, affecting the deposition of the strata of these formations. Thus, in some cases, as, for example, in the Carboniferous Limestone series, we find that the same fossils characterise the entire series from top to bottom—speaking roughly, at any rate—and that special kinds of fossils are not restricted to special horizons in the series. This apparent diffusion of the same kinds of fossils from the base to the summit of a series of beds perhaps two or three thousand feet in thickness, may of course be simply due to the fact that we have not sufficiently investigated the organic remains met with in the formation, and have not determined with sufficient precision the exact horizons at which each occurs. This is a work of time, and demands both great stratigraphical knowledge and also a wide and accurate acquaintance with the characters of the fossils themselves—two requirements rarely fulfilled in the same individual. Still there are reasons for believing that in certain formations, the common and characteristic fossils range from the top to the bottom of the series, so that it would not be possible to determine by means of the fossils the precise position in the series of any given bed. On the other hand, there are cases in which the fossils of a given formation may be divided into two principal groups. In the one group is comprised a series of common forms of life which may be regarded as characterising the formation *as a whole.* In the other group are included certain special fossils which are confined to particular parts of the formation, and which are characteristic of certain definite *horizons* or *zones* within the limits of the formation. All the great formations are to some extent capable of being broken up into minor rock-groups, characterised by special life-forms. Some of the differences in the kinds of fossils found in different parts of the same formation must, of course, be simply set down to the fact that different kinds of sediment imply changed conditions in the sea, and hence changes in the marine fauna.

If, for example, part of a formation consisted of limestone and part of sandstone, we should expect, beforehand, to find that each of these rock-groups would have some fossils not found in the other, since the two would have been formed under different conditions. Apart, however, from differences arising from causes of this nature, we meet with cases in which a formation, even if essentially homogeneous in its mineral nature, can be divided into *zones*, each of which is characterised by the possession of special groups of fossils. The most celebrated case of this subdivision of a formation by means of its fossils is that afforded by the *Lias*. This great and essentially argillaceous formation can be divided into a number of *zones*, each of which is characterised by possessing some special fossils, and particularly by some special Ammonite. These zones are extremely constant, and they are traceable wherever the formation is fully developed, and has been fully examined, in Europe; so that they enable us to effect a division of the formation into special horizons, which have no stratigraphical existence, and are not separated by any physical break, but which are of the utmost palæontological importance, and which can be rendered readily available in working out the stratigraphy of the formation. Similar "zones" are recognisable in the other Jurassic rocks and in the Cretaceous system; and it is tolerably certain that in time we shall be able to establish a similar, if less perfect, series of palæontological divisions in all the great formations.

The principal difficulty that we have to confront in dealing with these "zones," is to produce any plausible explanation accounting for the destruction of the special life-forms of the one zone and the appearance of those of the next zone. For the most part these zones are of very limited vertical extent, and they succeed each other in such a manner as totally to preclude the idea that the dying out of the old forms can have been in any way caused by any physical disturbance of the area. Perhaps the most probable view to adopt in the meanwhile is, that the formations in which distinct and limited life-zones can be recognised were deposited with extreme slowness, whereas those which show an essentially compact and homogeneous fauna from base to summit were deposited with comparative rapidity. Upon this view, a formation like the Lias is one formed by a process of very slow and intermittent sedimentation, the life-zones being separated by intervals, during which sedimentation must have been at a stand-still, but which were long enough to allow of more or less considerable biological changes, some forms dying out, or becoming modified, while other new ones came in. Upon this view, further, a formation like the Lias, though of comparatively small vertical extent, may represent as long a period of time as the whole of such a great formation as the Carboniferous, which appears to have been formed under conditions of comparatively rapid sedimentation.

Contemporaneity of Strata.

When groups of beds in different parts of the earth's surface, however widely separated from one another, contain the same fossils, or rather an assemblage of fossils in which many identical forms occur, they are ordinarily said to be "contemporaneous;" that is to say, they are ordinarily supposed to belong to the same geological period, and to have been formed at the same time in the history of the earth. They would therefore be unhesitatingly regarded as "geological equivalents," and would be classed as Silurian, Devonian, Carboniferous, and so on. It is to be remembered, however, that it is not necessary, to establish such a degree of equivalency between widely separated groups of strata, that the fossils of each should be to any great extent *specifically* identical. It is sufficient that, whilst some few species are identical in both, the majority of the fossils should be "representative forms," or, in other words, nearly allied species. It will be shown, however, that groups of strata *widely removed from one another in point of distance* can only exceptionally be "contemporaneous," in the *strict* sense of this term. On the contrary, in so far as we can judge from the known facts of the present distribution of living beings, the occurrence of *exactly* the same fossils in beds far removed from one another is *primâ facie* evidence that the strata are *not exactly* contemporaneous, but that they succeeded one another in point of time, though by no long interval geologically speaking.

Most of the facts bearing upon this question may be elicited by a consideration of such a widely extended and well-known formation as the Mountain Limestone or Sub-Carboniferous Limestone. This formation occurs in localities as remote from one another as Europe, Central Asia, North America, South America, and Australia; and it is characterised by an assemblage of well-marked fossils, amongst which *Brachiopods* belonging to the genus *Producta* may be specially singled out. Now, if we believe that the Carboniferous Limestone in all these widely distant localities was strictly contemporaneous, we should be compelled to admit the existence of an ocean embracing all these points, and, in spite of its enormous extent, so uniform in temperature, depth, and the other conditions of marine life, that beings either the same or very nearly the same inhabited it from end to end. We can, however, point to no such uniformity of conditions and consequent uniformity of life over any such vast area at the present day; and we have therefore no right to assume that this is the true explanation of the facts. Indeed this explanation would almost necessarily lead us to the now abandoned theory that each period in geological history was characterised by a special group of organisms spreading over the whole globe, and that there took place at the close of each period a general destruction of all existing forms of life, and a fresh creation of the new forms characteristic of the next period.

In our inability, then, to accept this view, we must seek for some other explanation of the observed facts. The most probable view, and the one which is supported most strongly both by what we see at the present day and by what we learn from numerous examples in past time, is this: The Carboniferous Limestone was not deposited all over the world in one given period, by one sea, or at exactly the same time; so that it cannot be said to be *strictly* "contemporaneous" wherever it is found. This would imply a uniformity of conditions over vast distances, such as exists nowhere at the present day, and such as we have no right to assume ever existed. On the contrary, the deposition of the Carboniferous Limestone must have first taken place in one comparatively limited area—say in Europe—where fitting conditions were present both for the animals which characterise it, and for the formation of beds of its peculiar mineral and physical characters. How wide this area may have been, signifies very little. It may have been as large as the area now covered by the Pacific, or larger, and yet it could not include all those localities in which strata of Carboniferous age with identical or representative fossils are already known to exist. Under any circumstances, some dispersion of the species of the original Carboniferous area must have been going on by the ordinary processes of migration from the commencement of the Carboniferous period, but this dispersion must have been greatly accelerated towards the close of the period of the deposition of the Carboniferous Limestone. At this time the conditions present in the original area must be supposed to have become unsuitable for the further existence in that area of the assemblage of animals which had been its inhabitants, or, at any rate, for a great many of them. The change from suitable to unsuitable conditions must, it is hardly necessary to say, have been an extremely slow and gradual one; and would doubtless be connected with the progressive shallowing of the sea, the diversion of old currents of heated water, or the incoming of new currents of cold water, or other physical changes tending to alter the climatic conditions of the area. What, then, would be the effect of such a change of conditions as we have supposed upon the animals inhabiting the area? *a.* Some of them would, doubtless, be sufficiently hardy and accommodating to bear up under the new state of things; and these would persist into the ensuing period, without any perceptible change, it might be, or more probably in the form of varieties or species allied to the old ones. In this case, therefore, we should get a certain number of species which would pass from the Carboniferous Limestone up into the Yoredale Series, the Millstone Grit or the Coal-measures; or, if we did not find any species exactly the same in all these groups, we should still find in the later groups some forms which would be varieties of those of the older, or which would be allied or representative species.

b. There would, in the second place, be a certain number of species which would be utterly unable to withstand the altered conditions of the area; and these would gradually die out and become wholly extinct. We should thus get a certain number of fossils which would be either exclusively confined to the Carboniferous Limestone in general, or which, perhaps, might not be found out of the Carboniferous Limestone of a single region, or even a single particular locality.

c. Lastly, some species would yield so far to the altered conditions of the area that they would "migrate," and seek elsewhere a more congenial home. This term is apt to convey false impressions; and it will be well here to consider what is meant by the "migration" of species or groups of animals. It is quite obvious that only animals like birds, mammals, insects, &c., which enjoy when grown up the power of active locomotion, can actually "migrate" in person, supposing they find themselves placed under unfavourable conditions. There are many animals, however, such as most shell-fish, corals, sea-urchins, &c., which have, when adult, either no power of changing their place, or at best a very limited one. Still in these cases even, though the *individual* has no means of removing his quarters to some more favoured spot, there may be a "migration" of the *species* from an unsuitable to a suitable locality. This is effected through the medium of the *young,* which have the power of choosing where they will settle, and are endowed with vigorous powers of locomotion. If, for example a bed of oysters should become placed under conditions unsuitable for the development of these molluscs, it is clear that the old oysters cannot change their location. The young oysters, however, swim about freely; and these will move away from the original bed till they find a place which will suit them. By a repetition of this process there may be in course of time a removal or "migration" of a species to almost any distance, irrespective of the fact that the adult is permanently rooted.

To return, then, to the case which we have been considering: When the conditions of life in the seas of the Carboniferous Limestone became unfavourable for the further existence of their fauna, some species would migrate to a more congenial area. In this way a greater or less number of the species characteristic of the Carboniferous Limestone would ultimately be transferred to some other area. Here they would mingle with the forms already inhabiting that area, perhaps more or less completely supplanting these, perhaps merely succeeding in maintaining a more or less precarious existence. In either case, their remains would be preserved in the sedimentary deposits of the new area. When, ages afterwards, we come to examine the crust of the earth geologically, we should find these identical and characteristic species of fossils in the rocks of the two areas, and we should say—"these rocks are contemporaneous." It is clear, however, that we should be wrong in so saying. The rocks in question would belong to the same geological period, but they would belong to different stages of the same period, and they would not be strictly contemporaneous. For deposits of this nature, believed to hold this relation to each other, the term of "homotaxeous" has been proposed, in place of the term "contemporaneous."

What has just been said about the Carboniferous rocks would apply with equal justice to all the great formations, and to many of the smaller rock-groups all over the world. The Silurian rocks of Europe, North America, South America, Australia, &c., contain very similar fossils, and are undoubtedly "homotaxeous." Nothing, however, that we see at the present day can justify us in believing that these widely separated deposits are strictly "contemporaneous," in the sense that they were deposited at *exactly* the same period of time. We should have to believe, if this conclusion is to be justified, that in Silurian times the ocean spread over a much larger area of the earth's surface than it does now, and that its temperature and depth were unnaturally uniform; and there are, perhaps, some who would accept this view. What has been said about the Silurian rocks as a whole applies with still greater force to certain of the minor subdivisions of the same, which contain many of exactly the same specific forms in parts of the globe very widely removed from one another. It is the very identity of the fossils, however, which proves that the beds in question, from their geographical position, cannot have been deposited at exactly the same time, though they doubtless belong to the same period, and may even be said to be related to one another, as far as the identical fossils are concerned, by lineal descent. Similar remarks might be made about the Devonian, Permian, Triassic, Jurassic, Cretaceous, and other formations; but it is not necessary further to multiply examples.

If we consider the present state of things upon the globe, we shall be further convinced of the justice of these views, which were first prominently brought forward in Britain by Professor Huxley. If we could suddenly remove the sea from the earth, we should find at various points of the earth's surface deposits of different kinds, now concealed from us by the ocean, or only partially known by dredgings or soundings. Thus we should find vast accumulations of calcareous matter, in the form of coral-rock and coral-reef, where now rolls the Pacific Ocean. In high northern and low southern latitudes we should find great deposits of sand and mud, with angular blocks of stone, the whole derived from the ice-clad regions of the poles. Over vast areas, again, in the deep Atlantic, we should find an impalpable chalky mud, or "ooze." All these different deposits are obviously and necessarily "contemporaneous," not only in the geological acceptation of the word, but in its most literal sense. In spite of this fact *they would not contain the same fossils;* and, indeed, they would be characterised by organic remains which would be wholly different in each case. The coral-reefs of the Pacific would be essentially characterised by the abundance of the remains of reef-building corals, though they would also present other tropical forms of life, especially Brachiopods and Echinoderms. The glacial mud of the Polar regions would contain the remains of Arctic molluscs, along with such other animals as delight in severe cold. Lastly, the ooze of the deep Atlantic would contain innumerable *Foraminifera*, along with siliceous Sponges, Sea-urchins, and Crinoids. We learn, therefore, from this, that contemporaneous deposits not only do not necessarily contain the same fossils, but that, if widely separated geographically, they may be characterised by wholly dissimilar assemblages of organisms.

It may happen, again, as pointed out by Sir Charles Lyell, that deposits belonging to different geographical and zoological provinces may, as regards space, be nearly approximated, and, as regards time, may be actually contemporaneous, and yet may not contain any fossils in common, or only a very few. If, for example, any sudden upheaval were to lay bare what is now the floor of the Red Sea, together with that of the Mediterranean, we should find the two areas to contain deposits actually synchronous as regards the time of their deposition, and very near to one another in point of distance, and yet containing, upon the whole, entirely distinct groups of organic remains. We learn, therefore, from this, that owing to the existence of geographical barriers, it is possible for contemporaneous deposits to be found in close contiguity, in a single region, and yet to contain very different fossils.

Again, we know from the researches of Professors Carpenter and Wyville Thomson and Mr Gwyn Jeffreys, that deposits may be formed, side by side, *in a single ocean*, and may yet differ from one another altogether, both in mineral characters and in their included fossils, though strictly contemporaneous in point of time. Thus, in parts of the deep Atlantic where the temperature of the bottom water is comparatively high, we have the calcareous deposit of the ooze, abounding in Foraminifera, Sponges, and Echinoderms. In certain other areas in the same ocean, and in comparatively close contiguity with the preceding, we have the temperature lowered by cold currents, and we find a sandy deposit in process of formation, with a fauna much more scanty than that of the ooze, and wholly distinct from it. We thus learn that sedimentary deposits may be strictly contemporaneous, and may be placed very near to one another in point of distance, and yet may contain very different fossils.

Lastly, synchronous deposits necessarily contain wholly

different fossils, if one has been deposited by fresh water, and the other has been laid down in the sea. The fresh-water deposits of one period are obviously contemporaneous with the marine formations of the same period, and they may not be far removed from one another in point of distance, but they must contain altogether different organic remains. The former will contain remains of the fresh-water and terrestrial animals of the period, and of these only; whilst the latter will principally, if not exclusively, be characterised by the remains of marine forms of life. In this way, there is some reason to believe, may be explained the differences between the fossils of the Old Red Sandstone and of the Devonian rocks, strictly so called. Both are believed to have been deposited in the same geological period, and to be truly "contemporaneous;" but they do not contain the same fossils. This may be readily explained, however, if we suppose the former to represent the fresh-water deposits of the Devonian period, or to have been laid down in an inland sea, whilst the latter is the true marine formation of the same period.

Under any circumstances, however, we must remember that the doctrine of "homotaxis," if rightly limited and defined, in no way diminishes the value of fossils as indications of the age of the formations in which they occur. If we give the term "contemporaneous" a purely geological sense, and endeavour to forget its literal signification as applying to events which have occurred at *precisely* the same moment of time, then it is just as good an epithet for the different deposits belonging to a given geological formation as is the term "homotaxeous." *All* the deposits which possess Carboniferous fossils, at whatever point of the earth's surface they may be situated, belong to the Carboniferous *period*, and are therefore *geologically* contemporaneous. All that is really implied by the doctrine of "homotaxis," rightly regarded, is that we cannot say that any great formation in any one country is the *precise* equivalent of the same formation in any country very widely removed in point of distance, in the sense that its deposition began and ended at *exactly* the same times; and therefore we cannot parallel the *sub-*

divisions of such formations with anything approaching to absolute precision. Regarded as a whole, however, the Carboniferous formation of America is the *geological* equivalent of the Carboniferous formation of Europe, and both belong to what geologists understand as the "Carboniferous period." As the same is true of all the great formations, in all parts of the world, it is clear that the principal advantage of the use of such a term as "homotaxis" is simply that we thereby avoid the employment of a word which common usage would wrongly interpret; and it is quite certain that we cannot abolish the idea of *geological* "contemporaneity," as demonstrated by the presence of identical or representative species of fossils; nor can we refuse to admit that formations containing such fossils, however far removed from one another in point of distance, must have been laid down within the limits of the same great "period" in the history of our earth.

We are now in a position very briefly to discuss the question of what may be called "geological continuity." It has already been stated that the entire series of Fossiliferous or Sedimentary rocks may be naturally divided into a certain number of definite rock-groups or "formations," each of which is characterised by the possession of a peculiar and characteristic assemblage of fossils, constituting, or rather representing, the "life" of the "period" in which the formation was deposited. The older geologists held, what probably every one would be tempted to think at first, that the close of each formation was characterised by a general destruction of the forms of life of the period, and that the commencement of each new formation was accompanied by the creation of a number of new animals and plants, destined to figure as the characteristic fossils of the same. This theory, however, not only invokes forces and processes which it can in no way account for, but overlooks the fact that most of the great formations are separated by lapses of time, unrepresented perhaps by any deposition of rock, or represented only in some particular area, and yet, perhaps, as great as, or greater than, the whole time occupied in the production of the formation itself.

Nowadays, most geologists hold that there was no such sudden destruction of life at the close of each great geological epoch, and no such creation of fresh forms at the commencement of the next period. On the contrary, they hold that there is a geological "continuity," such as we see in other departments of nature, and that the lines which we draw between the great formations merely mark periods of time in which no rocks were laid down, or the rocks deposited in which are at present unknown to us.

What are we to believe occurred at the close of any great geological period—say, the Cretaceous period? If we reject the view that the close of the period was marked by a sudden and universal extinction and destruction of the characteristic Cretaceous forms of life, there is only one other view which we can take. Confining our attention solely to those seas of the period of which alone we know enough for safe reasoning, we know that the close of the Cretaceous period in Europe was accompanied, or rather caused, by an upheaval of the Cretaceous area, and an obliteration of the Cretaceous sea. This upheaval was, of course, effected with extreme slowness, or, at any rate, not suddenly, and it must have completely changed the life-conditions or "environment" of the animals which swarmed in the Cretaceous seas. Some of these would doubtless be unable to accommodate themselves to their altered surroundings, and would simply die out. Others, we may presume, would migrate to some more favourable area, and some of these might accomplish their migration without undergoing any change. Most, however, of the forms which migrated, in the process of migration, and by reason of coming into contact with strange neighbours and untried conditions, would probably undergo more or less modification. Ultimately, therefore, many characteristic Cretaceous forms might be transferred to some sea far distant from their original home. Not only so, but some of the transferred species might have suffered so much modification that they would no longer be regarded as specifically identical with the original Cretaceous forms, but would be looked upon simply as allied or "representative" species, though really the lineal descendants of the animals of the Chalk.

It is perfectly clear that the process of rock-deposition which was going on in Europe towards the close of the Cretaceous period was not, and could not be, abolished by the elevation of the European area, and the obliteration of the Cretaceous sea, but was simply *transferred* to some other area. In this particular case, we do not happen to know where the new area of deposition may have been. It is quite certain, however, that in whatever area the Cretaceous animals took refuge, there rocks must have been deposited in course of time, as they are in all seas, though it does not in the least follow that the rocks of this new era should have the smallest likeness in mineral composition to the Cretaceous sediments. If we should at any time discover these rocks, it may pretty safely be predicted what we should find in them in the way of fossils. We should find, namely, some Cretaceous species, probably unchanged; with these there would be forms allied to the Cretaceous species, but differing from them to a greater or less extent; in addition, there would be a certain proportion of forms of life wholly unknown in the Cretaceous rocks; and lastly, there would be a conspicuous absence of certain characteristic species of the Chalk period. In other words, such deposits as we have been speaking of would contain an assemblage of fossils more or less intermediate in character between those of the true Cretaceous period and those of the lowest Tertiary beds (Eocene), which rest upon the Chalk, or they would present an intermixture of Cretaceous with Eocene types. In point of fact, we have fragments of such intermediate deposits (in the Mæstricht beds of Holland, the Pisolitic Limestone of France, the Faxoe Limestone of Denmark, and the Thanet Sands of Britain), and we find in them traces of such an intermixture. Moreover, when we come to examine the boundary-line between the Cretaceous and Tertiary in a region far removed from Europe—namely, in North America—we find that between these two formations, so widely separated in the Old World, we have some four thousand feet of strata (the so-called "Lignitic Series") containing such a complete intermixture of the forms of life characteristic of these two periods, that it has been a matter of lively controversy whether they should be regarded as the summit of the older or the base of the newer series of sediments.

We may pause here to consider how it is that we may never hope to find a complete series of deposits linking on one great formation to another, as, for example, the Chalk to the Eocene rocks. In the first place, only a limited portion of the earth has as yet been properly examined, and we have therefore no right to expect that we have as yet hit upon the area, or areas, to which the process of rock-forming was transferred at the close of the Cretaceous period proper in Europe. We have, however, the full right to expect that we shall ultimately find formations which will have to be intercalated in point of time between the White Chalk and the Eocene; and, as before said, traces of such are already known to us. In the second place, we have every reason to suppose that many of these intermediate deposits have been destroyed at some period subsequent to their formation by what is technically called "denudation," or, in other words, by the action of rain, rivers, ice, and the sea. In the third place, many of the missing deposits may have been concealed since their formation by the deposition upon them of other newer rocks; or they may be situated in areas which are at present covered by the ocean. Lastly, we must not forget that there may have been times in which great changes in life were actively progressing in areas in which there might be little or no contemporaneous deposition of rock, so that the extreme terms of a series might be preserved to us whilst all the intermediate links might have escaped record.

From these and similar causes, it is almost certain that we shall never be able to point to a complete series of deposits linking one great geological period, such as the Cretaceous, to another, such as the Eocene. Still, we may well have a strong conviction that such deposits must exist, or must have existed, as memorials of, at any rate, part of the time which elapsed between the close of the one formation and the commencement of the next. Upon any theory of "evolution," at any rate, it is certain that there can be no total break in the great series of the stratified deposits, but that there must have been a complete continuity of life, and a more or less complete continuity of deposition, from the Laurentian period to the present day. There was, and could have been, no such continuity in any one given area; but the chain could never have been snapped at one point and taken up at a wholly different one. The links must have been forged in different places, but the chain, nevertheless, remained unbroken. From this point of view, there would be little impropriety in saying that we are living in the Silurian period; but we could only say so in a very limited sense. While most geologists will readily admit that there must have been such an actual continuity of the great geological periods, from the earliest times up to the present day, it remains certain that we can never dispense with the division of the stratified series into definite rock-groups and life-periods. We can never hope to discover all the lost links of the geological chain, and the great formations will always be separated from one another by more or less evident physical or palæontological breaks, or by both combined. The utmost we can at present do is to arrive at the conviction that the lines of demarcation between the great formations only mark gaps in our knowledge, and that there can be in nature no *hiatus* in the long series of fossiliferous deposits.

The theory of "geological continuity," then, may in practice be carried so far as to be useless, or even injurious to the progress of science. This would seem to be the case with the attempt to show that we "are still living in the Cretaceous period," and that the ooze now forming at the bottom of the deep Atlantic is merely a continuation in point of time of the great and well-known formation of the White Chalk. The points of resemblance by which this is sought to be established are these: 1. The Atlantic ooze or "abyssal mud" is a whitish or greyish-looking mud, containing about sixty per cent of carbonate of lime, with from twenty to thirty per cent of silica, and a variable quantity of alumina. When dry, and especially if consolidated, it may fairly be compared in mineral composition to some varieties of Chalk or to Chalk-marl. 2. The abyssal mud of the Atlantic is to a very large extent composed of the microscopic shells of

Foraminfera, some of which are specifically identical with Cretaceous forms, whilst White Chalk is known to be very largely composed of the *débris* of these minute organisms. 3. The ooze contains siliceous sponges, in many respects comparable to the sponges which are so characteristic of the Cretaceous period. 4. The ooze contains Echinoderms, especially Sea-urchins and Crinoids, such as abounded in the Chalk period; whilst one of the latter is related to a Cretaceous type hitherto believed to be extinct.

On the other hand, as pointed out by Sir Charles Lyell, Prof. Prestwich, and other observers, the differences between the Atlantic ooze and the Chalk are, to say the least of it, quite as weighty as the resemblances, if not more so. Chalk is composed of from eighty to as much as ninety-nine per cent of carbonate of lime, and has therefore a very small proportion of any siliceous or aluminous impurity. Secondly, the occurrence of identical species of *Foraminifera* in the two formations amounts to very little; for it is well known that such lowly organised forms of life have an extraordinary power of persistence, surviving geological changes which are fatal to higher organisms. Moreover, it seems certain that the *Foraminifera* of the Atlantic ooze are principally derived from the *surface-waters* of the ocean, so that they prove nothing as to the depth at which the ooze was deposited. Thirdly, Dr Gwyn Jeffreys, one of the highest of authorities upon the Mollusca, has shown that the Molluscan fauna of the Chalk is essentially a *shallow-water* fauna, and certainly cannot be supposed to have inhabited any very great depth. Lastly, the most characteristic of the Chalk fossils, such as the various forms of *Cephalopoda* and Bivalve Molluscs, are entirely wanting in the Atlantic ooze.

Prof. Prestwich concludes that although it is probably true that "some considerable portion of the deep sea-bed of the mid-Atlantic has continued submerged since the period of our Chalk, and although the more adaptable forms of life may have been transmitted in unbroken succession through this channel, the immigration of other and more recent faunas may have so modified the old population that the original Chalk element is of no more importance than is the original British element in our own English people. As well might it have been said in the last century that we were living in the period of the early Britons, because their descendants and language still lingered in Cornwall, as that we are living in the Cretaceous period, because a few Cretaceous forms still linger in the deep Atlantic. Period in Geology must not be confounded with 'system' or 'formation.' The one is only relative, the other definite. A formation is deposited or takes place during a certain time, and that time is *the* period of *the* formation; but *a* geological period may include several formations, and is defined by the preponderance of certain orders, families, or genera, according to the extent of the period spoken of; and the passage of some of the forms into the next geological series does not carry the period with them, any more than would any particular historical epoch be delayed until the survivors of the preceding one had died out. Period is an arbitrary time-division. The Chalk or the 'London Clay' formations mark definite stratigraphical divisions. We may speak of the period of the London Clay, or we may speak of the Tertiary period. It merely refers to the 'time when' either were in course of construction. The occurrence of Triassic forms in the Jurassic series, of Oolitic forms in the Cretaceous series, and of Cretaceous forms in the Eocene, in no way lessens the independence of each series, although it may sometimes render it difficult to say where one series ceases and the other commences. The land and littoral faunas are necessarily more liable to change than a deep-sea fauna, because an island or part of a continent may be submerged, and all on it destroyed, while the fauna of the adjacent oceans would survive; and as we cannot suppose the elevation of entire ocean-beds at the same time, the maritime fauna of one period must be in part almost necessarily transmitted to the next."

In accordance, therefore, with the principles here laid down, we may conclude that it is not correct to say that we "are living in the Cretaceous period," in any other sense than one might say that we are living in the Silurian period, with this difference, that the Cretaceous period is much nearer to us in point of time than the Silurian, and that we can therefore trace a relationship between certain Cretaceous types and certain living forms that we cannot hope to establish in the case of Silurian fossils.

It is to be observed, lastly, that certain classes of animals are always likely to flourish in places and times in which favourable conditions are present, wholly irrespective of any genetic connection between successive faunæ. Thus the conditions present in the deep Atlantic are such as favour the existence of numerous *Foraminifera*, Sponges, *Echinoderms*, &c. Similar conditions existed in the seas in which the Chalk was deposited; and we need not, therefore, be surprised at the predominance of similar organisms in the Cretaceous period. In the same way, there are portions of the Carboniferous Limestone fairly comparable to the Chalk in mineral characters (making due allowance for difference of age), and containing forms of life which may be regarded as representative of the Cretaceous fauna—such as *Foraminifera*, smooth *Terebratulæ*, Crinoids, and Sea-urchins. The conditions, however, present in the deep Atlantic are not *exactly* similar to those under which the Chalk was deposited, for there are certain great classes, such as the *Cephalopoda*, which abounded in the Cretaceous seas, but which seem to have no representatives in the abyssal mud of the Atlantic.

DOCTRINE OF COLONIES.—It only remains in this connection to consider very briefly the doctrine of "colonies," laid down by M. Barrande, the eminent Bohemian palæontologist. It has been laid down as a law that when once a species disappears it never again makes its appearance in the geological record. This is unquestionably true, so long as we remember that it can only apply to cases in which a species has entirely and totally disappeared from the earth, and that it is often very difficult, or altogether impossible, to obtain evidence as to the exact time at which a given species has thus become actually extinct. There are plenty of cases in which a species seemingly disappears in a particular set of rocks, to reappear in some higher and later set of rocks in the same region, whilst its remains are wanting in all the intermediate deposits of the area. It also often occurs that a species, having disappeared in one region, is found in deposits of a later age in another area. The above-mentioned law, therefore, can obviously only hold good of cases in which a species has definitely and finally become extinct; and this implies an amount of knowledge on our part which we seldom or never possess. M. Barrande, however, has pointed out that there are other cases in which groups or species peculiar to one set of beds may appear in a temporary and sporadic manner in a much earlier set of beds, the two deposits thus characterised being separated by beds containing fossils peculiar to the earlier and older series. Thus, the Upper and Lower Silurian rocks of Bohemia are characterised by very distinct assemblages of fossils. It is found, however, that the Lower Silurian rocks contain in places a group of fossils characteristic of the Upper Silurian series. The beds containing this "colony" of Upper Silurian forms are succeeded by strata filled with Lower Silurian fossils; and it is only after several alternations of this kind that the Upper Silurian fauna comes in definitely and generally. These temporary appearances of a later fauna in the midst of an older fauna are termed by M. Barrande "colonies," and he explains their occurrence as follows: If we suppose the seas of the Bohemian area to have been peopled with Lower Silurian animals at a time when other portions of Europe were covered by a sea containing Upper Silurian animals, and suppose the former area to have been shut off from the latter by a land-barrier, we can readily understand how the "colonies" were produced. If, from any cause, a channel of communication were opened between the Bohemian area and the general area of Northern Europe, an immigration of species would take place from the latter into the former area. The Upper Silurian species of the latter area would thus be imported, in greater or less numbers, into the midst of the general Lower Silurian fauna of Bohemia, and would be preserved in the Lower Silurian rocks. If, however, the channel of communication were speedily closed, so that the new-comers could not be constantly reinforced by fresh immigrants, the "colonial" species would die out, and the general Lower Silurian fauna would again reign supreme. A reopening of the channel of communication would allow of a fresh immigration and the formation of a fresh "colony," and the process might be indefinitely repeated. Finally, however, we must suppose that the Bohemian area was permanently thrown open to immigration from the general European area, when the Upper Silurian fauna of the latter would succeed in permanently and completely displacing the old Lower Silurian fauna of the former region. The phenomenon, therefore, of "colonies" may be defined as "the coexistence of two general faunas, which, considered in their entirety, are nevertheless distinct;" and it is to be regarded as merely a case of migration under certain peculiar and exceptional circumstances.

It must be borne in mind that the phenomenon of "colonies," even if we regard it as established beyond contradiction in the Bohemian area, has never been recognised with any certainty elsewhere, and that it must, under any circumstances, be of *very* rare and exceptional occurrence. In fact, if "colonies" were at all common, it is quite clear that we should before now have met with plenty of instances of their existence, seeing that we have innumerable examples in which we can show that the migration of marine organisms from one area to another has taken place; and it is further clear that the occurrence of "colonies" as a common thing would greatly lower the general value of palæontological evidence as bearing upon the age of fossiliferous strata.

CHAPTER IV.

THE IMPERFECTION OF THE PALÆONTOLOGICAL RECORD.

As has been already pointed out, the series of the stratified formations is an imperfect one, and is likely ever to remain so. The causes of this "imperfection of the geological record," as it has been termed by Darwin, are various; but it is chiefly to be ascribed to our as yet incomplete knowledge of the geology of vast areas of the earth's surface, to denudation, and to the fact that many of the missing groups are buried beneath other deposits, whilst more than half of the superficies of the globe is hidden from us by the waters of the sea. The imperfection of the geological record necessarily implies an equal imperfection of the "palæontological record;" but, in truth, the record of life is far more imperfect than the mere physical series of deposits. As we are here chiefly concerned with the biological aspect of the question, we may advantageously consider some of the main causes of the numerous breaks and gaps in the palæontological record at some length.

I. Causes of the Absence of Certain Animals in Fossiliferous Deposits.—In the first place, even if the series of the stratified deposits had been preserved to us in its entirety, and we could point to the sedimentary accumulations belonging to every period of the earth's history, there would still be enormous deficiencies in the palæontological record, owing to the differences in the facility with which different animals may be preserved as fossils. This subject is sufficiently important to render it advisable to consider each of the primary groups of the animal kingdom separately from this point of view:—

a. Protozoa.—As regards the sub-kingdom of the *Protozoa*, the entire classes of the *Gregarinidæ* and *Infusorian Animalcules*, from their absence of hard parts, must ever remain unrepresented in a fossil condition. One or two of the latter, however, possess an integumentary covering capable under favourable circumstances of being preserved in rocks of recent age. The *Monera* present no structures capable of fossilisation; and the same may be said of the *Amœbea*, though one or two of the latter have a carapace which might possibly be preserved. The remaining Rhizopodous orders—viz., the *Foraminifera, Radiolaria*, and *Spongida*—almost invariably develop hard structures of lime or flint; and all these orders, therefore, have left abundant traces of their existence in past time.

b. Cœlenterata.—Amongst the Cœlenterate animals, the Fresh-water Polypes (*Hydra*), the Oceanic Hydrozoa, the Jelly-fishes (*Medusidæ*), the Sea-blubbers (*Lucernarida*), the Sea-anemones (*Actinidæ*), and the *Ctenophora* are destitute of hard parts which could be preserved as fossils. The Sea-blubbers, however, supply us with an instance of how a completely soft-bodied creature may leave traces of its past existence; for there is no doubt that impressions left by the stranded carcasses of these animals have been detected in certain fine-grained rocks (the Lithographic Slate of Solenhofen). On the other hand, the coralligenous Zoophytes or "corals" (comprising the *Zoantharia sclerodermata* and *sclerobasica*, and most of the *Alcyonaria*) possess hard parts capable of preservation, and the same is the case with most of the Hydroid Zoophytes. Accordingly, there are few more abundant fossils than corals; whilst the large extinct group of the *Graptolites* is generally placed in the vicinity of the Sea-firs (Sertularians).

c. Annuloida.—In this sub-kingdom the great class of the *Echinodermata* may be said to be represented more or less completely by all its orders. In the Sea-cucumbers (*Holothuroidea*), however, the calcareous structures so characteristic of the integuments of the other Echinoderms are reduced to their minimum; and accordingly, the evidence of the past existence of these creatures is of the most scanty description. The other great class of the *Annuloida* (viz., the *Scolecida*) comprises animals almost without exception destitute of hard parts, and which mostly live parasitically in the interior of other animals (*e.g.*, the Tape-worms, Suctorial-worms, Roundworms, &c.) We are therefore without any geological evidence of the former existence of *Scolecida*, though no doubt can be reasonably entertained but that the group dates back to a time long anterior to the present fauna.

d. Annulosa.—Many of the lower Annulose animals, such as Leeches, Earth-worms, and Errant Annelides, possess few or no structures by which we could expect to get direct evidence of their past existence. The last of these, however, have left ample traces of their former presence in the form of burrows or "tracks" upon the mud and sand of ancient sea-bottoms, and are known also by their horny jaws; while the so-called "Tubicolar" Annelides are well represented by their investing tubes. In the case of the higher *Annulosa*, another law steps in to regulate their comparative abundance as fossils. Most, in fact almost all, fossiliferous formations have been deposited in water; and of necessity, therefore, most fossils are the remains of animals whose habits are naturally aquatic. As most deposits, further, are not only aqueous, but are also marine, most fossils are those of sea-animals. It follows, therefore, that the remains of air-breathing animals, whether these be terrestrial or aerial, can only be preserved in an accidental manner, so to speak; except the animal inhabit water (as the Cetaceans do), or except in the rare instances in which old land-surfaces have been buried up by sediment, and thus partially kept for our inspection. In accordance with this law, the most important and abundant fossil Annulose animals are *Crustaceans;* since these not only have a resisting shell or "exoskeleton," but are also generally aquatic in their habits. The air-breathing classes of the *Myriapoda* (Centipedes and Millipedes), the *Arachnida* (Spiders and Scorpions), and the *Insecta* or true Insects, on the other hand, have been much less commonly and completely preserved, though many of them are perfectly capable of being fossilised. Almost all such remains, however, as we have of these three great classes, are the remains of isolated individuals, which may have been accidentally drowned; or else they occur in hollow trees, or in fragments of ancient soils, or in vegetable accumulations such as coal and peat. There is, however, a considerable number of aquatic insects (but exclusively in fresh water), and there are many insects the larvæ of which inhabit water, whether this be fresh or salt; so that instances of these occurring as fossils are not very infrequent.

e. Mollusca.—This sub-kingdom requires little notice, since the greater number of its members possess hard structures capable of being preserved in a fossil condition. Thus, the horny or calcareous polypidoms of many of the *Polyzoa*, the shells of the Brachiopods, the true Bivalves, and most of the *Gasteropoda*, the internal skeletons of the Cuttle-fishes, and the chambered shells of the Tetrabranchiate Cephalopods, all occur more or less abundantly as fossils. The entire class of the Tunicaries, however, presents (with one or two exceptions) no hard structures, and is hence not with certainty known by any fossil representative. Amongst the *Gasteropoda*, again, the Sea-slugs and their allies (*Nudibranchiata*) possess no shell, and are unknown to the palæontologist; whilst the shell of the Land-slugs is extremely minute, and has not been certainly recognised as fossil. Lastly, the air-breathing terrestrial Molluscs, from their habits, rarely occur as fossils; whilst those which inhabit rivers, ponds, and lakes are less largely represented than marine forms, owing to the preponderance of salt-water deposits over those of fresh water.

f. Vertebrata.—The majority of Vetebrate animals possess a bony skeleton, so that their preservation in a fossil state—so far as this point is concerned—is attended with no difficulty. Some of the fishes, however (such as the Lancelet, the Hag-fishes, and the Lampreys), have no scales, and either possess no "endoskeleton" or have one which is almost wholly cartilaginous. The only evidence, therefore, which could be obtained of the past existence of such fishes would be afforded by their teeth; but these are wanting in the Lancelet, and are very small in the Lampreys: so that we need not wonder that these fishes are unknown as fossils. The higher groups of the fishes, however, taking everything into consideration, may be said to be abundantly represented in a fossil condition by their scales, bones, teeth, and defensive spines.

The Amphibians are tolerably well represented by their bones and teeth, and, as regards one extinct order, by integumentary plates as well. They have also left many traces of their existence in the form of footprints. Most living Amphibians, however, frequent fresh waters, or spend a great part of their time upon the land; and hence their remains would not be apt to be preserved in marine deposits.

The abundance of Reptiles as fossils naturally varies much, according to the habits of the different orders. Of the living orders, the Chelonians (Tortoises and Turtles) are by no means rare; since many of them are habitual denizens of fresh water or of the sea, whilst all are provided with a hard integumentary skeleton. The Snakes are mainly represented by forms which frequented water, and especially by marine forms. The Lizards (*Lacertilia*) live mainly upon the land, and do not therefore abound as fossils; but some extinct forms (the Mosasauroids) were marine in their habits, and have consequently been pretty fully preserved. The *Crocodilia*, again, are so essentially aquatic in their habits, that their comparative frequency in aqueous deposits is no matter of wonder, especially if we recollect that many of the extinct members of this order seem to have frequented the sea itself. Of the extinct orders of Reptiles, the great *Ichthyosauri* and the *Plesiosauri* and their allies were marine in their habits, and their remains occur in what may fairly be called profusion. The Flying Reptiles, or *Pterodactyles*, would not seem to have any better chance of being preserved than Birds, if as good, yet their remains occur by no means very rarely in certain formations. The terrestrial *Deinosaurs* and *Dicynodonts*, again, come very much under the laws which regulate the preservation of Mammals as fossils; and their remains are chiefly, but not exclusively, to be found in fluviatile deposits.

As regards Birds, their powers of flight, as pointed out by Sir Charles Lyell, would save them from many destructive agencies, and the lightness of their bones would favour the long floating of the body in water, and thus increase the chances of its being devoured by predaceous animals. In accordance with these considerations the most abundant remains of Birds are referable to large wingless forms, to which the power of saving themselves from their enemies by flight was denied, whilst most of their bones were filled with marrow instead of air. Next in abundance after these come the remains of birds which frequent the sea-shore, lakes, estuaries, or rivers, or which delight in marshy situations.

Lastly, as regards Mammals, the record is far from being a full one, and from obvious causes. The great majority of Mammals live on land, and therefore are not likely to be buried in aqueous, and especially in marine, accumulations. That this cause is the chief one which has operated against the frequent preservation of Mammalian remains is shown by the fact that when we exhume an old land-surface, the remains of Mammals may be found in tolerable plenty. The strictly aquatic Mammals—such as Whales, Dolphins, and the like—are, of course, much more likely to have been preserved as fossils than the strictly terrestrial forms; but their want of integumentary hard structures places them at a disadvantage in this respect as compared with fishes. In a general way, we may conclude that the preservation of the terrestrial Mammals as fossils is due to the comparatively rare occurrence of a stray individual being killed whilst swimming a river or some other piece of water, or being mired in a bog, or to the bones of one that had died on land being washed into some stream, and thence into a lake or into the sea, by floods; but there are other cases for which a different explanation must be sought. The most abundant remains of Mammals have been found in deposits which have been laid down in lakes.

II. Unrepresented Time.—In the second place, we have seen that the *geological* record is very imperfect, and this of necessity causes vast gaps in our palæontological knowledge. In this connection we may briefly consider the evidence which we possess as to the immensity of the "unrepresented time" between some of the great formations, and no better example can be chosen than that of the Cretaceous and Eocene rocks, as developed in Europe. In considering such a case, the evidence may be divided into two heads, the one palæontological, the other purely physical, and each may be looked at separately.

The Chalk, as is well known, constitutes in Britain the highest member of the Cretaceous formation, and is the highest deposit there known as appertaining to the great Secondary or Mesozoic series. It is directly overlaid in various places by strata of Eocene age, which form the base of the great Tertiary or Kainozoic series of rocks. The question, then, before us is this, What evidence have we as to the lapse of time represented merely by the dividing-line between the highest beds of the Chalk and the lowest beds of the Eocene?

Taking the palæontological evidence first, it is found that out of five hundred species of fossils known in the Upper Cretaceous beds, only one *Brachiopod* and a few *Foraminifera* have hitherto been detected in the immediately overlying Eocene beds. These latter, on the contrary, are replete with organic remains wholly distinct from those of the Cretaceous beds. It may be said, therefore, that the very extensive assemblage of animals which lived in the later Cretaceous seas of Britain had entirely passed away and become a thing of the past, before a single grain of the Eocene rocks had been deposited. Now it is of course open to us to believe that the animals of the Chalk sea were suddenly extinguished by some natural agencies unknown to us, and that the animals of the Eocene sea had been in as sudden and as obscure a manner introduced *en masse* into the same waters. This theory, however, calls upon the stage forces of which we know nothing, and is contradicted by the whole tenor of the operations which we see going on around us at the present day. It is preferable, therefore, to believe that no such violent processes of destruction and re-peopling took place, but that the marked break in the life of the two periods indicates an enormous lapse of time. The Cretaceous animals, in consequence of the elevation of the British area at the close of the Cretaceous period, must have mostly migrated, some doubtless perishing, and others probably becoming modified in the process. When the British area became once more submerged beneath the sea, and became again a fitting home for marine life, an immigration into it would set in from neighbouring seas. By this time, however, the Cretaceous animals must have mostly died out, or must have become greatly changed in their characters; and the new immigrants would be forms characteristic of the Lower Eocene. How long the processes here described may have taken, it is utterly impossible to say, even approximately. Judging, however, from what we can observe at the present day, the palæontological break between the Chalk and the Eocene indicates a perfectly incalculable lapse of time; for all species change or die out slowly, marine species especially so; and we have here the disappearance of a large fauna almost in its entirety, and its replacement by another wholly distinct.

In the second place, to come to the physical evidence, the Eocene strata in Britain are seen to rest upon an eroded and denuded surface of Chalk, filling up "pipes" and winding hollows which descend far below the general surface of the latter. Not only so, but the base of the Eocene rocks is commonly composed of a bed of rolled and rounded flints, derived from the Chalk, affording incontestable proof that the Chalk had been greatly worn down and removed by denudation before the Eocene beds were deposited upon its surface. In short, the Eocene rocks repose "unconformably" upon the Chalk, and this, as is well known, indicates the following series of phenomena: Firstly, the Chalk was deposited in horizontal layers at the bottom of the Cretaceous sea. Secondly, at some wholly indefinite time after its deposition, after it had become more or less consolidated, the Chalk must have been raised by a gradual process of elevation above the level of the sea, during which it would inevitably suffer vast denudation. Thirdly, after another wholly indefinite period, the Chalk was again submerged beneath the sea, in which process it would be subjected to still further denudation, and an approximately level surface would be formed upon it. Fourthly, strata of Eocene age were deposited upon the denuded surface of the Chalk, filling up all the hollows and inequalities of its eroded surface (fig. 10).

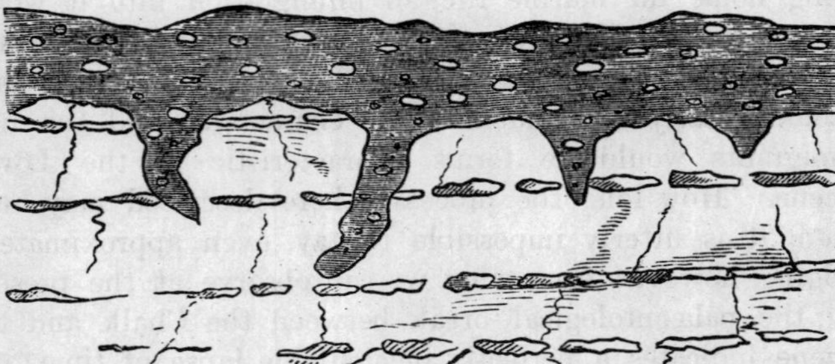

Fig. 10.—Section showing strata of Tertiary age a), resting upon a worn and denuded surface of White Chalk (b), the stratification of which is marked by lines of flints.

In the unconformability, then, between the Chalk and the Eocene rocks, we have unequivocal evidence—irrespective of anything that we learn from Palæontology—that the break between the two formations was one of enormous length. In Britain the interval of time thus indicated is not represented by any deposits; and in Europe generally there are but fragmentary traces of such. We may be quite sure, however, that during the time represented in Britain by the mere line of unconformability between the Chalk and the Eocene, there were somewhere deposited considerable accumulations of sediment. Whether we shall ever succeed in discovering these, or any part of these, is, of course, uncertain. We may be certain, however, that such deposits, if ever discovered, will prove to be charged with the remains of animals more or less intermediate in character between those of the Cretaceous and those of the Eocene period; and the huge gap now existing between these formations will thus be more or less completely bridged over. Indeed, in North America we actually find such a series of deposits, characterised by a transitional series of fossils, lying between the highest undoubted Cretaceous and the lowest unquestioned Tertiary strata. The deposits in question—the so-called "Lignitic Series"—are very thick, and the intermixture of Secondary and Tertiary types of life which they exhibit is so complete, that it has been found a matter of great difficulty to assign them definitely either to the Cretaceous or to the Eocene.

Amongst other well-known instances of more or less general unconformity in the stratified series, may be mentioned that between the Lower and Upper Silurian (not always present), that between the Lower and Upper Old Red Sandstone (also not universal), that between the Carboniferous and Permian rocks, that between the Permian and Triassic rocks (not universal), and that between the Lower and Upper Cretaceous rocks. All these physical breaks are accompanied by more or less extensive palæontological breaks as well. Other breaks which are rendered less important by the absence or scarcity of fossils, or which are as yet not thoroughly established, are those between the Lower and Upper Laurentian rocks, the Upper Laurentian and Huronian, and the Upper Cambrian and Lower Silurian.

It may not be out of place to point out that the unconformabilities here indicated must in no way be confounded with the common cases in which beds of one age rest unconformably upon beds far older than themselves. When, for example, we find beds of Carboniferous age reposing unconformably upon Silurian strata, this merely indicates that, in the particular locality under examination, the Devonian or Old Red Sandstone is amissing. This absence of a whole formation in any given region merely indicates that the area was dry land during the period of that formation, or that if any rocks of this age were deposited in this locality, they were removed by denudation before the higher group was laid down. The instances above spoken of, as where the Carboniferous rocks are succeeded unconformably by the Permian, though essentially of the same nature, are distinguished by an important point. In the former case we know what formation is wanting, and we can intercalate it from foreign areas, and thus complete the series. In the latter case we have two successive formations in unconformable junction, and we are not acquainted with any intermediate group of strata which could be intercalated from any other locality.

From the above facts, then, we learn that one of the chief causes of the imperfection of the palæontological record is to be found in the vast spaces of time which separate most of the great "formations," and which, so far as we yet know, are not represented by any formation of rock. In process of time we shall doubtless succeed in finding deposits to account for more or less of this "unrepresented time," but much will ever remain for which we cannot hope to find the representative sediments. It only remains to add that we have ample evidence within the limits of each formation, and wholly irrespective of any want of conformity, of such lengthened pauses in the work of deposition as to have allowed of great zoological changes in the interim, and to have thus caused irremediable blanks in the palæontological record. The work of rock-deposition is at best an intermittent process; the changes in a fauna, if slowly effected, are continuous. Thus there are scores of instances in which the fauna of a given bed, perhaps but a few inches in thickness, differs altogether from that of the beds immediately above and below, and is characterised by species peculiar to itself. In such cases we can only suppose, that though no physical break can be detected, the deposition of sediment was interrupted by pauses of incalculable length, during which no additional material was added to the sea-bottom, whilst time was allowed for the dying out of old species and the coming in of new. The incessant repetition of such intervals of unrepresented time throughout the whole stratified series is convincing proof that the palæontological record is, and ever must be, a mere excerpt from the biological annals of the globe.

III. Thinning out of Beds.—Another cause by which the continuity of the palæontological record is affected is what is technically called the "thinning out" of beds. Owing to the mode in which sedimentary rocks are produced, it is certain that there must be for every bed a point whence the largest amount of sediment was derived, and in the neighbourhood of which the bed will therefore be thickest. Thus, if we take a series of beds, such as sandstones and conglomerates, which are the product of littoral action, and are deposited in shallow water near a coast-line, it will be found that these gradually decrease in thickness, or "thin out," as we pass away from the coast in the direction of deep water. On approaching deep water, however, we might find that, though the sandstones were rapidly dying out, the thickness of the entire series might still be preserved, owing to the commencement now of some deep-water deposit, such as limestone. The beds of limestone would at first be very thin, but in proceeding still in the direction of deeper water, we should find that they would gradually expand till they reached a point of maximum thickness, on the other side of which they would gradually thin out. Each individual bed, therefore, in any group of stratified rocks, may be regarded as an unequal mass, thickest in the centre, and gradually tapering off or "thinning out" in all directions towards the circumference (fig. 11).

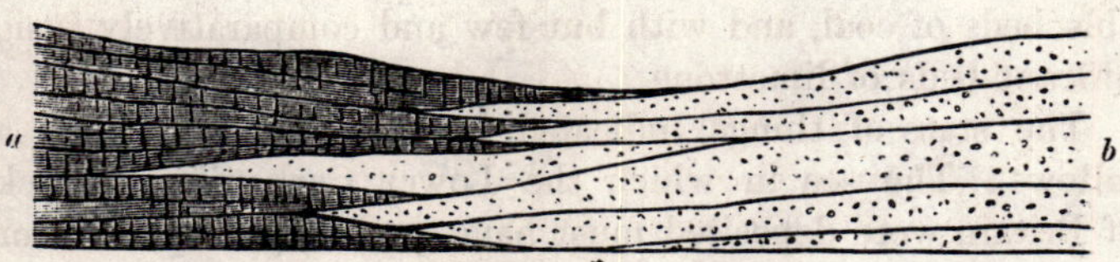

Fig. 11.—Diagram to show the "thinning out" of beds. *a*, Sandstones and Conglomerates; *b*, Limestones.

In a general way this holds good, not only for any particular bed, but for any particular aggregation or group of beds which we may choose to take. In the case, namely, of every group of beds, there must have been a particular point whither sediment was most abundantly conveyed, or where the other conditions of accumulation were especially favourable. At this point, therefore, the beds are thickest, and from this they thin out in all directions. It need scarcely be pointed out, indeed, that some such state of things is unavoidable in the case of every bed or group of beds, since no sea is boundless, and the sedimentary deposits of every ocean must come to an end somewhere.

An excellent example of the phenomena above described may be derived from the Lower Carboniferous rocks of Britain. Here we may start in South Wales and in Central England with the Carboniferous Limestone as a great calcareous mass over 1000 feet in thickness, and almost without a single intercalated layer of shale. Passing northwards, some of the beds of limestone begin to thin out, and their place is taken by strata of a different mineral nature, such as sandstone, grit, or shale. The result of this is, that by the time we have followed the Carboniferous Limestone into Yorkshire and Westmorland, in place of a single great mass of limestone, we have an equivalent mass of alternating strata of limestone, sandstone, grit, and shale, with one or two thin seams of coal—the limestones, however, still bearing a considerable proportion to the whole. Passing still further northwards, the limestones go on thinning out, till in Central Scotland, in place of the dense calcareous accumulations of Derbyshire, the Lower Carboniferous series consists of a great group of sandstones, grits, and shales, with thick and workable beds of coal, and with but few and comparatively insignificant beds of limestone.

The state of things indicated by these phenomena is as follows: The sea in which the Lower Carboniferous rocks of Britain were deposited must have gradually deepened from north to south. The land and coast-line whence the coarser mechanical sediments were derived must have been placed somewhere to the north of Scotland, and the deepest part of the ocean must have been somewhere about Derbyshire. Here the conditions for lime-making were most favourable, and here consequently we find the greatest thickness of calcareous strata, and the smallest intermixture of mechanical deposits.

The palæontological results of this are readily deducible. The entire Lower Carboniferous series of Britain was probably deposited in a single ocean, apparently destitute of land-barriers; and consequently, taken as a whole, the fauna of this series may be regarded as one and indivisible. The conditions, nevertheless, which obtained in different parts of this area were very different; and, as a necessary result, certain groups of animals flourished in certain localities, and were absent or but scantily represented in others. In the deeper parts of the area we have an abundance of Corals, with Crinoids, and at times *Foraminifera*. In the shallower parts of the area there is, on the other hand, a predominance of forms which affect water of no great depth. Still there is no difference in point of time between the deposits of different parts of the area; and in order to obtain a true notion of the Lower Carboniferous fauna, we must add the fossils derived from one portion of the area to those derived from another.

In many cases, however, we are acquainted with but one class of deposits belonging to a given period. We may have the deep-sea deposits of the period only, or we may know nothing but its littoral accumulations. In either case it is clear that there is an imperfection of the palæontological record; for we cannot have even a moderately complete record of the marine animals alone of a particular period unless we have access to a complete series of the deposits laid down in the seas of that period.

IV. DISAPPEARANCE OF FOSSILS.—The last subject which need be mentioned in connection with the imperfection of the palæontological record is that of the disappearance of fossils from rocks originally fossiliferous. This, as a rule, is due to "metamorphism"—that is to say, the subjection of the rock to a sufficient amount of heat to cause a rearrangement of its particles. When of at all a pronounced character, the result of metamorphism is invariably the obliteration of any fossils which might have been originally present in the rock. To this cause must be set down many great gaps in the palæontological record, and the irreparable loss of much fossil evidence. The most striking example which is to be found of this is the great Laurentian series, which comprises some 30,000 feet of highly metamorphosed sediments, but which, with one not absolutely certain exception, has as yet yielded no remains of life, though there is strong evidence of the former existence in it of fossils.

Another not uncommon cause of the disappearance of

organic remains from originally fossiliferous deposits is the percolation through them of water holding carbonic acid in solution. By this means fossils of a calcareous nature are dissolved out of the rock, and may leave no traces behind. This cause, however, can only operate to any extent in more or less loose and porous arenaceous deposits.

Lastly, "cleavage" may be mentioned as a common cause of the disappearance of fossils. The cleavage, however, must be very intense, if it actually prevents the recognition of the deposit as one in which fossils formerly existed, though cases are not uncommon in which this occurs through thousands of feet of strata. As a more general rule, however, it is not very difficult to determine whether a cleaved rock has ever contained fossils or not, though it may be quite impossible to make out the exact nature and character of the organic remains.

CHAPTER V.

CONCLUSIONS TO BE DRAWN FROM FOSSILS.

WE have already seen that geologists have been led by the study of fossils to the all-important generalisation that the vast series of the Fossiliferous or Sedimentary rocks may be divided into a number of definite groups or "formations," each of which is characterised by its organic remains. It may simply be repeated here that these formations are not properly and strictly characterised by the occurrence in them of any one particular fossil. It may be that a formation contains some particular fossil, or fossils, not occurring out of that formation, and that in this way an observer may identify a given group with tolerable certainty. It very often happens, indeed, that some particular stratum, or sub-group of a series, contains peculiar fossils, by which its existence may be determined in various localities. As before remarked, however, the great formations are characterised properly by the association of certain fossils, by the predominance of certain families or orders, or by an *assemblage* of fossil remains representing the "life" of the period in which the formation was deposited.

Fossils, then, enable us to determine the *age* of the deposits in which they occur. Fossils further enable us to come to very important conclusions as to the mode in which the fossiliferous bed was deposited, and thus as to the condition of the particular district or region occupied by the fossiliferous bed at the time of the formation of the latter. If, in the first place, the bed contain the remains of animals such as now inhabit rivers, we know that it is "fluviatile" in its origin, and that it must at one time have either formed an actual river-bed, or been deposited by the overflowing of an ancient stream. Secondly, if the bed contain the remains of shell-fish, minute crustaceans, or fish, such as now inhabit lakes, we know that it is "lacustrine," and was deposited beneath the waters of a former lake. Thirdly, if the bed contain the remains of animals such as now people the ocean, we know that it is "marine" in its origin, and that it is a fragment of an old sea-bottom.

We can, however, often determine the conditions under which a bed was deposited with greater accuracy than this. If, for example, the fossils are of kinds resembling the marine animals now inhabiting shallow waters, if they are accompanied by the detached relics of terrestrial organisms, or if they are partially rolled and broken, we may conclude that the fossiliferous deposit was laid down in a shallow sea, in the immediate vicinity of a coast-line, or as an actual shore-deposit. If, again, the remains are those of animals such as now live in the deeper parts of the ocean, and there is a very sparing intermixture of extraneous fossils (such as the bones of birds or quadrupeds, or the remains of plants), we may presume that the deposit is one of deep water. In other cases, we may find, scattered through the rock, and still in their natural position, the valves of shells such as we know at the present day as living buried in the sand or mud of the sea-shore or of estuaries. In other cases, the bed may obviously have been an ancient coral-reef, or an accumulation of social shells, like Oysters. Lastly, if we find the deposit to contain the remains of marine shells, but that these are dwarfed of their fair proportions and distorted in figure, we may conclude that it was laid down in a brackish sea, such as the Baltic, in which the proper saltness was wanting, owing to its receiving an excessive supply of fresh water.

In the preceding, we have been dealing simply with the remains of aquatic animals, and we have seen that certain conclusions can be accurately reached by an examination of these. As regards the determination of the conditions of deposition from the remains of aerial and terrestrial animals, or from plants, there is not such an absolute certainty. The remains of land-animals would, of course, occur in "sub-aerial" deposits—that is, in beds, like blown sand, accumulated upon the land. Most of the remains of land-animals, however, are found in deposits which have been laid down in water, and they owe their present position to having been drowned in rivers or lakes, or carried out to sea by streams. Birds, Flying Reptiles, and Flying Mammals might also similarly find their way into aqueous deposits; but it is to be remembered that many birds and mammals habitually spend a great part of their time in the water, and that these might therefore be naturally expected to present themselves as fossils in Sedimentary rocks. Plants, again, even when undoubtedly such as must have grown on land, do not prove that the bed in which they occur was formed on land. Many of the remains of plants known to us are extraneous to the bed in which they are now found, having reached their present site by falling into lakes or rivers, or being carried out to sea by floods or gales of wind. There are, however, many cases in which plants have undoubtedly grown on the very spot where we now find them. Thus it is now generally admitted that the great coal-fields of the Carboniferous age are the result of the growth *in situ* of the plants which compose coal, and that these grew on vast marshy or partially submerged tracts of level

Fig. 12.—Erect Tree containing Reptilian remains. Coal-measures, Nova Scotia. (After Dawson.)

alluvial land. We have, moreover, distinct evidence of old land-surfaces, both in the Coal-measures and in other cases (as, for instance, in the well-known "dirt-bed" of the Purbeck series). When, for example, we find the erect stumps of trees standing at right angles to the surrounding strata, we know that the surface through which these send their roots was at one time the surface of the dry land, or, in other words, was an ancient soil (fig. 12).

CONCLUSIONS AS TO CLIMATE.—In many cases fossils enable us to come to important conclusions as to the climate of the period in which they lived, but only a few instances of this can be here adduced. As fossils in the majority of instances are the remains of marine animals, it is mostly the temperature of the sea which can alone be determined in this way; and it is important to remember that, owing to the existence of heated currents, the marine climate of a given area does not necessarily imply a correspondingly warm climate in the neighbouring land. Land-climates can only be determined by the remains of land-animals or land-plants, and these are comparatively rare as fossils. It is also important to remember that all conclusions on this head are really based upon the present distribution of animal and vegetable life on the globe, and are therefore liable to be vitiated by the following considerations:—

a. Most fossils are extinct, and it is not certain that the habits and requirements of any extinct animal were exactly similar to, or even at all resembling, those of its nearest living relative.

b. When we get very far back in time, we meet with groups of organisms so unlike anything we know at the present day as to render all conjectures as to climate founded upon their supposed habits more or less uncertain and unsafe.

c. In the case of marine animals, we are as yet very far from knowing the exact limits of distribution of many species within our present seas; so that conclusions drawn from living forms as to extinct species are apt to prove incorrect. For instance, it has recently been shown that many shells formerly believed to be confined to the Arctic Seas have, by reason of the extension of Polar currents, a wide range to the south; and this has thrown doubt upon the conclusions drawn from fossil shells as to the Arctic conditions under which certain beds were supposed to have been deposited.

d. The distribution of animals at the present day is certainly dependent upon other conditions beside climate alone; and the causes which now limit the range of given animals are certainly such as belong to the existing order of things. But the establishment of the present order of things does not date back in many cases to the introduction of the present species of animals. Even in the case, therefore, of existing species of animals, it can often be shown that the past distribution of the species was different formerly to what it is now, not necessarily because the climate has changed, but because of the alteration of other conditions essential to the life of the species or conducing to its extension.

Still, we are in many cases able to draw completely reliable conclusions as to the climate of a given geological period, by an examination of the fossils belonging to the period. Among the more striking examples of how the past climate of a region may be deduced from the study of the organic remains contained in its rocks, the following may be mentioned: It has been shown that in Eocene times, or at the commencement of the Tertiary period, the climate of what is now Western Europe was of a tropical or subtropical character. Thus the Eocene beds are found to contain the remains of shells such as now inhabit tropical seas, as, for example, Cowries and Volutes; and with these are the fruits of palms, and the remains of other tropical plants. It has been shown, again, that in Miocene times, or about the middle of the Tertiary period, Central Europe was peopled with a luxuriant flora resembling that of the warmer parts of the United States, and leading to the conclusion that the mean annual temperature must have been at least 30° hotter than it is at present. It has been shown that, at the same time, Greenland, now buried beneath a vast ice-shroud, was warm enough to support a large number of trees, shrubs, and other plants, such as inhabit the temperate regions of the globe. Lastly, it has been shown, upon physical as well as palæontological evidence, that the greater part of the North Temperate Zone, at a comparatively recent geological period, has been visited with all the rigours of an Arctic climate, resembling that of Greenland at the present day. This is indicated by the occurrence of Arctic shells in the superficial deposits of this period, whilst the Musk-ox and the Reindeer roamed far south of their present limits.

CHAPTER VI.

DIVISIONS OF THE ANIMAL KINGDOM AND SUCCESSION OF ORGANIC TYPES.

It seems hardly necessary to remark that Palæontology, as a science, is based upon the kindred sciences of Zoology and Botany, and that no satisfactory acquaintance with the former can be arrived at without the previous acquisition of some knowledge of the latter. It cannot be pretended to teach here even the rudiments of these sciences, but there are a few points which may be noticed as having a special bearing upon the study of Palæontology.

Classification of the Animal Kingdom.—Leaving the vegetable kingdom till we come to speak of fossil plants, a few remarks may be made on the classification of the animal kingdom. Vast as is the number of known animals, all, whether living or extinct, may be classed under some five or six primary divisions or "morphological types," which are technically spoken of as the "sub-kingdoms." All the animals in any one sub-kingdom agree with one another in their structural type, or in the fundamental plan upon which they are constructed; and they differ from one another simply in the modifications of this common plan. No comparison, therefore, is possible between an animal belonging to one sub-kingdom, and one belonging to another, since their distinguishing characters are the result of the modification of two essentially different ground-plans. Hence it is possible to arrange the animals of any one sub-kingdom in something like a linear series, in which the lowest of the series most closely approaches the primitive or ideal form of the sub-kingdom, whilst the highest exhibits the greatest amount of complexity and specialisation of this type. But it is not possible to establish any such linear classification for the animal kingdom as a whole. Given an animal of a lower "sub-kingdom" than another animal, no amount of complexity, no specialisation of organisation, can raise the former above the latter. The one may be the result of the high evolution of a low morphological type, the other may be the result of the low evolution of a higher morphological type, but the superiority of the ground-plan gives the latter the higher place. We must therefore abandon the idea that it is possible to establish a linear classification of the animal kingdom.

The terms "class," "order," "genus," "sub-genus," "species," and "variety," are employed by the palæontologist in precisely the same sense, and with precisely the same limitations, as by the zoologist. We must notice, however, that a *palæontological* "species" has not always or necessarily the same value as that which a *zoological* species ought invariably to possess. This arises from the fact that the determination of fossil species is, almost without exception, based solely upon the characters of the *hard* parts of the animal—these, also, being often but imperfectly preserved. A fossil species, therefore, cannot, from the nature of things, be as thoroughly defined as a living one; and it is both possible and probable that variations in the form of the skeleton, especially if an integumentary one, may often depend upon mere individual, sexual, or local peculiarities, which could be at once discovered in the case of living forms, but which can hardly be detected as regards extinct types. Moreover, there is a *practical* inconvenience attending the use of the terms "variety" and "sub-genus" in palæontology, which is not found in zoology, owing to the very different nature of the working material of these two sciences. Many palæontologists, therefore, prefer, as we think rightly, to follow the general practice of giving distinct names to "varieties" and "sub-genera," thus practically raising them to the rank of "species" and "genera;" and this practice can hardly be injurious if accompanied with the well-understood reservation that this is done as a matter of convenience only, and that a somewhat wider and looser signification is to be given to the terms "species" and "genera" in palæontology than would be admissible in zoology. At the same time, this practice may be, and has been, carried too far; and in the case of very variable or "protean" species, it is certainly advisable to adhere to the plan usually adopted by British palæontologists—namely, to define the species by its central type, and to group the variable forms under this type as varieties.

The duration in time, or "vertical range," of fossil species varies greatly in different cases. Some species have an extraordinarily extended range, sometimes passing through two or three entire formations, and in such cases they generally exhibit numerous varieties. Others, again, are singularly restricted, and do not pass beyond the limits of a single subdivision of a formation, or sometimes even a single bed. In any case, when a species has once fairly died out, it never reappears again. As a general rule, it is the animals which have the lowest and simplest organisation that have the longest range in time, and the additional possession of microscopic or minute dimensions seems also to favour longevity. Thus some of the *Foraminifera* appear to have survived, with little or no perceptible alteration, from the Silurian period to the present day; whereas large and highly-organised animals, though long-lived as *individuals*, rarely seem to live long *specifically*, and have, therefore, usually a restricted vertical range. Some genera, as some species, are short-lived; whereas others extend through a succession of zoological periods with extraordinarily little modification. Among these "persistent types" may be specially mentioned the genus *Lingula* among the *Brachiopoda*, and *Nautilus* among the *Cephalopoda*, of which the former commenced in the Cambrian and the latter in the Silurian, and both of which are represented by living species. While the great majority of fossils are extinct, and while many of them are extremely unlike any existing forms, no fossil animal has hitherto been detected which cannot be referred to one or other of the existing *sub-kingdoms*. No fossil animals, indeed, possess peculiarities so great as to entitle them to be placed in any *class*, other than in one of the classes of recent forms. On the other hand, the differences between some of the ancient types of life and the existing ones are so great, that palæontologists have been compelled to construct new *sub-classes, orders*, and *genera* for their reception. Moreover, many fossil animals are not only very different from living ones, but they are often "comprehensive" or "transitional" in their characters. In other words, fossil animals are often "comprehensive types," and combine in themselves characters now only found separate, thus serving as "transitional links" between groups which are at present widely removed from one another. For example, the reptiles and the birds form at the present day two widely separated classes, but some fossil birds exhibit peculiarities of a distinctly reptilian character, and some fossil reptiles approximate to the birds in structure; and the fossil forms thus partially fill up the gap which now exists between these two great divisions of the animal kingdom.

Again, many fossil animals exhibit what are termed "generalised" characters. If, namely, we construct for ourselves a "general" or "ideal" *type* for any great group of animals—a *type* which shall possess all the *essential* characters of the group, without its non-essential ones—then we find that the fossil animals of the same group are generally nearer to this type than are its living representatives. Moreover, the older representatives of any given group are usually nearer to the ideal type of the group—or are more "generalised"—than are the later representatives of the same group. All zoologists, however, admit that the process of development in any *individual* animal is one in which there is a gradual progress from the general to the special, the embryo being nearer to the general type of the group to which it belongs than the adult is. In other words, the embryo animal is more *generalised* than the adult, and the process of development is one of *specialisation*. Admitting this, it follows that the fossil forms belonging to any given group, in so far as they are "generalised" in their characters, may fairly be said to be "embryonic" types; and as the oldest forms of any given group are usually the least specialised, so they are likewise the most "embryonic." It must be borne in mind, however, that if we speak of fossil animals as being "embryonic types," we can only do so on the distinct understanding that it is not thereby implied that they were in any way *degraded* forms, or that they were at all less perfectly constructed or less thoroughly adapted for their surroundings than their modern representatives.

The following synoptical table gives briefly the leading divisions of the animal kingdom, and the chief characters of these:—

Tabular View of the Chief Divisions of the Animal Kingdom.

INVERTEBRATE ANIMALS.

SUB-KINGDOM I.—PROTOZOA.

Animal simple or forming colonies, usually very minute; the body composed of the structureless, jelly-like, albuminous substance called "sarcode;" not divided into regular segments; having no nervous system; no regular circulatory system; usually no mouth; no definite body-cavity or digestive system, or at most but a short gullet.

Class A. Gregarinidæ.—Minute Protozoa which inhabit the interior of insects and other animals, and which have not the power of throwing out prolongations of their substance (pseudopodia). No mouth.

Class B. Rhizopoda (Root-footed Protozoa).—Protozoa which are simple or compound, and have the power of throwing out and retracting prolongations of the body-substance (the so-called "pseudopodia"). No mouth, in most, if not in all.

Order 1. *Monera.—Ex.* Protogenes.

Order 2. *Amœbea.—Ex.* Proteus Animalcule (Amœba).

Order 3. *Foraminifera.—Ex.* Lagena, Nodosaria, Globigerina.

Order 4. *Radiolaria.—Ex.* Thalassicolla, Polycystina.

Order 5. *Spongida.—Ex.* Fresh-water Sponge (Spongilla), Venus's Flower-basket (Euplectella).

Class C. Infusoria (Infusorian Animalcules).—Protozoa with a mouth and short gullet; destitute of the power of emitting pseudopodia; furnished with vibratile cilia or contractile filaments; the body usually composed of three distinct layers.

Order 1. *Ciliata.—Ex.* Bell-animalcule (Vorticella), Paramœcium.

Order 2. *Flagellata.—Ex.* Peranema.

Order 3. *Suctoria.—Ex.* Podophrya.

SUB-KINGDOM II.—CŒLENTERATA.

Animals whose alimentary canal communicates freely with the general cavity of the body; body composed essentially of two layers or membranes, an outer layer or "ectoderm," and an inner layer or "endoderm." No circulatory system or heart, and in most no nervous system. Skin furnished with minute stinging organs or "thread-cells." Distinct reproductive organs in all.

CLASS A. HYDROZOA.—Walls of the digestive sac not separated from those of the general body-cavity, the two coinciding with one another. Reproductive organs external.

Sub-class I. HYDROIDA (Hydroid Zoophytes).
- Order 1. *Hydrida.—Ex.* Fresh-water Polype (Hydra).
- Order 2. *Corynida.—Ex.* Pipe-coralline (Tubularia).
- Order 3. *Sertularida.—Ex.* Sea-firs (Sertularia).

Sub-class II. SIPHONOPHORA (Oceanic Hydrozoa).
- Order 4. *Calycophoridæ.—Ex.* Diphyes.
- Order 5. *Physophoridæ.—Ex.* Portuguese Man-of-war (Physalia).

Sub-class III. DISCOPHORA (Jelly-fish).
- Order 6. *Medusidæ.—Ex.* Trachynema.

Sub-class IV. LUCERNARIDA (Sea-blubbers).
- Order 7. *Lucernaridæ.—Ex.* Lucernaria.
- Order 8. *Pelagidæ.—Ex.* Pelagia.
- Order 9. *Rhizostomidæ.—Ex.* Rhizostoma.

Sub-class V. GRAPTOLITIDÆ (extinct).

Sub-class VI. HYDROCORALLINÆ.—*Ex.* Millepora, Stylaster.

CLASS B. ACTINOZOA.—Stomach opening below into the body-cavity, which is divided into a number of compartments by a series of vertical partitions or "mesenteries." Reproductive organs internal.

- Order 1. *Zoantharia.*—Tentacles simply rounded, in multiples of five or six.—*Ex.* Sea-anemones (Actinidæ), Star-corals (Astræidæ), Brain-corals (Meandrina), Madrepores (Madreporidæ).
- Order 2. *Alcyonaria.*—Tentacles fringed, in multiples of four.—*Ex.* Dead-man's-toes (Alcyonium), Organ-pipe Coral (Tubipora), Sea-rods (Virgularia), Sea-pens (Pennatula), Red Coral (Corallium), Heliopora, Heliolites.
- Order 3. *Rugosa* (extinct).
- Order 4. *Ctenophora.*—Animal oceanic, swimming by means of bands of cilia or "ctenophores."—*Ex.* Pleurobrachia, Venus's Girdle (Cestum).

SUB-KINGDOM III.—ANNULOIDA.

Animals in which the alimentary canal is completely shut off from the general cavity of the body, and in which there is a distinct nervous system. A true blood-circulatory system may or may not be present. In all there is a peculiar system of canals, which usually communicate with the exterior, and which constitute what is called the "water-vascular system." The body of the adult is never composed of a succession of definite rings, or provided with successive pairs of appendages disposed symmetrically on the two sides of the body.

The Annuloida are divided into two great classes :—

A. ECHINODERMATA.—Integument composed of numerous calcareous plates joined together, or leathery and having grains, spines, or tubercles of calcareous matter developed in it. Water-vascular system (ambulacral system) mostly employed in locomotion, and generally communicating with the exterior. Adult generally more or less star-like or "radiate" in shape ; young mostly showing more or less complete "bilateral symmetry," that is, showing similar parts on the two sides of the body. Nervous system radiate.

- Order 1. *Crinoidea* (Sea-lilies).—*Ex.* Feather-star (Comatula), Medusa-head Crinoid (Pentacrinus), Stone-lily (Encrinus).
- Order 2. *Blastoidea* (extinct).
- Order 3. *Cystoidea* (extinct).
- Order 4. *Ophiuroidea* (Brittle-stars).—*Ex.* Sand-stars (Ophiura), Brittle-stars (Ophiocoma).
- Order 5. *Asteroidea* (Star-fishes).—*Ex.* Cross-fish (Uraster), Sun-star (Solaster), Cushion-star (Goniaster).
- Order 6. *Echinoidea* (Sea-urchins). — *Ex.* Sea-eggs (Echinus), Heart-urchins (Spatangus).
- Order 7. *Holothuroidea* (Sea-cucumbers).—*Ex.* Trepangs (Holothuria).

B. SCOLECIDA.—Body usually flattened, or cylindrical and worm-like ; integument soft, without lime. Water-vascular system not assisting in locomotion. Nervous system consisting of one or two ganglia or little masses, and not disposed in a radiate manner.

- Order 1. *Tæniada.*—*Ex.* Tape-worm (Tænia).
- Order 2. *Trematoda* (Suctorial worms).—*Ex.* Liver-fluke (Distoma).
- Order 3. *Turbellaria.*—*Ex.* Planarians (Planaria), Ribbon-worms (Nemertes).
- Order 4. *Acanthocephala* (Thorn-headed worms).—*Ex.* Echinorhynchus.
- Order 5. *Gordiacea* (Hair-worms).—*Ex.* Gordius.
- Order 6. *Nematoda* (Thread-worms).—*Ex.* Round-worm (Ascaris), Guinea-worm (Filaria), Vinegar-eel (Anguillula).
- Order 7. *Rotifera* (Wheel-animalcules).—*Ex.* Builder-animalcule (Melicerta), Flexible Creeper (Notommata).

SUB-KINGDOM IV.—ANNULOSA.

Animal composed of numerous definite segments or "somites," arranged longitudinally, one behind the other. Nervous system always present, consisting typically of a double chain of nervous masses, or ganglia, which are placed along the lower surface of the body, and form a collar around the gullet. Limbs (when present) turned toward that side of the body on which the main masses of the nervous system are situated.

DIVISION A. ANARTHROPODA.—Locomotive appendages, when present, not distinctly jointed or articulated to the body.

CLASS I. GEPHYREA.—*Ex.* Spoon-worms (Sipunculus).

CLASS II. ANNELIDA (Ringed-worms).

Order 1. *Hirudinea.*—*Ex.* Leeches (Sanguisuga, Hirudo).

Order 2. *Oligochæta.*—*Ex.* Earth-worms (Lumbricus), Water-worms (Nais).

Order 3. *Tubicola.*—*Ex.* Tube-worms (Serpula).

Order 4. *Errantia.*—*Ex.* Sand-worms and Sea-centipedes (Nereis), Lob-worm (Arenicola), Sea-mouse (Aphrodite).

CLASS III. CHÆTOGNATHA (Arrow-worms).—*Ex.* Sagitta.

DIVISION B. ARTHROPODA.—Locomotive appendages jointed or articulated to the body.

CLASS I. CRUSTACEA.—Respiration aquatic, mostly by gills. Two pairs of antennæ. Limbs more than four pairs in number, carried upon the thorax, and generally upon the abdomen also.

Order 1. *Ichthyophthira.*—*Ex.* Lernæa.

Order 2. *Rhizocephala.*—*Ex.* Peltogaster.

Order 3. *Cirripedia.*—*Ex.* Barnacles (Lepas), Acorn-shells (Balanus).

Order 4. *Ostracoda.*—*Ex.* Water-fleas (Cypris).

Order 5. *Copepoda.*—*Ex.* Cyclops.

Order 6. *Cladocera.*—*Ex.* Branched-horned Water-fleas (Daphnia).

Order 7. *Phyllopoda.*—*Ex.* Brine-shrimp (Artemia).

Order 8. *Trilobita* (Extinct).

Order 9. *Merostomata.*—*Ex.* King-crabs (Limulus).

Order 10. *Læmodipoda.*—*Ex.* Whale-louse (Cyamus).

Order 11. *Isopoda.*—*Ex.* Wood-lice (Oniscus), Slaters (Ligia).

Order 12. *Amphipoda.*—*Ex.* Sandhopper (Talitrus), Fresh-water Shrimp (Gammarus).

Order 13. *Stomapoda.*—*Ex.* Locust-shrimp (Squilla).

Order 14. *Decapoda.*—*Ex.* Lobster (Homarus), Cray-fish (Astacus), Shrimps (Crangon); Hermit-crabs (Pagurus); Crabs (Cancer, Carcinus), Land-crabs (Gecarcinus).

CLASS II. ARACHNIDA.—Respiration aerial, by pulmonary chambers or air-tubes (tracheæ) in the higher forms. Antennæ converted into jaws. Head and thorax amalgamated. Four pairs of legs. Abdomen without limbs.

Order 1. *Podosomata* (Sea-spiders).—*Ex.* Pycnogonum.

Order 2. *Monomerosomata.*—*Ex.* Mites (Acarus), Water-mites (Hydrachna), Ticks (Ixodes).

Order 3. *Adelarthrosomata.*—*Ex.* Harvest-spiders (Phalangidæ), Book-scorpions (Chelifer).

Order 4. *Pedipalpi.*—*Ex.* Scorpions (Scorpio).

Order 5. *Araneida.—Ex.* House-spiders (Tegenaria), Field-spiders (Epeira).

Class III. Myriapoda.—Respiration aerial, by tracheæ (air-tubes) or by the skin. Head distinct; remainder of body composed of nearly similar segments. Legs more than eight pairs in number, and borne partly upon the abdomen. One pair of antennæ.

Order 1. *Chilopoda.—Ex.* Centipedes (Scolopendra).

Order 2. *Chilognatha.—Ex.* Millipedes (Iulus).

Order 3. *Pauropoda.—Ex.* Pauropus.

Order 4. *Onychophora.—Ex.* Peripatus.

Class IV. Insecta.—Respiration aerial, by tracheæ. Head, thorax, and abdomen distinct. One pair of antennæ. Three pairs of legs, and generally two pairs of wings on the thorax. No locomotive limbs on the abdomen.

Order 1. *Anoplura.—Ex.* Lice (Pediculus).

Order 2. *Mallophaga* (Bird-lice).

Order 3. *Collembola.—Ex.* Podura.

Order 4. *Thysanura.—Ex.* Lepisma.

Order 5. *Hemiptera.—Ex.* Plant-lice (Aphides), Field-bug (Pentatoma), Cochineal Insects (Coccus).

Order 6. *Orthoptera.—Ex.* Locusts (Acrydium), Grasshoppers (Gryllus), Crickets (Acheta), Cockroach (Blatta).

Order 7. *Neuroptera.—Ex.* White Ants (Termes), Dragon-flies (Libellulidæ), May-flies (Ephemeridæ).

Order 8. *Aphaniptera.—Ex.* Fleas (Pulex).

Order 9. *Diptera.—Ex.* Gnats (Culex), Crane-flies (Tipula), House-flies and Flesh-flies (Musca).

Order 10. *Lepidoptera* (Butterflies and Moths).

Order 11. *Hymenoptera.—Ex.* Bees (Apidæ), Humble-bees (Bombidæ), Wasps (Vespidæ), Ants (Formicidæ), Saw-flies (Tenthredinidæ).

Order 12. *Strepsiptera.—Ex.* Stylops.

Order 13. *Coleoptera* (Beetles).

SUB-KINGDOM V.—MOLLUSCA.

Animal soft-bodied, generally with a hard covering or shell. Nervous system consisting of a single ganglion or of scattered pairs of ganglia. A distinct heart and breathing-organ, or neither.

The Mollusca may be divided into the two following primary divisions, containing the following classes :—

A. Molluscoida.—Nervous system consisting of a single ganglion or of a principal pair of ganglia. No heart, or an imperfect one.

Class I. Polyzoa.—Animal always forming compound growths or colonies. No heart. The mouth of each zoöid surrounded by a circle or crescent of ciliated tentacles.—*Ex.* Sea-mats (Flustra), Lace-coral (Fenestella).

CLASS II. TUNICATA.—Animal simple or compound, enclosed in a leathery or gristly case. An imperfect heart.—*Ex.* Sea-squirts (Ascidia).

CLASS III. BRACHIOPODA.—Animal always simple; the body enclosed in a bivalve shell. Mouth furnished with two long fringed processes or "arms."—*Ex.* Lamp-shells (Terebratula).

B. MOLLUSCA PROPER.—Nervous system consisting of three principal pairs of ganglia. Heart well developed, consisting of at least two chambers.

CLASS IV. LAMELLIBRANCHIATA (Bivalve Shell-fish).—No distinct head; no teeth. Body enclosed in a shell which is "bivalve," or composed of two distinct pieces. One or two leaf-like gills on each side of the body.—*Ex.* Oyster (Ostrea), Scallop (Pecten), Mussel (Mytilus).

CLASS V. GASTEROPODA.—A distinct head and toothed tongue. Shell absent in some, but mostly present, and usually consisting of a single piece ("univalve"). Locomotion effected by creeping about on the flattened under-surface of the body ("foot"), or by swimming by means of a fin-like modification of the same.—*Ex.* Whelks (Buccinum), Limpets (Patella), Sea-lemons (Doris), Land-snails (Helix), Slugs (Limax).

CLASS VI. PTEROPODA.—Animal oceanic, swimming by means of two wing-like appendages, one on each side of the head. Size minute.—*Ex.* Cleodora.

CLASS VII. CEPHALOPODA.—Animal with eight or more arms, placed in a circle round the mouth. Mouth armed with jaws, and a toothed tongue. Two or four plume-like gills. In front of the body, a muscular tube ("funnel") through which is expelled the water which has been used in respiration. An external shell in some, an internal skeleton in others.—*Ex.* Calamaries (Loligo), Cuttle-fishes or Poulpes (Octopus), Paper-Nautilus (Argonauta), Pearly Nautilus (Nautilus).

VERTEBRATE ANIMALS.

SUB-KINGDOM VI.—VERTEBRATA.

Body composed of a number of definite segments arranged longitudinally, or one behind the other. The main masses of the nervous system are placed on the dorsal aspect of the body, and are completely shut off from the general body-cavity. The limbs (when present) are turned away from that side of the body on which the main nervous masses are situated, and are never more than four in number. In most cases a back-bone, or "vertebral column," is present in the fully-grown animal.

CLASS I. PISCES (Fishes).—Breathing-organs in the form of gills.

Heart usually of two chambers, rarely of three. Blood cold. Limbs, when present, converted into fins.

Order 1. *Pharyngobranchii.*—*Ex.* Lancelet (Amphioxus).

Order 2. *Marsipobranchii.*—*Ex.* Lamprey (Petromyzon), Hag-fish (Myxine).

Order 3. *Teleostei* (Bony Fishes).—*Ex.* Eels (Murænidæ), Herrings (Clupeidæ), Salmon and Trout (Salmonidæ), Cod and Haddock (Gadidæ), Flat-fishes (Pleuronectidæ), Perch (Percidæ), Mackerel (Scomberidæ).

Order 4. *Ganoidei.*—*Ex.* Bony Pike (Lepidosteus), Paddle-fish (Spatularia), Sturgeon (Sturio).

Order 5. *Elasmobranchii.*—*Ex.* Sharks (Carcharidæ), Dog-fishes (Scylliadæ), Saw-fishes (Pristis), Rays and Skates (Raiidæ).

Order 6. *Dipnoi.*—*Ex.* Mud-fish (Lepidosiren), Ceratodus.

Class II. Amphibia (Amphibians).—Breathing-organs in the young in the form of gills alone, afterwards lungs, either alone or associated with gills. Skull joined to the backbone by two articulating surfaces ("condyles"). Limbs never converted into fins. Heart in the young of two chambers only, in the adult of three chambers. Blood cold.

Order 1. *Labyrinthodontia* (extinct).

Order 2. *Ophiomorpha.*—*Ex.* Cæcilia.

Order 3. *Urodela* (Tailed Amphibians).—*Ex.* Water-newts (Triton), Salamanders (Salamandra), Axolotl (Siredon), Mud-eel (Siren).

Order 4. *Anoura* (Tailless Amphibians).—*Ex.* Frogs (Rana), Tree-frogs (Hyla), Toads (Bufo), Surinam Toads (Pipa).

Class III. Reptilia (Reptiles).—Respiratory organs in the form of lungs, never in the form of gills. Heart three-chambered, rarely four-chambered, the pulmonary and systemic circulations always connected together directly, either in the heart itself or in its immediate neighbourhood. Blood cold. Skull jointed to the backbone by a single articulating surface or "condyle." Each half of the lower jaw composed of several pieces. Appendages of the skin in the form of scales or plates.

Order 1. *Chelonia.*—*Ex.* Turtles (Cheloniidæ), Soft Tortoises (Trionycidæ), Terrapins (Emydidæ), Land Tortoises (Testudinidæ).

Order 2.—*Ophidia.*—*Ex.* Vipers (Viperidæ), Rattlesnakes (Crotalidæ), Sea-snakes (Hydrophidæ), Boas and Pythons (Boidæ).

Order 3. *Lacertilia.*—*Ex.* Lizards (Lacerta), Iguanas (Iguanidæ), Monitors (Varanidæ), Chameleons (Chamæleontidæ).

Order 4. *Crocodilia.*—*Ex.* Crocodiles, Alligators, Gavials.

Order 5. *Ichthyopterygia* (extinct).—*Ex.* Ichthyosaurus.

Order 6. *Sauropterygia* (extinct).—*Ex.* Plesiosaurus.

Order 7. *Pterosauria* (extinct).—*Ex.* Pterodactylus, Pteranodon.

Order 8. *Anomodontia* (extinct).—*Ex.* Dicynodon.

Order 9. *Deinosauria* (extinct).—*Ex.* Iguanodon.

Order 10. *Theriodontia* (extinct).—*Ex.* Cynodraco.

CLASS IV. AVES (Birds).—Respiratory organs in the form of lungs, never in the form of gills. Lungs connected with air-receptacles placed in different parts of the body. Heart four-chambered. Blood warm. Skull connected with the backbone by a single articulating surface or "condyle." Each half of the lower jaw composed of several pieces. Appendages of the skin in the form of feathers. Cavities of the chest and abdomen not separated by a complete partition (diaphragm). Fore-limbs converted into wings. Animal oviparous.

A. Sub-class RATITÆ.

Order 3. *Cursores* (Runners).—*Ex.* Ostrich (Struthio), American Ostrich (Rhea), Emeu (Dromaius), Cassowary (Casuarius), Apteryx, Dinornis.

B. Sub-class CARINATÆ.

Order 1. *Natatores* (Swimmers).—*Ex.* Penguins (Spheniscidæ), Gulls (Laridæ), Ducks (Anatidæ), Geese (Anserinæ), Flamingos (Phænicopteridæ).

Order 2. *Grallatores* (Waders).—*Ex.* Rails (Rallidæ), Water-hens (Gallinulæ), Cranes (Gruidæ), Herons (Ardeidæ), Storks (Ciconinæ), Snipes and Woodcock (Scolopacidæ), Plovers, Oyster-catchers, and Turnstones (Charadriidæ).

Order 3. *Rasores* (Scratchers).—*Ex.* Grouse, Ptarmigan, Partridges, Pheasants, Turkey, Guinea-fowl, Domestic Fowl, Pea-fowl (Gallinacei); Doves, Pigeons, Ground-pigeons (Columbacei).

Order 4. *Scansores* (Climbers).—*Ex.* Cuckoos (Cuculidæ), Woodpeckers (Picidæ), Parrots, Cockatoos, Parrakeets (Psittacidæ), Toucans (Rhamphastidæ), Trogons (Trogonidæ).

Order 5. *Insessores* (Perchers).—*Ex.* Crows, Magpies, and Jays (Corvidæ), Starlings (Sturnidæ), Finches, Grosbeaks, Larks (Fringillidæ), Thrushes, Blackbirds, Orioles (Merulidæ), Creepers and Wrens (Certhidæ), Humming-birds (Trochilidæ), Swallows and Martins (Hirundinidæ), Swifts (Cypselidæ), King-fishers (Alcedinidæ).

Order 6. *Raptores* (Birds of Prey).—*Ex.* Owls (Strigidæ), Falcons and Hawks (Falconidæ), Eagles (Aquilina) Vultures (Vulturidæ), American Vultures (Cathartidæ), Secretary-bird (Gypogeranus).

C. Sub-class SAURORNITHES.

Order *Saururæ* (extinct).—Archæopteryx.

D. Sub-class ODONTORNITHES.

Order 1. *Odontolcæ* (extinct).—Hesperornis.

Order 2. *Odontotormæ* (extinct).—Ichthyornis, Apatornis.

CLASS V. MAMMALIA (Mammals or Quadrupeds).—Respiratory organs in the form of lungs, which are never connected with air-sacs placed in different parts of the body. Heart four-chambered. Blood warm. Skull united to the backbone by two articulating surfaces or "condyles." Each half of the lower jaw composed of a single piece. Appendages of the skin in the form of hairs. Young nourished by means of a special fluid—the milk,—secreted by special glands—the mammary glands. Animal viviparous.

A. NON-PLACENTAL MAMMALS.—The young not provided with a placenta.

Order 1. *Monotremata.—Ex.* Duck-mole (Ornithorhynchus), Spiny Ant-eater (Echidna).

Order 2. *Marsupialia.—Ex.* Kangaroos (Macropodidæ), Kangaroo-bear (Phascolarctos), Phalangers (Phalangistidæ), Opossums (Didelphidæ), Tasmanian Devil (Dasyurus).

B. PLACENTAL MAMMALS.—The young provided with a placenta.

Order 3. *Edentata.—Ex.* Sloths (Bradypodidæ), Armadillos (Dasypodidæ), Hairy Ant-eaters (Myrmecophagidæ), Scaly Ant-eaters (Manis).

Order 4. *Sirenia.—Ex.* Manatee (Manatus), Dugong (Halicore).

Order 5. *Cetacea.—Ex.* Whalebone-whales (Balænidæ), Sperm-whales (Physeteridæ), Dolphins and Porpoises (Delphinidæ).

Order 6. *Ungulata* (Hoofed Quadrupeds).—*Ex.* Rhinoceros; Tapir; Horse, Ass, and Zebra (Equidæ); Hippopotamus; Hogs and Peccaries (Suida); Camels and Llamas (Camelidæ); Giraffe; Stags, Elk, Reindeer (Cervidæ); Antelopes (Antilopidæ); Sheep and Goats (Ovidæ); Oxen and Buffaloes (Bovidæ).

Order 7. *Dinocerata* (extinct).—*Ex.* Dinoceras.

Order 8. *Tillodontia* (extinct).—*Ex.* Tillotherium.

Order 9. *Toxodontia* (extinct).—*Ex.* Toxodon.

Order 10. *Hyracoidea.—Ex.* Hyrax.

Order 11. *Proboscidea.—Ex.* Elephants (Elephas).

Order 12. *Carnivora.—Ex.* Seals (Phocidæ), Bears (Ursidæ), Racoons (Procyon), Badgers (Melidæ), Weasels and Otters (Mustelidæ), Civets and Genettes (Viverridæ), Dogs, Wolves, and Foxes (Canidæ); Hyænas (Hyænidæ), Cats, Lynxes, Leopards, Tigers, Lions (Felidæ).

Order 13. *Rodentia.—Ex.* Hares and Rabbits (Leporidæ), Porcupines (Hystricidæ), Beavers (Castoridæ), Mice and Rats (Muridæ), Dormice (Myoxidæ), Squirrels and Marmots (Sciuridæ).

Order 14. *Cheiroptera.—Ex.* Common Bats (Vespertilionidæ), Horseshoe-bats (Rhinolophidæ), Vampire-bats (Phyllostomidæ), Fox-bats (Pteropidæ).

Order 15. *Insectivora—Ex.* Moles (Talpidæ), Shrew-mice (Soricidæ), Hedgehogs (Erinaceidæ).

Order 16. *Quadrumana.—Ex.* Aye-aye (Cheiromys), Lemurs (Lemuridæ), Spider-monkeys (Ateles), Howlers (Mycetes), Macaques (Macacus), Baboons (Cynocephalus), Gibbons (Hylobates), Orang (Simia), Gorilla and Chimpanzee (Troglodytes).

Order 17. *Bimana.*—Man (Homo Sapiens).

General Succession and Progression of Organic Types.

Whilst admitting the impossibility of arranging the animal kingdom upon any linear plan, no doubt obtains as to the fact that some of the fundamental "morphological types," or plans upon which animals have been constructed, are higher than others. Every one admits, for example, that the Vertebrate type is higher than the Molluscan or the Articulate type, an admission which is not affected by the fact that the highest Molluscs and Articulates are superior in point of organisation to the lowest Vertebrates. In the same way, within the limits of each sub-kingdom, every one admits that some of the groups are higher than others. Every one, for example, would admit that a Mammal is a superior animal to a Fish. It follows from this that a certain general arrangement of the animal kingdom, as a whole, is possible, upon the comparative basis of the morphological type of the sub-kingdoms. Similarly a general and more exact arrangement of the classes and orders of each sub-kingdom may be made by the degree of perfection in which the type of the sub-kingdom is carried out in each.

No generalisation of Palæontology seems to stand on a firmer basis than that which asserts that there has been a general succession of organic types; and that the appearance of the lower forms of life has in the main preceded that of the higher forms in point of time. In other words, it is one of the generalisations of Palæontology that there has not only been a *succession*, but also a *progression*, of organic types in proceeding from the earliest fossiliferous deposits up to the present day. Whilst this *general* law remains, as we believe, unassailable, there are some important considerations which must not be lost sight of. In the first place, it is very doubtful if we are as yet acquainted with the *absolute* time of the first appearance upon the globe of even one of the sub-kingdoms. Future discoveries, therefore, are almost certain to push back still further into the remote vistas of the past the point of time at which each morphological type first made its appearance upon the globe. Still, there is little likelihood that the *relative* times of appearance of the great groups, as compared with one another, will be affected by the researches of the future. It remains almost certain that we shall find that the lower types were followed in point of time by the higher.

In the second place, we find all the primary types in existence before the close of the Silurian period; and he would be rash indeed who would dogmatically deny that they might all have been present in the earlier Cambrian period. This, at first sight, might seem almost to negative the above generalisation, but it does not affect its value if fairly examined. The lower sub-kingdoms of the Invertebrate animals appeared so early that their origin is lost in the mists of antiquity, and we can say nothing positively as to the time when each came into existence. The Cambrian deposits are underlaid by the vast series of the Laurentian deposits, representing an incalculable lapse of time. These ancient sediments, with one exception, have hitherto proved barren of life, owing to the intense metamorphism to which they have been subjected, and they consequently yield no evidence bearing on the question in hand. They serve to show us, however, by their presence alone, that we must in the meanwhile leave the Invertebrate sub-kingdoms out of account altogether as bearing upon the question of the succession and progression of organic types. We do not know when these sub-kingdoms commenced, and hence we have no right to assert either that they were all introduced simultaneously, or that they came into being successively. We may be sure, however, of one thing—they did not commence at the points where now we find their earliest traces. There remains, then, only the sub-kingdom of the Vertebrate animals which can reasonably be appealed to as evidence on this question. The stratified series is long enough to render it certain that it contains traces of the first appearance of, at any rate, the higher classes of these, though we doubtless are ignorant of the absolute moment at which each appeared. If, therefore, it can be shown that there has been a progression as far as this sub-kingdom is concerned, then there would, by analogy, be the greatest probability that a similar progression has taken place in all the sub-kingdoms.

So far as our present knowledge goes, it would appear that there *is* such a progression in the Vertebrate sub-kingdom. The classes of Vertebrates make their appearance, on the whole, in the order indicated by their zoological position, the lowest first and the highest last. Not only does this hold good for the classes of the Vertebrates, but the same *general* statement may be made as to the orders of each class. Where apparent exceptions occur, a reasonable explanation can be given, or our knowledge can be shown to be defective. Space will not allow a discussion of this question, but a single example may be taken. So far as we know at present, the earliest remains of Vertebrate animals are those of Fishes—the lowest class of the sub-kingdom—and these appear in the Upper Silurian rocks for the first time. Granting the probability that Fishes may some day be found in the Lower Silurian rocks, or even in Cambrian deposits, there still seems no likelihood that they will be deprived by any future discoveries of their position as being the earliest of their sub-kingdom. The oldest remains of Fishes, however, are by no means those which would be expected, but belong to two of the higher orders of the class. This seeming anomaly, however, disappears when we consider that the two lowest orders of Fishes possess almost no structures by which we can reasonably expect to find them recorded in a fossil state. They may therefore have been in existence long before the Ganoids and Placoids of the Upper Silurian rocks, and we have no right to assume that they were not. As to the remaining great order of Fishes (the Teleostean Fishes), it is certain that their appearance was much later, and they are generally regarded as inferior to the Ganoids and Placoids in zoological position. This, however, is a matter of opinion, and reasons are not wanting for regarding them as the highest of their class.

It only remains to add that nothing further is contended for here than the general fact of there having been a progression of morphological types, the lowest presenting themselves first, the highest being the last to appear upon the scene. It is by no means contended that the Ganoid Fishes of the Upper Silurian rocks were in any way degraded members of their order, or inferior in point of organisation to the Ganoids of the present day. On the contrary, there is reason to think that many types early presented a development more varied than that exhibited by their successors. It is simply contended that, *on the whole*, there has been a zoological progression as we ascend from the Cambrian period to the present day. It is also to be remembered, that though the commencement of the Invertebrate sub-kingdoms may be unknown to us, a similar progression can be in many cases shown as regards the *orders* and *classes* of these, even more completely than in the case of the Vertebrate sub-kingdom.

Lastly, the evidence of Palæontology points, in the main, to the operation of some general law of *evolution*, whereby the later forms of life have been derived from the older ones. That this law has acted along with, and has sometimes been counteracted by, some other and as yet obscure law regulating the appearance of new types, seems equally certain; but it is not necessary to enter upon this complex and imperfectly understood question in this place. We are dealing here primarily with facts, and in the following pages we shall meet with unmistakable evidence of the operation of some law of evolution, while we shall, at the same time, find ourselves confronted with phenomena which, in the present state of our knowledge, appear to be irreconcilable with the *universal* or *exclusive* action of this law. It would be an easy solution of the difficulty to adopt the course of definitely accepting some hard-and-fast theory upon this subject, and to bring forward prominently all the known facts favouring this theory, while we left in comparative abeyance the facts pointing in other directions, or explained them away by more or less probable assumptions. Upon the whole, however, it seems preferable to enter upon the study of the actual facts of the science of Palæontology, as far as may be, unfettered by preconceptions and unpledged to theories; while we may, at the same time, safely accept the doctrine of evolution, in the shape presented to us by the master-mind of Darwin, as an invaluable, indeed an indispensable, *working hypothesis.*

PART II.

PALÆOZOOLOGY

PART II.

CHAPTER VII.

SUB-KINGDOM I.—PROTOZOA.

SUB-KINGDOM I. PROTOZOA.—*Animal simple or composite, generally of very minute size, composed of a structureless or slightly differentiated, jelly-like, albuminoid substance (termed "sarcode"), showing no composition out of definite parts or segments, having no definite body-cavity, presenting no traces of a nervous system, and having either no alimentary apparatus, or but a very rudimentary one.*

TABLE OF THE DIVISIONS OF THE PROTOZOA.

CLASS A. GREGARINIDÆ.—Parasitic Protozoa, which are destitute of a mouth, and do not possess the power of emitting processes of their body-substance (pseudopodia).

CLASS B. RHIZOPODA.—Protozoa, which are destitute of a mouth, and have the power of emitting extensile and contractile processes of the body-substance (pseudopodia).

Order 1. *Monera.*—*Ex.* Protogenes.
Order 2. *Amœbea.*—*Ex.* Amœba.
Order 3. *Foraminifera.*—*Ex.* Nummulites.
Order 4. *Radiolaria.*—*Ex.* Haliomma.
Order 5. *Spongida.*—*Ex.* Spongilla.

CLASS C. INFUSORIA (Infusorian Animalcules).—Protozoa mostly with a mouth, and rudimentary digestive canal; destitute of the power of emitting pseudopodia; furnished with vibratile cilia or contractile filaments; the body usually with a distinct cuticle covering a layer of firm sarcode.

Regarded palæontologically, we may eliminate from the *Protozoa* the entire class of the *Gregarinidæ*, with the Rhizopodous orders of the *Monera*[1] and *Amœbea*, no trace of the past existence of which has yet been obtained, or, from their soft-bodied nature, is ever likely to be. For all practical purposes the same may be said of the large and universally-distributed class of the Infusorian Animalcules.[2] Some of these, however, possess horny or membranous cases, which might possibly be preserved in a fossil state; and Ehrenberg has found in the flints of the Chalk certain microscopic bodies, which he regarded as being the protective carapaces of *Peridinium* and allied forms of Flagellate Infusoria. With this doubtful exception, however, no Infusorian animalcule has ever been detected in the fossil state, though the class has doubtless existed from the most remote antiquity. There remain, then, only the three Rhizopodous orders of the *Foraminifera*, *Radiolaria*, and *Spongida*, all of which secrete hard structures, and all of which are more or less extensively represented as fossils, so that they demand our attention separately and in detail.

I.—FORAMINIFERA.

The *Foraminifera* may be defined as *Rhizopoda in which the body is protected by a shell or "test," which is composed of carbonate of lime, or which may consist of particles of sand cemented together by some animal cement, or may be simply horny (chitinous). The animal may be simple, or may repeat itself indefinitely by budding, and the body-substance gives out long and thread-like processes (pseudopodia) which interlace with one another to form a network, and often coalesce at their bases to form a continuous layer of sarcode outside the shell.*

[1] The "coccoliths" are sometimes regarded as being referable to the *Monera*; but they will be considered here as belonging to the vegetable kingdom, and they will be briefly described in speaking of fossil *Algæ*.

[2] "Fossil Infusoria" are often spoken of as forming more or less extensive deposits in the earth's crust, but the organisms so named are really *Diatoms* and *Polycystina*.

The pseudopodia reach the exterior either by perforations in the walls of the shell, or simply by the mouth of the latter (fig. 13, *c b*).

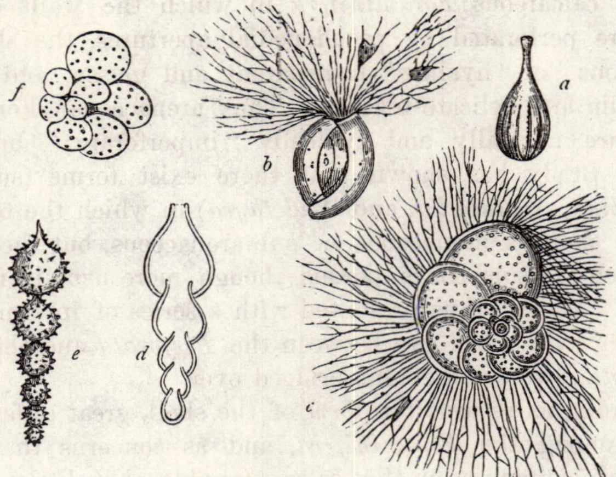

Fig. 13.—Morphology of Foraminifera. *a*, *Lagena vulgaris*, a monothalamous Foraminifer; *b*, *Miliola* (after Schultze), showing the pseudopodia protruded from the oral aperture of the shell; *c*, *Discorbina* (after Schultze), showing the nautiloid shell with the foramina in the shell-wall giving exit to pseudopodia; *d*, Section of *Nodosaria* (after Carpenter); *e*, *Nodosaria hispida*; *f*, *Globigerina bulloides*.

From a palæontological point of view the only part of a Foraminifer with which we have to deal is the shell or "test," and there are several points to notice in this connection. Firstly, as regards the actual composition of the shell, it is in the majority of cases calcareous, or composed of carbonate of lime, but it is rarely chitinous, and it is not uncommonly "arenaceous"—that is, composed of particles of sand cemented together by some animal substance or by carbonate of lime. With the horny or chitinous *Foraminifera* (*Gromidæ*) we have nothing to do here, as they have never been, and are never likely to be, detected in a fossil condition.

In the second place, the *Foraminifera* may be divided into two primary groups—the *Perforata* and *Imperforata*—according as the walls of the test are or are not perforated by pores or foramina, through which the pseudopodia reach the surface. In the imperforate calcareous *Foraminifera* the substance of the shell is "porcellanous," homogeneous, and opaque-white when viewed by reflected light. On the other hand, in those of the calcareous *Foraminifera* in which the walls of the test are perforated by pseudopodial apertures, the shell is "vitreous" or "hyaline," transparent and glassy, and often of a thin and delicate texture. The "arenaceous" Foraminifera are normally and typically "imperforate;" but Mr Henry Brady has shown that there exist forms (such as *Valvulina*, *Nodosinella*, and *Endothyra*) in which the texture of the shell is arenaceous or sub-arenaceous, but the walls of which are sometimes porous, though more usually imperforate. We are thus presented with a series of intermediate forms by which the gap between the *Perforata* and the *Imperforata* is to some extent bridged over.

Thirdly, as regards the *form* of the shell, great differences exist among the *Foraminifera*, and as concerns the mere external configuration, this is so variable that little or no value can be attached to it in classification. Moreover, in the two great series of the Perforate and the Imperforate *Foraminifera* it is common to find parallel or "isomorphic" groups. That is to say, we meet with two series of forms, repeating each other's peculiarities and variations in *form*, but the shell in the one series being perforate, while in the other it is imperforate.

The simplest form among the *Foraminifera* is that of a single spheroid of sarcode, capable of secreting for itself a hard covering, as in the flask-shaped *Lagena* (fig. 13, *a*) or the globular *Orbulina* (fig. 14). Forms such as these are said to be "unilocular" or "monothalamous," the test consisting of but a single chamber, not subdivided by partitions or "septa." In the more complex *Foraminifera*, the sarcode of the body undergoes a subdivision into partially separated segments, which may be produced by a process of budding, or, perhaps, by the occurrence of constrictions in the growing protoplasm, and each of these segments becomes more or less completely divided off from its neighbours, or enclosed by a

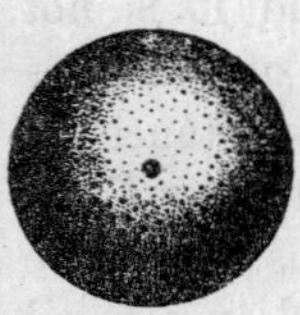

Fig. 14.—*Orbulina universa.* A simple Foraminifer from the Pliocene strata (Sub-Apennine beds) of Italy. (D'Orbigny.)

wall of shell. In these "multilocular" or "polythalamous" *Foraminifera*, therefore, the shell ultimately comes to consist of a series of chambers, separated by partitions of the test, and filled with sarcode. The partitions, however, or "septa," between the different chambers, are perforated by one or more apertures, through which pass connecting bands, or "stolons," of sarcode; so that the sarcode occupying the different chambers is united into a continuous and organic whole. Each segment may give out its own pseudopodia through perforations in its investing wall (fig. 13, *c*), or the pseudopodia may be simply emitted from the mouth of the shell by the last segment only (fig. 13, *b*). In any case the direction in which the segments are developed is governed by a determinate law, and differs in different species, the form ultimately assumed by the shell depending wholly upon this. The forms, however, assumed by the shells of *Foraminifera* are extremely variable, even within the limits of a single species, and it would be impossible to notice even the chief types in this place. There are, however, two or three important variations which may be noticed. If the buds are thrown out from the primitive spherule in a linear series so as to form a shell composed of numerous chambers arranged in a straight line, we get such a type as *Nodosaria* (fig. 13, *e*). When the new chambers are added in a spiral direction, each being a little larger than the one which preceded it, and the coils of the spiral lying in the same plane, we get such a form as *Discorbina* (fig. 13, *c*), or *Robulina* (fig. 15). These are the so-called "nautiloid" *Foraminifera*, from the resemblance of the shell, in figure, to that of the Pearly Nautilus. From this resemblance the nautiloid *Foraminifera* were originally placed in the same class as the *Ammonites* (*Cephalopoda*), but their true position was shown by the examination of their soft parts. In the typical nautiloid shell the convolutions

Fig. 15.—*Cristellaria* (*Robulina*) *echinata*, a "nautiloid" Foraminifer. (D'Orbigny.)

of the spiral all lie in one plane; but in other cases, as in *Rotalia* (fig. 16), the shell becomes turreted or top-shaped, in consequence of the coils of the spiral passing obliquely round a central axis.

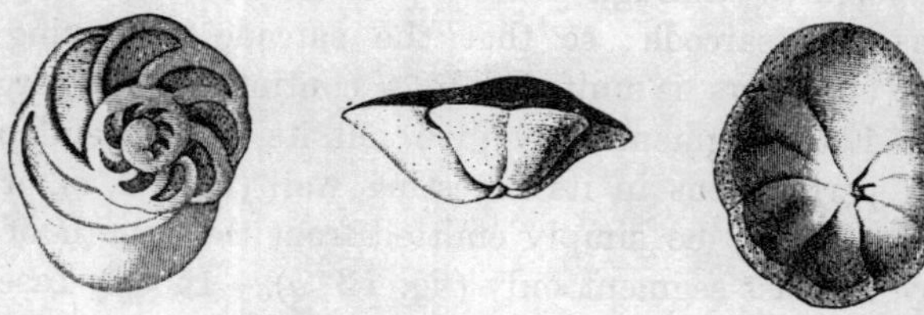

Fig. 16.—*Rotalia Boueana.* (D'Orbigny.)

In a few types of the *Foraminifera* (*e.g.*, in the *Dactyloporidæ*, fig. 19) the successive chambers of the multilocular test have no direct communication with one another, and simply cohere by their walls. In the majority of the compound shells, the successive chambers are so produced, that the septum between any two of them is formed solely by the anterior wall of the older chamber, which thus constitutes the posterior wall of the newer one (fig. 13, *e*). In the highest types of the compound *Foraminifera*, however, each segment is provided with its own proper wall of shell, each segment, as it is produced, forming for itself a posterior wall which applies itself to the anterior wall of the preceding segment, so that each septum ("septal plane") is composed of *two* lamellæ, as seen in fig. 17, A (Carpenter). Moreover, "in the higher types of the hyaline or vitreous series we frequently meet with an 'intermediate' or 'supplemental' skeleton, formed by a secondary or exogenous deposit upon the outer walls of the chambers, by which they receive a great accession of strength. This deposit not only fills up what would otherwise be superficial hollows at the junctions of the chambers (fig. 17, A, *d*), or (as in *Polystomella*) at the umbilical depression, but often forms a layer of considerable thickness over the whole surface, thus separating each whorl from that which encloses it; and it is sometimes prolonged into outgrowths that give a very peculiar variety to the ordinary contour, as in some varieties of *Rotalia* and *Poly-*

stomella, but most characteristically in *Calcarina* (fig. 17, B), and the stellate form of *Tinoporus.* This intermediate or

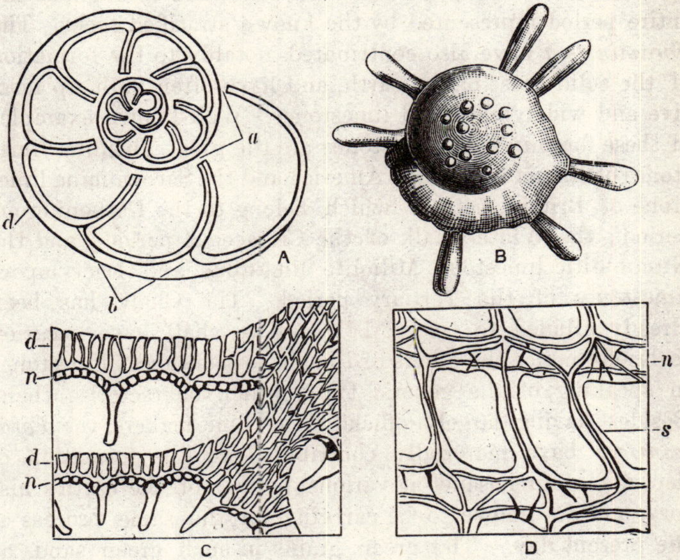

Fig. 17.—A, Diagram of one of the higher forms of the vitreous *Foraminifera*, showing the double nature of the septa (*b*), the stolon-passages between successive chambers (*a*), and the supplemental skeleton (*d*); B, Test of *Calcarina Spengleri*, magnified twelve diameters, showing the spines formed by the supplemental skeleton; C, Part of a section of the test of *Calcarina*, magnified fifty diameters, showing the tubulated "proper walls" of the chambers (*n*), and the canal-system of the intermediate skeleton (*d*); D, Part of the test of *Nummulina lævigata*, highly magnified, showing the canal-system of the septa (*s*), and marginal cord (*n*). (After Carpenter.)

supplemental skeleton, wherever developed to any considerable extent, is traversed by a set of 'canals,' which are usually arranged upon a systematic plan, and are sometimes distributed with considerable minuteness" (Carpenter). The canals of this system are doubtless filled in the living state by prolongations of the sarcode, which serve to keep up the vitality of the intermediate skeleton. This intermediate skeleton, with its canal-system, is largely developed in many of the highest and largest of the types of the Hyaline *Foraminifera,* and very specially so in the ancient *Eozoön,* if this be rightly regarded as a Foraminifer.

As regards the range of the *Foraminifera in time,* representatives of this group are found in almost all formations in which calcareous rocks are developed, and if we admit *Eozoön* to be a member of this group, then the order dates from the Laurentian, and has been continued throughout the entire period represented by the known stratified rocks. The *Foraminifera* have also contributed notably to the formation of the solid crust of the earth, and have often built up massive and widely extended limestones. Well-known examples of these foraminiferal limestones are the great Fusulina limestones of Russia and North America and the Saccammina limestone of Britain, both of which belong to the Carboniferous period; the White Chalk of the Cretaceous period; and the Nummulitic limestone, Miliolite limestone, and Dactylopora limestones of the Tertiary period. The Chalk has been already alluded to (see p. 15), and we shall have occasion to briefly notice the others of the above-mentioned limestones in speaking of the generic types which characterise them. Besides having largely officiated as lime-makers, the *Foraminifera* have materially contributed to the formation of deposits of greensand at various periods of the earth's history, and are known to be carrying on the same process at the present day. The green grains in such green sands as those of the Cretaceous period (as first shown by Professor Ehrenberg for similar green grains in the Lower Silurian of Russia), seem to be often really *casts* of *Foraminifera* in glauconite (silicate of iron and potash), from which the calcareous shell has been dissolved away. Similar green sands, similarly composed in part of internal casts of *Foraminifera*, are now being laid down in various of the warmer seas of the globe. All the recent *Foraminifera* (with the exception of the chitinous *Gromidæ*) are exclusively marine in habit, and all the extinct members of the group were doubtless inhabitants of the sea. Like many other lowly organised forms, the *Foraminifera* have been very "persistent" types of life. Various of the Palæozoic genera have descended to us unchanged from the Palæozoic period; and the prevalent forms in the Chalk are hardly different from those of the Atlantic "ooze." Upon the whole, Dr Carpenter concludes that "there is no evidence of any fundamental modification or advance of the Foraminiferous type from the Palæozoic period to the present time." Lastly, the *Foraminifera* are not altogether reliable tests as to the depth of water in which the deposits containing them were laid down. As a rule, they abound principally in warm and shallow seas. The "Globigerina ooze" of the deep Atlantic and Pacific occurs mainly at great depths, but though doubtless partly composed of forms which really lived at those depths, it is principally made up of the shells of *Foraminifera* which live at or near the surface of the sea. The White Chalk—the ancient analogue of the Atlantic "ooze"—may therefore have been laid down in any depth of water, since its prevalent types of *Foraminifera* were probably mainly surface-forms.

Classification of the Foraminifera.—The classification of the *Foraminifera* has proved a matter of considerable difficulty. The older arrangements were unnatural, as being based wholly on the form of the shell, a point in which the *Foraminifera* show a most marvellous variability. For this reason the artificial systems proposed by D'Orbigny and Max Schultze have now been generally abandoned, and their place has been taken by the schemes of classification put forward independently and almost simultaneously by Professor Von Reuss upon the Continent, and by Dr Carpenter, Mr Parker, and Professor T. Rupert Jones in this country. Both these arrangements agree in the essential feature that they divide the *Foraminifera* into two great primary divisions, in accordance with the nature of the shelly investment. In the one division (*Imperforata*), the test is not perforated by pseudopodial apertures, and it may be either "arenaceous" or "porcellanous." In the other division, the test is perforated by more or less numerous pseudopodial foramina, and to this division the name of *Perforata* is applied. The following tables exhibit the arrangements proposed by Carpenter, Parker, and Rupert Jones, on the one hand, and Reuss, on the other hand; the former being the most natural, and the one most widely adopted:—

A. Classification of the Foraminifera, according to Carpenter, Parker, and Rupert Jones.

Sub-Order I. Imperforata.—Test membranous, calcareous, or arenaceous, not perforated by pseudopodial foramina.

Family 1. *Gromida.*
„ 2. *Miliolida.*
„ 3. *Lituolida.*

Sub-Order II. Perforata.—Test perforated by pseudopodial foramina, generally calcareous.

Family 1. *Lagenida.*
„ 2. *Globigerinida.*
„ 3. *Nummulinida.*

B. Classification of the Foraminifera according to Reuss.

I. Foraminifera with a Non-perforate Test.

A.—With arenaceous tests.

1. *Lituolidea.*
2. *Uvellidea.*

B.—With compact, porcellanous, calcareous tests.

1. *Squamulinidea.*
2. *Miliolidea.*
3. *Peneroplidea.*
4. *Orbitulitidea.*

II. Foraminifera with a Perforate Test.

A.—With a glassy, finely porous, calcareous test.

1. *Spirillinidea.*
2. *Ovulitidea.*
3. *Rhabdoidea.*
4. *Cristellaridea.*
5. *Polymorphinidea.*
6. *Cryptostegia.*
7. *Textilaridea.*
8. *Cassidulinidea.*

B.—With an exceedingly porous, calcareous test.

1. *Rotalidea.*

C.—With a calcareous shell, traversed by a ramified canal-system.

1. *Polystomellidea.*
2. *Nummulitidea.*

With regard to the classification of the *Foraminifera,* the author may advantageously quote some remarks on this subject made by Mr Henry Bowman Brady, F.R.S., one of the highest living authorities on this group of organisms; since they not only have a most important bearing upon the special point in question, but forcibly express the principles which should guide the philosophic naturalist in his systematic treatment of all such variable forms of life: "A purely artificial classification is ill adapted to the conditions presented by a class of organisms like the *Foraminifera*, largely made up of groups of which the modifications run in parallel lines. This 'isomorphism,' demonstrated chiefly by the labours of Messrs Parker and Jones, whilst it is the source of most of the difficulties the systematist has to contend with, is, at the same time, the key to the natural history of the order. It exists not merely between a single series, say in one of the larger divisions, with a single series in another, but often amongst several series even of the same family. It not unfrequently happens that a member of one group presents a greater similarity to its isomorph in another group with which it has a relationship, than it does to any other member of its own group. Take a familiar illustration —suppose the fingers of the two hands to represent the modifications ('species') of two such parallel types of *Foraminifera:* the thumb of one hand resembles more closely the thumb of the other hand than it does any other of the fingers on its own. In other words, the extreme member of one series resembles more closely its isomorph in the other series than it does its own nearer relations, and so on through the remaining members of the respective groups. Under conditions like these, artificial subdivision, based upon minor morphological characters, is certain to infringe the order of nature. Its tendency is to separate forms closely allied, and in many cases to place together such as have no close affinity."

The principal fossil groups of *Foraminifera* deserve a brief consideration, but in the short summary of these which follows—as in the case of similar summaries which will subsequently be given—it must be understood that nothing further is proposed than to select for notice and characterisation those *leading types* of each great group of fossils which may seem to demand mention on the ground of their being *common*, or in other respects, geologically or zoologically, of peculiar *importance*. For anything like a complete list of the known structural types of each group, or the characters of the recorded genera, the specialist will consult special treatises; and it does not appear to be necessary for the wants of ordinary students to do more than to supply a brief statement of the conspicuous characters—especially the *differential* characters—of the more widely distributed and more important types in each group. Nor can even this limited characterisation of leading types be carried out with equal fulness in the case of all groups of fossils, or upon any absolutely uniform plan. In the case, however, of Invertebrate fossils, as being those with which the palæontologist is more especially called upon to deal, the families of each group will, where possible, be defined, and some of the chief generic types will be noticed. The subjoined engraving, representing some of the principal type-forms of the *Foraminifera*, is from a drawing kindly made for the author by his friend, Mr Henry Brady, F.R.S., who has so greatly contributed to our knowledge of this difficult group of organisms.

IMPERFORATE FORAMINIFERA.—Among the *Imperforata*, we have the three families of the *Gromida*, *Miliolida*, and *Lituolida*, of which the first needs no notice, as being quite unknown in the fossil condition.

In the family of the *Miliolida*, the test is opaque, porcellanous, unilocular, or multilocular, and extremely variable in shape; the oral aperture being simple and undivided, or being formed by numerous pores. The family, as far as known at present, is not represented in the Palæozoic period, but ranges from the Trias to the Recent period inclusive. One of the simplest forms of this group is *Cornuspira* (fig. 18, *a*), in which the shell is a simple unchambered spiral, like the shell of a *Planorbis*. The genus is represented in the early Tertiary, and is found under living forms in our seas. *Nubecularia* is a much older type, beginning in the Trias, and its test, extraordinarily variable in shape, is parasitic upon shells and other foreign bodies. In *Miliola*, again (fig. 18, *b*, representing the sub-generic form *Quinqueloculina*), the shell is still extremely variable in form, but it consists typically of a series of chambers wound round an axis, in such a manner

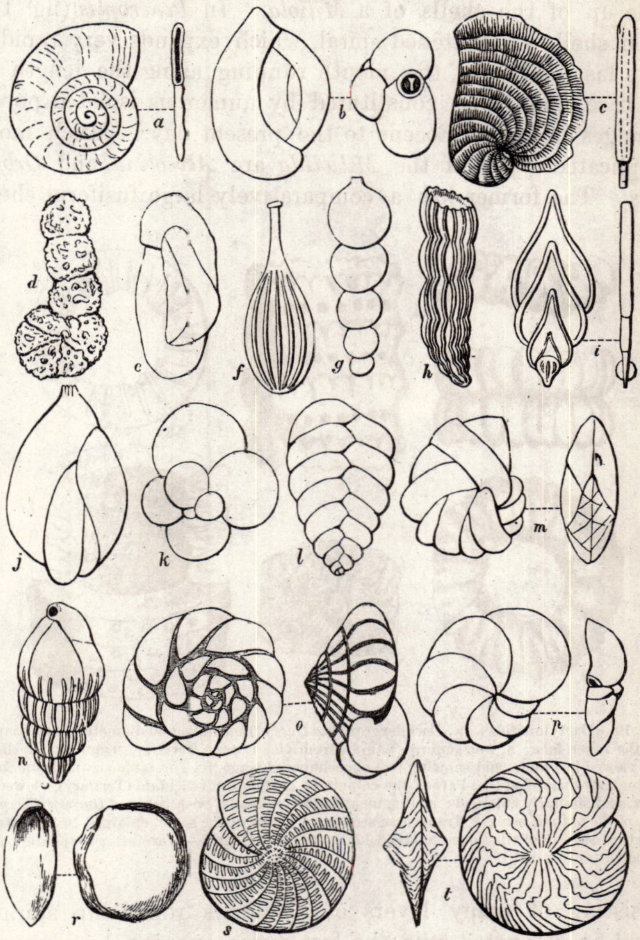

Fig. 18.—Types of Foraminifera. *a*, *Cornuspira foliacea*; *b*, *Quinqueloculina seminulum*; *c*, *Peneroplis pertusus*; *d*, *Lituola agglutinans*; *e*, *Trochammina pusillus*; *f*, *Lagena sulcata*; *g*, *Nodosaria radicual*; *h*, *Marginulina raphanus*; *i*, *Frondicularia Archiaciana*; *j*, *Polymorphina lactea*; *k*, *Globigerina bulloides*; *l*, *Textularia sagittula*; *m*, *Cassidulina lævigata*; *n*, *Bulimina Buchiana*; *o*, *Rotalia Beccarii*; *p*, *Truncatulina lobatula*; *r*, *Archædiscus Karreri*; *s*, *Polystomella crispa*; *t*, *Amphistegina Lessoni*. All the figures are greatly enlarged, the real diameters varying from 1-100 to 1-10 inch. (H. B. Brady.)

that each embraces half the entire circumference. This genus dates from the Jurassic (Lias), and is well represented in recent seas. It abounded in Eocene times, one of the Tertiary limestones of the Paris basin being known as the "Miliolite limestone," in consequence of its being largely made up of the shells of a *Miliola*. In *Peneroplis* (fig. 18, *c*) the shell is a flattened spiral, which expands very rapidly in its last half turn, the mouth running along the length of the base, and being constituted by numerous isolated pores. It ranges from the Eocene to the present day. Much more complicated types of the *Miliolida* are *Alveolina* and *Orbitolites*. The former has a comparatively large fusiform shell,

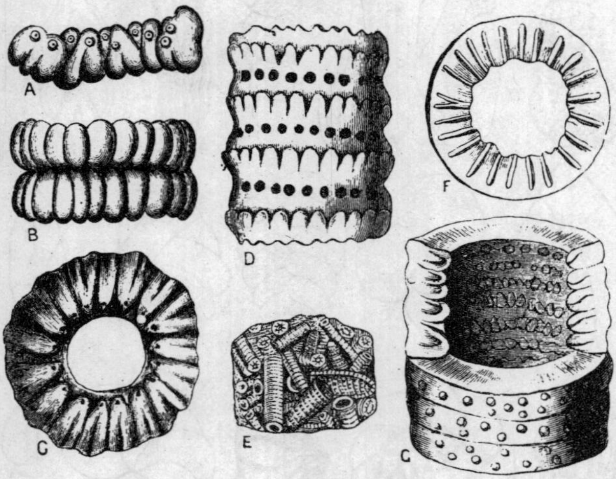

Fig. 19.—Dactyloporidæ. A, *Dactylopora eruca* (recent) magnified 30 diameters, and viewed from the inner face; B, *Dactylopora annulus*, from the Eocene Tertiary, magnified 40 diameters, viewed in profile, and showing two superimposed rings; C, The same viewed from above and similarly magnified; D, Part of the column of *Dactylopora reticulata* (Tertiary), viewed in profile, and similarly enlarged; E, Fragment of Muschelkalk, with tubes of *Gyroporella cylindrica*, of the natural size; F, Transverse section of a tube of the same, enlarged 10 diameters; G, Vertical section of the same, enlarged 12 diameters. (Figs. A—D are after Carpenter; figs. E—G are after Gümbel.)

consisting of many layers of chambers rolled up spirally round an elongated axis, the last series opening by a row of pores; and it dates from the Cretaceous, and has largely contributed to the formation of various of the Tertiary limestones. The latter is coin-shaped, sometimes more than half an inch in diameter, and very complex as regards the arrangement of its chambers. The genus is especially abundant in the Eocene Tertiary, but it dates from the Lias, and occurs plentifully in the recent seas.

The last of the *Miliolida* which need notice are the aberrant types, which may be grouped together in a sub-family under the name of *Dactyloporidæ*.[1] The simplest forms of this curious group, such as *Dactylopora eruca* (fig. 19, *a*), consist of a series of isolated but coherent chambers, each with a single opening, disposed in a half ring, and either free, or more commonly parasitic on shells, and found in tropical seas at the present day. The fossil forms are similar in structure to the simple type just alluded to, but they form complete rings, and these rings are superimposed upon one another so as to form longer or shorter columns, closed at their lower end but open above (fig. 19, B—G). Each ring is quite independent of the others, the orifices of the constituent chambers all opening into the central cavity (fig. 19, C); but in some of the fossil forms the place of the chambers is taken by canals, which, like the former, do not communicate with one another (fig. 19, F). The various members of the *Dactyloporidæ* range from the Trias to the present day, and they have a special interest, from the fact that certain forms of them (*Gyroporella*, fig. 19, E) constitute vast masses of limestone in the Trias of the Bavarian and Tyrolese Alps.

Coming next to the family of the *Lituolida*, we have to deal with imperforate *Foraminifera*, mostly with arenaceous tests, but sometimes sub-arenaceous in texture, and sometimes composed of purely calcareous particles embedded in a calcareous cement. In *Lituola* itself, the type of the group, (fig. 18, *d*), the test is generally crosier-shaped, sometimes nautiloid, usually with a rough exterior, and composed of sand-grains agglutinated together. The genus ranges from the Carboniferous to the present day. An essentially Carboniferous type is *Endothyra*, in which the shell is exactly like that of a *Rotalia* in shape, and which is found abund-

[1] Of late years high authorities have brought forward strong evidence to prove that most or all of the *Dactyloporidæ* are really calcareous *Algæ*. The question cannot be discussed here in detail; but until a final decision has been given by specialists in the department of the *Foraminifera*, it seems safest to retain the group in its present position.

antly in the Mountain Limestone of Britain. It forms in America entire beds of the Carboniferous Limestone (fig. 20). In *Trochammina* (fig. 18, *e*) the test is usually spiral, consisting of one or many chambers, free or attached, and, though sandy, with a smooth surface. It ranges from the Carboniferous to the present day. *Valvulina* (fig. 21) also generally has a spiral shell which may be free or attached, and is normally thick-walled, imperforate, and sandy. Sometimes, however, the shell is porous and smooth, and in other cases the sandy coating seems to be a mere incrustation on a calcareous and perforate shell, so that *Valvulina* may be regarded as a transitional type between the great series of the imperforate and perforate *Foraminifera*. The

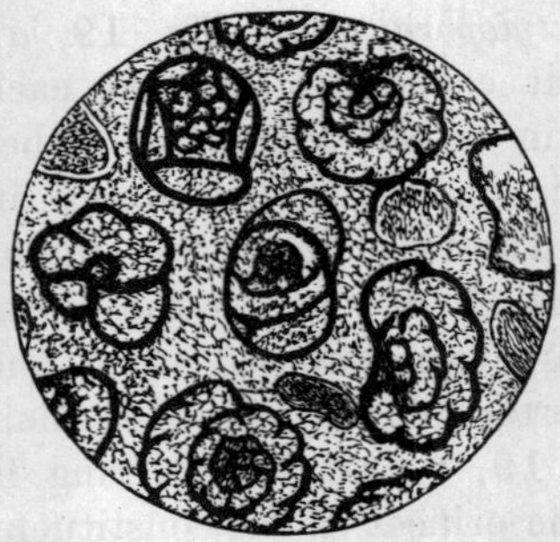

Fig. 20.—Section of Carboniferous Limestone from Spergen Hill, Indiana, U.S., showing numerous large-sized *Foraminifera* (*Endothyra*) and a few oolitic grains magnified. (Original.)

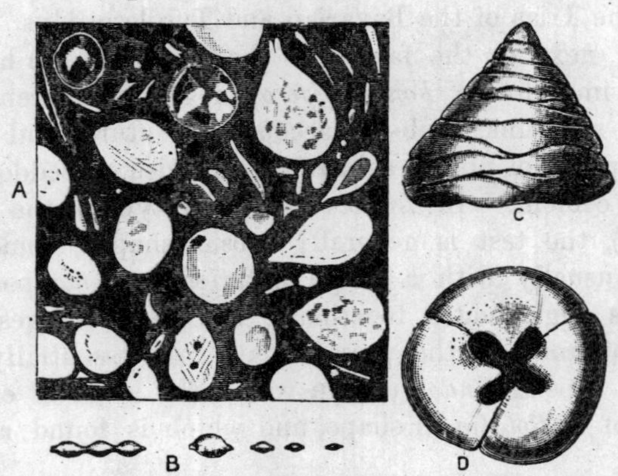

Fig. 21.—A, Slice of limestone with *Saccammina Carteri*, enlarged 5 diameters; B, Spheres of the same, of the natural size, exhibiting variations; C, *Valvulina palæotrochus*, in profile; and D, the same viewed from below, enlarged 45 diameters. All from the Carboniferous. (After Brady.)

genus makes its first appearance in the Carboniferous of Britain, is abundant in the Tertiaries, and is represented in our recent seas. Of the remaining types of the *Lituolida*, the genus *Saccammina* merits special mention as being the only Foraminifer which in Britain actually forms a limestone. It consists of free spherical, pyriform, or fusiform chambers (fig. 21), sometimes separate, sometimes united end to end in twos or threes, with thick, internally labyrinthic walls. The central chamber communicates with the exterior by a single aperture, and the average length of the chambers of the British Carboniferous species (*Saccammina Carteri*, Brady) is as much as 1-8th inch. It forms beds of limestone in the Carboniferous of the South of Scotland and North of England; but the genus is not known to occur again till we meet it in the Post-Pliocene, and, in a living state, in the North Sea. The genus has also been found recently in the Lower Silurian rocks of Scotland. In the *Nodosinella* of the Carboniferous we have another curious type, closely resembling the well-known *Nodosaria* in *form*, but having a sub-arenaceous, imperforate test. A still more singular form is the *Stacheia* of the Carboniferous, in which the test is also sub-arenaceous and imperforate, but grows parasitically upon foreign bodies, in the shape of a crust composed of "an acervuline mass of chamberlets" (Brady). Lastly, we must place here the extraordinary and colossal extinct forms which have been described under the names of *Parkeria*[1] (Carpenter) and *Loftusia* (Brady). Both of these are arenaceous in texture, and both have a very complex and truly "labyrinthic" internal structure. *Parkeria* occurs in the Upper Greensand of Britain, in the form of spheres, which are sometimes over an inch in diameter; while *Loftusia* is found in the Eocene Tertiary of Persia, and has a fusiform shell which may attain a length of between two and nearly three inches.

PERFORATE FORAMINIFERA.—The forms included under this head have a calcareous shell more or less freely perforated by pseudopodial apertures, and they form a great series, of which only a few of the most important forms can be noticed here.

[1] According to Mr Carter, *Parkeria* is a Hydrozoön allied to the recent *Hydractinia*.

The first family under this section—that of the *Lagenida*—comprises "hyaline" or "vitreous" *Foraminifera*, with a calcareous shell, the walls of which are pierced by numerous *minute* pores, and are usually more or less strikingly thin and glassy. In the compound forms of this group the successive chambers have their posterior walls formed by the front wall of the preceding segment, so that the septa are always single, instead of being double, and there is never any "intermediate" skeleton. The family may be divided into two series, *Lagena* itself being the type of the one, while *Nodosaria* is the type of the other. In *Lagena* (fig. 18, *f*) the shell is simple, flask-shaped, unilocular, with a single prominent aperture. The genus commences in the Carboniferous, with a few rare forms, is further developed in the Secondary and Tertiary, and is well represented at the present day. *Polymorphina* (fig. 18, *j*) is allied to *Lagena*, but it is multilocular, the chambers being usually arranged in a double series. It is represented in the Trias, and survives under common types at the present day. In the series of which *Nodosaria* is the type, we have perforate Foraminifera consisting of a succession of chambers, each of which is essentially similar to a *Lagena*, arranged in a series, which is usually nearly or quite straight, though sometimes spirally involuted. In *Nodosaria* itself (fig. 18, *g*) the chambers are simple, and are disposed in a straight line. It ranges from the Permian to the present day. *Dentalina*, ranging from the Carboniferous onwards, is fundamentally like *Nodosaria*, but the shell is bent like a bow. *Vaginulina* comprises forms similar to *Nodosaria*, but laterally compressed, and begins in the Trias. *Marginulina* (fig. 18, *h*) is slightly curved, or is sometimes crosier-shaped, and also starts in the Trias. *Frondicularia* (fig. 18, *i*) has the shell flattened out and leaf-like, and likewise makes its first appearance at the summit of the Trias. Lastly, *Cristellaria* (with *Robulina*) comprises forms more or less spirally inrolled or crosier-shaped, which extend from the Chalk to the present day, and have a very wide development both individually and specifically.

In the second family of the Perforate *Foraminifera*—that

of the *Globigerinida*—are included by Dr Carpenter all those hyaline or vitreous *Foraminifera* which have their shell-substance *coarsely* perforated for the exit of the pseudopodia. There is sometimes, though not usually, a "supplemental" skeleton, and the chambers generally communicate with one another by a larger or smaller crescentic aperture, and not by circular pores. Three simple or unilocular types—viz., *Orbulina, Ovulites,* and *Spirillina,* are known; of which the first (fig. 14) is the most important. It has a spherical shell, with numerous large-sized pores distributed among the smaller ones; and though its distribution at the present day is universal, its earliest appearance seems to be in the Miocene Tertiary. In *Ovulites*[1] the one-chambered ovate test possesses an aperture at both ends. It is found in the Eocene and Miocene. In *Spirillina,* again, the test is coiled into a flat spiral; and, likewise commencing in the Eocene, it is continued to the present day. The *type* of the *Globigerinida,* however, is *Globigerina* itself (fig. 18, *k*), in which there is a polythalamous shell consisting of globose segments arranged in a turbinate spiral, or irregularly disposed. The chambers do not communicate with one another directly, but each opens by a special aperture into a deep central or umbilical depression. In some forms (as also in *Orbulina*) the test, when perfect, appears to be covered with long and extremely delicate spines. *Globigerina* dates from the Cretaceous (Trias ?), and is extremely abundant. It is of special interest, as being the principal constituent of the "ooze" found at great depths in the larger oceans at the present day; while its shells form an equally large portion of the White Chalk (see p. 15).

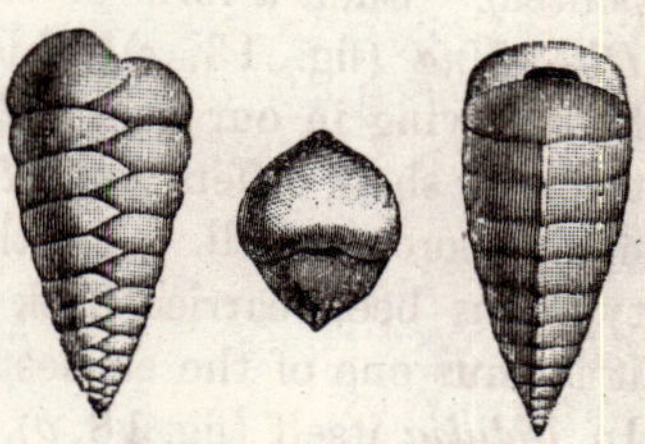

Fig. 22.—*Textularia Meyeriana.* D'Orbigny.

The remaining members of the *Globigerinida* fall into two

[1] Recent researches point to the view that *Ovulites* is a detached segment of a calcareous *Alga,* but further evidence is required before this conclusion is finally adopted.

great sections, typified respectively by *Textularia* and *Rotalia*. In *Textularia* itself (figs. 18, *l*, and 22) the test is generally conical or wedge-shaped, and consists of numerous chambers arranged in two alternate parallel series. *Bigenerina* is much the same as *Textularia*, except that the last-formed segments are disposed in a single and not a double series, and both make their first appearance in the Carboniferous, the latter being a common type in many formations, and being specially abundant in the Chalk. *Bulimina* (fig. 18, *n*), dating from the Trias onwards, consists of spheroidal segments which progressively increase in size, and form an oblique spiral; while *Cassidulina* (fig. 18, *m*), ranging from the Miocene to the present day, though truly biserial, is more or less completely rolled up, and may thus be regarded as an involute *Textularia*. Lastly, *Chrysidalina*, dating from the Chalk, is like *Textularia*, but is triserial.

In the Rotaline series, the shell is typically composed "of a succession of coarsely porous or globigerine segments, arranged in a turbinoid spire, and communicating with each other by a crescentic aperture situated at the junction of the septal plane with the free surface of the convolution" (Carpenter). Such a form of shell is exhibited, for example, by *Discorbina* (fig. 13, *c*), which dates from the Chalk, and is found living in our seas. *Pulvinulina*, with a spiral, usually trochoid shell, differs from *Discorbina* in having a much more finely porous shell. By the researches of Mr Brady, this type has been carried back to the Carboniferous period; and it is thus one of the earliest representatives of the Rotalines. In *Rotalia* itself (fig. 18, *o*), the test is also spiral and turbinoid, but its structure is more complex than in the preceding, the shell-substance being compact and very finely porous; while each chamber is enclosed by a complete wall of its own, and there are canal-like spaces between the two lamellæ forming each septum. In these respects, *Rotalia* closely approaches the Nummuline type. The earliest *Rotaliæ* appear in the Chalk, but the genus attains its maximum in the Tertiary period, and is well represented at the present day. The approximation to the Nummuline type is further manifested by *Calcarina* (fig. 17, B and C), in which the shell is spiral and discoidal, with spur-like marginal appendages, and with a well-developed "supplemental skeleton" and "canal-system." The genus has been shown by Brady to commence in the Carboniferous. In *Planorbulina* the shell is composed of numerous segments, at first spirally and then cyclically disposed. It dates from the Tertiary period, but the forms which are included under the sub-generic name of *Truncatulina* (fig. 18, *p*) commence in the Carboniferous. *Tinoporus*, dating from the Chalk, is in some respects intermediate between *Calcarina* and *Planorbulina*, its general form being like the former, while the irregular and partly cyclical arrangement of its chambers recalls the latter. There is also sometimes a "supplemental skeleton" and "canal-system." We may just mention, also, the genus *Polytrema*, though not yet known in the fossil state, since it has some curious resemblances to some forms of corals and *Polyzoa*. It forms crusts, or, more commonly, branched outgrowths, parasitically attached to foreign bodies; and it consists of numerous intercommunicating irregular chambers, the walls of which are penetrated by an extensive system of capillary canals. *Polytrema* seems to be the representative in the Rotaline series of the singular genus *Stacheia* among the *Imperforata*. Lastly, the genus *Involutina*, from the Lias, is usually placed among the Rotalines, though it presents some peculiarities which would remove it from this series, or would even place it altogether outside the section of the Perforate *Foraminifera*.

Finally, we have the family of the *Nummulinida*, comprising the most complex and the most highly organised of all the *Foraminifera*. In the forms included under this head, the shell is compound, the successive chambers are enclosed each in its proper wall (as diagrammatically shown in fig. 17, A), there is almost always a well-developed "intermediate" or "supplemental" skeleton which renders the shell strong and compact, and which is perforated by a "canal-system," originating in the spaces between the two lamellæ of which each septum is composed; while the shell-substance is pierced by close-set and extremely fine tubules, the septa alone wanting these, so that contiguous chambers usually communicate by but one large aperture. The form of the shell is typically a discoidal spiral or a cycloidal disc.

There is a relationship of a decided character between the higher Rotalines and the Nummulinida, as exhibited by forms like *Rotalia* itself, and *Calcarina* on the one hand, and by *Polystomella* and *Amphistegina* on the other hand. In *Polystomella* (fig. 18, *s*) the shell is lenticular, discoidal, composed of successive chambers, which are prolonged into wing-like ("alar") prolongations, which extend inwards to the centre, thus concealing the earlier turns of the spire from view, while the centre itself is occupied by a solid calcareous boss, penetrated by irregular canals. The "canal-system" is extraordinarily developed and very complex. Some of the simpler types of *Polystomella* are grouped together under the name of *Nonionina;* and the genus seems to make its first appearance in the Upper Chalk, being well represented in the Tertiaries, and surviving to the present day.

Amphistegina still more closely approaches the Rotalines, with which it has sometimes been grouped. Its shell is spiral and discoidal (fig. 18, *t*), usually more or less inequilateral, each chamber being saddle-shaped, and sending forth "alar" prolongations which reach nearly to the centre, where is placed a solid boss. The shell-substance, with exception of the septa and the central boss, is penetrated by numerous close-set, parallel, extremely minute tubules, but the "canal-system" is only imperfectly developed. Brady has shown that the genus occurs in the Carboniferous; but with this exception it is Tertiary and Recent.

Another very ancient, and more anomalous, type of the Nummuline group is the *Archædiscus* of Mr Brady (fig. 18, *r*), which occurs also in the Carboniferous Limestone. In this curious form the test is "convoluted, rounded, more or less unsymmetrical; formed of a non-septate tube coiled upon itself in a constantly varying direction; the shell-wall transversed by very numerous parallel minute tubuli" (Brady).

In the genus *Nummulina* itself (fig. 23) the shell is coin-shaped, of large size, sometimes as big as a florin, or larger, composed of numerous chambers arranged on one plane in a regular spiral. Each chamber is saddle-shaped, the internal or "alar" prolongations of each extending to the centre, so that each revolution completely encloses and conceals from view all the preceding ones. The successive chambers communicate by means of arched fissures, which perforate each septum, close to the periphery of the previous turn of the

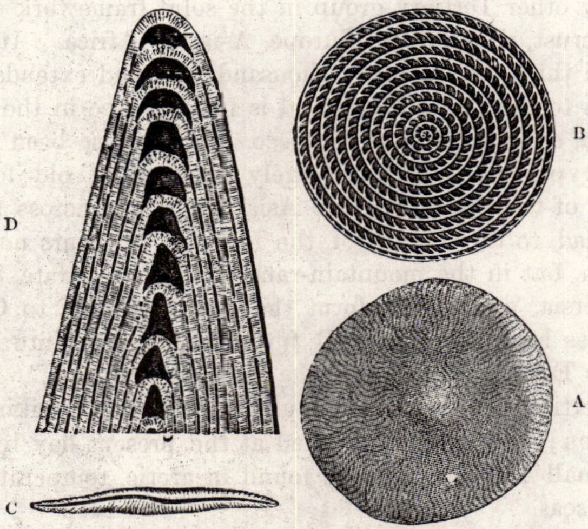

Fig. 23.—*Nummulina nummularia*. A, The shell viewed from above; B, The same, horizontally bisected; The same vertically bisected; D, Vertical section of part of the shell, highly magnified, showing the chambers of the median plane, the alar prolongations, and the tubuli of the shell-substance. Eocene Tertiary.

spire, while secondary and irregular pores in the septa discharge the same function. The general shell-substance is traversed by extremely minute parallel tubuli (fig. 23, D); and there is a supplemental skeleton (forming the so-called "marginal cord"), which, together with the septa, is penetrated by a well-developed and ramified "canal-system" (see fig. 17, D). By the researches of Mr Henry Brady, we know now that the range of the genus *Nummulina* in time must be carried back to the Carboniferous, one small form having been detected in the Mountain Limestone of Belgium. A few "Nummulites" have also been detected in strata of Jurassic and Cretaceous age, but the maximum development of the genus is recorded in the early Tertiary period (Middle

Eocene). At this period in the earth's history we find the Nummulites existing in extraordinary profusion, and building up the wide-spread and massive series of calcareous deposits which are known as the "Nummulitic Limestone." According to Sir Charles Lyell, "the Nummulitic Limestone, with its characteristic fossils, plays a far more conspicuous part than any other Tertiary group in the solid framework of the earth's crust, whether in Europe, Asia, or Africa. It often attains a thickness of many thousand feet, and extends from the Alps to the Carpathians, and is in full force in the north of Africa, as in Algeria or Morocco. It has also been traced from Egypt, where it was largely quarried of old for the building of the Pyramids, into Asia Minor, and across Persia, by Bagdad, to the mouths of the Indus. It occurs not only in Cutch, but in the mountain-ranges which separate Scinde from Persia, and which form the passes leading to Cabul; and it has been followed still further eastwards into India, as far as Eastern Bengal and the frontiers of China." In the later Tertiary period, the genus underwent a striking degeneration; and it is represented at the present day by only a few small forms, which are found in arctic, temperate, and tropical seas.

Very closely allied to *Nummulina*, and of equal or even greater geological importance, is the genus *Fusulina*, the typical forms of which (fig. 24) are spindle-shaped in figure, and may be compared to a Nummulite drawn out at its umbilici. According to Brady, however, some species of *Fusulina* are discoidal and symmetrical, and thus not distinguishable in *form* from *Nummulina*; while in other species the test is spherical. In internal structure, and especially in the minute tubulation of the shell-substance, the genus approaches *Nummulina*, but a regular interseptal "canal-system" appears to be wanting, and the chambers are broken up into chamberlets. Most of the *Fusulinæ* are of considerable size, often from a third to a half of an inch in length, and they often constitute massive beds of limestone,

Fig. 24.—*Fusulina cylindrica.* Carboniferous, Russia.

which have been justly paralleled with the Nummulitic Limestone of the Eocene. Thus they form whole beds of the Carboniferous Limestone in Russia, Central Europe, Armenia, India, China, Japan, and the United States. Though pre-eminently Carboniferous, they occur also in the Permian.

The remaining types of the *Nummulinida*, with the exception of the much-disputed *Eozoön*, can be merely alluded to here. The genus *Orbitoides* is extremely like *Nummulina* in external appearance and form, and has been often mistaken for it, but it differs considerably in its internal structure, and especially in the fact that its mode of growth is cyclical instead of spiral, and the place of the "alar prolongations" of the chambers of the latter is taken by a multitude of chamberlets. The genus appears first at the summit of the Cretaceous, but it undergoes, along with its ally *Nummulina*, an extraordinary development in the early Tertiary period, and it forms immense masses of Eocene limestone in the Southern United States, the West Indies, and in various parts of the Old World. A nearly allied genus is *Cyclocly-peus*, which is also coin-shaped, and is strictly cyclical in its mode of growth. It occurs in the Miocene Tertiary, and the only known recent types attain an extraordinary size (over two inches in diameter). *Operculina*, again, is much more closely related to *Nummulina* proper in its internal structure, though it differs in *form*, owing to the fact that the chambers of the spirally-inrolled shell have no "alar prolongations," and thus approximate to the Rotaline type. The genus commences in the Upper Cretaceous, but is particularly developed in the Eocene of the South of Europe and Africa. Lastly, *Heterostegina* (Tertiary and Recent) differs from *Operculina* chiefly in having the principal chambers broken up into chamberlets by secondary septa.

Finally, if we admit that it is truly a fossil, we must include here the singular body which is known as *Eozoön Canadense*. Upon the true nature of this body a long controversy has been carried on, into which it would be impossible and out of place to enter here. It is sufficient to say that while the highest living authorities upon this special group of organisms regard *Eozoön Canadense* as an aberrant Nummuline Foraminifer, there are other observers who look upon it as a purely mineral and inorganic structure.[1] If, however, we accept the Foraminiferal nature of *Eozoön* as, at any rate, highly probable, we are presented here with a type of extraordinary interest, not only from its intrinsic peculiarities, but also as the most ancient representative of the group of *Foraminifera*, and, indeed, as the oldest fossil which has yet been exhumed from the earth's crust.

The structure known as *Eozoön* is found in various localities in the Lower Laurentian limestones of Canada, in the form of isolated masses or spreading layers, which are composed of thin alternating laminæ, arranged more or less concentrically (fig. 25). The laminæ of these masses are usually

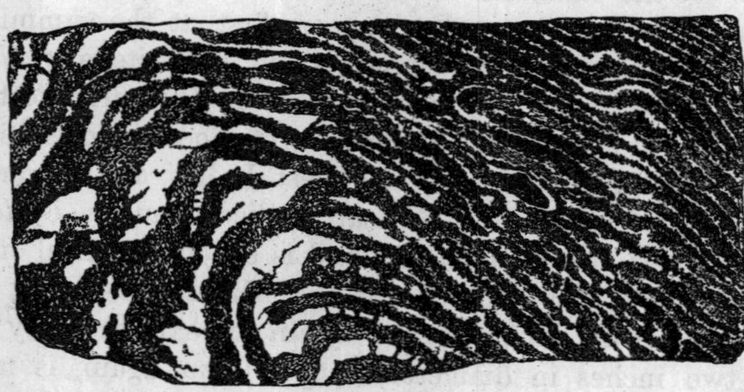

Fig. 25.—Fragment of *Eozoön*, of the natural size, showing alternately laminæ of loganite and dolomite. (After Dawson.)

of different colours and composition; one series being white, and composed of carbonate of lime—whilst the laminæ of the second series, alternate with the preceding, are green in colour, and are found by chemical analysis to consist of some silicate, generally serpentine or the closely related "loganite," or white pyroxene. In some instances, however, all the laminæ are calcareous, the concentric arrangement still remaining visible in consequence of the fact that the laminæ

[1] Since the above was written, Professor Möbius has published an elaborate treatise upon *Eozoön*, and has arrived at the conclusion that it is not truly organic. It would not appear, however, so far as the author is able to judge, that the arguments of Möbius are by any means decisive; and it may safely be concluded that the last word on *Eozoön* has yet to be spoken.

are composed alternately of lighter and darker coloured limestone.

When first discovered, the masses of *Eozoön* were supposed to be of a mineral nature; but their striking general resemblance to the undoubted fossils which will be subsequently spoken of under the name of *Stromatopora* was recognised by Sir William Logan, and specimens were submitted for minute examination, first to Principal Dawson, and subsequently to Dr W. B. Carpenter. After a careful microscopic examination, these two distinguished observers came to the conclusion that *Eozoön* was truly organic, and in this opinion they were afterwards corroborated by other high authorities (Mr W. K. Parker, Professor Rupert Jones, Mr H. B. Brady, Professor Gümbel, &c.) Stated briefly, the structure of *Eozoön*, as exhibited by the microscope, is as follows:—

The concentrically-laminated mass of *Eozoön*, as described by Dr Carpenter and Principal Dawson, is composed of numerous calcareous layers, representing the original skeleton of the organism (fig. 26, *b*). These calcareous layers serve to separate and define a series of chambers arranged in successive tiers, one above the other (fig. 26, A, B, C); and they are perforated not only by passages (fig. 26, *c*), which serve to place successive tiers of chambers in communication, but also by a system of delicate branching canals (fig. 26, *d*). Moreover, the central and principal portion of each calcareous layer, with the ramified canal-system just spoken of, is bounded both above and below by a thin lamina which has a structure of its own, and which

Fig. 26.—Diagram of a portion of *Eozoön* cut vertically. A, B, C, Three tiers of chambers communicating with one another by slightly constricted apertures; *a a*, The true shell-wall, perforated by numerous delicate tubes; *b b*, The main calcareous skeleton ("intermediate skeleton"); *c*, Passage of communication ("stolon-passage") from one tier of chambers to another; *d*, Ramifying tubes in the calcareous skeleton. (After Carpenter.)

may be regarded as the proper shell-wall (fig. 26, *a a*). This proper wall forms the actual lining of the chambers, as well as the outer surface of the whole mass; and it is perforated with numerous fine vertical tubes (fig. 27, *a a*), opening into the chambers and on the surface by corresponding fine pores. From the resemblance of this tubulated layer to similar structures in the shell of the Nummulite, it is often spoken of as the "Nummuline layer." The chambers are sometimes piled up one above the other in an irregular manner; but they are more commonly arranged in regular tiers, the separate chambers being marked off from one another by projections of the wall in the form of partitions, which are so far imperfect as to allow of a free communication between contiguous chambers. In the original condition of the organism, all these chambers, of course, must have been

Fig. 27.—Portion of one of the calcareous layers of *Eozoön*, magnified 100 diameters. *a a*, The proper wall ("Nummuline layer") of one of the chambers, showing the fine vertical tubuli with which it is penetrated, and which are slightly bent along the line *a' a'*; *c c*, The intermediate skeleton, with numerous branched canals. The oblique lines are the cleavage planes of the carbonate of lime, extending across both the intermediate skeleton and the proper wall. (After Carpenter.)

filled with living matter; but they are found in the present state of the fossil to be generally filled with some silicate, such as serpentine, which not only fills the actual chambers, but has also penetrated the minute tubes of the proper wall and the branching canals of the intermediate skeleton. In some cases the chambers are simply filled with crystalline carbonate of lime. When the originally porous fossil has been permeated by a silicate, it is possible to dissolve away the whole of the calcareous skeleton by means of acids, leaving an accurate and beautiful cast of the chambers and the tubes connected with them in the insoluble silicate.

From the point of view that *Eozoön* is truly *Foraminiferal*, it must be regarded as a gigantic member of the *Nummulinida*, which must have grown in reef-like masses. It also has decided affinities to the Rotaline genera *Polytrema* and *Calcarina*, resembling the former in its irregular mode of growth, while it approaches the latter in intimate structure. The test in *Eozoön* is distinctly of a Nummuline type, as shown by its possessing a *minutely* porous or tubular "proper wall" to the sarcode-chambers, while there is also a largely developed "intermediate" or "supplemental" skeleton, penetrated by a "canal-system;" but it differs from all the known *Nummulinida* in its indefinite and often "acervuline" mode of increase. The minute structure of the test will be readily understood by comparing figs. 26 and 27 with fig. 17 c, the latter representing a much-enlarged view of part of the test of *Calcarina*. On the other hand, Professors King and Rowney, Mr Carter, and others, maintain that *Eozoön* is inorganic, and that its so-called "proper wall" is really nothing more than fibrous serpentine.

Eozoön Canadense occurs in the crystalline metamorphic limestones of the Lower Laurentian in Canada, and it has also been detected in the same country in similar limestones believed to be of the age of the Upper Laurentian or Huronian. An allied form (species ?) has been found in rocks supposed to be Laurentian in Newfoundland; Dr Gümbel has described a third form from crystalline limestone belonging to the "Hercynian gneiss formation" (Lower Cambrian or Huronian ?) of Bavaria; while similar structures are stated to occur in the serpentinous marbles of Connemara in Ireland (which are thought to be of Lower Silurian age).

Lastly, Dr Dawson has given the name of *Archæosphærinæ* to small spherical masses of serpentine, sometimes single, sometimes united together in small numbers, which he finds in the Laurentian limestones of Canada, and which he states to be surrounded by a tubulated calcareous shell, resembling the "proper wall" of *Eozoön*. He is of opinion that these bodies are either detached chamberlets of *Eozoön*, or that they are independent organisms, allied to *Eozoön*, but of a simpler type.

Receptaculites.

Before leaving the *Foraminifera*, we must briefly consider the curious fossils grouped together under the name of *Receptaculites* (figs. 28 and 29), which appear to constitute an aberrant type of the *Foraminifera*. If truly referable to this group of animals, *Receptaculites* is not only highly abnormal in point of magnitude, being sometimes as much as a foot in diameter, but its actual structure is quite anomalous. The genus includes large fossils, which are usually discoid, basin-shaped, funnel-shaped, cylindrical or globular in shape, and which consist of a large central cavity, probably filled with sarcode in the living condition (fig. 28), surrounded by a thick wall of complicated structure. In the cup-shaped forms (fig. 29, A) this central cavity is widely open above; but in the globular forms, according to Billings, it communicates with the exterior by but a small aperture situated on the upper surface (fig. 28). The wall bounding the central chamber is composed of (1) an external integument, (2) an internal integument, and (3) an intermediate space crossed perpendicularly by more or less closely approximated tubular pillars (fig. 28, and fig. 29, c). Both the outer and inner

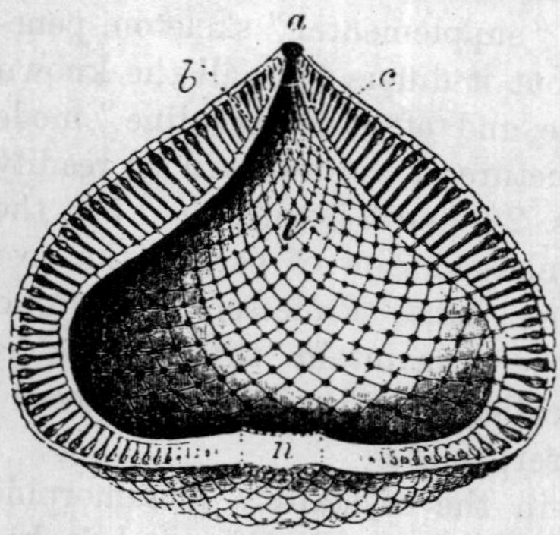

Fig. 28.—Diagram of the structure of *Receptaculites*, as it would be shown by a vertical section of a perfect specimen. *a*, The aperture at the summit; *b*, The inner integument; *c*, The outer integument; *n*, The usual position of the nucleus; *v*, The great internal cavity. The unshaded bands running from the outer to the inner integument represent the pillars. (After Billings.)

integuments of the wall are composed of numerous rhomboidal calcareous plates (fig. 29, B), closely fitting together, and arranged in obliquely-curved intersecting rows, giving fragments of the fossil very much the appearance of the engine-turned case of a watch. According to Billings, the points of junction of each quartette of plates in the inner integument are pierced by foramina (see fig. 28), which allow of a communication between the central cavity and

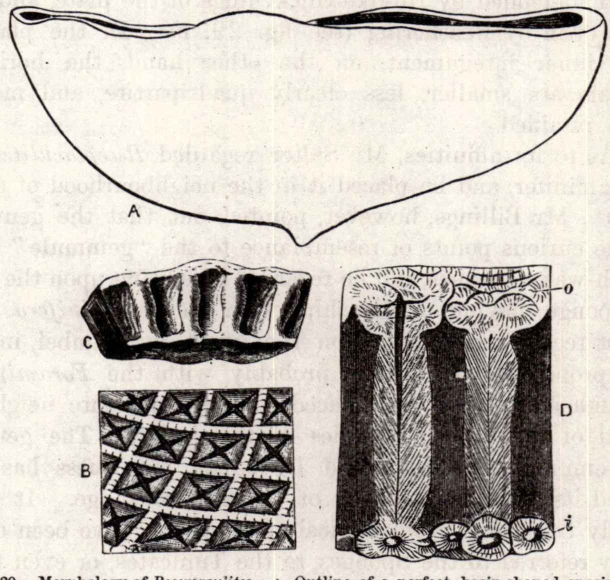

Fig. 29.—Morphology of *Receptaculites*. A, Outline of a perfect, basin-shaped specimen of *Receptaculites Neptuni*, viewed in profile, of the natural size; B, Part of the outer integument of the same, enlarged two diameters, and so far weathered as to show the canals radiating from the summit of the radial pillars; C, Side view of a fragment of the same, showing the radial pillars uniting the inner and outer integuments; D, Vertical section of the same, magnified, showing the outer and inner integuments (*o* and *i*), and the central canals of the pillars. From the Devonian of Germany. (After Gümbel.)

the intercolumnar spaces of the wall; but no such apertures could be detected by Gümbel in the specimens examined by him; and he thinks that this communication is effected by means of minute, microscopic canals between the edges of contiguous plates. The last-named observer has also shown that the radial pillars of the wall are thick and nearly solid, but are perforated by a central tube or canal (fig. 29, D), while their substance is composed of calcareous fibres arranged in a feather-like manner. Each pillar, further, is attached to the centre of one of the plates of both the outer and inner integuments, and the central tube of the pillar opens into a system of horizontal canals which penetrate the substance of these plates. In the plates of the outer integument there are four of these canals, springing from the main tube of the pillar, directed towards the four angles of the plate, defended by rib-like thickenings of the plate, and often laid open by weathering (see fig. 29, B). In the plates of the inner integument, on the other hand, the horizontal canals are smaller, less clearly quadripartite, and more or less ramified.

As to its affinities, Mr Salter regarded *Receptaculites* as a Foraminifer, and he placed it in the neighbourhood of *Orbitolites.* Mr Billings, however, pointed out that the genus has some curious points of resemblance to the "gemmule" of the fresh-water sponges, and he regarded it as being upon the whole a sponge, having relationships with the *Foraminifera.* The most recent researches upon the genus, by Gümbel, indicate its proper position to be probably with the *Foraminifera,* though it can hardly be placed in the immediate neighbourhood of any of the families of this order. The genus is pre-eminently Silurian and Devonian, but Suess has indicated its existence in rocks of Carboniferous age. It seems likely that some problematical types, which have been doubtfully referred to the Sponges, to the Tunicates, or even to the Cystideans, may really belong to the same family with *Receptaculites.* This may be the ultimate destination of the various singular bodies described by Billings as *Pasceolus,* by Eichwald as *Cyclocrinus,* by Salter as *Nidulites,* and by Pengelly as *Sphærospongia,* the true nature of all these being still uncertain; but the little that need be said about these problematical forms will be given when treating of the Cystideans. In any case, the Silurian genera described under the names of *Ischadites* and *Tetragonis* are certainly the same as *Receptaculites.*

CHAPTER VIII.

SUB-KINGDOM I.—PROTOZOA (*Continued*).

RADIOLARIA AND SPONGIDA.

II.—Radiolaria.

Under the head of *Radiolaria* are grouped together at the present day various, mostly microscopic, Protozoans which *typically possess a siliceous skeleton, the parts of which are often more or less radiate, the sarcode of the body being differentiated into a central mass, surrounded by a membranous capsule, and an outer layer usually containing cell-like bodies, while the pseudopodia are long, filamentous, and ray-like* (fig. 30). Though the typical Radiolarians are distinguished by the above-mentioned characters, some of the forms which must be included here are devoid of certain of these features. Thus the so-called *Heliozoa* have no central capsule, and only occasionally possess skeletal structures. In other cases, though the general type of the group is retained, the skeleton is wholly wanting, while the nature of the skeleton when present varies greatly. From the last-mentioned point of view, the Radiolarians are divided into four groups. In the first of these there is no skeleton at all; we have therefore nothing to do with these as fossils. In the second group are forms in which there is a skeleton, but this consists merely of scattered spicules, which lie wholly outside the central capsule of the body; and these also are unknown as fossils. In the third group are forms with a skeleton of radial spicules, but these are now disposed as a symmetrical whole, and lie partly inside the central capsule as well as outside it. In this group are forms like *Acanthometra* (fig. 30, *a*); but though capable of preservation, these forms, like

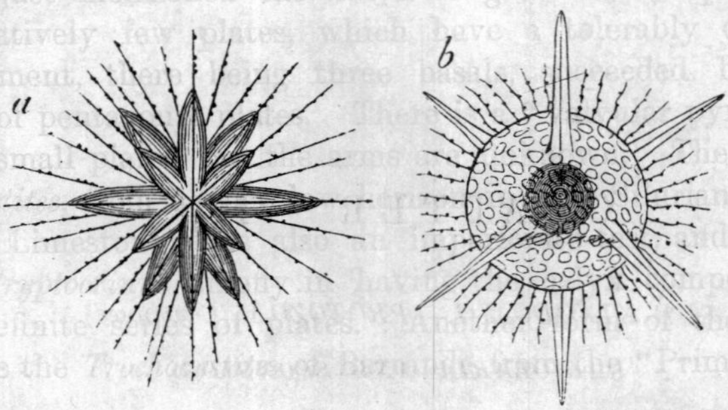

Fig. 30.—Recent Radiolaria. *a*, *Acanthometra*; *b*, *Actinomma* (*Haliomma*), one of the *Polycystina*, showing the siliceous test and radiating pseudopodia.

the preceding, do not occur in the fossil state. Lastly, we have a group in which there is a siliceous, fenestrated or perforated, coherent shell or "test," which is usually furnished with projecting spines. In this group are the microscopic marine organisms, well known for their beautiful glassy skeletons, to which the name of *Polycystina* was given by Ehrenberg.

Many forms of the *Polycystina* have been preserved in the fossil condition, and the distribution of the *Radiolaria* in time, so far as known, is thus identical with that of this particular section of the order. The earliest[1] undoubted remains of *Polycystina* occur in the Jurassic series, and several well-marked types have been detected in the Chalk. All the remaining fossil *Polycystina* are referable to the Tertiary period; and the two most famous localities in which they occur are Barbadoes and the Nicobar Islands. In the former island, in particular, is found a great deposit of sandstone and marls, which rises to heights of over 1000 feet above the sea-level, and which is fundamentally composed of the siliceous skeletons of *Polycystina* (fig. 7). The "Barbadoes

[1] Some forms of *Polycystina* have been indicated as occurring in the Trias. Some Carboniferous fossils have also been referred to this order, but these supposed ancient Radiolarians appear to be really of a vegetable nature.

earth," as this deposit is generally called, is thus a Tertiary representative of the "Radiolarian ooze," which has been shown by Sir Wyville Thomson to cover large areas of the ocean-bed, up to the enormous depth of 4500 fathoms, and which is likewise principally made up of the shells of Polycystines. As is the case, however, with the *Globigerinæ*, the Polycystines are principally surface-forms, inhabiting the open ocean, so that a deposit formed of their shells can in no way be regarded as indicating necessarily the depth of the sea in which it was laid down.

There are many known recent types of *Polycystina* which have not yet been detected in a fossil state; and the principal fossil forms are the following: One of the simplest,

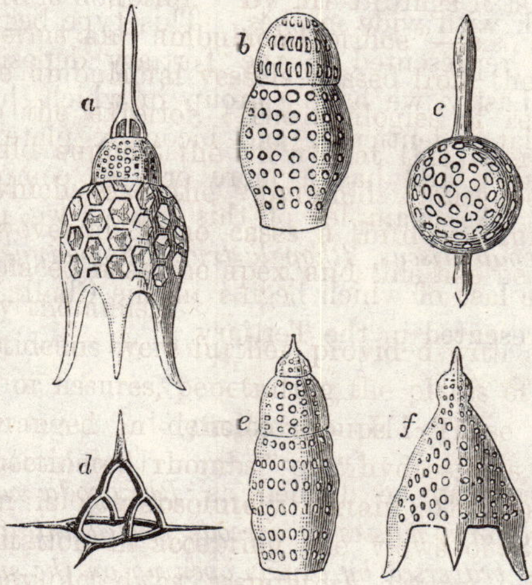

Fig. 31.—Types of Polycystina. *a*, *Podocyrtis Schomburgki*; *b*, *Dictyomitra Mongolfieri*; *c*, *Haliomma dixiphos*; *d*, *Dictyocha Messanensis*; *e*, *Eucyrtidium elegans*; *f*, *Lychnocanium lucerna*. *d* is living, and is after Hæckel; the remaining are Tertiary, and are after Ehrenberg. All the figures are greatly enlarged.

and it is also the most ancient, of the fossil types is *Cœnosphæra* (sometimes referred to the *Thalassicollida*), which is found in the Jurassic and Cretaceous, and has survived to the present day. Of the typical Polycystines, *Haliomma* (fig. 31, *c*), *Heliodiscus*, *Actinomma* (fig. 30, *b*), and *Didymocyrtis*, represent those forms in which the skeleton consists of two, three, or more porous spherical shells, included concentrically within one another, the smaller within the larger, and united by radial bars. Most of the fossil forms, however, belong to the type in which the shells are in the shape of a porous siliceous trellis-work, which may be quite undivided, or is partially marked off into two or more compartments by longitudinal or transverse constrictions. The two poles of the shell are quite unlike one another, and the central membranous capsule of the living animal is enclosed within the closed apical pole. As examples of this type, we may select the genera *Podocyrtis*, *Eucyrtidium*, *Lychnocanium*, and *Dictyomitra* (see fig. 31), all of which are found in the Tertiary, and the last of which is represented in the Chalk. In another group, of which *Dictyocha* (fig. 31, *d*) is the type, the skeleton is composed of irregular bars of flint united into a loose network with wide meshes. This type begins in the Chalk, and is represented in the Tertiary deposits and in recent seas. Lastly, we have a group in which the skeleton consists of a flat or lenticular and biconvex plate, which is sometimes double, and has a more or less complex internal structure. As examples of this group, we may select *Astromma*, *Trematodiscus*, *Rhopalastrum*, *Stephanastrum*, and *Stylodictya*, the last of which begins in the Chalk, while the others are represented in the Tertiary.

III.—Spongida.

The Sponges may be defined as *Rhizopoda composed of numerous amœbiform masses of sarcode united into a composite mass, which is traversed by canals opening on the surface, and is almost always supported by an internal skeleton or framework of horny fibres or of calcareous or siliceous spicula.*

The only portion of the Sponges with which the palæontologist is concerned, is the skeleton. Whatever the nature of the skeleton may be, it is so arranged that its parts surround two sets of apertures which open on the surface of the sponge, and which are connected with one another by a system of canals ramifying in its deeper portions. Of the apertures which penetrate the substance of the sponge in every direction, one set consists of large chimney-like openings, which are called "oscula," or "exhalant apertures." There may be only a single osculum, or many may be present. The other set consists of very much smaller openings, which are always very numerous, and which are termed the "pores," or "inhalant apertures." The pores and oscula are connected by a system of canals excavated in the substance of the sponge, and a constant circulation of water can be kept up through the whole mass, the former serving for the incoming currents, the latter for the outgoing.

The Sponges are by far the largest of the *Protozoa*, and, as above defined, they consist of a soft basis of living protoplasm, which, with hardly an exception, is supported by certain skeletal structures, which vary in composition and arrangement, and are more or less capable of being preserved in a fossil condition. As the soft parts of the Sponges seem to be essentially identical, and as it is only the supporting framework or skeleton which is capable of undergoing petrifaction, we may divide Sponges according to the nature of their hard parts into the following three groups:—

1. THE HORNY SPONGES (*Keratoda*), in which the skeleton is composed of a substance allied to horn, and consists of innumerable fibres matted and felted together, so as to give rise to a very variably-shaped mass. The fibres may be solid or hollow, and the skeleton may consist wholly of these, or may be more or less extensively strengthened by means of variously-shaped microscopic spicules of flint (fig. 32). The horny framework of Sponges such as these is obviously incapable of preservation in the fossil condition; unless we suppose (what has not been proved to occur) that it may be replaced, during the process of petrifaction, by flint or carbonate of lime; while the spicules which are often present,

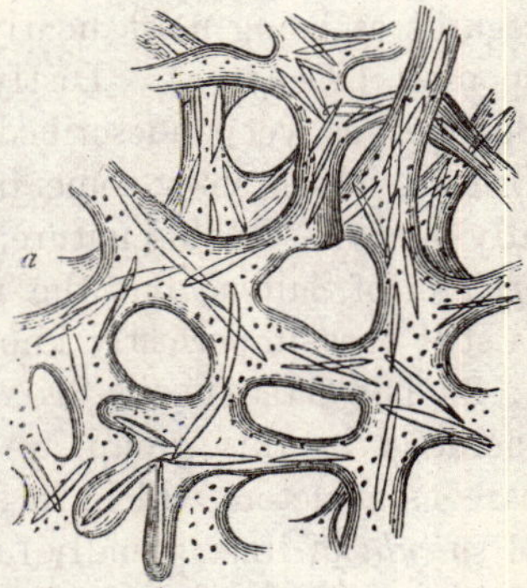

Fig. 32.—Fragment of the skeleton of a horny sponge (after Bowerbank), greatly enlarged, showing interlacing horny fibres with spicula.

though doubtless capable of preservation, and though probably often really present in the rocks, can be with difficulty detected, from their minute size, and can hardly be said to be known *with certainty* except in the Secondary and Tertiary deposits. Many fossil Sponges have, it is true, been referred by different observers to the section of the Horny Sponges, but it is now certain that most of these are certainly truly Siliceous Sponges, while the others are equally referable to other groups.

An exception to the above statement must be made in favour of the aberrant group of living Sponges known as the *Clionidæ*. In all formations, from the Lower Silurian onwards, we meet with shells and corals mined with winding tunnels or borings, which have a circular cross-section. These tunnels are usually regarded as being the work of Sponges belonging to the living genus *Cliona* (*Vioa*), or to forms closely allied to this; and in many instances this reference is doubtless correct. It must not be forgotten, however, that it is very difficult, or impossible, to distinguish, in the case of fossils, between the borings made by Sponges and those produced by Annelides or by carnivorous Gasteropods. Geinitz has also described from the Permian rocks a sponge to which he gives the name of *Spongillopsis*, and which he regards as being most nearly allied to the living fresh-water sponges (*Spongilla*). Lastly, we meet with the remains of Sponges, as yet undescribed, in the Lower Silurian rocks of Britain, which show some indications of having been originally horny. Of this nature, perhaps, is the cup-shaped *Protospongia* of Salter; but the minute structure of this old type is still very imperfectly known.

2. THE CALCAREOUS SPONGES (*Calcispongiæ*).—The Sponges included in this group are invariably furnished with a calcareous skeleton, which, in all the *living* species, is composed of *spicula* of lime, usually fusiform or triradiate in shape, and always entirely free and independent of each other. No living member of the *Calcispongiæ*, then, is possessed of a *continuous* skeleton, and the calcareous spicules which constitute the sole skeletal elements are microscopic in their dimensions; so we might have anticipated beforehand that we should find few representatives of this group of Sponges

in past time. Of the fossil forms, indeed, hardly any can be said to be true *Calcispongiæ*, if we restrict this name to Sponges with a skeleton composed of *free calcareous spicula*, and some doubt attaches to even the very few forms which one might be disposed to place here. Two ancient genera—viz., *Astræospongia* and *Amphispongia*—may, however, be more specially mentioned, as being, perhaps, ancestral types of the modern *Calcispongiæ*. In *Astræospongia* (fig. 33, *a*), we have

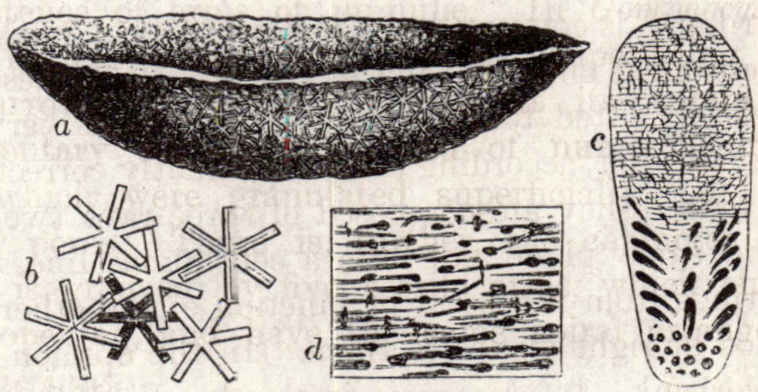

Fig. 33.—*a*, Side-view of a specimen of *Astræospongia meniscus*, of the natural size, Upper Silurian; *b*, Spicules of the same, enlarged (after Roemer); *c*, A split specimen of *Amphispongia oblonga*, of the natural size, Upper Silurian; *d*, Part of the upper portion of the same, enlarged. (Original.)

a cup-shaped or discoid sponge, found in the Upper Silurian and Devonian, and composed of irregularly disposed, free, six-rayed, calcareous spicula, without any definite canal-system. The rays of the spicules are all in one plane, and their size is much greater than that of the spicules of any living member of the *Calcispongiæ*, but there is no sufficient reason for regarding them as having been anything but calcareous to begin with, since they occur in beds in which the other fossils have undergone no change.

The genus *Amphispongia* (fig. 33, *c*) is another curious type, which approximates to the living *Calcispongiæ* more closely than the preceding. It is only known from certain soft sandstones, of Upper Silurian age, in the Pentland Hills, near Edinburgh; and, like almost all the other fossils in the same beds, its original calcareous skeleton has been dissolved away, but we are not thereby justified in concluding that its hard parts were primitively composed of anything but lime. The general form of this sponge is that of a somewhat clavate mass, between one and two inches long, and, owing to its peculiar state of preservation, the spicules which formed the original skeleton are now only represented by vacant spaces or cavities, while the actual canals and intervals between these are filled with the sandy matrix of the rock. The lower part of the sponge (fig. 33, *c*) consists of a series of radiating canals, separated by a series of calcareous bars or bundles of spicules, the former now filled with the surrounding matrix, while the latter, being dissolved out, are represented by hollow tubes. The upper part of the sponge, on the other hand, is composed of a matted mass of small spicules, now represented only by cavities (fig. 33, *d*). These spicules are so closely fitted together that their form is very difficult to make out; but, according to the apparently correct observations of Salter, they are composed of three rays, two of which lie in the same plane and form a continuous line, while the third springs from the point of junction of the other two, and is directed at right angles to them. In the opinion of Salter and Bowerbank, *Amphispongia* is closely allied to the living genus *Grantia*, and, under any circumstances, there can be little doubt as to its being truly an ancient type of the *Calcispongiæ*.

While all the living *Calcispongiæ* possess a skeleton composed of *disconnected spicules*, there is no *à priori* reason why we should not meet with fossil forms of an essentially similar nature, but having a *continuous* or *vermiculate* skeleton, composed of calcareous spicules primitively free, but ultimately anchylosed so as to form a single and connected framework. Many fossil sponges have been supposed to belong to this now unrepresented category of *Calcispongiæ* with a continuous skeleton, but most of them (including most of the forms formerly known as *Petrospongiadæ*) have been shown by recent microscopic researches to be truly siliceous sponges.[1] There still remain, however, some fossils which were beyond all question calcareous to begin with, and which cannot, with our present knowledge, be assigned to a definite place in the zoological series, unless we regard them, provisionally at any

[1] Since the above was written, Zittel has published a memoir in which he refers a large number of Triassic, Jurassic, and Cretaceous Sponges to a special group of *Calcispongiæ* in which the skeleton is fibrous and continuous, instead of being composed of separate spicules. The principal types referred to this group are the genera *Peronella* and *Corynella*. The apparently fibrous skeleton of these forms is stated to be really composed of spicula.

rate, as a group of Calcareous Sponges with a continuous skeleton—thus giving us a link between the *Spongida* and the *Foraminifera.* Of the fossils referred to, we need only speak more particularly here of the abundant Palæozoic forms which have usually been grouped together by palæontologists under the name of *Stromatopora*, and of the singular forms composing the genus *Archæocyathus.* The genus *Stromatopora*—using this term in a purely general sense—comprises a great number of Silurian and Devonian (possibly also Carboniferous) fossils, which form hemispherical, globular, or irregular masses, varying from an inch or less up to many feet in diameter, and which are always composed essentially of closely approximated calcareous laminæ (fig. 34) arranged concentrically round one or more centres, and often demonstrably

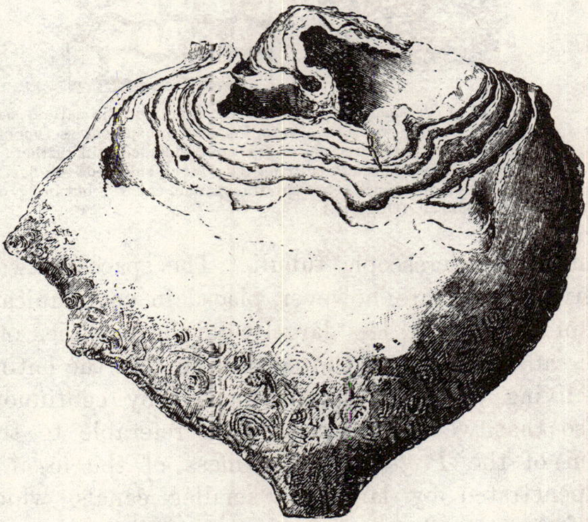

Fig. 34.—A small and perfect specimen of *Stromatopora rugosa* (Hall). From the Memoirs of the Geological Survey of Canada.

attached to foreign bodies. Sometimes they form thinner or thicker crusts, growing parasitically on shells or corals, or spreading out as horizontal expansions. The concentric laminæ, which are the essential feature of this group of fossils (and which strongly call to mind the appearance of masses of *Eozoön*), are separated by wider or narrower interspaces, which are more or less completely subdivided by vertical pillars or imperfect partitions (fig. 35). We might readily be disposed to regard these singular forms as aberrant and gigantic *Foraminifera*, but in no case has it been satisfactorily proved that the calcareous walls of the fossil are

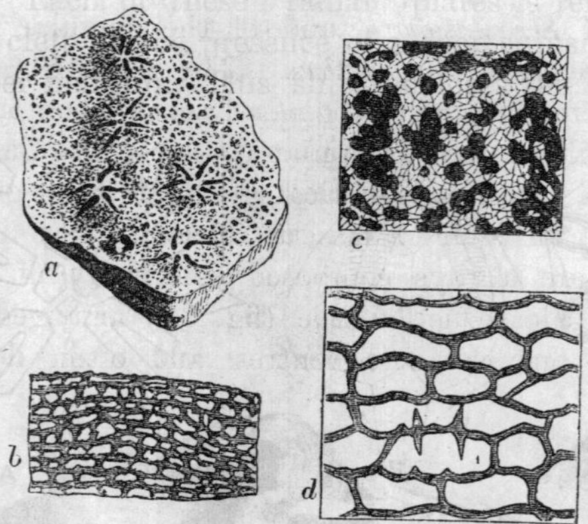

Fig. 35.—*a*, Fragment of *Stromatopora granulata* (Devonian), of the natural size, showing the upper surface, with stellate water-canals; *b*, Vertical transparent section of the same, magnified; *c*, Horizontal transparent section of the same enlarged still further; *d*, Vertical section of another Stromatoporoid (*Clathrodictyon cellulosum*), enlarged. In figs. *b* and *c* the skeleton is dark, and the light spaces represent transparent calcite; but in fig. *d* the latter represent an infilling of silica. (Original.)

penetrated by microscopic tubuli. The spaces between the successive laminæ are, however, placed in communication by means of a more or less largely developed series of pores, and we can hardly avoid the conclusion that the entire fossil in its living condition was permeated by continuous sarcode; so that we must regard it as referable to the subkingdom of the *Protozoa*. The mass of the fossil is also often penetrated by larger or smaller canals, which can hardly have served any other purpose than that of conveying water to different parts of the organism (fig. 36, *a*), and which may fairly be compared with the "aquiferous canals" of the Sponges. The surface of the mass, also, often exhibits conical elevations, or papillæ, which are perforated at their summits by the apertures of water-canals, and from which there radiate branched and vermiculate grooves, these characters strongly reminding us of some of the living Sponges (see fig. 35). It should be added that many palæontologists are disposed to take a different view of the affinities of the "Stromatoporoids," and to place them among the *Hydrozoa*, in the sub-class of the "Hydrocorallines." This view seems to be based principally upon the curious Devonian genus *Caunopora*, which differs from *Stromatopora* proper in many important respects, and the structure of which—in spite of many able researches—has still to be fully worked out.

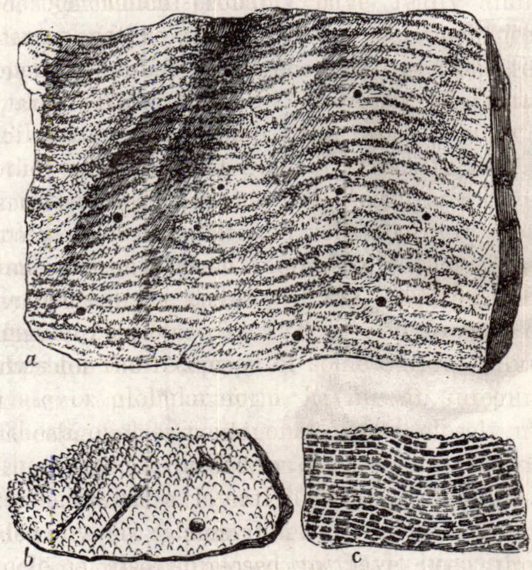

Fig. 36.—*a*, Part of the under surface of *Stromatopora tuberculata*, showing the wrinkled basement membrane and the openings of water-canals, of the natural size; *b*, Portion of the upper surface of the same, enlarged; *c*, Vertical section of a fragment, magnified to show the internal structure. Corniferous Limestone, Canada (Original.)

Upon the whole, therefore, we must at present regard the position of *Stromatopora* and its allies as uncertain.

In the genus *Archæocyathus* (fig. 37), we have a group of singular fossils which have been described by Mr Billings from the summit of the Cambrian and the base of the Lower Silurian (Potsdam Sandstone and Calciferous Sand-rock), in North America, and have been subsequently investigated by Principal Dawson. The general form of the fossils herein included is that of a hollow cone or hollow cylinder, enclosing a large cup-shaped cavity, and tapering towards one extremity, which was presumably fixed to some foreign body.

Specimens appear to have reached a very large size, a length of two or three feet and a diameter of three or four inches being sometimes attained. The sponge consists of an outer wall, usually perforated with numerous small irregular apertures, and a thin inner wall pierced with many openings (fig. 37, *a*). The space between the outer and inner wall is subdivided by a number of vertical radiating partitions, thus simulating the structure of one of the septate corals. The genus, however, is certainly not a coral, and we have the curious feature of the occurrence of numerous branching, cylindrical, or fusiform siliceous spicula within the substance of the organism. In the same geological horizon, and also in higher strata, occurs the somewhat allied genus *Calathium*, in which the skeleton also assumed a turbinate form. According to Dawson, the vertical laminæ or septa in the upper portion of the cup of *Archæocyathus* are not only perforated by numerous round apertures—thus allowing contiguous chambers to communicate freely with one another,—but they are themselves traversed by branching delicate canals running in their substance; and he regards the genus as in some respects allied to *Eozoön*. Lastly, it is possible that, in the Carboniferous genus *Palæacis*, often referred to the Perforate Corals, we have in reality a type of calcareous Sponges, with a vermiculate skeleton.

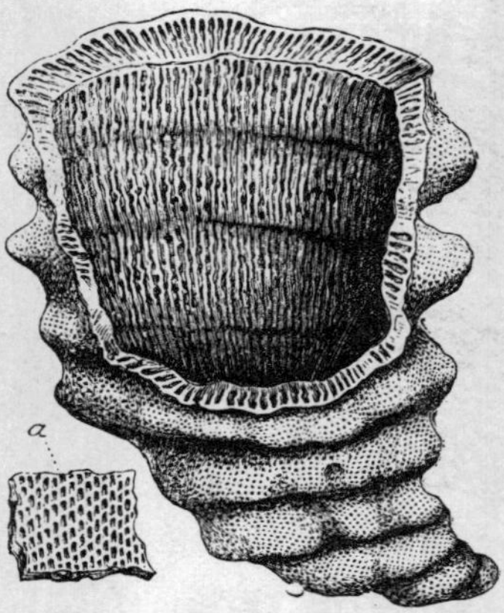

Fig. 37.—Restoration of the lower part of *Archæocyathus Minganensis*. *a*, The pores of the inner wall of the cup. (After Billings.)

3. The Siliceous Sponges (*Silicispongiæ*). — In this group are included those Sponges in which the skeleton is made up of siliceous spicula or fibres. The skeleton may be a discontinuous or continuous one, and in the living forms the sarcode contains free siliceous spicules of microscopic size and very variable form. These "flesh-spicules," though

often met with in the rocks, are of comparatively little use to the palæontologist so far as enabling him to classify the fossil forms is concerned, since it is rarely possible to refer them to the Sponge to which they originally belonged. They are, however, of great use in the determination of the living types. The spicules of the true skeleton, on the other hand, are usually united to one another by a sort of articulation, or become cemented together by a deposit of glassy silica; so that the skeleton forms a more or less continuous framework, admirably adapted for preservation in a fossil condition.

Until of late years very little was known, with any precision, as to the structure, affinities, or real nature of a great many fossil Sponges, which are now recognised as belonging to the group at present under consideration. This arose partly from the fact that the value of the microscope in palæontology had not been recognised, and partly from the fact that the structure of the living types has only recently been at all fully understood; while the state of preservation in which these fossils often occur was such as almost inevitably to lead to misconceptions as to their nature, and to give rise to difficulties which are even yet not fully cleared up. It was known, namely, that Sponges with a siliceous skeleton were of common occurrence in various formations, and especially in the Jurassic and Cretaceous rocks; but it was generally supposed that in these cases the skeleton had been originally composed of *lime* or of *horn,* which had in the process of fossilisation been dissolved away and *replaced* by flint. These Sponges, in fact, were supposed to have undergone *silicification* —a change to which we know that fossils are very often subjected. In the face of the now recognised fact that the minute structure of these fossil forms agrees perfectly with that of *living* siliceous Sponges, and differs wholly from that of any living types of calcareous or horny Sponges, we cannot doubt that their skeleton was primarily composed of flint. We are thus compelled to believe that in many instances the original siliceous skeleton has been more or less completely dissolved away, the space which it originally occupied in the rock being left permanently vacant, or being simultaneously or subsequently filled up with peroxide of iron, or with crystalline carbonate of lime. Though this dissolution of a flinty skeleton, with or without replacement by lime, is at variance with all preconceived ideas on this subject, and though it is very difficult to give any adequate or satisfactory explanation of the way in which it is effected, the researches of Prof. Zittel and Mr Sollas hardly allow us to doubt its actual occurrence in nature. It is impossible here to pursue this intricate and still controverted subject further; but it may be pointed out that in this dissolution of the skeletons of siliceous Sponges (and of other flinty organisms) by waters percolating through sediments in course of formation, we find a sufficient source for supplying the amorphous and gelatinous silica of the chalk-flints. A similar origin may with all probability be ascribed to at least some of the chert-nodules so common in many formations.

As regards the *classification* of the Siliceous Sponges, naturalists now universally accept the division of the group into the two primary sections of the *Hexactinellidæ* and *Lithistidæ*, first proposed by Oscar Schmidt, and subsequently very variously subdivided by different authorities. It is not necessary to consider the minor subdivisions here, but we may define the two primary sections above named, and glance at some of the leading types of each, taking these in geological rather than zoological order.

(A.) *Section Hexactinellidæ.*—Siliceous Sponges, the fundamental elements in the skeleton of which are *six-rayed spicules, the rays of which are almost invariably at right angles to each other.* In the centre of each spicule is an axial canal, consisting of three tubes cutting each other at right angles. The spicules may be only united by sarcode, or they may be fused together by amorphous silica, in either case being so disposed as to form a trellis-work (fig. 38, B) with rectangular or polyhedral meshes. Besides the true "skeleton-spicules," the sarcode contains (in the living forms at any rate) numerous detached "flesh-spicules," which are also fundamentally six-armed, but which give off secondary branches, so as to form a "rosette."

Until of late years, even the living *Hexactinellidæ* were very little known, the most familiar example of the group being the beautiful Venus's Flower-basket (*Euplectella*). Now we not only know of a long series of living Sponges with a flinty skeleton built upon the plan above defined, but we

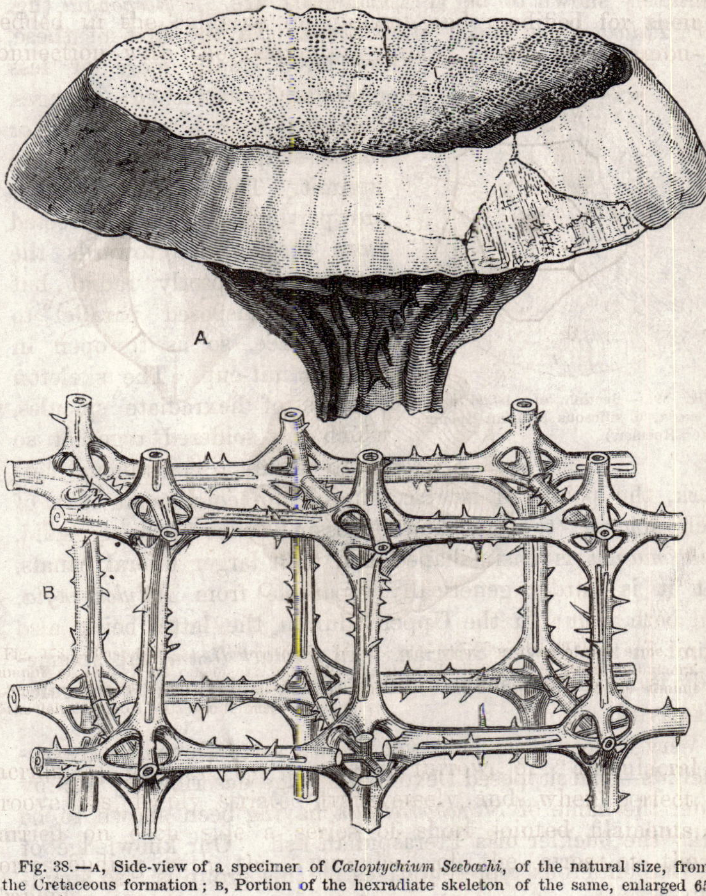

Fig. 38.—A, Side-view of a specimen of *Cœloptychium Seebachi*, of the natural size, from the Cretaceous formation; B, Portion of the hexradiate skeleton of the same, enlarged 65 times. (After Zittel.)

know that the group enjoyed a great development in past time, and that it can be traced back to the Silurian period. All the recent *Hexactinellidæ* are inhabitants of the more profound depths of the ocean, and are in no case strictly shallow-water forms. It is probable, therefore, that the fossil forms indicate that the deposits in which they occur were laid down in a considerable depth of water.

Of the Silurian Sponges, the only forms which have been definitely shown to be Hexactinellids are *Astylospongia* (fig. 39), *Palæomanon*, and *Protachilleum*. In the first of these, we have small, more or less globular, unattached sponges, furnished with a cup-shaped or funnel-shaped depression at the summit. The aquiferous canals are partly radial and directed from the surface towards the centre, and partly radial but vertically disposed parallel to the surface, so as to open in the summit-cup. The skeleton consists of hexradiate spicules, which are soldered together so as to form a continuous net-work, the points of intersection of the component rays of each spicule (the so-called "crossing-nodes") being solid. *Palæomanon* is basin-shaped, and with larger lateral canals, but it is hardly generically separable from *Astylospongia*, and both occur in the Upper Silurian, the latter being also found in the Lower Silurian. In *Protachilleum*, the sponge-body is mushroom-shaped and stalked, and there is no sum mit-cavity.

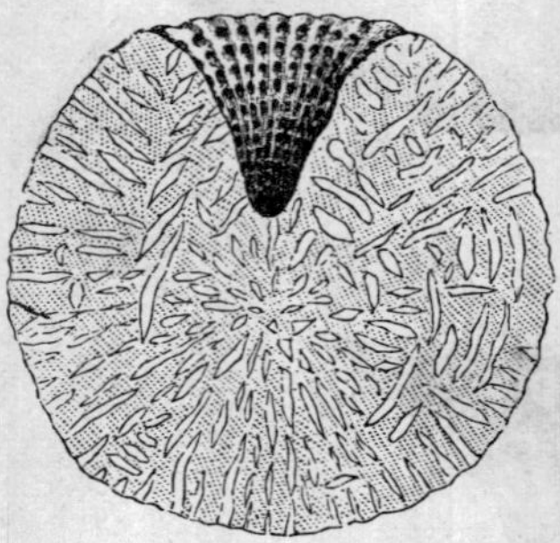

Fig. 39. — Section of *Astylospongia præmorsa*, a siliceous Silurian Sponge. (After Roemer.)

Very little, indeed, is as yet known as to Devonian Hexactinellids—the supposed Devonian Sponge described by McCoy under the name of *Steganodictyum* having been shown to be really the buckler of a Pteraspidian fish. Our knowledge of the Carboniferous representatives of this group is also singularly defective; but we have the extremely interesting fact that the deposits of this age contain, in no inconsiderable numbers, the remains of ancient members of the now living genus *Hyalonema*. In this genus, comprising the so-called "Glass-rope Zoophytes," there is a comparatively small sponge-body, rooted to the mud of the sea-bottom by a longer or shorter rope of delicate siliceous fibres. In addition to this skein of "anchoring fibres," there are branched spicules, which are four-armed or five-armed in the recent forms, but some of which in the fossil forms are hexradiate. A Carboniferous species has been described as *Hyalonema Smithii;* and its "rope" was long known to palæontologists under McCoy's title of *Serpula parallela,* being supposed to be formed of the parallel tubes of one of the Tubicolous Annelides. The root-fibres of *Hyalonema Smithii* are of large size, with a minute central canal, and terminating in anchoring hooklets, their sides being smooth. Silurian species of *Hyalonema,* still undescribed, are also known to occur.

Nothing is at present known of Permian Hexactinellids (unless we refer here the imperfectly examined *Bothroconis*

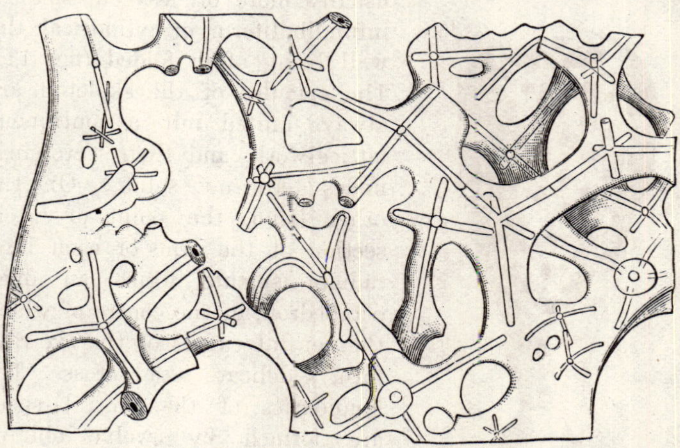

Fig. 40.—Portion of the skeleton of *Tremadictyon reticulatum,* enlarged 50 diameters, from the Jurassic (after Zittel). The original spicules are soldered into a continuous trellis-work by a coating of silica; but their position and hexradiate form is shown by their axial canals. The "crossing-nodes," or points of intersection of the arms of each spicule, are solid.

of King). No members of the group, also, have been hitherto detected in the Trias, and they are scarce or wanting in the lower portion of the Jurassic series. In the Upper Jurassic, however, we meet with a great number of Hexactinellid Sponges. Of these, *Craticularia, Verrucocœlia, Tremadictyon* (fig. 40), *Sporadopyle,* and *Sphenaulax,* possess a skeleton built upon the same type as the living

Eurete and *Farrea*; *Pachyteichisma* and *Trochobolus* are early forms of the great family of the *Ventriculitidæ*; *Cypellia*, *Staurodcrma*, and others represent the extinct family of the *Staurodermidæ*; and *Stauractinella* belongs to the group of Hexactinellids in which the skeleton-spicules are only united by sarcode, so that they do not form a continuous network.

In the Cretaceous deposits, and especially in the Chalk itself, the Hexactinellids are very largely and abundantly represented. Of the family of the *Euretidæ* we have now few forms (*Craticularia*, *Verrucocœlia*, &c.); but the great family of the *Ventriculitidæ* (employing this term here in a general sense for the groups allied to the *Ventriculitidæ* proper) undergoes a marvellous expansion. The shape of the sponge-body in this family is very variable, but is usually more or less cup-shaped, infundibuliform, or cylindrical, the wall being often folded (fig. 41). The spicules of the skeleton are always united into a continuous lattice-work, and their "crossing-nodes" are not solid. On the other hand, the point of intersection of the arms of each hexradiate spicule forms an open octahedron, in the centre of which the central canals of the six rays form a delicate axial cross. The boundaries of the central space are formed by twelve oblique uniting beams, the whole forming an elegant octahedron, which is known as the "lantern" (see fig. 38, B). The most important of the Cretaceous genera of the *Ventriculitidæ*—using the term in the above wide sense—are *Ventriculites* (fig. 41), *Cephalites*, *Cœloptychium*, *Callodictyon*, *Marshallia*, *Pleurope*, *Plocoscyphia*, *Etheridgia*, *Camerospongia* (fig. 42), and *Tremabolites*. Very few Tertiary representatives of the *Ventriculitidæ* have hitherto been

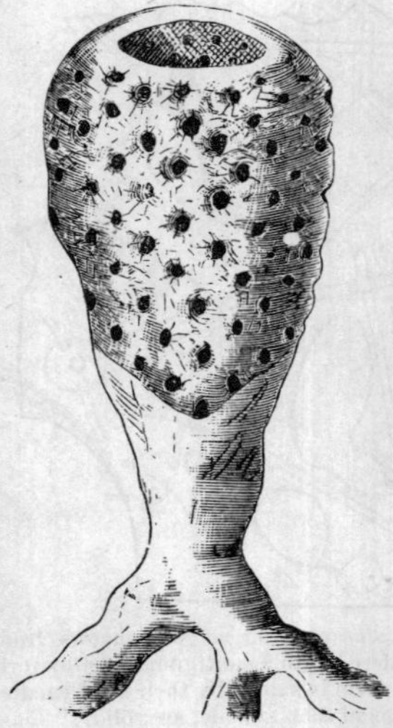

Fig. 41.—*Ventriculites simplex*. White Chalk, Britain.

discovered, but the family survives in the existing genus *Myliusia*. Of the remaining Cretaceous Hexactinellids, we may just mention *Coscinopora* (fig. 43), with its cup-shaped skeleton affixed to foreign bodies by ramified roots. The

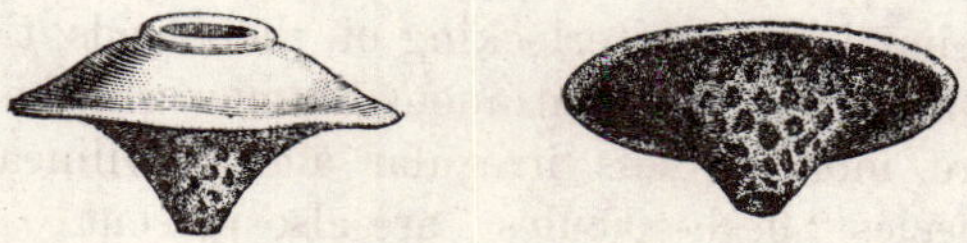

Fig. 42.—*Camerospongia fungiformis.* Cretaceous.

lattice-work of the skeleton is irregular, and the crossing-nodes of the spicules are partly solid, and partly furnished with a "lantern." In the nearly allied *Guettardia* the wall is deeply folded in a stellate manner, and the crossing-nodes are solid. Lastly, we find in the Cretaceous the first representative of the curious modern genus *Aphrocallistes*.

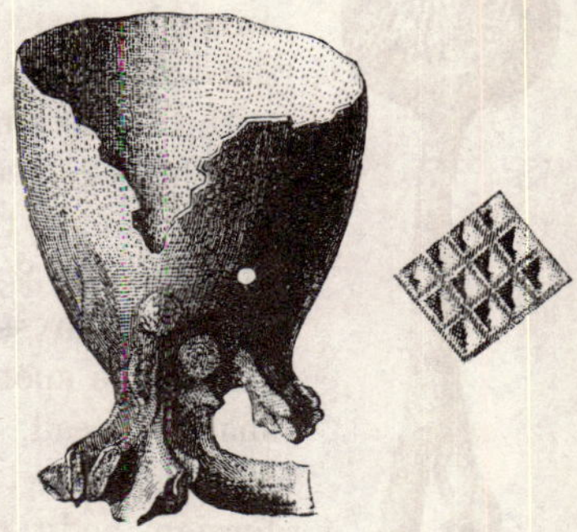

Fig. 43.—*Coscinopora cupuliformis,* and a portion of its surface enlarged. Cretaceous.

In the Tertiary period comparatively few Hexactinellids make their appearance, doubtless owing to the fact that so many of the formations of this age are shallow-water deposits. In the Miocene of Oran, in North Africa, however, an abundance of Hexactinellids (*Craticularia, Aphrocallistes,* &c.) have been detected.

At the present day, we find an abundance of Hexactinellid Sponges, of which the most striking forms are the exclusively recent *Euplectellidæ*, in which the skeleton-spicules are cemented together into a ladder-like trellis-work, and there is a single "osculum" provided with a porous lid.

(B.) *Section Lithistidæ.*—In this group are comprised Siliceous Sponges in which the spicules are fundamentally quadriradiate, three of the four arms being so disposed as to come together at an angle of 120°, their extremities being divided into processes by means of which contiguous spicules interlock with one another; while the fourth arm lies in a different plane and forms a cylindrical shaft from which the other three spring. The spicules are not united by a siliceous cement; but by the interlocking of their ends the skeleton forms a more or less continuous framework, the meshes of which are more or less irregular and curvilinear. In the recent species "flesh-spicules" are also present.

The fossil *Lithistidæ*[1] have not yet been so extensively worked out, as is the case with the fossil Hexactinellids; but they are said by Zittel to be represented in the Silurian (*Aulocopium*, *Calathium* (?), and *Eospongia* (?)). The same observer has likewise indicated their abundant occurrence in the Jurassic period; but it is not till we reach the Cretaceous period that we are confronted with the maximum development of this group. Of the Cretaceous Lithistids, by far the most important is the familiar and widely distributed genus *Siphonia*, the structure of which has been investigated by Mr Sollas. In this genus the sponge-body consists of a variably-shaped head, supported upon a longer or shorter stem, and attached thereby to some foreign body (fig. 44), but the stem may be wanting, when the sponge is attached by diverging root-fibres. The exhalant water-canals open at the summit of the sponge, usually by oscula situated within a cup-shaped apical cavity, while the inhalant canals open by "pores" on the lateral surfaces. The skeleton-spicules (fig. 45, B) are furnished with three diverging arms, are bifurcated, and furnished

Fig. 44.—*Siphonia ficus*, a Cretaceous Lithistid Sponge.

[1] Since the above was written, Professor Zittel has published a detailed memoir on the *Lithistidæ* (Neues Jahrbuch für Mineralogie, &c., 1878), in which this difficult group is systematically worked out. The general results above stated are, however, not affected by this, though our knowledge of the fossil forms is immensely increased.

with tubercles and intervening depressions, by means of which they are interlocked into a rigid framework. The agreement in shape between the skeleton-spicules of *Siphonia*, and those of the living Lithistid genus *Discodermia* (fig. 45, A), is so close, as shown by Mr Sollas, that it will

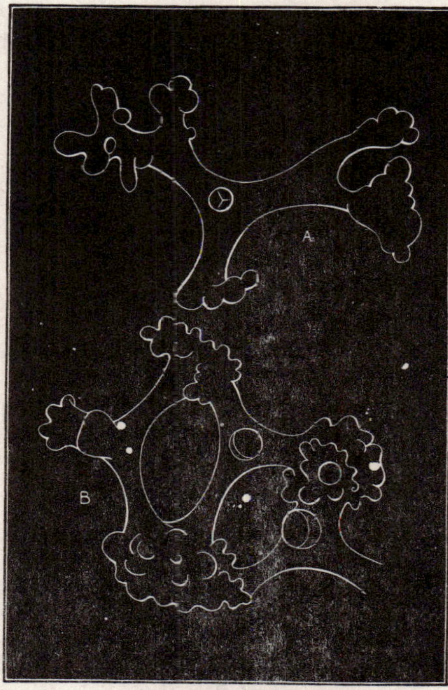

Fig. 45.—A, A single skeleton-spicule of the living Lithistid *Discodermia polydiscus*, magnified 60 diameters; B, Small portion of the skeleton of *Siphonia pyriformis*, similarly enlarged. (After Sollas.)

be perhaps with difficulty that the latter can be retained as generically distinct from the Cretaceous type. Other well-known Cretaceous Lithistids belong to the genera *Chenendopora*, *Jerea*, *Chonella*, *Verruculina*, *Astrocladia*, *Amphithelion*, &c. In the Tertiary period, lastly, numerous Lithistid Sponges have been detected, but none of these are sufficiently important to demand special consideration.

LITERATURE.

[In the following brief bibliography of works which may advantageously be consulted by the student of the fossil Protozoa, as in all subsequent lists of a similar nature, it is hardly necessary to say that only a very limited number of the most important and easily accessible treatises and memoirs have been selected for mention.]

Foraminifera.

1. "Introduction to the Study of the Foraminifera." ('Ray Society.') 1862. Carpenter.
2. "Carboniferous and Permian Foraminifera" (with a general Introduction). 'Monographs of the Palæontographical Society,' 1876. H. B. Brady.
3. "Handbuch der Palæontologie," vol. i. pp. 61-114, 1876. Zittel.
4. "Mikrogeologie." Ehrenberg. 1854.
5. "Foraminifères Fossiles du Bassin Tertiaire de Vienne." D'Orbigny.
6. "Entwurf einer Systematischen Zusammenstellung der Foraminiferen." 'Sitzungsber. d. K. Akad. Wiss. Wien,' 1861. Reuss.
7. "Monograph of the Foraminifera of the Crag," 1868. Rupert Jones, Parker, and Brady.
8. "Beiträge zur Kenntniss der Organisation und Systematischen Stellung von Receptaculites." C. W. Gümbel. 'Abhandl. der K. Bayerischen Akad. der Wiss.' Bd. XII. 1876.
9. "Die Sogenannten Nulliporen" (Lithothamnium und Dactylopora). Zweiter Theil. 'Abhandl. der K. Bayerischen Akad. der Wiss.' Bd. XI. 1874.
10. 'The Dawn of Life.' J. W. Dawson. 1876.
11. "On the Occurrence of Organic Remains in the Laurentian Rocks of Canada." Sir W. E. Logan. 'Quart. Journ. Geol. Soc.,' xxi. 45-50. 1865.
12. "On the Structure of Certain Organic Remains in the Laurentian Limestones of Canada." J. W. Dawson. 'Quart. Journ. Geol. Soc.,' xxi. 51-59. 1865.
13. "Additional Note on the Structure and Affinities of Eozoön Canadense." W. B. Carpenter. 'Quart. Journ. Geol. Soc.,' xxi. 59-66. 1865.
14. "Supplemental Note on the Structure and Affinities of Eozoön Canadense." W. B. Carpenter. 'Quart. Journ. Geol. Soc.,' xxii. 219-228. 1866.
15. "On the So-Called Eozoönal Rock." King & Rowney. 'Quart. Journ. Geol. Soc.,' xxii. 185-218. 1866.

RADIOLARIA.

16. "Mikrogeologie," 1854. Ehrenberg.
17. "Die Radiolarien," 1862. Hæckel.
18. "Fortsetzung der Mikrogeologischen Studien," &c. 'Abhandl. d. K. Akad. Berlin,' 1875. Ehrenberg.
19. "Handbuch der Palæontologie," 1877. Zittel.

SPONGIDA.

20. "Grundzüge einer Spongien-fauna der Atlantischen Gebietes," 1876. Oscar Schmidt.
21. "Die Kalk-Schwämme" Hæckel. 1872.
22. "Anatomy and Classification of Sponges." 'Annals of Nat. Hist.,' 1875. Carter.
23. "Vitreous Sponges." 'Annals Nat. Hist.,' 1868. Wyville Thomson.
24. "Ventriculitidæ of the Chalk." 'Annals Nat. Hist.,' 1847-48. Toulmin Smith.
25. "Beiträge zur Systematik der Fossilen Spongien." 'Neues Jahrbuch für Mineralogie,' &c., 1877. Karl Zittel. (Trans. in 'Ann. Nat. Hist.,' 1877.)
26. "Ueber Cœloptychium. Ein Beitrag zur Kenntniss der Organisation Fossilen Spongien." 'Abhandl. der K. Bayerischen Akad. der Wiss.' Bd. XII. 1876. Karl Zittel.
27. "Die Spongitarien des Nord-deutschen Kreide-gebirges." 'Palæontographica,' 1864. F. A. Roemer.
28. "Palæontologie de la Province d'Oran. 5th Fasc. Spongiaires," 1872. Pomel.
29. "On the Hexactinellidæ and Lithistidæ." 'Ann. Nat. Hist.,' 1873. Carter.
30. "On a Carboniferous Hyalonema." 'Ann. Nat. Hist.,' 1877. Prof. Young and Mr John Young.
31. "Structure and Affinities of the Genus Siphonia." 'Quart. Journ. Geol. Soc.' Vol. XXXIII. 1877. Sollas.
32. "On Cliona." 'Ann. Nat. Hist.,' 1849. Albany Hancock.
33. "On the Minute Structure of Stromatopora and its Allies." 'Proc. Linn. Soc.,' 1878. Nicholson and Murie.
34. "Ueber die Natur der Stromatoporen," 1867. Von Rosen.
35. "Ideen über die Verwandschaftsverhältnisse der Hexactinelliden." 'Zeitschr. für Wiss. Zool.,' 1876. Marshall.
36. "Studies on Fossil Sponges" (Lithistidæ). Karl Zittel. 'Ann. and Mag. Nat. Hist.,' Ser. V. Vol. II. 1878. (Trans. by W. S. Dallas from 'Abhandl. d. K. Bayer. Akad. der Wiss.' Bd. XIII. 1878.)

CHAPTER IX.

SUB-KINGDOM II.—CŒLENTERATA.

FOSSIL HYDROZOA.

THE sub-kingdom *Cœlenterata* (Frey and Leuckhart) may be considered as a modern representative of the *Radiata* of Cuvier. From the *Radiata*, however, the *Echinodermata* and *Scolecida* have been removed to form the *Annuloida*, the entire sub-kingdom of the *Protozoa* has been taken away, and the *Polyzoa* have been relegated to their proper place amongst the *Mollusca*. Deducting these groups from the old *Radiata*, the residue, comprising most of the animals commonly known as Polypes or Zoophytes, remains to constitute the modern *Cœlenterata*.

The *Cœlenterata* may be defined as *animals whose alimentary canal communicates freely with the general cavity of the body ("somatic cavity"). The substance of the body is made up of two fundamental membranes—an outer layer, called the "ectoderm," and an inner layer, or "endoderm." Peculiar urticating organs, or "thread-cells," are usually present; and, generally speaking, a radiate condition of the organs is perceptible, especially in the tentacles with which most are provided. In all the Cœlenterata distinct reproductive organs have been shown to exist.*

The sub-kingdom *Cœlenterata* is divided into the two great classes of the *Hydrozoa* and *Actinozoa*, and the following table indicates the characters and principal subdivisions of these :—

Table of the Divisions of the Cœlenterata.

Class A. Hydrozoa.—The walls of the digestive sac not separated from those of the general body-cavity, the two coinciding with one another. Reproductive organs in the form of external processes of the body-wall.

Sub-class I. Hydroida (Hydroid Zoophytes).
- Order 1. *Hydrida.*—*Ex.* Hydra.
- Order 2. *Corynida.*—*Ex.* Tubularia.
- Order 3. *Thecaphora.*—*Ex.* Sertularia, Campanularia.
- Order 4. *Thecomedusæ.*—*Ex.* Stephanoscyphus.

Sub-class II. Siphonophora (Oceanic Hydrozoa).
- Order 5. *Calycophoridæ.*—*Ex.* Diphyes.
- Order 6. *Physophoridæ.*—*Ex.* Physalia.

Sub-class III. Discophora (Jelly-fishes).
- Order 7. *Medusidæ.*—*Ex.* Ægina.

Sub-class IV. Lucernarida (Sea-blubbers).
- Order 8. *Lucernariadæ.*—*Ex.* Lucernaria.
- Order 9. *Pelagidæ.*—*Ex.* Pelagia.
- Order 10. *Rhizostomidæ.*—*Ex.* Rhizostoma.

Sub-class V. Graptolitidæ (Graptolites).

Sub-class VI. Hydrocorallinæ.—*Ex.* Millepora, Stylaster.

Class B. Actinozoa.—Animal with a differentiated digestive sac opening below into the body-cavity, but separated from it by an intervening "perivisceral space," which is divided into compartments by a series of radiating vertical partitions or "mesenteries," to the faces of which the reproductive organs are attached.
- Order 1. *Zoantharia.*—*Ex.* Sea-anemones, Star-corals, Brain-corals.
- Order 2. *Alcyonaria.*—*Ex.* Sea-pens, Fan-corals, Sea-shrubs, Red-coral, *Heliopora, Heliolites.*
- Order 3. *Rugosa.*—*Ex.* Cyathophyllum.
- Order 4. *Ctenophora.*—*Ex.* Venus's Girdle.

With the exception of one or two Graptolites (*Dendrograptus*), which are known in the Upper Cambrian, the *Cœlenterata* are first well represented in the Lower Silurian; and when we consider that we find the two great classes thoroughly differentiated, and existing under many and varied types, at this early period, we are forced to conclude that the first appearance of the sub-kingdom must really have been at a much more ancient epoch. The Cœlenterates are extremely abundant and important as fossils, some large groups being wholly or largely unrepresented at the present day; but owing to the fact that other large groups (such as the *Lucernarida*, the Oceanic Hydrozoa, and the *Ctenophora*) are almost or quite without hard parts, and therefore only capable of preservation in the fossil condition under very exceptional circumstances, the geological history of the sub-kingdom is very imperfect.

Fossil Hydrozoa.

Of the living orders of *Hydrozoa*, the Fresh-water Polypes (*Hydrida*) and the Oceanic Hydrozoa (*Calycophoridæ* and *Physophoridæ*) have left no traces of their former presence, as might have been anticipated from their want of hard structures. The order of the *Medusidæ* and the sub-class *Lucernarida* (Jelly-fishes and Sea-blubbers) are equally destitute of hard parts, and their absence from the palæontological record might have been confidently predicted. Curiously enough, however, traces of both groups have been detected in the fine-grained lithographic slates of Solenhofen, Pappenheim, and Eichstadt. Of the *Medusidæ*, the two living families of the *Æquoridæ* and *Trachynemidæ* have been recognised by their impressions; and an ancient member of the order *Rhizostomidæ* (*Hexarhizites*) represents the *Lucernarida* in the same formation. With these exceptions, however, the only living orders of *Hydrozoa* which have fossil representatives are the *Corynida* and *Thecaphora*, both of which possess a chitinous or horny integumentary skeleton. In neither case, however, can the evidence be said to be wholly free from suspicion, except in the instance of *Hydractinia* and its immediate allies.

I. Corynida or Tubularida (Pipe-corallines).—*Animal simple, consisting of a single polypite; or compound, consisting of several polypites united to one another by a common flesh or cœnosarc. The cœnosarc generally secretes a hard chitinous outer covering or "polypary;" but the separate polypites are never protected by cup-like expansions of the polypary.* As a rule, the polypary extends only to the bases of the polypites (fig. 46, A); but in one case it is continued far enough upwards to include the bases of the tentacles (fig. 46, B). Generally the polypary is more or less plant-like, and is attached to some foreign body by its base; sometimes it forms a crust-like investment to shells (*Hydractinia*); some-

Fig. 46.—Recent Corynida. A, Portion of the colony of *Perigonimus minutus*, with polypites and reproductive buds, enlarged about 25 diameters; B, A single polypite of *Bimeria vestita*, greatly enlarged, showing the extension of the polypary upon the bases of the tentacles. (After Allman.)

times it is rooted in the sand of the sea-bottom. Usually the polypary is horny in texture, but forms of *Hydractinia* with a calcareous polypary have been described by Mr Carter.

The most important genus of *Corynida* which has been certainly detected in the fossil condition is *Hydractinia*, which is still represented by living species. The recent *Hydractiniæ*, as a rule, are horny as regards the texture of the skeleton, and form crusts attached to the outer surface of shells. By age, these crusts come to be composed of successive close-set, vertically superimposed laminæ, and the shell upon which they originally grew is commonly more or less dissolved away and replaced by the substance of the parasite. In an African *Hydractinia*, described by Mr Carter, the skeleton is *calcareous*, but not essentially different to the horny forms in minute structure. Several fossil forms of *Hydractinia* are now known, two of them (*H. cretacea* and *H. Vicaryi*) being from the Upper Cretaceous system. Of these ancient types, the latter is described as being siliceous, but it is more probable that its skeleton was originally calcareous, and that it has been *silicified*. In the Miocene Tertiary another species occurs; and in the Pliocene (Coralline Crag of Suffolk) we have another species, in which the skeleton is calcareous, Mr Carter's discovery of a living calcareous *Hydractinia* rendering it probable that the fossil form possessed a skeleton primitively composed of carbonate of lime. The genera *Thalaminia* (Jurassic and Cretaceous) and *Sphæractinia* (Jurassic) have been founded by Steinmann for forms supposed to be allied to *Hydractinia*.

According to Mr Carter, the fossils which have been usually described under the name of *Stromatopora* (see p. 137), together with the large arenaceous *Foraminifera* described by Dr Carpenter under the title of *Parkeria*, are really closely related to *Hydractinia*, and are truly fossil *Hydrozoa*. So far as *Stromatopora* and its allies are concerned, the author is unable at present to accept this view, which appears to be founded upon resemblances of analogy rather than of real and fundamental likeness. It is not impossible, however, that Lindström is correct in regarding the Silurian genus *Labechia*—which has generally been regarded as a coral—as a close ally of *Hydractinia*. In this ancient and singular type we have flattened calcareous expansions (fig. 47), the upper surface of which is studded by blunt spines. These spines are seen in vertical sections (fig. 47, c), to be the summits of perpendicular pillars, the spaces between which are occupied by vesicular calcareous plates. The chief points in which *Labechia* differs from *Hydractinia* are the apparent absence of any apertures on the surface, and the fact that the former does not grow in the form of crusts enveloping shells, but in the shape of expansions attached to a foreign body at one point only.

Two other fossil genera, viz., *Palæocoryne* and *Corynoides*, have been referred to the *Corynida*, but in neither case is the reference free from doubt. *Palæocoryne* (fig. 48) is a minute organism which was discovered by Dr Martin Duncan and Mr Jenkins growing attached to the margins of Lace-corals (*Fenestellæ*) in the Carboniferous rocks of Scotland.

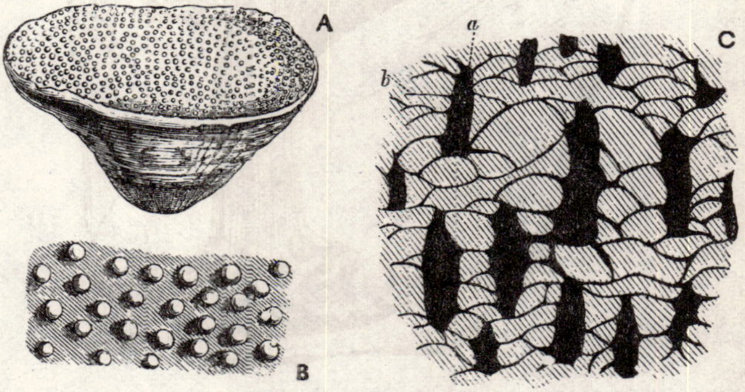

Fig. 47.—*Labechia conferta*, Edw. & H. A, A small specimen, of the natural size. B, A piece of the upper surface of the same, enlarged. C, Portion of a vertical section under a low microscopic power: *a*, The calcareous columns, represented as opaque; *b*, The vesicular tabulæ, filled up with calcite.

Its base is expanded, with finger-like processes of attachment. From the base rises a short robust stem, which is marked with flutings and superficial granulations. The stem terminates in a single polypite, the mouth of which is surrounded by a single whorl of slender processes or "tentacles," in the centre of which is the mouth. The entire polypary, as above described, is "calcareous, dense, and ornamented." In one living form only (viz., *Bimeria*, fig. 46, B), is the polypary continued along the tentacles and upper part of the body of the polypite, and in this case the polypary is simply of the consistence of parchment. This peculiarity, therefore, with the possession of a *calcareous* polypary, renders the reference of *Palæocoryne* to the *Corynida* a matter of question. According to the views of Prof. Young and Mr

John Young, indeed, the fossils known as *Palæocoryne* are really peculiar processes belonging to the Polyzoön (*Fenestella*) upon which they grow.

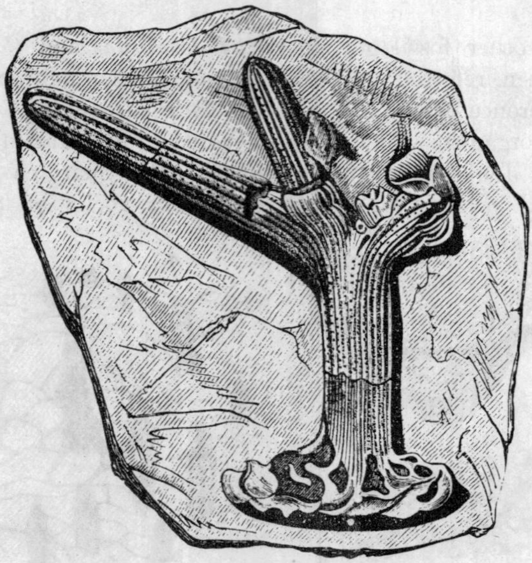

Fig. 48.—*Palæocoryne radiatum*, enlarged fifteen diameters. (After Duncan and Jenkins.)

The genus *Corynoides* was proposed by the author for some singular fossils from the Lower Silurian rocks of Scotland. Each consists of a cylindrical corneous tube (fig. 49), tapering towards the base, where it is furnished with two small spines, and expanding above into a species of toothed cup. *Corynoides* consists of a single polypite, and in this respect may be compared with some living *Corynida*. It would seem, however, not to have been attached to any foreign body—as all living *Corynids* are — and its true affinities are thus rendered uncertain.

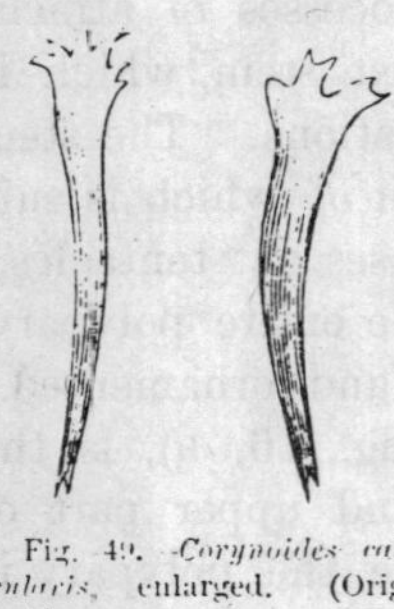

Fig. 49. *Corynoides calicularis*, enlarged. (Original.)

II. Thecaphora (or *Sertularida* and *Campanularida*).—*Animal compound, rooted and plant-like, consisting of numer-*

ous polypites united by common flesh or cœnosarc. The cœnosarc is more or less branched (fig. 50, *a*), *and secretes a strong chitinous investment or "polypary." The polypites are also protected within "hydrothecæ," or little cup-like expansions derived from the polypary. The process of reproduction is carried on by the development of the reproductive elements within horny urn-like sacs, which are of larger size than the "hydrothecæ," and are known as "ovarian capsules" or "gonothecæ" (often called "gonophores"). Type of the order the Sea-fir* (*Sertularia*, fig. 50).

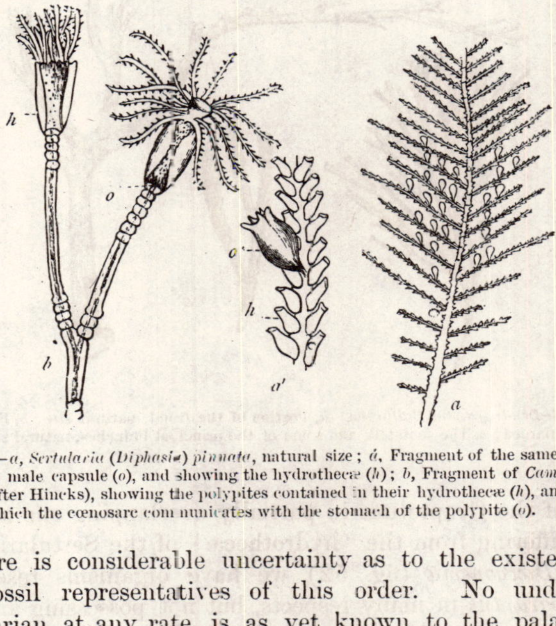

Fig. 50.—*a*, *Sertularia* (*Diphasia*) *pinnata*, natural size; *a'*, Fragment of the same enlarged, carrying a male capsule (*o*), and showing the hydrothecæ (*h*); *b*, Fragment of *Campanularia neglecta* (after Hincks), showing the polypites contained in their hydrothecæ (*h*), and also the point at which the cœnosarc communicates with the stomach of the polypite (*o*).

There is considerable uncertainty as to the existence of any fossil representatives of this order. No undoubted Sertularian, at any rate, is as yet known to the palæontologist; but there are several genera which may with more or less probability be referred to this place. The most important of these—as being those in which the reference is probably correct—are certain forms usually referred to the *Graptolitidæ*, of which the genera *Dendrograptus* and *Dictyonema* may be noticed in particular. The forms referred to *Dendrograptus* are exclusively confined to the Upper Cambrian and Lower Silurian formations. They consist of plant-like spreading and branched growths, which are furnished with a strong footstalk (fig. 51). In all probability the organism was attached by the base of the footstalk to some foreign body, but no actual demonstration of this has as yet been obtained. The branchlets carry upon one side a series of little chitinous cups or "cellules," each of which must have contained a polypite, and which agree with the similar struc-

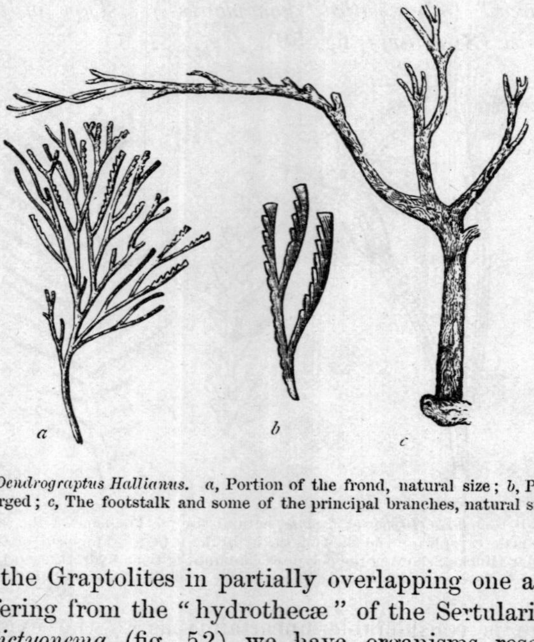

Fig. 51.—*Dendrograptus Hallianus.* *a*, Portion of the frond, natural size; *b*, Portion of a branch, enlarged; *c*, The footstalk and some of the principal branches, natural size. (After Hall.)

tures of the Graptolites in partially overlapping one another; thus differing from the "hydrothecæ" of the Sertularians.

In *Dictyonema* (fig. 52) we have organisms resembling *Dendrograptus* in many respects, but not possessing any footstalk. The frond is branched and plant-like, and is fan-shaped or funnel-shaped in form. It is not certainly known whether the organism was attached by its base or not; but there is the strongest probability in favour of its having been fixed. The branches radiate from the base, running nearly parallel with one another, and often bifurcating. They are united to one another at short intervals by numerous, irregular, slender, transverse processes or dissepiments, and they bear small horny cups or "cellules" like those of the

Graptolites. *Dictyonema* ranges from the Upper Cambrian to the Middle Devonian. The genus bears a close superficial resemblance to the *Fenestellæ* or Lace-corals (belonging to the *Polyzoa*); but the latter have a calcareous skeleton, and have no "cellules." Besides the above-mentioned genera, *Callograptus* and *Ptilograptus* may with great probability be referred to the *Sertularida;* as may, perhaps, be the obscure

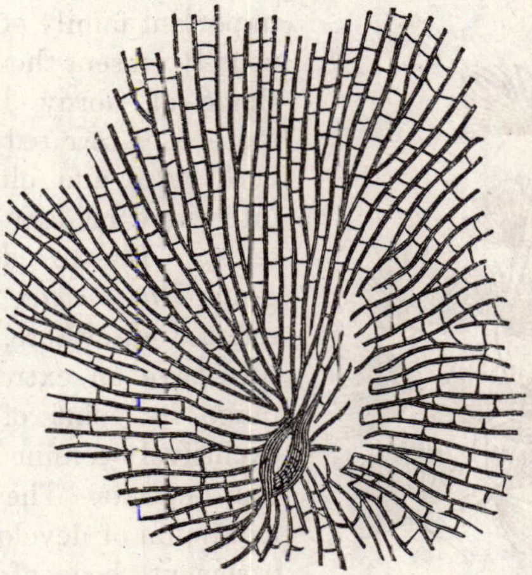

Fig. 52.—*Dictyonema retiforme*, Hall. (After Hall.)

fossils *Buthograptus* and *Thamnograptus.* All these genera are Silurian in age.

OLDHAMIA.—The singular fossils described under the genus *Oldhamia* may be noticed here, as they have been referred to the *Hydrozoa;* though their true nature is altogether uncertain. *Oldhamia* occurs in certain green and purple grits of Lower Cambrian age, at Bray Head, in Wicklow, Ireland, where the fronds are found in great abundance, matted together, and spreading over the surfaces of the strata. A species of *Oldhamia* is also said to occur plentifully in the Potsdam Sandstone (Upper Cambrian) of Wisconsin, in North America. *Oldhamia antiqua*, the commonest species, consists of a central thread-like axis from which spring bundles or umbels of short radiating branches (fig. 53), at regular intervals. Each branch "is formed of a series of articulations marking the position of minute cells" (E. Forbes). *Oldhamia* has been variously referred to the Sertularian Zoophytes, to the *Polyzoa*, and to the vegetable kingdom. The most probable conjecture, perhaps, would refer the genus to the calcareous sea-weeds (Salter).

III. Sub-class Graptolitidæ (Graptolites — Rhabdophora, Allman).—The Graptolites form a very large and important family of fossils which usually present themselves in the shape of horny linear bodies, toothed or serrated upon one or both sides, and often combined into more or less complex systems. If we disregard the genus *Dictyonema*, which is best referred elsewhere, the Graptolites have an extremely definite range in point of time, being exclusively confined to the Silurian deposits. They attain their maximum of development in the basement-beds of the Silurian (Quebec group of Canada and Skiddaw Slates of England), are abundantly represented in the higher portion of the Lower Silurian, and die out altogether before the close of the Upper Silurian period.

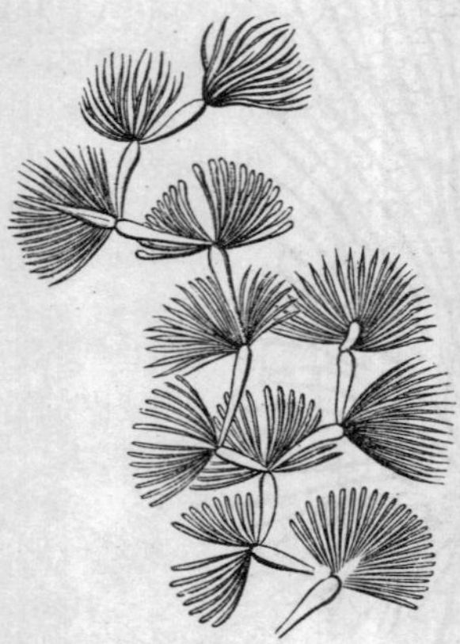

Fig. 53.—*Oldhamia antiqua*, natural size (after Salter). Cambrian.

Excluding the genera *Dictyonema*, *Dendrograptus*, *Ptilograptus*, and *Callograptus*, the *Graptolitidæ* may be defined by the possession of a compound polypary, consisting of a tubular chitinous investment enclosing the cœnosarc, giving origin to numerous cup-like "cellules" or "hydrothecæ," each of which protected a polypite. The polypary was free, and was not attached to any foreign body; and the polypites were not separated from the cœnosarc by any partition. Lastly, the polypary was almost always strengthened by a chitinous rod or fibre, which is termed the "solid axis," and which is analogous to the chitinous rod described by Dr Allman in the singular Polyzoön, *Rhabdopleura.*

From the above definition, it will be seen that the Graptolites agree with the living Sertularians in possessing a corneous polypary, which not only invests the cœnosarc, but is expanded into little cups or "hydrothecæ," within which each polypite is protected. The Graptolites, however, differ from the Sertularians in the fact that the polypary was unattached, and apparently free-floating, whilst it has not, except in a few cases, anything like the plant-like appearance of the latter. Further, the hydrothecæ of the Graptolites, except in the genus *Rastrites,* always more or less overlap one another; whereas those of the Sertularians are not in contact. Lastly, no Sertularian exhibits any structure which can be compared with the "solid axis" of the Graptolites.

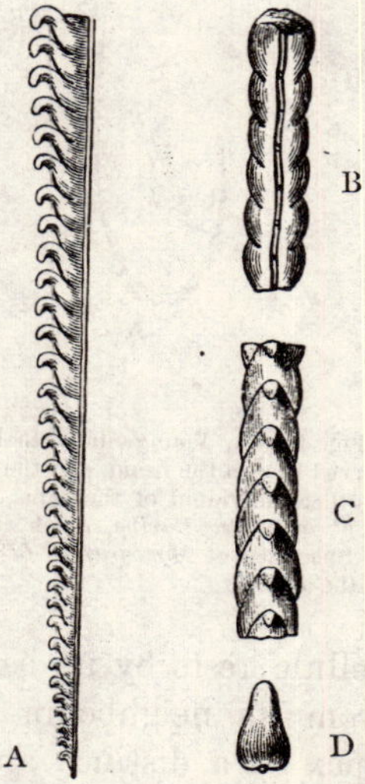

Fig. 54.—Morphology of *Monograptus priodon.* A, *Monograptus priodon*, Bronn, preserved in relief—lateral view slightly enlarged; B, Dorsal view of a fragment of the same species—considerably enlarged; C, Front view of a fragment of the same, showing the mouths of the cellules—much enlarged; D, Transverse section of the same. All from the base of the Coniston Flags. (Original.)

Taking such a simple Graptolite as *Monograptus priodon* (fig. 54), or *M. sagittarius* (fig. 55, A), as the type of the sub-class, the polypary is seen to consist of three elements, which are known as the "solid axis," the "common canal," and the "cellules." The "solid axis" is a cylindrical fibrous rod which gives support to the corneous and flexible polypary. The term "solid" is probably a misnomer; for it was almost certainly hollow, and filled with living material. It appears to be absent in *Retiolites Geinitzianus,* but some uncertainty rests upon this point. As a very general rule, it is prolonged as a longer or shorter naked rod beyond one or both ends of the polypary, and either extension may be more or less dilated. Its basal prolongation, with or without an accompanying extension of the common canal, is termed the "radicle," or "initial point," as marking the organic base of the frond.

The "common canal" is the tube in which the cœnosarc was enclosed; but it commonly appears, in compressed specimens, merely as a vacant space between the "cellules" and the solid axis. The common canal gives origin, by a process of budding, to the "cellules" or "hydrothecæ," which are little horny cups for the reception of the polypites. Each

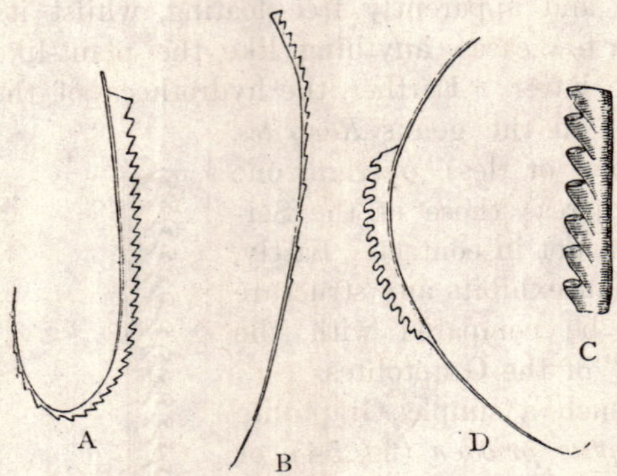

Fig. 55.—A, Young individual of *Monograptus sagittarius*, His., showing the slender curved base of the frond, and the extension of the axis beyond its opposite end; B, Base of another individual of the same, in which there is an extremely long "radicle;" C, Fragment of *M. sagittarius*, much enlarged to show the cellules—from a specimen in relief; D, Specimen of *Monograptus Clingani*, Carr., showing the distal and proximal extensions of the axis.

cellule rests by its base upon the common canal, is separated from its neighbours by "cell-partitions," and opens at its apex by a distinct aperture or "cell-mouth," through which the polypite could exsert its tentaculate head.

The reproductive process appears, in some cases at any rate, to have been carried on by the formation at certain seasons of horny capsules, of much greater size than the cellules, within which the generative elements were matured. In some cases these "ovarian vesicles" have been found actually attached to the fronds of Graptolites. In other cases, as described by the writer, we find numerous bell-shaped horny capsules, termed "*Dawsoniæ*" (fig. 56), each with a little spine at its summit, scattered through the rock in which the Graptolites occur, but only doubtfully attached to the fronds of the latter. These we may infer to have been "ovarian vesicles;" but they differ from the bodies so called in the Sertularians in becoming detached from the parent colony.

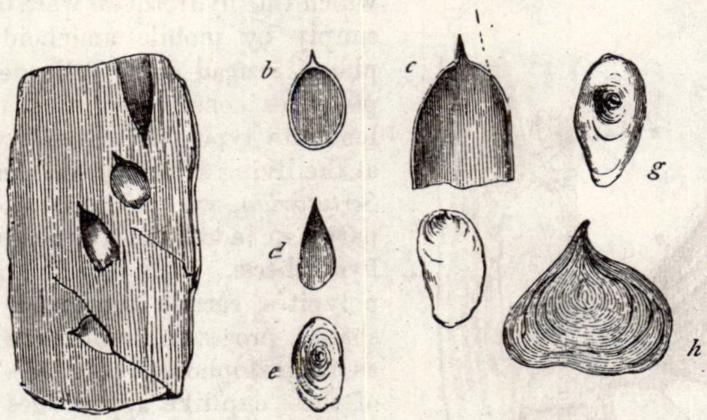

Fig. 56.—Supposed "ovarian capsules" or reproductive buds of Graptolites.

As regards the *affinities* of the Graptolites, opinions widely differ, though the view now generally adopted by palæontologists is that they constitute an aberrant and ancient type of the *Hydrozoa*. In the actual structure of the polypary, as is easily seen by a comparison with the

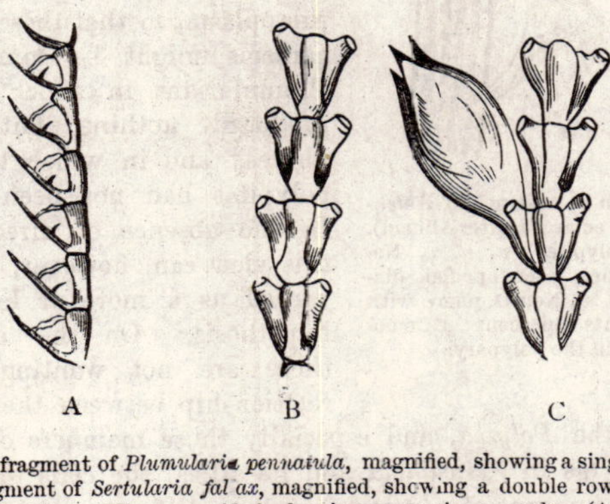

Fig. 57.—A, A fragment of *Plumularia pennatula*, magnified, showing a single row of hydrothecæ; B, A fragment of *Sertularia fallax*, magnified, showing a double row of hydrothecæ; C, Fragment of *Sertularia fallax*, magnified, showing an ovarian capsule. (After Johnston.)

subjoined illustration (fig. 57), representing, on an enlarged scale, the shape and arrangement of the hydrothecæ among the *Sertularida*, the Graptolites certainly closely approach the recent Sea-firs, though the latter possess no "solid axis," and the hydrothecæ do not overlap, while the polypary is always fixed. In the unquestionable fact, also, that the reproductive elements in some Graptolites (if not in all) are matured in special chitinous receptacles, we have an undoubted approach to the Sertularians, with their "ovarian capsules" (fig. 57, c). By Professor Allman, our highest living authority on the *Hydrozoa*, the Graptolites are regarded as referable to the sub-kingdom of the *Hydrozoa*, but as presenting us with an ancient and degraded type of this class, in which the hydrothecæ were occupied simply by mobile amœboid protoplasm, instead of by fully-developed polypites constructed upon the Cœlenterate type. If, namely, we look at the living *Plumularians* among the *Sertularida*, we find that the polypary, in addition to the ordinary hydrothecæ, with their contained polypites, carries a number of cup-shaped processes, which are known as "nematophores" (fig. 58, *n*). Each of these cup-like appendages is filled with protoplasmic matter, which has the power of emitting amœboid prolongations or filaments, strictly comparable to the "pseudopodia" of the Rhizopods. Upon Dr Allman's view, the "cellules" of the Graptolites were similarly occupied by amœboid protoplasm, so that these extinct organisms might be compared with Plumularians in which the colony produced nothing but "nematophores," and in which the ordinary polypites had not been developed. In the absence of direct evidence, this view can, however, only be regarded as a more or less probable hypothesis. On the other hand, there are not wanting points of relationship between the Graptolites and some of the *Polyzoa*, and especially those members of the latter class in which (as in *Vesicularia* and its allies) the cells of the colony communicate by means of a common tube. A further point of affinity between these two groups of organisms is established by the presence in *Rhabdopleura*—an unquestionable marine Polyzoön—of a hollow chitinous axial tube, which may in many respects be compared with the "solid axis" of the Graptolites. In other points, however, *Rhabdopleura* is entirely unlike any known Graptolite, and especially so in its general form, and in the fact that it is fixed to solid bodies; and, upon the whole,

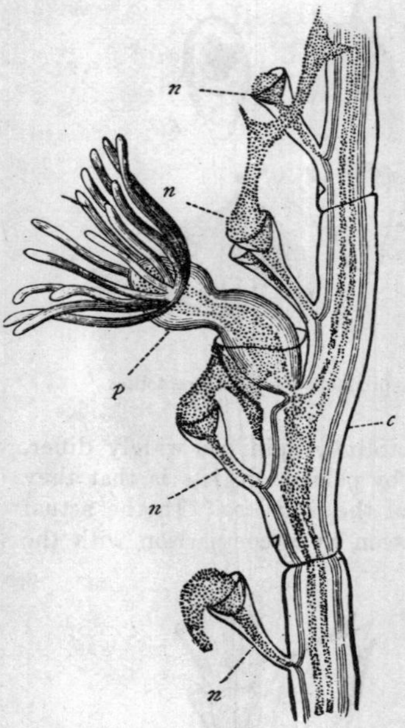

Fig. 58.—Portion of a branch of *Antennularia antennina*, enlarged (after Allman). *p*, One of the polypites; *n*, *n*, *n*, Nematophores emitting pseudopodial filaments of sarcode; *n'*, Nematophore with its sarcodic contents quiescent; *c*, Cœnosarc enclosed within the polypary.

the balance of evidence is unquestionably in favour of a reference of the *Graptolitidæ* to the *Hydrozoa*.

Two leading types may be distinguished amongst the Graptolites, which are termed respectively "monoprionidian" and "diprionidian." The monoprionidian Graptolites, such as *Monograptus priodon* (fig. 54), are distinguished by the fact that the polypary, whether simple or branched, possesses but a single row of cellules or "hydrothecæ." In the diprionidian forms, on the other hand, as in *Diplograptus* (fig. 63), the polypary possesses a row of cellules on each side. It is noticeable that the diprionidian Graptolites, with hardly an exception, are confined to the *Lower* Silurian rocks; whilst the monoprionidian forms range from the base of the Silurian to the summit of the *Upper* Silurian series.

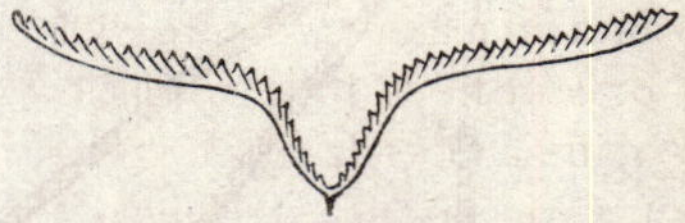

Fig. 59.—*Didymograptus V-fractus.* Lower Silurian (Skiddaw Slates).

Numerous genera of Graptolites, as here restricted, are known to science; but it will be sufficient to give the diagnostic characters of a few of the commonest and more important types. In the genus *Monograptus* (figs. 54, 55), the polypary is simple, linear, possessing but a single row of cellules on one side, and commencing by an attenuated, usually curved base. Species of this genus are found from near the base of the Lower Silurian series to the very summit of the Upper Silurian deposits.

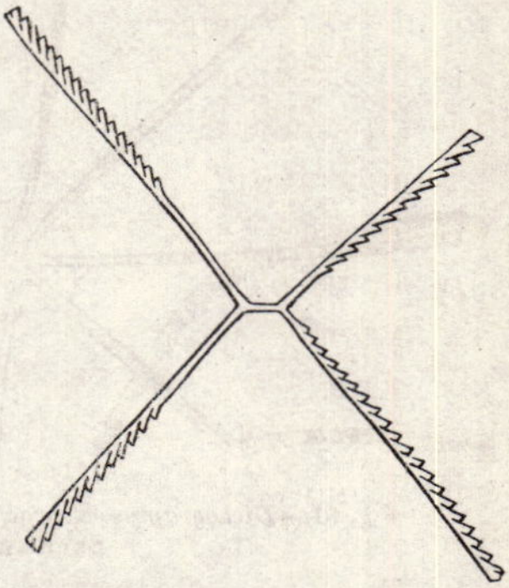

Fig. 60.—*Tetragraptus quadribrachiatus* (after Hall). Lower Silurian (Skiddaw and Quebec groups).

In the genus *Didymograptus* (fig. 59), the polypary consists of two simple monoprionidian branches, which spring from a common point, which is almost invariably marked by a small spine-like "radicle." The genus attains its maximum in the Quebec group of Canada and Skiddaw Slates of England, and is well represented in the succeeding portions of the Lower Silurian (Llandeilo rocks); but no species of the genus is known as late as the Upper Silurian period.

In the genus *Tetragraptus* (fig. 60), the polypary consists of four simple monoprionidian branches, springing from a central non-celluliferous connecting process, which bifurcates at each end. The celluliferous branches do not subdivide, and the base may be enveloped in a peculiar corneous "disc," as will be immediately described in the genus *Dichograptus*. The species of *Tetragraptus* are exclusively confined to the Skiddaw and Quebec groups (Lower Silurian).

In the genus *Dichograptus* there are more than four usually eight) simple monoprionidian branches, which arise from the same number of divisions of a non-celluliferous basal process. In many cases the divisions of the basal connecting process (fig. 61), are enveloped in a species of corneous "disc" or plate, which is believed to have been composed of two laminæ. The functions of this disc are doubtful; but it has been compared with the "float" or buoy of the *Physophoridæ*, an order of the Oceanic Hydrozoa.

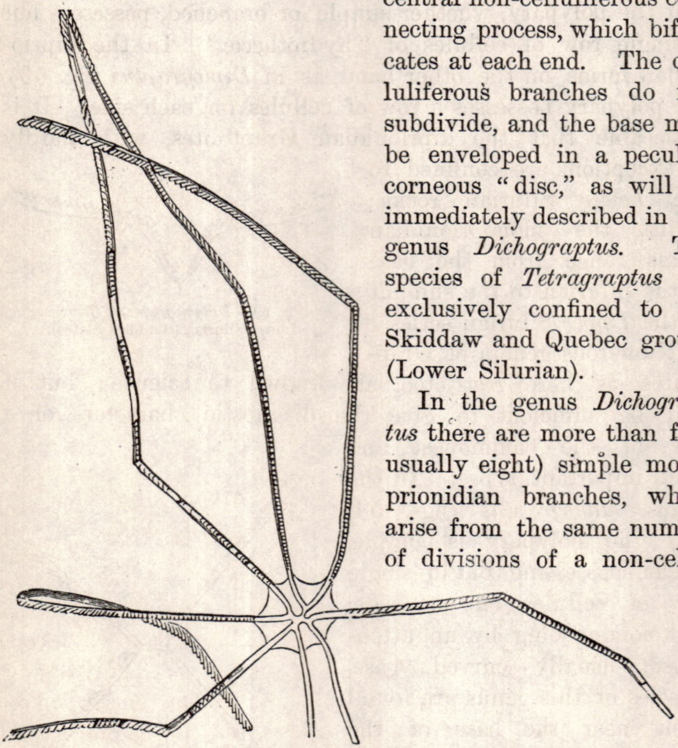

Fig. 61.—*Dichograptus octobrachiatus*, showing the central disc (after Hall). Skiddaw and Quebec groups.

This genus is likewise restricted to the earlier portion of the Lower Silurian period.

In the genus *Rastrites* (fig. 62), the polypary consists of a slender axial tube, giving off on one side a series of linear tubular cellules or "hydrothecæ," which are free throughout their entire length. The genus differs from all the other Graptolites, in the fact that the cellules do not overlap one another, but are free through their whole length, whilst it is not certain if a true "solid axis" is always present. In

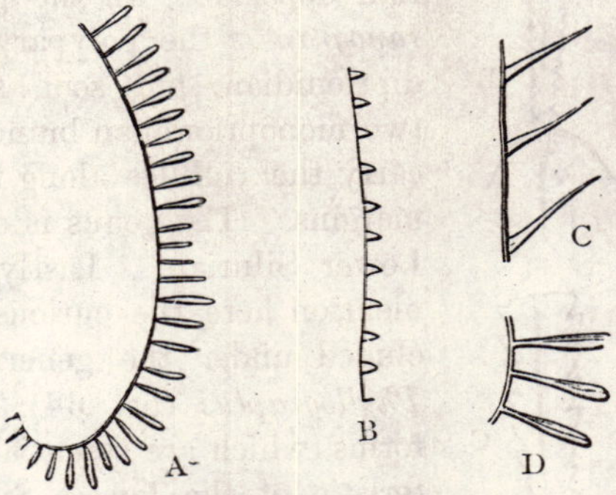

Fig. 62.—Morphology of Rastrites. A, *Rastrites peregrinus*, Barr., from the Mudstones of the Coniston Series—enlarged; B, *Rastrites capillaris*, Carr., from the Upper Llandeilo Shales of Dumfriesshire—enlarged; C, Fragment of *Rastrites Linnæi*, Barr., from the Coniston Mudstones—enlarged; D, Fragment of *R. peregrinus*, greatly enlarged, showing the impressed line running up the centre of each cellule. (Original.)

Britain and North America the species of *Rastrites* are exclusively confined to the Lower Silurian rocks, but in Bohemia they pass up into the lowest beds of the Upper Silurian.

In the genus *Diplograptus* (fig. 63), the polypary consists of two simple monoprionidian stipes, firmly united to one another, back to back. The frond, therefore, is "diprionidian," or carries cellules on both sides. The solid axis is usually prolonged beyond the base of the polypary as a longer or shorter process or "radicle," which is often flanked by lateral spines. The solid axis is also almost invariably prolonged beyond the opposite or "distal" end of the polypary as a naked rod. In the nearly-allied genus *Climaco-*

graptus, the structure is much as above described, but the cellules have such a structure that their mouths appear to be sunk below the general surface of the polypary, forming a row of rounded or quadrangular openings on each side. Both *Diplograptus* and *Climacograptus* range in Britain and North America from the base to the summit of the Lower Silurian series; but in Bohemia they rise into the lower portion of the Upper Silurian deposits. In the genus *Dicranograptus* the polypary is at first diprionidian, but soon splits into two monoprionidian branches which carry the cellules along their outer margins. The genus is exclusively Lower Silurian. Lastly, we may mention here the curious forms included under the generic title of *Phyllograptus* (fig. 64). In these forms, which are essentially characteristic of the lowest Silurian deposits (Skiddaw and Quebec groups), the polypary is leaf-like in shape, and consists of *four* rows of cellules placed back to back, thus resembling two *Diplograpti* intersecting each other at right angles. In consequence of the peculiar structure of the polypary, the *Phyllograpti* are sometimes spoken of as the "*tetraprionidian*" Graptolites.

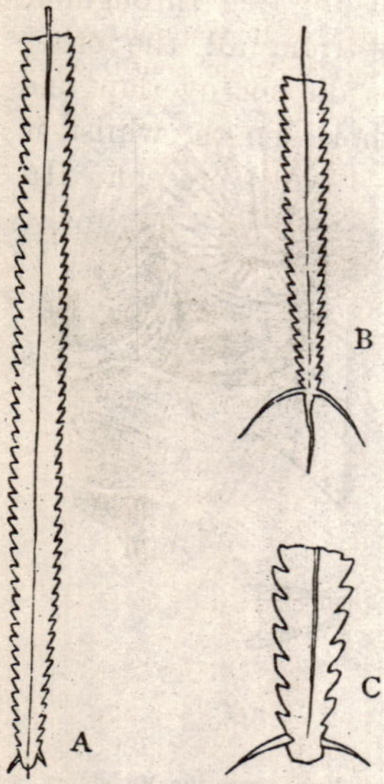

Fig. 63.—A, *Diplograptus pristis*, His., slightly enlarged, showing the normal condition of the base; B, Another example of the same, slightly enlarged, showing a long radicle, and long lateral spines; C, Another of the same, enlarged, showing lateral spines, succeeded proximally by a small bulb, but showing no true radicle. (Original.)

IV. Sub-class Hydrocorallinæ.—This name has recently been proposed by Mr Moseley for two groups of marine animals which produce a regular skeleton of carbonate of lime, often of large size, and which have been generally referred to the Corals (*Actinozoa*). One of these groups comprises the well-known *Millepora* (fig. 65), which is found contributing so largely to the formation of coral-reefs in the

West Indies and Pacific. The calcareous skeleton of *Millepora* is mostly in the form of foliaceous or laminar expansions, stony in texture, with a smooth surface studded with minute apertures of two sizes, the larger of these being much the fewest (fig. 65, C). The larger openings are the mouths of tubes (fig. 65, B, *p*, *p*), which are divided by transverse calcareous partitions into a number of compartments, only the most superficial of these being actually tenanted by the living animal. The smaller tubes are similarly septate or "tabulate," and the general tissue of the skeleton (fig. 65, C) is composed of calcareous trabeculæ traversed by a series of ramifying and anastomosing cœnosarcal canals, which place the tubes occupied by the zoöids in direct communication.[1]

Fig. 64.—Group of individuals of *Phyllograptus typus*, from the Quebec group of Canada (after Hall). One of the four rows of cells is hidden on the under surface.

From the presence of transverse partitions, or "tabulæ," in its tubes, *Millepora* was generally placed amongst the so-called "Tabulate Corals," with the typical forms of which it has no affinity. Though its skeleton is abundantly obtained in the regions where it occurs, the living animal has been rarely observed. The late Professor Agassiz was the first to examine *Millepora* in its living condition, and he was led to the conclusion that the genus was unequivocally referable to

[1] As regards the living animal, Mr Moseley has shown that the colony of *Millepora* consists of two kinds of zoöids, differing in size, in structure, and in function. The larger zoöids occupy the larger tubes of the corallum, and have a mouth, surrounded by four tentacles and opening into a digestive cavity. The smaller and more numerous zoöids surround the larger ones, are long and slender in form, carry numerous clavate tentacles, and are destitute of any mouth. They perform the functions of prehension for the colony, and supply food to the nutritive zoöids.

the *Hydrozoa.* This conclusion has been confirmed by the more recent and more complete researches of Mr Moseley. At the present day, *Millepora* contributes largely to the formation of coral-reefs; but little is known of its distribution in past time. The genus has been detected in the Tertiaries, and allied forms (*Porosphæra*) occur in the Chalk. The Tertiary genus *Axopora* is apparently allied to *Millepora*, but the tubes inhabited by the larger zoöids are traversed

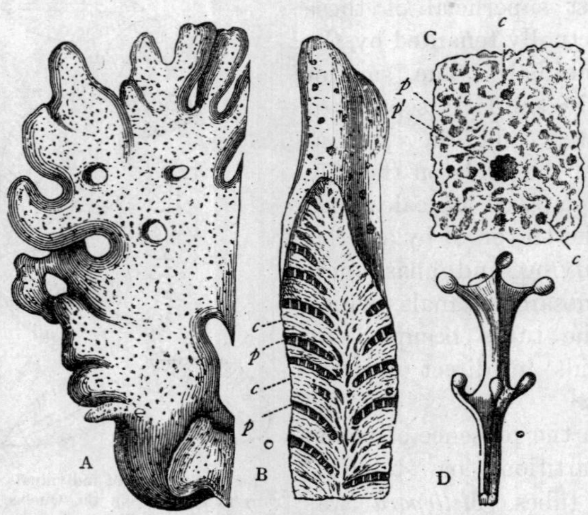

Fig. 65.—A, Portion of a mass of *Millepora alcicornis*, of the natural size; B, Portion of the same, cut open vertically to show the larger tabulate tubes (*p, p*), and the spongy cœnosarcal skeleton (*c, c*)—enlarged; C, Small portion of the surface, enlarged to show the larger and smaller openings (*p* and *c*) inhabited by the different zoöids, and the reticulated calcareous tissue of the skeleton; D, Part of a tentacular polypite, enlarged, showing two whorls of knobbed tentacles. (A, B, and C are after Milne-Edwards and Haime; D is after Martin Duncan and Major-General Nelson.)

by a large fasciculate "columella" or central rod. According to Mr Carter, the ancient and widely distributed forms referred to *Stromatopora* and to allied genera are really closely related to *Millepora*, but this conclusion cannot be accepted without further evidence.

We must also mention here the extraordinary forms known as the *Stylasteridæ*, which have been shown by Mr Moseley to be true Hydroids, producing a calcareous corallum. But brief notice, however, can be given to the group, since there is at present only one known fossil example of it (*Disticho-*

pora antiqua of the Tertiary of France). The family includes several genera (*Stylaster*, *Allopora*, *Cryptohelia*, &c.), all of which secrete a branched calcareous skeleton, so closely resembling some undoubted corals (such as *Oculina*) that the group has always been regarded as referable to the *Oculinidæ*. Taking *Stylaster* (fig. 66) as the type of the

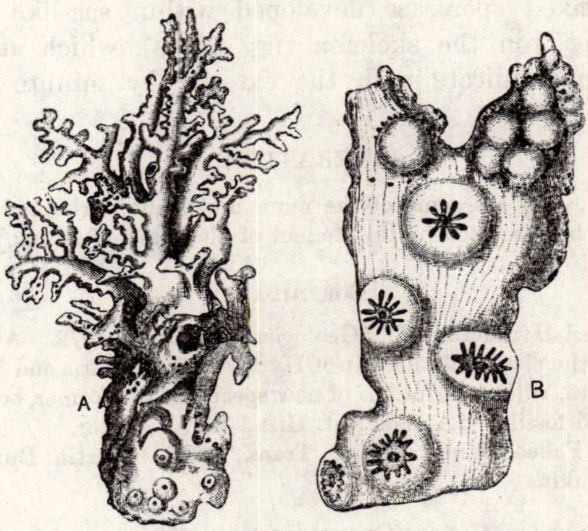

Fig. 66.—A, Portion of the skeleton of *Stylaster sanguineus*, of the natural size; B, Small portion of a branch of the same, enlarged, showing the calices and ampullæ. Living in the Australian seas. (After Milne-Edwards and Haime.)

group, we find the skeleton to be a branched calcareous structure, studded at intervals with cup-like depressions, each of which exhibits a central chamber, occupied axially by a styliform rod ("columella"), and surrounded by a series of secondary chambers, separated from one another by short partitions ("septa"), which appear to be formed by a folding of the wall. Though the general appearance thus produced — especially as regards the existence of "calices" and "septa" — is distinctly that of the ordinary compound corals, yet Mr Moseley has shown that the animal forming this skeleton is truly a composite Hydroid. The colony consists of two sets of zoöids, of which the larger and perfect ones inhabit the central chambers of the calices, while the smaller chambers, marked off by the septa, are occupied by rudimentary and imperfect zoöids, resembling tentacles in shape, and destitute of a mouth. The cavities of the zoöids are placed in communication with one another by a complex system of canals, ramifying in the cœnosarcal skeleton; and the true Hydrozoal character of these coral-like forms is shown by the fact that the reproductive organs are situated outside the bodies of the ordinary zoöids, being in the form of fixed sporosacs developed within sac-like cavities ("ampullæ") in the skeleton (fig. 66, B), which at certain periods communicate with the exterior by minute pores.

LITERATURE.

[The following are some of the more important works and memoirs which may be consulted by the student of the fossil *Hydrozoa*.]

Corynida.

1. "Fossil Hydractiniæ." 'Geological Magazine,' 1872. Allman.
2. "On the close Relationship of Hydractinia, Parkeria and Stromatopora, with descriptions of new species of the former, both recent and fossil." 'Annals Nat. Hist.,' 1877. Carter.
3. "On Palæocoryne." 'Phil. Trans.,' 1869. Martin Duncan and Jenkins.

Graptolitidæ.

4. "Graptolites de Bohême." Barrande. 1850.
5. "Ueber Graptolithen." Scharenberg. 1851.
6. "Die Graptolithen." 'Versteinerungen der Grauwacken-formation.' Geinitz. 1852.
7. "Graptolites of the Quebec Series." 'Descript. of Canadian Organic Remains.' Decade ii. Hall. 1865.
8. "British Graptolites." 'Siluria,' 4th Edition, Appendix. Carruthers. 1867.
9. "Monograph of the British Graptolitidæ." Part i., General Introduction. Nicholson. 1872.
10. "Morphology and Affinities of Graptolites." 'Annals Nat. Hist.,' 1872. Allman.
11. "Notes on British Graptolites." 'Geol. Mag.,' 1873. Lapworth.

Hydrocorallinæ.

12. "Contributions to the Natural History of the United States." Louis Agassiz. (The animal of *Millepora* is figured, vol. iii., pl. xv., and the genus is referred to the Hydrozoa.)
13. "Observations critiques sur la classification des Polypiers paléozoiques." Compt. Rend. t. lxxx., 1875. Dollfus.

14. "Notes on Two Species of Millepora," &c. 'Phil. Trans.,' 1876. Moseley. (This memoir also contains a note on the structure of a Stylaster.)

15. "Structure of a Species of Millepora occurring at Tahiti." 'Annals Nat. Hist.,' 1876. Moseley.

16. "Preliminary Note on the Structure of the Stylasteridæ." 'Annals Nat. Hist.,' 1877. Moseley.

17. "On the Actinozoan Nature of Millepora alcicornis." 'Annals Nat. Hist.,' 1876. Nelson and Martin Duncan.

18. "Recherches sur les Polypiers; Cinquième Mémoire. Monographie des Oculinides." 'Annales des Sci. Nat.,' 1850. Milne-Edwards and Haime. (Descriptions and figures of Stylasterids.)

CHAPTER X.

FOSSIL ACTINOZOA.

Of the living groups of the *Actinozoa* (see Table, p. 153), the *Ctenophora* and the Sea-anemones (*Zoantharia malacodermata*), from their absence of hard parts, are unknown in a fossil condition. The remaining groups—viz., the *Zoantharia sclerobasica*, *Zoantharia sclerodermata*, *Alcyonaria*, and *Rugosa*—secrete a hard skeleton, which is known by the general name of the "coral" or "corallum." All these groups, therefore, are known as fossils; but they are of very unequal importance. The *Zoantharia sclerobasica* are known by very few fossil representatives, and require to be little more than mentioned. The *Alcyonaria*, also, with the exception of the remarkable group of the *Heliolitidæ*, are of little geological importance. The *Zoantharia sclerodermata* and *Rugosa*, on the other hand, have left very numerous and interesting traces of their former existence—the latter being almost altogether extinct,—and both will require to be noticed at some length. Regarded as a whole, the class of the *Actinozoa* appears, so far as we yet know, to have commenced its existence in the Lower Silurian period, and to have attained its maximum of development at the present day.

Order I.—Zoantharia.

Tentacles simple, rounded; soft parts in multiples of five or six.

Sub-order 1. *Zoantharia malacodermata.—Ex.* Sea-anemone.
Sub-order 2. *Zoantharia sclerobasica.—Ex.* Antipathes.
Sub-order 3. *Zoantharia sclerodermata.—Ex.* Reef-building Corals.

A. ZOANTHARIA MALACODERMATA.—Though, from their soft nature, unknown in a fossil condition, the Sea-anemones merit a brief description here, as they may be taken as the types of the order, and as the somewhat complicated structure of the sclerodermic coral will thereby be rendered much more intelligible.

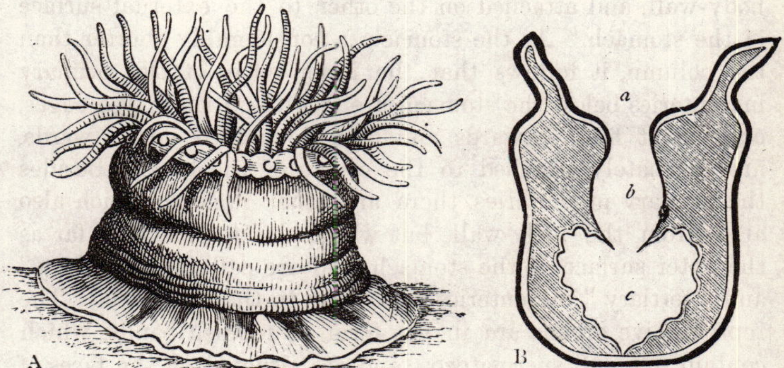

A B

Fig. 67.—A, *Actinia mesembryanthemum*, one of the Sea-anemones (after Johnston); B, Section of the same, showing the mouth (*a*), the stomach (*b*), and the body-cavity (*c*).

The body of a Sea-anemone (fig. 67) is a truncated cone, or a short cylinder, termed the "column," and is of a soft, leathery consistence. The two extremities of the column are termed respectively the "base" and the "disc,"—the former constituting the sucker, whereby the animal attaches itself at will, whilst the mouth is situated in the centre of the latter. In a few cases (*Cerianthus* and *Peachia*) the centre of the base is perforated, but the object of this arrangement is unknown. Between the mouth and the circumference of the disc is a flat space, without appendages of any kind, termed the "peristomial space." Round the circumference of the disc are placed numerous tentacles, usually retractile, arranged in alternating rows, and amounting to as many as 200 in number in the common *Actinia*. The tentacles are tubular prolongations of the ectoderm and endoderm, containing diverticula from the somatic chambers, and sometimes having apertures at their free extremities. The mouth leads directly into the stomach, which is a wide membranous tube, opening by a large aperture into the general body-cavity below, and extending about half-way between the mouth and the base. The wide space between the stomach and column-wall is subdivided into a number of compartments by radiating vertical lamellæ, termed the "primary mesenteries," arising on the one hand from the inner surface of the body-wall, and attached on the other to the external surface of the stomach. As the stomach is considerably shorter than the column, it follows that the inner edges of the primary mesenteries below the stomach are free; and these free edges, curving at first outwards and then downward and inwards, are ultimately attached to the centre of the base. Besides the primary mesenteries, there are other lamellæ which also arise from the body-wall, but which do not reach so far as the outer surface of the stomach, and are called "secondary" and "tertiary" mesenteries, according to their breadth. The reproductive organs are in the form of reddish bands, which contain ova and spermatozoa, and are situated on the faces of the mesenteries.

B. ZOANTHARIA SCLEROBASICA.—The members of this group are all composite organisms, consisting of numerous polypes, each of which has essentially the structure of a small Sea-anemone, united together by a common organised medium or "cœnosarc" (fig. 68). Each polype has, with

Fig. 68.—Part of a living stem of *Antipathes anguina*, of the natural size. (After Dana.)

rare exceptions, six tentacles, and the entire organism is supported by an internal skeleton or "corallum." The coral is horny, and it is what is called "sclerobasic"—that is to say, it forms an internal axis, over which the cœnosarc is spread, much as the bark encloses the wood of a tree. As the polypes are sunk in the cœnosarc, and as this simply forms a rind for the coral, it follows that the polypes are *outside* the corallum. In other words, the polypes take no part in the secretion of the corallum; but this is deposited solely by the cœnosarc or common flesh by which the polypes are connected together.

The corallum is sometimes simple and unbranched, but is more commonly branched and plant-like, and its surface, though occasionally smooth, is usually covered with minute spines, being thus readily distinguished from the grooved and striated sclerobasis of the Gorgonidæ.

The *Zoantharia sclerobasica* are not known as occurring in either the Palæozoic or Mesozoic period. They appear for the first time in the Miocene Tertiary (*Leiopathes vetusta*).

C. Zoantharia Sclerodermata (Hexacoralla).—This group includes most of the so-called "corals," and is of very high geological importance. All the members of this group secrete a skeleton or "corallum," and this is necessarily the only part of the animal with which the palæontologist has to deal; so that it becomes necessary to enter into its structure at some length. The animal itself in the *Zoantharia sclerodermata*, in its essential structure, resembles a Sea-anemone; but it very often has the power of repeating itself by budding (gemmation) or cleavage (fission), so that from a simple it becomes a compound organism. It may therefore consist of a single "polype," or of many similar polypes united by a common flesh or "cœnosarc." The corallum is what is called "sclerodermic," its essential peculiarity being that it is secreted by the polype or polypes. The sclerodermic coral, in fact, is an actual calcification of part of the tissues of the polype. When, therefore, we have a *simple* coral, produced by a simple member of this group (as in fig. 69), we have clearly to do with nothing but skeletal structures produced in the interior of the polype itself. When, on the other hand, we have a *compound* sclerodermic coral to deal with, we have usually more than this. We have, namely, two parts or elements of the coral to consider: 1. The parts of the coral secreted by each individual polype; and, 2. The parts secreted by the cœnosarc which unites all the polypes into an organic whole. A compound coral may be theoretically regarded, therefore, as consisting of a greater or less number of simple corals, such as the preceding, united together by a greater or less quantity of calcareous matter secreted by the cœnosarc. The entire compound corallum consists, therefore, of a greater or less number of "corallites" bound together by a calcareous basis, which is secreted by the cœnosarc, and is called the "cœnenchyma." In practice, however, this theoretical view of the subject is not always

Fig. 69.—*Caryophyllia borealis.* A simple sclerodermic coral, twice the natural size. Recent. (After Sir Wyville Thomson.)

borne out. The compound corallum may, and often does, consist of a number of corallites produced by budding or cleavage from a primitive corallite, having their outer walls amalgamated, or more or less completely free, but not sunk in any general cœnenchyma. In other cases, the cœnenchyma, though not actually absent, is very much reduced in quantity.

To comprehend the more intimate structure of a sclerodermic coral, we may take a simple corallum, such as figured in fig. 69. Typically, such a coral is conical in shape, sometimes discoid, sometimes cylindrical, but in all cases possessing an external "wall" or "theca," with an internal included space. The theca may be very imperfect, often porous or cribriform ("*Perforate Corals*"), or it may be strengthened by a secondary calcareous investment, termed the "epitheca." The "theca" encloses a larger or smaller space which is known as the "visceral chamber," is variously subdivided below, and superiorly presents itself as a shallower or deeper cup-shaped depression, termed the "calice." Within the calice is contained, in the living state, the stomach-sac of the polype; and the visceral chamber *below the calice* is subdivided into a number of vertical compartments—the "interseptal loculi"—this subdivision being effected by means of calcareous partitions or "septa," which spring from the inner wall of the theca, and are directed inwards towards the centre. The "septa" correspond with the "mesenteries" of the living animal, with which they agree in number and size. Some of them—the so-called "primary septa"—are much wider than the others, and may extend far enough inwards to meet in the centre of the visceral chamber (fig. 70, A). Others of the septa fall short of the centre by a

Fig. 70.—Diagram of the arrangement of the septa in the *Zoantharia sclerodermata* and *Rugosa*. A, Transverse section of a simple sclerodermic coral (*Turbinolia*), showing the theca, with its projecting ridges or "costæ" outside, the visceral chamber and radiating septa inside, and the columella in the centre; B, Transverse section of a simple Rugose coral (*Cyathophyllum*), showing the wall, costæ, and septa.

greater or less distance, and are known as the "secondary" and "tertiary" septa, according to their width. In the centre of the visceral chamber there may or may not be an axial calcareous rod, known as the "columella" (fig. 70, A). The structure of the columella varies in different cases, but it extends, typically, from the floor of the visceral chamber to the bottom of the calice, into which it projects for a greater or less distance, and the primary septa are often closely connected with it.

The *number* of the septa (when present at all) varies, but there are never less than six of these structures, and however great the number may be, a "*hexameral*" arrangement of the septa can be usually more or less clearly demonstrated in the corallum of the *Zoantharia sclerodermata*. Hence the name of *Hexacoralla* often applied to this group of corals. While the rule among the *Zoantharia sclerodermata* is that the septa are arranged in *six systems* (see fig. 70, A), and are, however numerous, some multiple of six, there are cases in which no such hexameral arrangement is demonstrable.

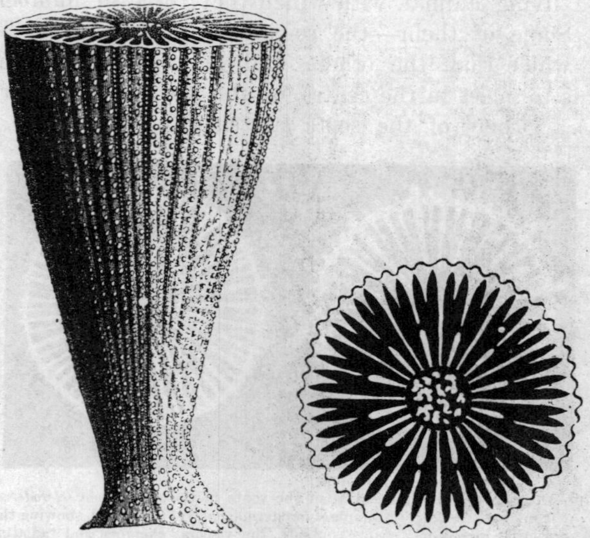

Fig. 71.—*Caryophyllia* (*Cyathina*) *Bowerbanki*, from the Gault (Cretaceous). The left-hand figure represents a specimen imperfect above, and enlarged, showing the tuberculated costæ. The right-hand figure is a magnified cross-section, showing the septa and pali. (After Milne-Edwards and Haime.)

As connected with the septa, we may also notice here the structures which are known as "*pali*." These are "small processes which exist between certain septa and the columella (fig. 71). They generally arise from the base of the visceral cavity, or close to it, and pass upwards, united by one edge to the columella, and by the other to the inner end or margin of the septa. When there is no columella, they are adherent to the septa, present a free edge to the cavity in the axis of the corallum, and arise with the septa" (Duncan). They are only occasionally present.

The septa, further, may, for all practical purposes, be regarded as being continued in the *Zoantharia sclerodermata through* the wall, so as to become visible on the exterior. The outer surface of the theca thus comes to be covered with a series of vertical ridges or ribs, which are termed the "costæ" (figs. 70 and 71). The costæ vary much as to the distance by which they are separated from one another, and as to their breadth, their solidity, and their ornamentation with spines, tubercles, or teeth.

Theoretically, the "interseptal loculi" are vacant spaces or vertical compartments, bounded laterally by the septa, and extending from the lower and lateral surfaces of the theca to the floor of the calice. In practice, however, the continuity of the interseptal loculi is liable to be more or less interfered with by the development of the structures known as "*synapticulæ*," "*dissepiments*," and "*tabulæ*." The "synapticulæ" are transverse calcareous bars which stretch across the interseptal loculi, and form a kind of trellis-work, uniting the opposite faces of adjacent septa. They are characteristic of the *Fungidæ*. The "dissepiments" are commonly present in a great many corals, and have the form of incomplete, approximately horizontal plates, which stretch between adjacent septa, and break up the interseptal loculi into secondary compartments or cells. Lastly, the "tabulæ" may be regarded as highly developed dissepiments, and like them, are approximately horizontal, as a rule, at any rate. They differ from the dissepiments in the fact that they cut across the interseptal loculi at the same level. When fully developed (fig. 72), they are transverse plates, which extend completely across the visceral chamber, and divide it into a series of storeys placed one above the other, the only living portion of the coral being above the last-formed tabula. Tabulæ are found in various of the *Zoantharia sclerodermata*, in some of the *Alcyonaria*, and in a great many of the *Rugosa*.

The above is the essential structure of the typical form of a simple sclerodermic coral, and it is easy to see that it is produced by the calcification, or conversion into carbonate of lime, of the *lower* portion of a polype similar in structure to an ordinary Sea-anemone. The " theca " of the coral corresponds to, and is secreted by, the " column-wall " or general wall of the body of the polype. The " septa," again, correspond with the " mesenteries," and, like them, are " primary," " secondary," or " tertiary," according as they reach the centre or fall short of it by a greater or less distance. We must remember, however, that it is only the *inferior* portion of the body of the polype which is thus calcified. The tentacular disc and mouth are placed at some distance above the upper margin of the theca, and the digestive sac occupies the calice;

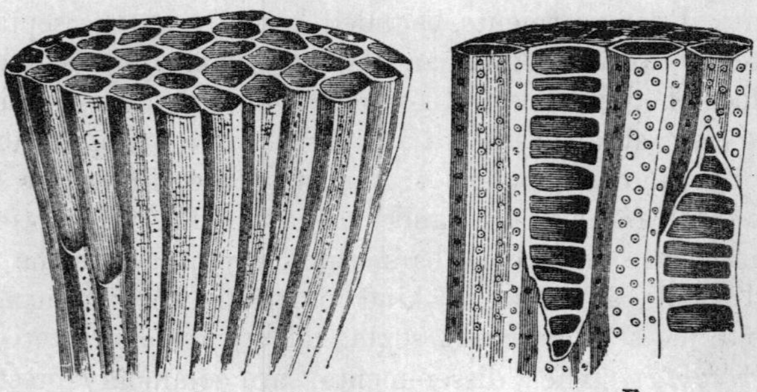

A B

Fig. 72.—A, Portion of the corallum of *Favosites favosa*, of the natural size; B, Portion of four corallites of *Favosites Gothlandica*, enlarged, showing the tabulæ.

whilst the whole of the space comprised within the theca is lined by the endoderm, and the whole of its outer surface is covered by the ectoderm.

Having now considered the general structure of a simple sclerodermic corallum, as produced by a single polype, we must glance for a moment at that of a *compound* corallum of the same group. Such a corallum is the aggregate skeleton produced by a *colony* of polypes, each of which is essentially similar to a Sea-anemone in structure, and it varies in size and form according to the characters of the colony by which it is produced. Such a colony (fig. 73) consists of a number of polypes, which may spring directly from one another, or which may be united by a common flesh or " cœnosarc," and corresponding differences are found in the resulting corallum.

In the former instance, as previously remarked, the compound corallum consists of an assemblage of separate "corallites," as the skeletons of the individual polypes are called, these being united with one another directly and in various ways.

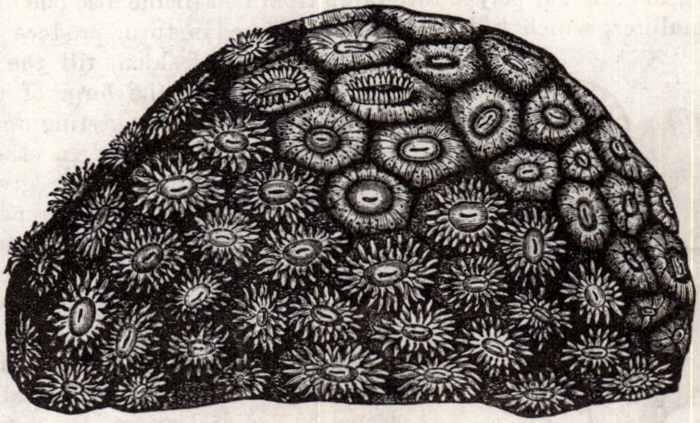

Fig. 73.—*Astræa pallida*, a compound sclerodermic coral, in its living condition. (After Dana.)

In the latter instance the corallum consists of a number of "corallites," and of a common calcareous basis or tissue, which unites the various corallites into a whole, is secreted by the cœnosarc, and is known as the "cœnenchyma."

The compound coralla are, of course, primitively simple, and they become composite either by budding or by cleavage of the original polype. The following are the principal methods in which this increase is effected; and in considering this subject briefly, it will be as well to take into account not only the *Zoantharia sclerodermata*, but also the *Rugosa*, the modes of increase in the two groups being very similar :—

1. *Lateral* or *parietal gemmation.*—In this mode of increase the original polype throws out buds from some point on its sides between the base and the circle of tentacles, and these buds on becoming perfect corallites may repeat the process. This is one of the commonest modes of growth amongst the recent corals, and it gives rise chiefly to dendroid or tree-like corals.

2. *Basal gemmation.*—In this method the original polype gives forth from its base a rudimentary cœnosarc, from which new buds are thrown up, and which may have the form of root-like prolongations or of a continuous horizontal expansion. The resulting coralla are usually massive or incrusting, and the youngest corallites are, of course, those placed on the periphery of the colony.

3. *Calicular gemmation.* —This consists in the production of buds from the calicine disc of the parent corallite, which may or may not continue to grow thereafter, whilst the new corallites thus produced generally repeat the process. This mode of growth is exceedingly rare amongst the *Zoantharia sclerodermata*, and is never typically exhibited; but it is a characteristic feature in many of the *Rugose* corals. In many of these (fig. 74), the original polype throws up from its calicine disc one or more new corallites, which kill the parent. These, in turn, produce others after a similar fashion, till the entire corallum assumes the form of an inverted pyramidal mass resting upon the original budding polype. In other *Rugose* corals the calicine disc gives off but a single bud, which may repeat the process indefinitely till the corallum presents the appearance of a succession of inverted cones placed one above the other.

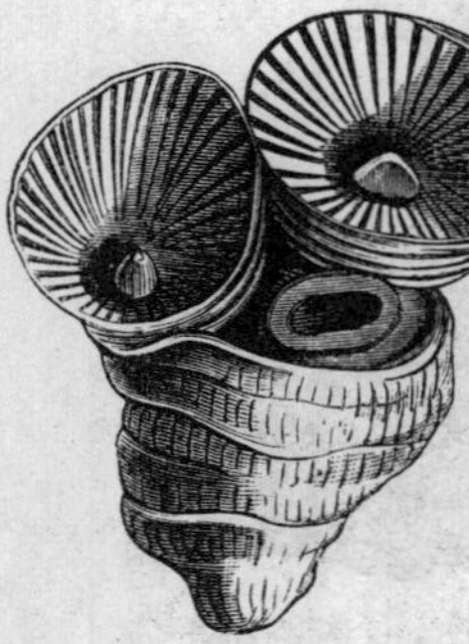

Fig. 74.—Calicular gemmation as seen in *Lonsdaleia floriformis*. Carboniferous.

4. *Fission.* — This process in the coralligenous *Actinozoa* is usually effected by "oral cleavage," the divisional groove commencing at the oral disc, and deepening to a greater or less extent, the proximal extremity always remaining undivided. According to Dana, in fission a new mouth is formed in the disc near the old mouth, and a new stomach is formed for the new mouth, round which the new tentacles are then developed. This, therefore, is not, strictly speaking, a subdivision into halves; since one half carries off the old mouth and stomach. More rarely, fission "is effected by the separation of small portions from the attached base of the primitive organism, whose form and structure they subsequently, by gradual development, tend to assume."

"The coral-structures which result from a repetition of the fissiparous process are of two principal kinds, according as they tend most to increase in a *vertical* or in a *horizontal* direction. In the first of these cases the corallum is *cæspitose*, or tufted, convex on its distal aspect, and resolvable into a succession of short diverging pairs of branches, each resulting from the division of a single corallite." In the second case the coral becomes *lamellar*. "Here the secondary corallites are united throughout their whole height, and disposed in a linear series, the entire mass presenting one continuous theca." Both these forms of corallum "are liable to become *massive* by the union of several rows or tufts of corallites throughout the whole or a portion of their height. An illustration of this is afforded by the large *gyrate* corallum of *Meandrina*, over the surface of whose spheroidal mass the calicine region of the combined corallites winds in so complex a manner as at once to suggest that resemblance to the convolutions of the brain which its popular name of Brain-stone Coral has been devised to indicate" (Greene).

As to their habitat and *distribution in space*, all the living *Zoantharia sclerodermata* are inhabitants of the sea, and there is no reason to suppose that any of the fossil forms were other than marine. At the present day, also, they attain their maximum development in warm seas; and this was probably the case in past times too. In existing seas, further, as has been specially insisted on by Professor Martin Duncan, we find two great groups of the Sclerodermic *Zoantharia*—viz., those which inhabit tolerably deep water, and those which build the great masses of coral which are known as "coral-reefs." The deep-sea corals, though often attaining, as individuals, a considerable size, and though often compound, never form massive aggregations or "reefs." This is due to the fact that, when composite, the separate corallites are not united together by a lax cellular cœnenchyma, so that the colony cannot increase to an indefinitely large size. The deep-sea corals seem to have existed in all the great geological periods, from the Silurian upwards. The chief genera of this group at the present day are *Caryophyllia*, *Balanophyllia*, *Flabellum*, *Desmophyllum*, and *Sphenotrochus*, amongst the simple forms; and *Lophohelia*, *Amphihelia*, *Dendrophyllia*, and *Astrangia*, amongst the compound forms.

The reef-building corals, when simple, are provided with special structures which enable the polypes to grow rapidly. The great majority of the reef-builders, however, are compound, and owe the large size to which they attain to the fact that the corallites are mostly united by a loose cellular cœnenchyma. The chief genera of reef-building *Zoantharia* in Mesozoic, Kainozoic, and Recent times, belong to the families of the *Astræidæ*, *Poritidæ*, and *Madreporidæ*, though the *Oculinidæ* and *Fungidæ* also contribute to the formation of reefs.

In the Palæozoic period, if true "reefs" can be said to have existed at all, they were built up essentially by Rugose corals. In Mesozoic times, however, true coral-reefs existed towards the close of the Trias in Western Europe, and largely in Oolitic times both in Western Europe and in England. In the earlier portion of the Tertiary period, again, vast coral-reefs were formed in Central and Southern Europe, in Egypt, Syria, and Arabia, and in parts of India.

As to the *distribution in time* of the *Zoantharia sclerodermata*, it is difficult to speak with precision, as much doubt obtains as to the true systematic position of many ancient forms often referred to this group. It may be certainly affirmed, however, that the group attained no strikingly predominant position during the whole of the Palæozoic epoch; that it underwent a great development in the Secondary and Tertiary periods; and that it has, perhaps, reached its maximum at the present day.

The *Zoantharia sclerodermata* were divided by Milne-Edwards and Haime into the four sections of the *Aporosa*, *Perforata*, *Tabulata*, and *Tubulosa*, of which the two first are large and natural divisions, while the two latter are of doubtful affinities and uncertain value. We shall, however, briefly consider the characters, geological distribution, and leading types of each of these sections.

I.—Aporosa.

The *Aporose Zoantharia* possess a corallum composed of more or less compact calcareous tissue, the "theca" or wall surrounding the visceral chamber being *complete*, and *rarely perforated by apertures or pores*. The septa are well developed, and usually constitute complete lamellæ; and though dissepiments or synapticulæ are present, tabulæ very rarely exist.

Taken as a whole, the *Aporosa* are an essentially Secondary and Tertiary group, being represented during both these periods by an immense variety of types. In the Palæozoic period, there is still some doubt as to the precise position and structure of some of the corals which have at one time or another been regarded as ancient forms of the *Aporosa*. The Silurian genus *Palæocyclus*, usually referred to the *Fungidæ*, appears to belong rather to the *Rugosa*; but it is possible that the genus *Columnaria* (*Favistella*), from the same formation, should be looked upon as a tabulate form of the *Astræidæ*. By Prcf. Martin Duncan, the *Battersbyia* of the Devonian, and the *Heterophyllia* of the Carboniferous, are considered old types of the *Astræidæ*. Lastly, some of the forms placed in the badly-characterised genus *Petraia*, of the Silurian and Devonian, may very probably turn out to be *Turbinolidæ*, and this may possibly be the place of the Silurian *Duncanella*.

The *Aporosa* are divided into the following families:—

1. *Turbinolidæ*.—In this family the corallum may be simple or compound, but in the latter case it is without a cœnenchyma. The *interseptal loculi are open from top to bottom*, and are not crossed by dissepiments or synapticulæ; and the septa are mostly granulated on their sides. Leaving the doubtful ancient forms (*Petraia* and *Duncanella*) out of sight, the family makes its first certain appearance in the Lias (*Thecocyathus*). In the Cretaceous numerous forms are known, and in the Eocene Tertiary a still greater development of this type takes place, after which the family begins to decline in numbers to the present day.

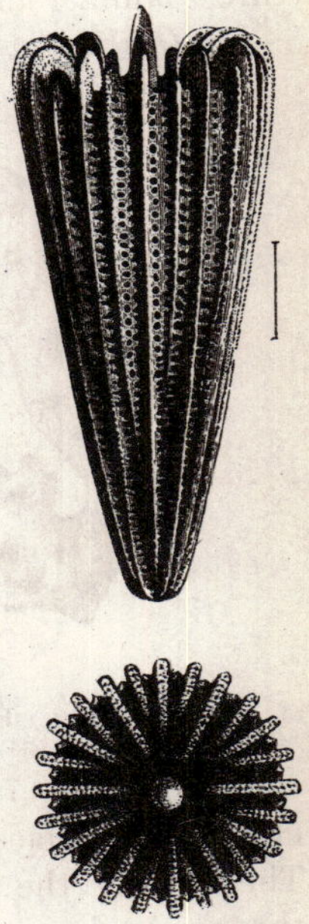

Fig. 75. — *Turbinolia sulcata*. The upper figure shows the exterior of the theca with the costæ. The lower figure shows the calice, with the columella and primary and secondary septa. Eocene.

In *Turbinolia* itself (fig. 75) the corallum is simple and conical, with a styliform columella, but without pali. The costæ are very prominent, and the spaces between them are marked with rows of small dimples, which look like perforations in the wall, but which really do not penetrate to the visceral chamber. The genus is characteristic of the Eocene period. *Flabellum*, ranging from the Eocene Tertiary to the present day, is nearly allied to *Turbinolia*, but the corallum is compressed, so as to produce an elliptic form of calice, and the wall is covered with a thin epitheca. *Caryophyllia*, ranging from the Cretaceous to the present day, is another close ally of *Turbinolia*, from which it differs principally in the possession of a crown of "pali" (fig. 71); while the widely distributed *Trochocyathus* of the Jurassic, Cretaceous, and Tertiary formations, possesses more than one circle of these structures.

2. *Pseudoturbinolidæ.*—In this family we have only the extinct genus *Dasmia*, of the Cretaceous and Tertiary, in which the corallum is in most respects similar to that of the *Turbinolidæ* proper, but each septum is composed of three laminæ united externally by a single costa.

3. *Oculinidæ.*—The corallum in this family is always

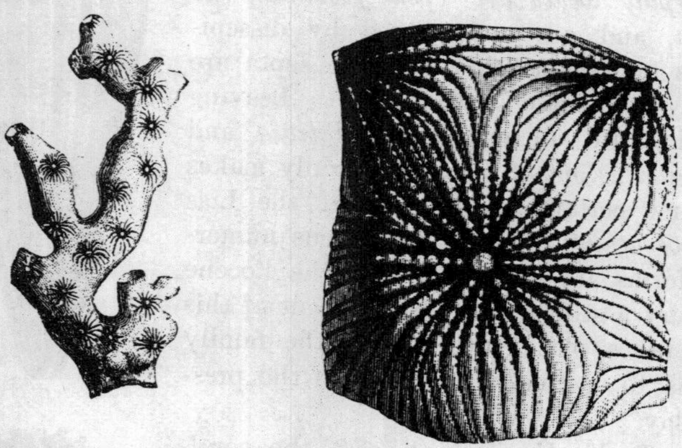

Fig. 76.—*Synhelia Sharpeana.* Cretaceous.

compound (fig. 76), with an abundant and compact cœnenchyma, its surface smooth or striated, but never echinulate. The wall of the corallites is imperforate, not distinct from the cœnenchyma, the lower portion of the visceral chamber becoming filled up with advancing age. A few dissepiments are present, but no synapticulæ.

The *Oculinidæ* appear for the first time in the Oolitic rocks (*Euhelia, Enallohelia*), and are also represented in the Cretaceous (*Synhelia*, fig. 76, and *Diblasus*). In the Eocene Tertiary we meet with *Oculina* itself, with its arborescent corallum and nearly smooth cœnenchyma. The well-known living genera *Lophohelia* and *Amphihelia* are found in the late Tertiaries, and *Diplohelia* is Eocene and Miocene. From the researches of Verrill, it would appear that *Pocillopora* (Miocene to Recent), formerly referred to the *Tabulata*, should be placed here, in which case we have a member of this family exhibiting "tabulæ" in a well-marked form, these structures being occasionally present in *Lophohelia* also. Probably the genus *Seriatopora* (which has been said to occur even in the Palæozoic rocks) should likewise be referred to this family. On the other hand, the limits of the group have been contracted in another direction by the removal of *Stylaster* and allied genera to the *Hydrozoa*.

4. *Astræidæ.*—In this, the most important of all the families of the *Zoantharia sclerodermata*, the corallum may be simple or compound, usually increasing in the composite

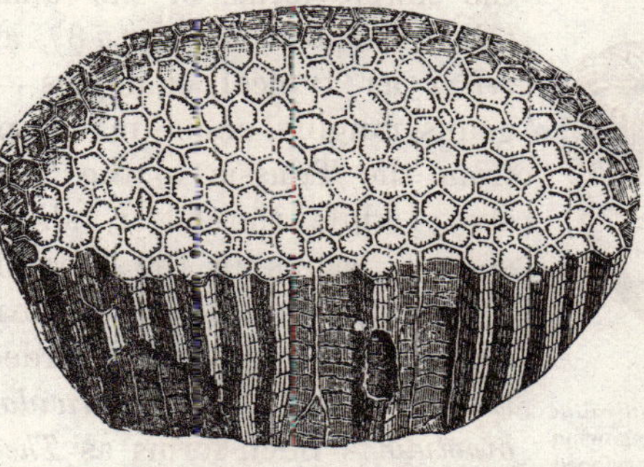

Fig. 77.—*Columnaria* (?) *Halli*, showing the corallites partitioned off into storeys by tabulæ. Silurian.

forms (fig. 78) by fission. There is no, or little, cœnenchyma, but an abundance of dissepiments, without synapticulæ or tabulæ.

Little is known of the distribution of *Astræidæ* in the Palæozoic period, but, as before said, Dr Martin Duncan, one of the highest of authorities on this subject, regards the Devonian *Battersbyia* and the Carboniferous *Heterophyllia* as ancestral and aberrant types of the family (*Palastræidæ*). If it were not, also, for the presence of tabulæ, we should place the Silurian genus *Columnaria* in this family, and as it is now admitted that tabulæ have little or no systematic significance, it is difficult to see how this old form can be excluded from the *Astræidæ.* In the *Columnariæ* proper (= *Favistella*) the septa are very well developed, but in other forms often referred here (fig. 77) the septa are rudimentary.

Leaving the Palæozoic period, we find a great development of *Astræidæ* to take place towards the close of the Trias where the family is represented by numerous and varied types; a still further expansion takes place in the Oolites; very numerous forms are met with in the Cretaceous, and though there is some decrease in the Tertiaries, this great family still holds its ground as the most important group of the "reef-building" corals.

Of the many genera of this family, only two or three of the most important can be so much as alluded to here. Of the simple forms of the family, we may take *Montlivaltia* (fig. 78), and *Trochosmilia* as typical examples—the former genus ranging, under many specific forms, from the Triassic to the Tertiary inclusive; while the latter, also with many species, begins in the Jurassic, and continues to the later Tertiaries. These simple types may be regarded as transitional between the *Astræidæ* and *Turbinolidæ.* Such forms as *Thecosmilia* (fig. 79), again, may be compared to a colony of *Montlivaltiæ*, the separate corallites being bound together by a strong "epitheca," and united into a tufted corallum. The species of the genus are numerous, and are found in the Jurassic, Cretaceous, and Tertiary periods. In a third group of *Astræidæ* we have very numerous and important forms, which agree with the last mentioned in being compound, but in which the corallum consists of numerous closely-approximated corallites, produced by fission, and giving rise, as a rule, to massive "astræiform" colonies. Of these "star-corals," *Astræa* itself may be taken as the type, though not known to have existed earlier than the Tertiary. In the Secondary period, however, we have a vast development of

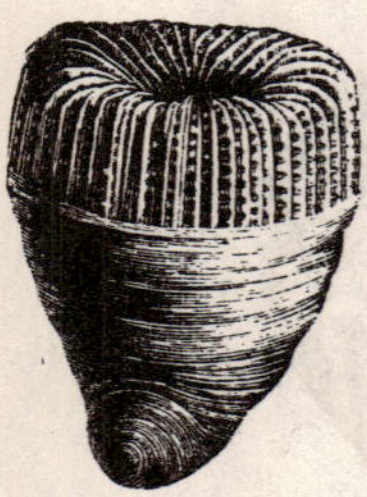

Fig. 78.—*Montlivaltia caryophyllata*, showing the greatly-developed epitheca covering the lower part of the coral. Great Oolite.

forms more or less closely related to the living *Astræa*, such as *Prionastræa, Isastræa* (fig. 80), *Septastræa, Convexastræa, Thamnastræa, Heliastræa,* &c. Lastly, not to mention others,

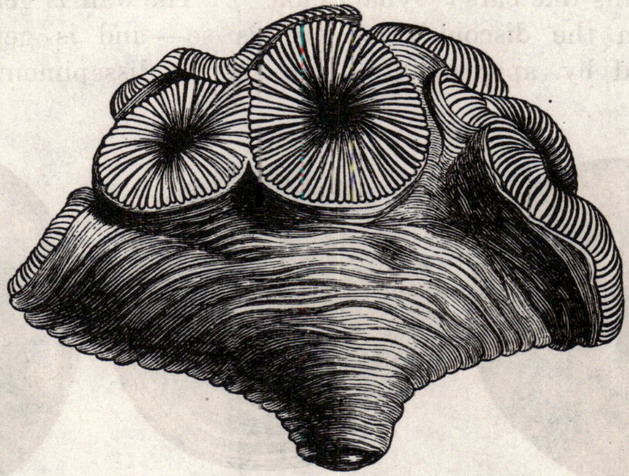

Fig. 79.—*Thecosmilia annularis.* Coral-rag, England.

we have a group of forms very similar to the preceding, but having the calices of the separate corallites more or less completely confluent with one another. As examples of this

Fig. 80.—*Isastræa oblonga*; portion of a small polished slab, of the natural size, and a few calices enlarged. Jurassic (Portland Oolite).

group, we may take *Meandrina* (the "Brain-corals"), *Latimeandra*, and *Diploria*.

5. *Fungidæ.*—In this group, the corallum is simple or compound, usually discoidal or laminar, the calices shallow

and open in the simple forms, confluent in the compound forms, with complete imperforate septa, the edges of which are dentate; while the interseptal loculi are crossed by numerous trellis-like bars ("synapticulæ"). The wall is generally basal—in the discoid forms always so—and is generally perforated by apertures. There are no dissepiments nor tabulæ.

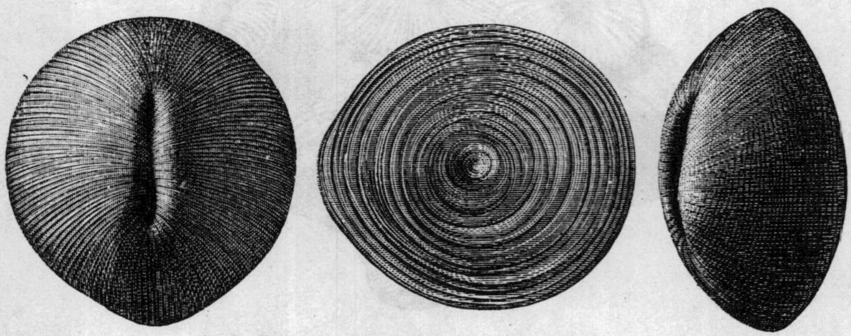

Fig. 81.—*Cyclolites elliptica*, a simple type of the *Fungidæ*, viewed from above, from below and from the side. Cretaceous.

If we except the Silurian *Palæocyclus* (which appears to be truly a *Rugose* Coral), the *Fungidæ* are not known to have existed prior to the Jurassic, in which they are represented by numerous forms (*Comoseris, Protoseris, Anabacia*, &c.) Numerous forms are known in the Cretaceous, and there is also a considerable number of Tertiary species. As an example of the family we may take the genus *Cyclolites* (fig. 81), which ranges from the Cretaceous to the Miocene Tertiary inclusive. In this genus the corallum is simple and discoid, with a concentrically-striated basal epitheca and numerous delicate septa. *Micrabacia*, of the Cretaceous, is like *Cyclolites*, but has no epitheca, and has its basal wall perforated; while in the Jurassic *Anabacia*, also very similar to the preceding, the basal wall is imperfect, so that the under side of the disc is covered by the projecting septa.

6. *Pseudofungidæ*.—This, the last family of the *Aporosa*, merely requires mention, as affording a connecting link between the *Fungidæ* and *Astræidæ*, the corallum agreeing with the former in having a perforated basal plate, but

resembling the latter in having no synapticulæ, and in possessing dissepiments. The only genus of this family is *Merulina*, and it is not known to possess any fossil representatives.

II.—Perforata.

In the *Perforate* section of the *Zoantharia sclerodermata*, the calcareous tissue of the corallum is more or less porous, often spongy or reticulate, *the "wall" of corallites being always perforated by more or fewer apertures.* The septa may be well developed, but they are usually more or less porous, and are sometimes represented by mere calcareous trabeculæ. There may be imperfect dissepiments, and sometimes there are well-developed tabulæ, but in general the visceral chamber is open from top to bottom, the interseptal loculi being continuous.

Taken as a whole, the Perforate Corals must be regarded as an essentially Tertiary and Recent group. The only Palæozoic Corals which can at present be referred to the *Perforata*, with anything like certainty, are the *Protaræa* of the Lower Silurian, and the *Calostylis* of the upper division of the same formation. The Lower Silurian genus *Columnopora* may possibly be one of the *Poritidæ*, and there is considerable probability that the true place of the great and important family of Palæozoic Corals termed the *Favositidæ* is truly in this section, but we shall provisionally consider these among the *Tabulata*. In deposits later than the Silurian, the only forms which have been referred here—namely, *Pleurodictyum* and *Palæacis*—are certainly wrongly placed, the former being founded upon casts of a member of the *Favositidæ*, while the latter is probably a sponge, and is apparently not a coral. In the Permian rocks, the Trias, and the Lias, no Perforate Corals are known; and in the Oolitic series we have only the aberrant genus *Microsolena* (including *Dendraræa*), an ancient type of the *Poritidæ*. In the Cretaceous series, however, we have examples of all the existing families of this section, and in the Tertiary period we find a great development of Perforate Corals, the group apparently reaching its maximum at the present day.

The Perforate Corals are divided into the following three families—if we omit the *Favositidæ*, the true place of which is still uncertain:—

1. *Eupsammidæ.*—In this family the corallum may be simple or compound, the *wall* being always perforated and granular, while the septa, though comparatively well developed and lamellar, are generally also perforated. There is a spongy columella, and the interseptal loculi are open, or crossed by but few dissepiments.

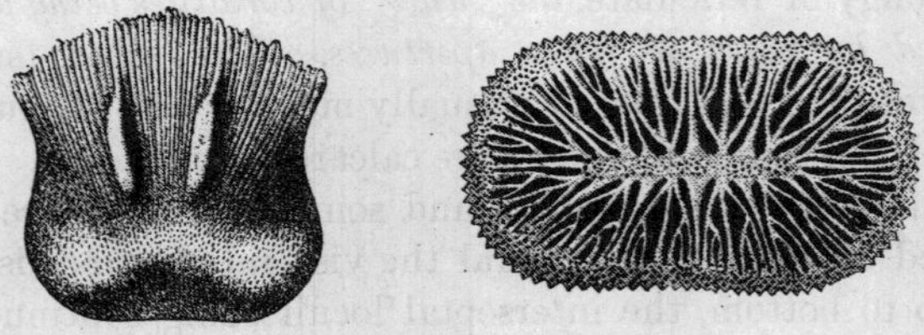

Fig 82.—*Endopachys Maclurii*, viewed in profile and from above. Eocene Tertiary.

The most ancient type of the *Eupsammidæ* is the Upper Silurian *Calostylis*, but with this exception the earliest known representatives of this family occur in the Cretaceous (*Stephanophyllia*), and there is a considerable expansion of the group in the Eocene Tertiary. In *Eupsammia* itself, the corallum is simple, free, and turbinate in shape; and *Endopachys* (fig. 82) is essentially the same, except that it is much compressed, and its keeled base is continued into two wing-like expansions. *Balanophyllia*, ranging from the Eocene to the present day, is also simple, but the corallum is fixed; while the Cretaceous and Tertiary *Stephanophyllia* is free, simple, and discoid, with an open circular calyx. *Dendrophyllia*, again, a well-known recent type, is composite, the corallum increasing by lateral gemmation, so as to assume a dendroid or shrub-like form. It begins in the Eocene Tertiary.

2. *Madreporidæ.*—The members of this family are distinguished by possessing a composite corallum, increasing by gemmation, the various corallites being united by an abundant and spongy cœnenchyma. The *walls* of the corallites are not distinct from the cœnenchyma, and are porous,

and the septa are often well developed. There are no synapticulæ, and usually no dissepiments, but there may be tabulæ. The family makes its first undoubted appearance in the Cretaceous (*Actinacis*), and is largely represented in the Tertiaries, and by living forms in the "coral-reef region" of the present era.

The genus *Madrepora* itself, with its lobate, ramose, or fasciculate corallum, and its loose and delicately echinulate cœnenchyma, appears for the first time in the Eocene Tertiary, and survives to the present day; while the range of the allied genus *Astræopora* is essentially similar.

3. *Poritidæ.*—In this family the corallum is entirely made up of reticulated calcareous tissue ("sclerenchyma").

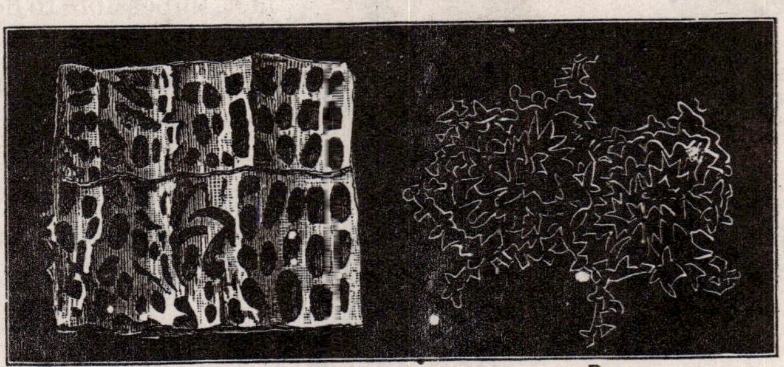

A B

Fig. 83.—*Alveopora spongiosa*, one of the recent *Poritidæ* (after Dana). A, Some of the corallites cut vertically and enlarged, showing the tabulæ and the perforated walls; B, View of the calices from above, enlarged.

The septa are not lamellar, but are composed of styliform processes, which constitute by their junction a sort of trellis-work, and the *walls* are similarly constructed and are not distinct from the cœnenchyma, when this is present. There are a few dissepiments, but generally no tabulæ.

The oldest known types of the normal *Poritidæ* appear in the Silurian, where the family is represented by the curious *Protaræa* and the nearly allied *Stylaræa*, both of which are believed to be closely related to the Tertiary genus *Litharæa*. In the Oolitic rocks we meet only with the singular genus *Microsolena* (fig. 84), but in the Cretaceous we find *Porites* itself. In the Tertiary rocks, again, the family is well represented, principally by the still existing *Porites*, *Alveo-*

pora (fig. 83), and *Rhodaræa.* The genus *Alveopora* is one of special interest, as its corallites are provided with well-developed tabulæ (fig. 83, A). The closely-allied *Favositipora* of Mr Saville Kent, found both in the Devonian and in existing seas, not only possesses tabulæ, but is in other respects extremely similar to some of the *Favositidæ.* It seems, in fact, impossible to doubt that the place of the large and principally Palæozoic family just mentioned is in the vicinity of the *Poritidæ,* though, in deference to long received and still current systems, we shall here retain it in the provisional group of the "Tabulate Corals."

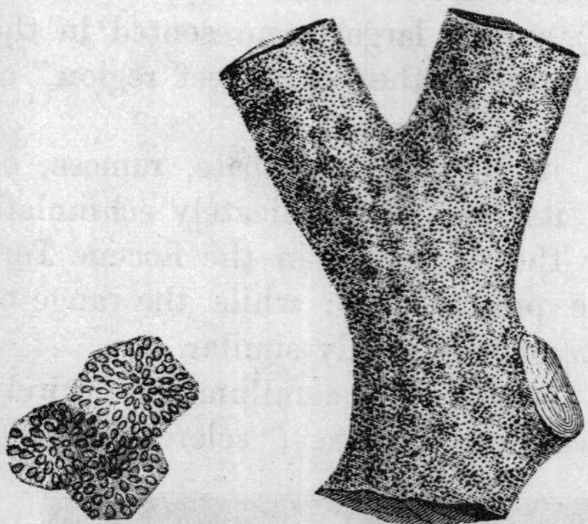

Fig. 84.—Fragment of *Microsolena (Dendraræa) ramosa,* and three of the calices of the same, enlarged. Jurassic.

III.—Tabulata.

The group of the "Tabulate Corals," as founded by Milne-Edwards and Haime, included a large number of corals, in many respects very unlike each other, but characterised by the rudimentary condition or absence of the septa, conjoined with the presence of well-developed *tabulæ* dividing the visceral chamber into so many distinct stories. It is now known, however, that the presence of tabulæ cannot be regarded as a point of any great classificatory value; and the researches of naturalists, and especially of Verrill and Moseley, into the structure of living forms have shown that the various recent "Tabulate Corals" are of the most diverse nature. Thus it has been shown that *Pocillopora* (and probably *Seriatopora* also) is a true *Aporose* Zoantharian. On the other hand, the living *Heliopora* is not a Zoantharian at all, but a genuine *Alcyonarian;* and this discovery removes to the *Alcyonaria* the great group of extinct "Tabulate Corals" of which *Heliolites* is the type. Lastly, *Millepora*, as originally affirmed by Agassiz, has been shown to be likewise not a Zoantharian, but to be referable to the *Hydrozoa*.

In spite of these great deductions, there still remain some extinct groups of corals which may, in the meanwhile, be retained to form the section *Tabulata*, though their true affinities and systematic position are matters of great doubt.

1. *Favositidæ*.—The first of these groups comprises the well-known "honeycomb" corals, and is almost exclusively Palæozoic in its range. In all the members of this family the corallum is compound, the septa are rudimentary or absent, the tabulæ are extremely well developed, and the *walls of the corallites are perforated more or less freely with apertures or "mural pores"* (fig. 72). In their perforated walls and general structure the *Favositidæ* make such a near approach to such living genera as *Alveopora* and

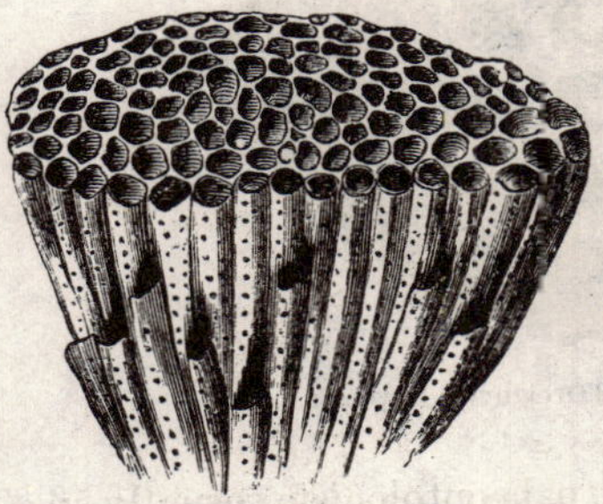

Fig. 85.—Portion of a mass of *Favosites Gothlandica*, of the natural size. Upper Silurian and Devonian of Europe and America. (Original.)

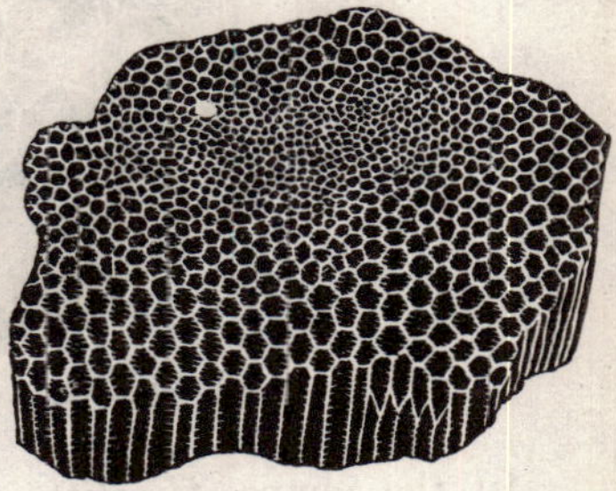

Fig. 86.—Fragment of *Favosites* (*Emmonsia*) *hemispherica*, of the natural size. Upper Silurian and Devonian of America. (After Billings.)

Favositipora, that it seems certain that they will ultimately be removed from the "*Tabulata*," and will be placed in the *Zoantharia Perforata*, in or near the family of the *Poritidæ*.

The type-genus of this family is *Favosites* itself, with many species in the Silurian, Devonian, and Carboniferous rocks. The corallum in this genus is massive or branched, composed of more or less prismatic, closely approximated corallites, which are often enveloped basally by a common epitheca, and have their walls pierced by one or more rows of regular apertures ("mural pores"). The septa are absent or spiniform, and the tabulæ, though usually complete, are sometimes imperfect (*Emmonsia*, fig. 86). The genus *Michelinia* (fig. 87) possesses a corallum very like that of the massive forms of *Favosites;* but the epitheca is often furnished with root-like prolongations, the tabulæ are arched and somewhat vesicular, and the mural pores are numerous and usually irregularly distributed. The genus is essentially Devonian and Carboniferous.

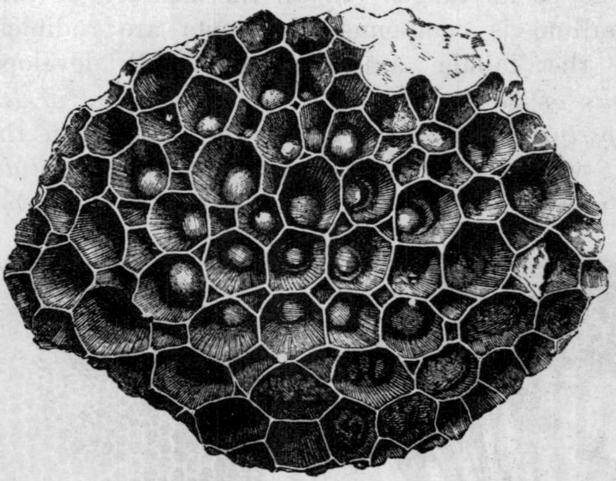

Fig. 87.—*Michelinia convexa* (D'Orbigny). Devonian.

In the genus *Alveolites*, again, palæontologists usually include a number of branching or massive corals, which agree with *Favosites* in most respects, but which have short *oblique* corallites, with sub-triangular or crescentic calices (fig. 88). Nearly allied to *Alveolites* are the genera *Cœnites* and *Pachypora*, of the Silurian and Devonian. Of the remaining genera of the *Favositidæ*, *Striatopora* (fig. 89), of the Silurian and Devonian, is remarkable for the form of its calices; and *Koninckia* is specially noticeable as being found in strata as young as the Cretaceous.

2. *Chætetidæ.*—In this family we have a great number of Palæozoic corals, in which there is a compound corallum

(fig. 90), composed of closely approximated corallites, destitute of septa, provided with well-developed tabulæ, and differing from the *Favositidæ* chiefly in the fact that the *walls of the corallites are imperforate.* Though ranging from

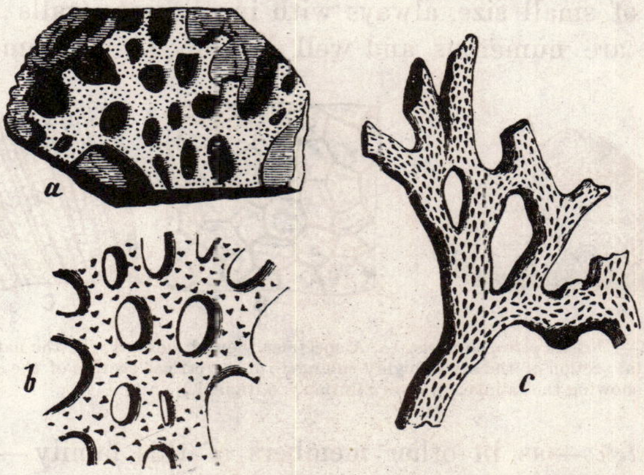

Fig. 88.—*a*, Fragment of *Alveolites ramulosa*, of the natural size; *b*, Portion of the same enlarged, showing the calices; *c*, Fragment of *Alveolites Billingsi*, of the natural size. Devonian. (Original.)

the Lower Silurian to the Permian, inclusive, we have no Mesozoic or Tertiary representatives of this family, so far as is certainly known. Though presenting a striking *general* resemblance to the *Favositidæ*, it is very doubtful if any close relationship exists between the *Chætetidæ* and the former. The family, indeed, may possibly be really *Alcyonarian*, though at present the evidence would rather point to its ultimate removal to the *Polyzoa*. At any rate, it should be remembered that we have Mesozoic, Tertiary, and Recent *Polyzoa* (such as *Heteropora*), which can hardly be distinguished from the *Chætetidæ* except by their not possessing tabulæ.

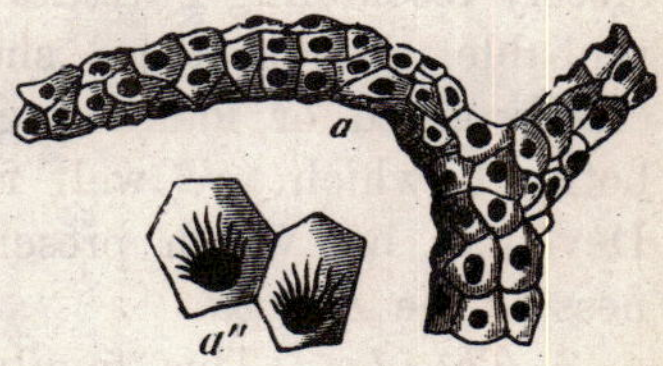

Fig. 89.—Fragment of *Striatopora flexuosa* of the natural size, and two calices enlarged. Upper Silurian. (After Hall.)

The chief genus of this family is *Chætetes* itself (provisionally including under this name the forms known by the name of *Monticulipora*); and it is widely represented by numerous and varied forms in the Silurian, Devonian, Carboniferous, and Permian rocks. The corallum is sometimes massive, sometimes branched, sometimes laminar, and sometimes encrusting; the corallites are prismatic, generally of small size, always with imperforate walls; and the tabulæ are numerous and well developed. In some species

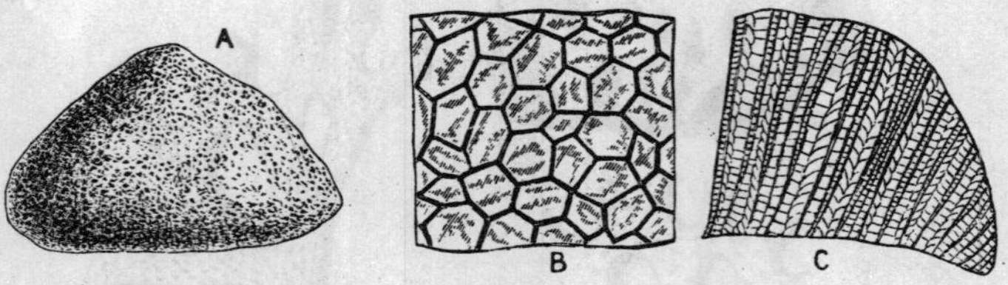

Fig. 90.—*Chætetes petropolitanus.* A, A specimen viewed sideways, of the natural size; B, A horizontal section of the same, highly enlarged; C, A vertical section of the same, greatly enlarged, showing the tabulæ. Lower Silurian. (Original.)

of *Chætetes*—as in other members of this family—some of the corallites are of larger size than the others; this probably indicating that the colony was composed, in its living condition, of two distinct and different sets of zoöids.

The genera *Prasopora, Dania, Dekayia,* and *Constellaria,* are nearly allied to *Chætetes,* and are all Silurian. *Beaumontia,* closely resembling *Favosites* in *form,* is Carboniferous. It is probable, also, that we should include in this family the closely allied or identical genera, *Fistulipora* and *Callopora,* both of which are well represented in the Silurian and Devonian, but which present many striking points of likeness to the *Polyzoa.*

3. *Thecidæ.*—This family includes only the single genus *Thecia,* confined to the Silurian period. The corallum is compound, septa are present, and tabulæ are well developed. The precise affinities of this genus are still obscure, but there is a considerable probability that it should really be regarded as an *Alcyonarian,* and placed in the neighbourhood of *Heliolites.*

4. *Halysitidæ.*—In this family we have the most typical of the "Tabulate Corals," or, at any rate, those which appear most likely to hold their ground as a separate division of the *Zoantharia.* The corallum is always compound, rudimentary

septa (typically twelve in number), are mostly present, there are well-developed tabulæ, and the walls are imperforate.

In *Halysites*, the "Chain-coral," which may be taken as the type of this family, the corallum (fig. 91), consists of long tubular corallites, united to one another in such a way as to form vertical plates or expansions, which, in turn, are so disposed as generally to form by reticulation a loosely netted mass. The tabulæ are horizontal. The "Chain-corals" are characteristically Upper Silurian forms, but they occur also in the Lower Silurian.

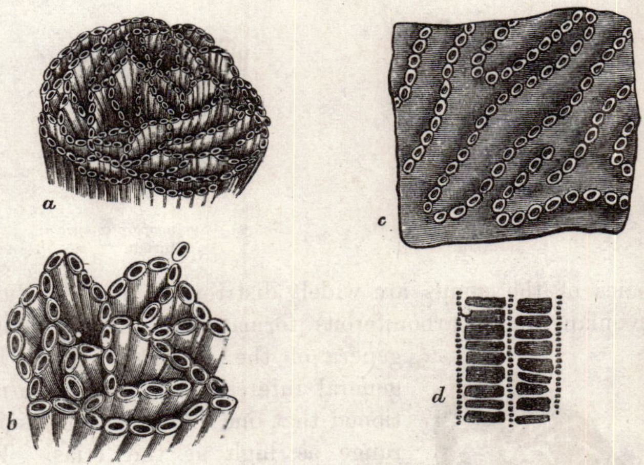

Fig. 91.—*a*, *Halysites catenularia*, small variety, of the natural size; *b*, Fragment of a large variety of the same, of the natural size; *c*, Fragment of limestone with the tubes of *Halysites agglomerata*, of the natural size; *d*, Vertical section of two tubes of the same, showing the tabulæ, enlarged. Niagara Limestone (Upper Silurian), Canada. (Original.)

The genus *Syringopora* (figs. 92-95), is in many respects closely allied to *Halysites*, though very different in external aspect. The corallum is fasciculate, the corallites being cylindrical, lengthy, and united by hollow, tubular, horizontal connecting-processes; so that though the wall is imperforate, the visceral chambers of contiguous corallites are placed in communication. Septa, though very rudimentary, are not wholly absent, and the tabulæ have the form of funnel-shaped plates invaginated into one another. Young forms, and the basal portions of old colonies, closely resemble the Tubulose genus *Aulopora,* from which they can only be separated by the presence of infundibuliform tabulæ. The

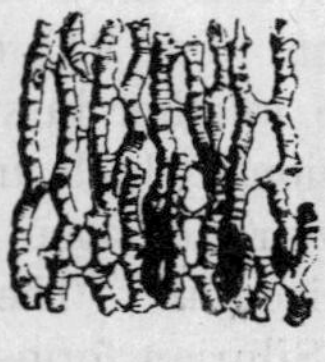

Fig. 92.—*Syringopora retiformis.* A Silurian Tabulate Coral.

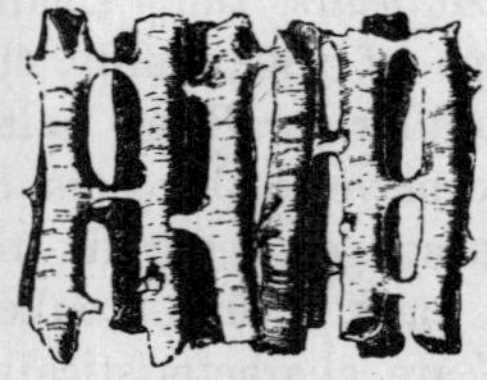

Fig. 93.—*Syringopora verticillata.* Silurian.

Fig. 94.—*Syringopora Dalmani.* Silurian.

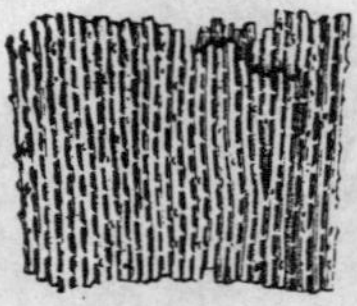

Fig. 95.—*Syringopora compacta.* Silurian.

species of the genus are widely distributed in the Silurian, Devonian, and Carboniferous formations. The remaining genera of the *Halysitidæ* have little general interest; but it may be mentioned that one (*Fletcheria*) is said to range as high as the Trias. With this exception, the family seems to die out in the Carboniferous. The curious Silurian genus, *Tetradium,* widely distributed in North America, may also be noted here, as affording a link between the *Halysitidæ* and the *Chætetidæ.*

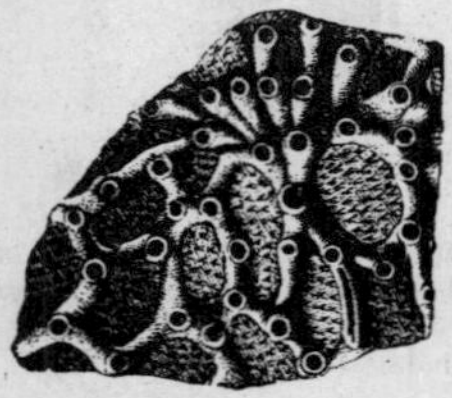

Fig. 96.—*Aulopora serpens.* Devonian.

IV.—Tubulosa.

This is a small group of corals, including the genera *Aulopora* and *Cladochonus* (*Pyrgia*), to which, perhaps, *Stomatopora* should be added. The corallum may be simple or compound (fig. 96), the corallites being pyriform, or trumpet-shaped, without tabulæ, and having the septa indicated by mere striæ on the wall (fig. 97, B). The family is entirely Palæozoic, and its systematic position is wholly doubtful. Save for the alleged absence of tabulæ, its nearest ally appears to be *Syringopora*.

In *Aulopora*, a genus which ranges from the Silurian to the Carboniferous, the corallum (fig. 97, A) is compound, and grows parasitically upon foreign bodies. The corallites are

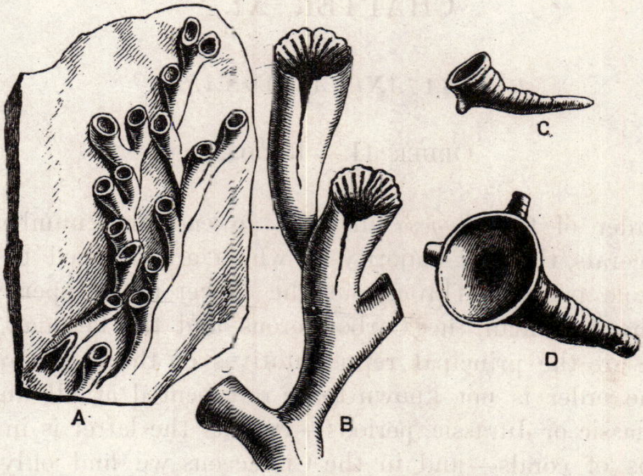

Fig. 97.—A. Portion of *Aulopora tubæformis*, of the natural size; and B, Portion of the same, enlarged (after Goldfuss). Devonian. C, *Cladochonus (Pyrgia) Michelini*, of the natural size, and enlarged (after Milne-Edwards and Haime). Carboniferous.

tubular or pyriform, produced by lateral gemmation, and furnished with a strong imperforate wall. In the Carboniferous genus *Cladochonus (Pyrgia)*, on the other hand, the corallum is typically simple, and resembles a free corallite of *Aulopora*.

CHAPTER XI.

RUGOSA AND ALCYONARIA.

ORDER II.—RUGOSA.

THE order of the *Rugosa* includes an enormous number of fossil corals, the vast majority of which are confined to the Palæozoic period. Throughout the Lower and Upper Silurian, the Devonian, the Carboniferous, and the Permian, the *Rugosa* are the principal representatives of the *Cœlenterata*; but the order is not known to be represented at all during the Triassic or Jurassic periods—rich as the latter is in the remains of corals—and in the Cretaceous we find only the singular little *Holocystis* of the Lower Greensand. In the great series of the Tertiary deposits, again, there has been discovered but one Rugose genus—viz., the *Conosmilia* of the later Tertiaries of Australia. Lastly, at the present day we find only two living genera (*Haplophyllia* and *Guynia*) which have any title to be regarded as Rugose Corals. While, therefore, we may well admit that our knowledge of the history of the Rugose Corals since the close of the Permian period is extremely imperfect and fragmentary, still it remains certain that the group is an essentially Palæozoic one, and that it underwent a very marked diminution before the commencement of Mesozoic time.

As regards their *general* characters, the *Rugosa* agree with the *Zoantharia sclerodermata* in possessing a well-developed sclerodermic corallum, with a true *theca*, and generally presenting well-developed *septa*, though these are usually combined with *tabulæ* as well. The corallum may be simple or compound, but in the latter case a true cœnenchyma is not present.

Seeing that there is this striking general likeness between the corallum of the *Rugosa* and that of the *Zoantharia sclerodermata*, and seeing, further, that the only *living* forms which have any claim to be regarded as Rugose Corals are unquestionable and undoubted *Actinozoa*, it would seem that there ought to be little difficulty in deciding as to the systematic position of the *Rugosa*, unless it were in separating them satisfactorily from the *Zoantharia*. Of late years, however, an opinion—first started by the late Prof. Agassiz—has gained considerable acceptance, according to which the *Rugosa* are to be considered as really belonging to the *Hydrozoa*. So far as any actual evidence in support of this view goes, it can only be said that we really do know now of living *Hydrozoa* which do secrete a calcareous skeleton (*Millepora* and the *Stylasteridæ*). The skeleton of these forms, however, does not present any special resemblance to that of the *Rugosa*, and its characters, indeed, are such that its makers—till their soft parts were investigated—were always referred to the *Zoantharia*. It is clear, therefore, that the existence of coralligenous *Hydrozoa* cannot be used as an argument for removing the *Rugosa* from the *Actinozoa*.

On the other hand, the resemblances between the corallum of the *Rugosa* and that of the *Zoantharia sclerodermata* are so numerous and so weighty, that it is difficult to imagine that the coralla in the two cases were secreted by different methods, or bore dissimilar relations to the soft parts of the animals producing them. Thus, in both groups alike the simple form of corallum (fig. 98) consists of an outer wall or "theca," enclosing a central space or "visceral chamber," which is ordinarily divided into a series of compartments by vertical partitions or "septa;" in both alike the "visceral chamber" may be partitioned off into storeys by horizontal plates or "tabulæ;" in both alike the interseptal loculi are liable to be more or less subdivided by "dissepiments;" and in both alike the axial rod, known as the "columella," may be developed. In both groups alike, moreover, the corallum is often composite, and may be regarded as a variously formed aggregate of "corallites," each of these subordinate elements of the colony being essentially similar in structure to the typical simple corallum.

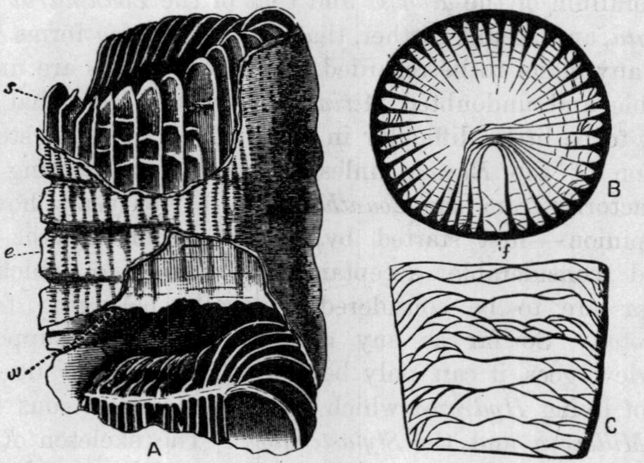

Fig. 98.—Morphology of the *Rugosa*. A, Fragment of *Zaphrentis gigantea*, showing the septa (*s*), with the sparse dissepiments crossing the interseptal loculi, the epitheca (*e*), and the thin proper wall (*w*); B, Transverse section of *Zaphrentis Guerangeri*, showing the septa and dissepiments, the central area occupied solely by the tabulæ, and the "fossula" (*f*); C, Longitudinal section of the last, showing the arrangement of the tabulæ. (A is after Edwards and Haime; B and C are after James Thomson.)

On the other hand, there are various points in which the corallum of the *Rugosa* differs from that of the *Zoantharia sclerodermata*, and some of the more important of these differences may be briefly alluded to here. In the first place, the "septa" appear to be primitively developed in *four systems*, so that the corallum is fundamentally constructed upon a *tetrameral*, instead of an *hexameral* type. In some cases, as in *Stauria* (fig. 100, A), this quadripartite disposition of the septa is very conspicuous, since there are four pre-eminently large septa, which form a cross in the centre of the calice. In the genus *Anisophyllum* (fig. 100, B and C) there are *three* of these pre-eminently developed septa. In the second place, the septa usually present themselves in the adult as of two sizes only, a larger and a smaller; and their arrangement is very generally rendered irregular by the presence of a singular vacant space, which is known as the "fossula" or "fossette" (fig. 98, B, and fig. 99). This space appears to take the place of one of the four principal septa, and usually presents itself as a more or

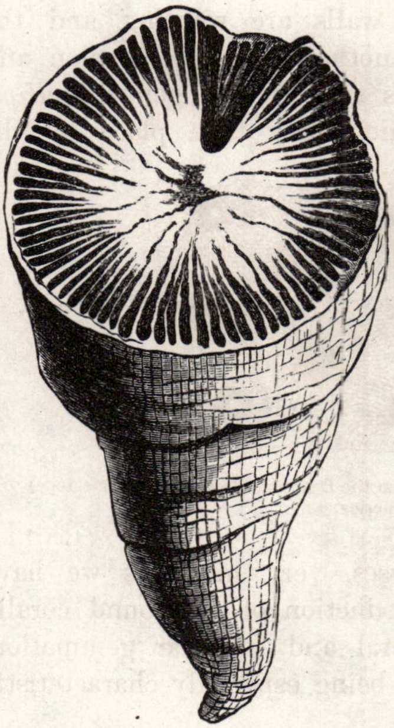

Fig. 99.—*Zaphrentis cornicula*, the walls of the calice broken away, and showing the "fossula," of the natural size. Devonian. America. (Original.)

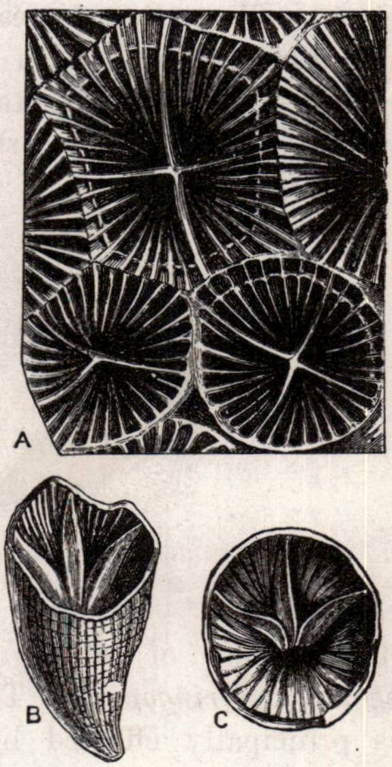

Fig. 100.—A, A few calices of *Stauria astræiformis*, enlarged—Silurian; B, *Anisophyllum Agassizi*, slightly enlarged; C, Calice of same, viewed from above—Devonian. (After Milne-Edwards and Haime.)

less conspicuous depression or groove in the calice. Sometimes there may be two small lateral fossulæ, and in other cases (*Omphyma*) there are four shallow fossulæ arranged in a crucial manner. In the third place, the corallum of the *Rugosa* generally exhibits tabulæ in conjunction with well-developed septa; whereas in the *Zoantharia sclerodermata*, if the tabulæ are conspicuously developed, the septa are rudimentary or wanting, and *vice versa*. The tabulæ of the *Rugosa* may be "complete," passing completely across the visceral chamber from side to side (fig. 98, C), or they may be confined to a larger or smaller central area. Lastly, the corallites of the compound corallum of the *Rugosa* are never connected by a true cœnenchyma. Sometimes the corallites are placed in close contact, so that the corallum becomes "massive," and then they are usually united by a fusion of their walls. Sometimes the walls are wanting, and the corallites are united to one another by the extension and confluence of their septa, as is seen in the genus *Phillipsastræa* (fig. 101), or the union may take place by the

Fig. 101.—*Phillipsastræa Verneuilli.* From the Devonian (Corniferous Limestone) of N. America.

development of lateral processes, very much as we have seen in *Syringopora.* The production of compound coralla is principally effected by lateral and calicular gemmation, the latter process (see p. 186) being especially characteristic of the Rugose Corals.

The divisions of the Rugose Corals which were laid down by Milne-Edwards and Haime, and which have subsequently been generally adopted, are as follows :—

1. *Stauridæ.*—In this family the corallum may be simple or compound, the septa are well developed, conspicuously arranged in four systems, and both dissepiments and tabulæ are present.

In *Stauria* (fig. 100, A), which is the type of the family, there is a compound astræiform corallum, four of the principal septa forming a cross in each calice. The increase of the corallum is effected by calicular gemmation, and there is no columella. The genus is wholly Silurian in its range. The genus *Holocystis* (fig. 102) is closely allied to *Stauria*, but the corallites are united by their costæ, and a styliform columella is present. It is remarkable as being the sole representative of the *Rugosa* at present known from deposits of Secondary age, occurring, as it does, in the Lower Greensand (Cretaceous). The remaining members of the *Stauridæ* possess a simple corallum. In *Metriophyllum*, of the Devonian, the septa are arranged in four groups, separated by as many fossulæ; and in the Permian *Polycœlia* the septa are divided by four principal ones into as many systems. The Tertiary genus *Conosmilia* may perhaps be associated with the above, but its systematic position is not free from doubt.

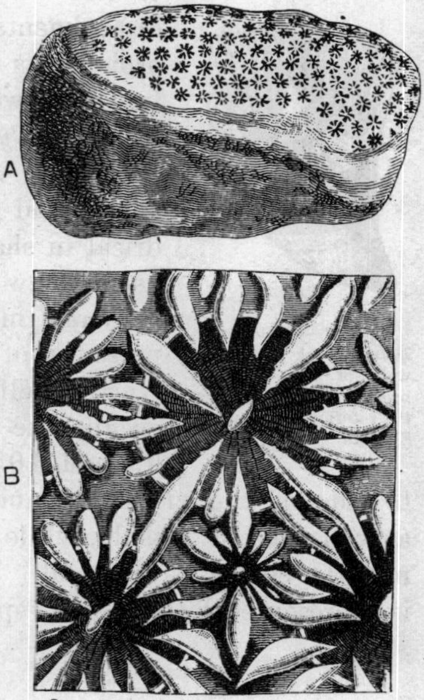

Fig. 102.—A, Small mass of *Holocystis elegans*, of the natural size; B, A few calices of the same, enlarged. Cretaceous. (After Milne-Edwards and Haime.)

2. *Cyathaxonidæ.* — In this family the corallum is simple, the septa are well developed, and the interseptal loculi are open, and are not partitioned off by dissepiments or tabulæ. In many respects this family is very nearly allied to the family of the *Turbinolidæ*, among the *Zoantharia sclerodermata*, but the septa have a tetrameral arrangement. *Cyathaxonia*, the type-genus, has a styliform columella (fig. 103), and ranges from the Silurian to the Carboniferous.[1] The living genera, *Haplophyllia* and *Guynia*, are referred here, but no Secondary or Tertiary members of the family are known.

3. *Cyathophyllidæ.*—In this, by far the largest and most important family of the Rugose Corals, the corallum may be simple or compound, the septa are not arranged in a con-

[1] Recent researches have shown that some of the forms usually referred to *Cyathaxonia* (such as *C. Dalmani*) really possess tabulæ, and are in other respects entitled to be regarded as forming a separate genus, for which the name of *Lindströmia* has been proposed.

spicuously quadripartite manner, except in a few cases; tabulæ are always present, and the interseptal loculi are more or less broken up by the development of dissepiments. The family is wholly Palæozoic in its distribution, and it is divided into the two great tribes of the *Zaphrentinæ* and *Cyathophyllinæ*.

Fig. 103.—*Cyathaxonia Dalmani*. A portion of wall of the theca is broken, in order to show the interior of the calice. Silurian.

In the *Zaphrentinæ*, the corallum is simple and free, conical, discoidal, or cylindrical in shape, with complete tabulæ, and usually few dissepiments. The septa are rendered more or less irregular by the presence of a septal fossula. In *Zaphrentis* itself, which is the type of the group, the corallum is turbinate (figs. 99 and 105), the tabulæ pass from side to side of the visceral chamber (fig. 98, c), and there is a well-marked fossula, while the septa extend inwards to near the centre of the coral. This large and important genus is represented by numerous species in the Silurian, Devonian,

Fig. 104.—*Petraia calicula*. Upper Silurian.

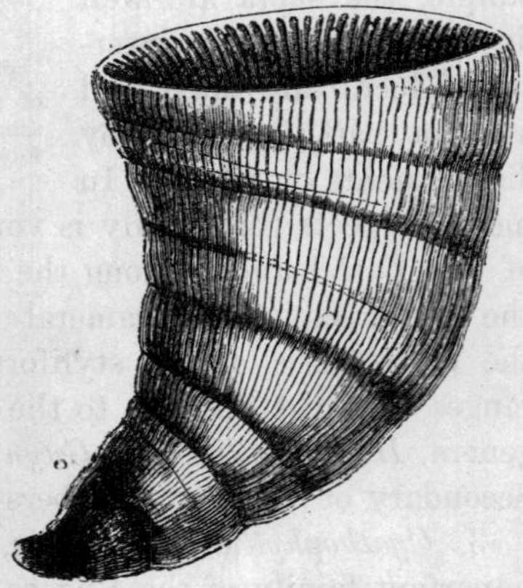

Fig. 105.—*Zaphrentis Stokesi*. Upper Silurian.

and Carboniferous rocks. The Silurian and Devonian genus *Petraia* (fig. 104) has often been placed in the neighbourhood of *Zaphrentis*, but there is considerable uncertainty as to the forms which are really to be included under this head, and if the typical members of the genus are destitute of tabulæ and dissepiments, it must then be rather referred to the *Turbinolidæ*. In the genus *Amplexus*, which ranges from the Silurian to the Carboniferous, the structure of the corallum is essentially similar to that of *Zaphrentis*, but the septa are much less developed, and are so short as to leave the central portion of the tabulæ smooth and bare. In *Lophophyllum*, of the Devonian and Carboniferous, the corallum essentially resembles *Zaphrentis;* but there is a flattened columella. In the Devonian *Anisophyllum*, again, there are three septa pre-eminently developed (fig. 100, B and C), and in *Hallia* (also Devonian), there is one such predominant septum, towards which a number of the septa are inclined. Lastly, in the Silurian *Streptelasma*, the tabulæ are less developed than in *Zaphrentis*, and some of the septa are prolonged inwards to the centre of the visceral chamber in the form of twisted plates.

In the *Cyathophyllinæ*, in the second place, the septa are more or less regularly radiate in their arrangement (fig. 106),

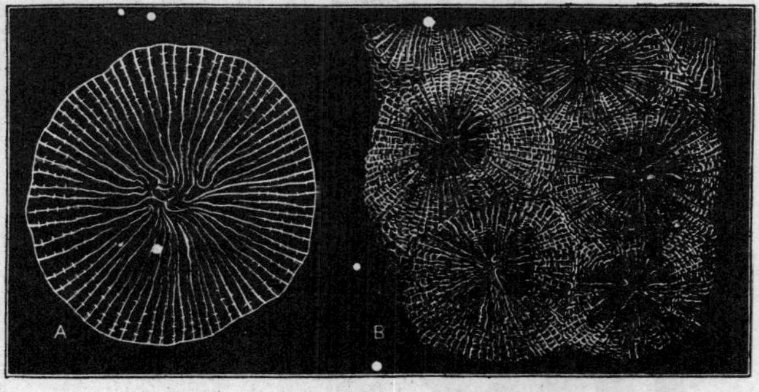

Fig. 106.—A, Cross-section of *Heliophyllum Halli*, of the natural size—Devonian ; B, Cross-section of a fragment of *Cyathophyllum regium*, of the natural size—Carboniferous.

and there is often no septal fossula. The corallum may be simple or compound, the tabulæ are confined to a more or less extensive central area, and there is often an external zone of vesicular tissue formed by the great development of the dissepiments in this region.

In *Cyathophyllum*, the type-genus of the family, the corallum may be simple or compound, and the septa are well developed, some of them extending to the centre of the visceral chamber, where they are twisted together to form a spurious columella (fig. 106, B). The genus ranges from the Silurian to the Carboniferous. In *Heliophyllum* (figs. 106, A, and 107) the septa are provided with singular dissepimental outgrowths, which appear as so many spines or teeth on the free edges of the septa within the calice, and which give them a characteristically cross-barred appearance in transverse sections.

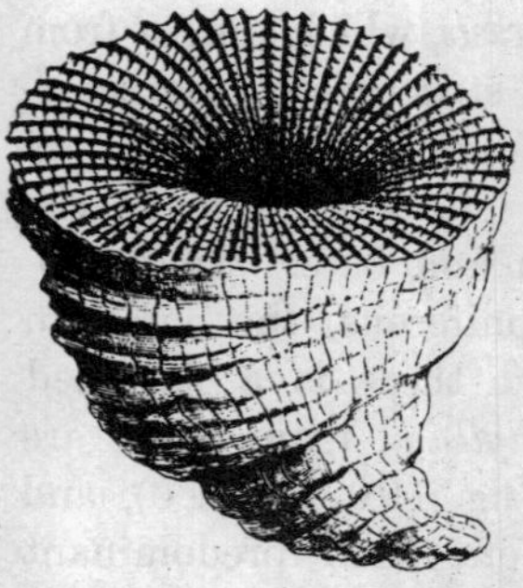

Fig. 107.—A young form of *Heliophyllum Halli*, viewed from one side. Devonian. Of the natural size. (Original.)

The genus is abundantly represented in the Devonian. *Palæocyclus*, which is exclusively Silurian, may be placed here in the meanwhile, though it is typically discoid in form, and has other peculiarities as well. The Silurian genus *Omphyma* is closely related to the simple forms of *Cyathophyllum*, but the septa are divided into four groups by as many shallow depressions or fossulæ, and the coral-

Fig. 108.—*Strombodes pentagonus*. A Silurian Rugose Coral.

Fig. 109.—*Strombodes gracilis*. Silurian.

lum is attached by root-like prolongations of the epitheca. In the genus *Acervularia*, again, of the Silurian and Devonian, we have forms in many respects resembling the compound species of *Cyathophyllum*, but differing in the fact that each corallite possesses a central circular space, invested by a secondary and interior wall. *Strombodes* (figs. 108 and 109), of the Silurian, is very like *Acervularia*, but the walls of the corallites are imperfectly developed, and the tabulæ are funnel-shaped and invaginated in one another. *Phillipsastræa*, again, approaches *Strombodes* in appearance, but the calices are not distinctly circumscribed (fig. 101); the walls of the corallites are deficient; and the corallites are united by the fusion of their septa. The genus is Devonian and Carboniferous.

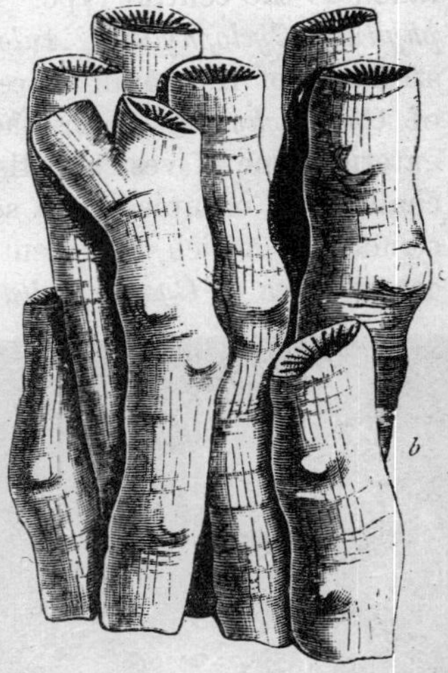

Fig 110.—Fragment of a mass of *Lithostrotion irregulare*, of the natural size. Carboniferous. (After De Koninck.)

As the type of another group of the *Cyathophyllinæ*, we may select the great genus *Lithostrotion* (fig. 110), so highly characteristic of the Carboniferous deposits in almost all parts of the world. The corallum in this genus is compound and fasciculate or massive, composed of cylindrical or prismatic corallites, which may or may not be in close contact with one another. The corallites possess a central tabulate area, which is traversed by a well-developed styliform columella. The genus *Diphyphyllum*, ranging from the Silurian to the Carboniferous, closely resembles the fasciculate forms of *Lithostrotion*, but the corallites have no columella. The Devonian genus *Eridophyllum*, again, differs from *Diphyphyllum*, principally in the fact that the corallites are united by horizontal connecting-processes. Lastly, the genus *Lonsdaleia* (figs. 74 and 111 A), of the Carboniferous rocks, presents some resemblance to the massive and astræiform species of *Lithostrotion*, but the corallites have a secondary or inner wall, enclosing a central area, and the columella is formed of twisted lamellæ.

Lastly, we must mention here a great group of Cyathophylline Corals, which are characteristically Carboniferous in their range, and of which *Clisiophyllum* (fig. 111, B) may be taken as the central type. In the forms in question (*Clisiophyllum, Cyclophyllum, Aulophyllum, Rhodophyllum*, &c.) the structure of the corallum resembles that of the simple forms of *Cyathophyllum* in the fact that there is a well-developed exterior zone of vesicular tissue; but the axis of the visceral chamber is occupied by a series of more or less complicated structures, which represent a modification of the tabulate central area of *Cyathophyllum*, rather than a true columella.

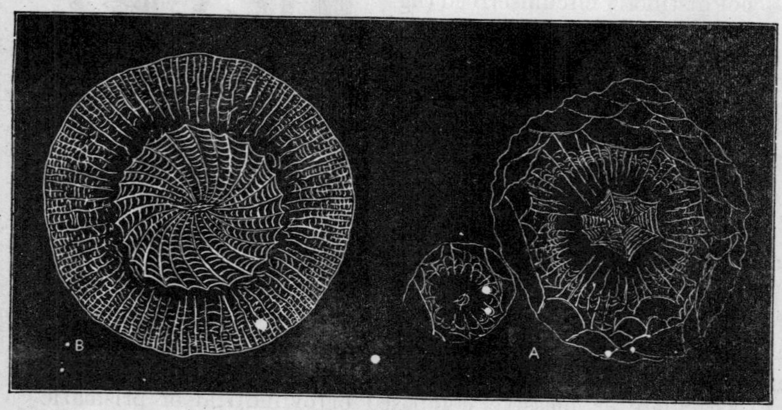

Fig. 111.—A, Cross-section of two corallites of *Lonsdaleia duplicata*, Lower Carboniferous, enlarged; B, Cross-section of the corallum of *Clisiophyllum Keyserlingi*, Lower Carboniferous, of the natural size. (After James Thomson and the Author.)

4. *Cystiphyllidæ.*—In this, the last and most aberrant family of the Rugose Corals, the skeleton is usually simple, though occasionally compound; and the septa are rudimentary, existing only as so many vertical striæ or ridges within the calice (fig. 112). The outer wall is complete, but the entire visceral chamber is filled with small lenticular vesicles formed by a combination of dissepiments and tabulæ. A distinct septal fossula is sometimes present. The family is Palæozoic, and is confined to the Silurian and Devonian periods.

In *Cystiphyllum* (fig. 112), the type-genus of the family, the corallum is usually simple, and the vesicular tissue of which it is made up, is often disposed in funnel-shaped layers. In most of the species the calice is open, but in one form (*C. prismaticum*) the calice is closed by a lid or operculum, consisting of four or more valves. In the genus *Goniophyllum*, of the Upper Silurian, a lid of four valves was present, and the extraordinary Devonian genus *Calceola* (fig. 113), long referred to the *Brachiopoda*, has been shown to be a coral of this family in which the calice is closed by an operculum consisting of a single piece. In this connection, it is worthy of notice that some of the living Alcyonarian Corals (species of *Primnoa*, *Paramuricea*, and others) exhibit also a more or less complete operculum. The calices of *Cryptohelia pudica* (one of the Hydroid group of the *Stylasteridæ*) are also protected by a calcareous lamina in front of each.

Fig. 112.—*Cystiphyllum vesiculosum*, showing a succession of cups produced by budding from the original coral. One side of the calice is broken away, and shows the internal structure. Of the natural size. Devonian, America and Europe. (Original.)

Before leaving the *Cystiphyllidæ*, a few words may be said as to the singular and problematical fossils termed *Beatricea*, which, so far, are only known as occurring in the Lower Silurian of North America. In these extraordinary forms (fig. 114) we have bodies of great size, often many feet in length, of more or less cylindrical form, and principally composed of a calcareous vesicular tissue, essentially similar to that of *Cystiphyllum.* In the centre is what appears to represent a tabulate area, such as we meet with in many of the *Cyathophyllidæ.* If truly corals, these huge fossils must be regarded as very aberrant members of the *Cystiphyllidæ;* but there is some reason for thinking that *Beatricea* is perhaps really founded upon peculiar forms of *Stromatoporoids.*

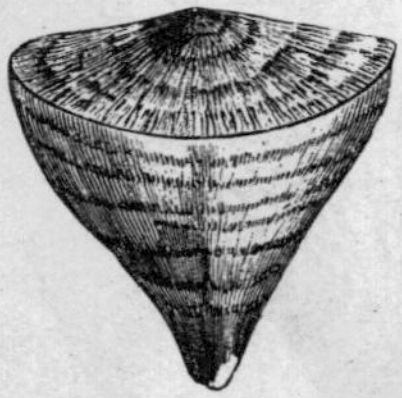

Fig. 113.—*Calceola sandalina.* An operculate Rugose Coral. Devonian.

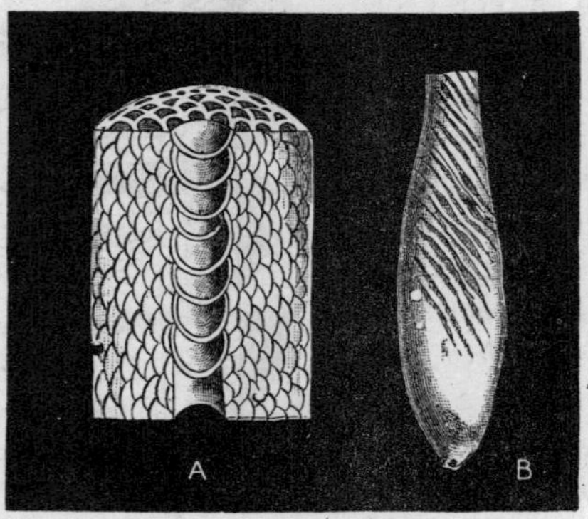

Fig. 114.—*Beatricea undulata.* A, Diagram showing the internal structure as exhibited by a longitudinal section; B, Portion of the base of a specimen from the Hudson River formation (Lower Silurian), the real length of the portion figured being about two feet. (After Billings.)

Order III.—Alcyonaria.

The Alcyonarian Zoophytes are *Actinozoa* in which *the polypes possess eight tentacles, which are fringed on their sides with lateral pinnæ or papillæ,* hence the name of *Octocoralla* often applied to the order. Almost all the members of the order are composite, the tubular polypes being united by a cœnosarc, through which ramify canals by which their body-cavities are placed in communication.

Of the living groups of the *Alcyonaria,* the "Organ-pipe Corals" (*Tubiporidæ*) have a well-developed sclerodermic compound corallum, but they are unknown in a fossil condition. In the *Alcyonidæ* the skeleton consists simply of calcareous spicules scattered in the soft parts, and the family is almost unknown in the fossil condition. A species of *Alcyonium*, however, has been recognised in the Pliocene (Red Crag) of Britain. In the *Pennatulidæ* (the "Sea-rods" and "Sea-pens"), besides detached spicules, there is a skeleton in the form of a horny or calcareous rod, supporting the soft colony. The Silurian genus *Protovirgularia* was believed to belong to the *Pennatulidæ*, but it is certainly not of this nature, and is probably a Graptolite. The family has, indeed, no certain fossil representatives till we reach the latest Secondary or earliest Tertiary deposits. In the Pisolitic Limestone (late Cretaceous) of France, we have the genus *Pavonaria*; while *Graphularia* occurs in the Eocene, and the Miocene Tertiary has yielded examples of *Graphularia*, *Virgularia*, and *Cœlographula*. In the family of the *Gorgonidæ* ("Sea-shrubs") there is a branched sclerobasic corallum, the surface of which is grooved or sulcate. The corallum may be horny or calcareous, or it may be composed of alternating calcareous and corneous segments (as in *Isis* and *Mopsea*). The earliest representatives of the *Gorgonidæ*—so far as we have any sufficient evidence—are found in deposits of the age of the Eocene Tertiary, the genera *Mopsea* and *Websteria* occurring in the London clay. The genus *Corallium*, embracing the living Red Coral, has been quoted from the Jurassic and Cretaceous, and undoubtedly occurs in the Miocene Tertiary; and deposits of the same age have yielded species of *Isis*, *Gorgonia*, and *Melithœa*.

So far as all the preceding forms are concerned, it will be seen that the *Alcyonaria*, though widely distributed in existing seas, are of little geological importance, and are, moreover, comparatively of modern origin. Mr Moseley, however, has shown that the living corals of the genus *Heliopora* are truly referable to the *Alcyonaria*, and not to the *Zoantharia*, in which they had been previously placed; and he has further shown that the large and ancient group of fossil corals of which *Heliolites* is the central type, is essentially similar to *Heliopora* in structure. We thus are presented with a new family of the *Alcyonaria*—that of the *Helioporidæ*—which has played no inconsiderable part in geological history. The *Helioporidæ* were formerly placed in the "Tabulate" section of the *Zoantharia sclerodermata*, and possess *a well-developed sclerodermic corallum, composed of tabulate tubes of two sizes, the larger ones being furnished with rudimentary septa.* In the living *Heliopora cœrulea* (fig. 115) the corallum is composite and sclerodermic, and is

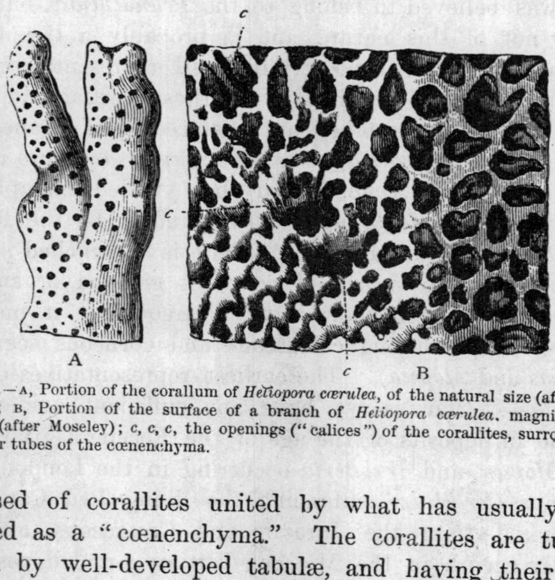

Fig 115.—A, Portion of the corallum of *Heliopora cœrulea*, of the natural size (after Milne-Edwards); B, Portion of the surface of a branch of *Heliopora cœrulea*, magnified eight diameters (after Moseley); *c, c, c*, the openings ("calices") of the corallites, surrounded by the smaller tubes of the cœnenchyma.

composed of corallites united by what has usually been regarded as a "cœnenchyma." The corallites are tubular, crossed by well-developed tabulæ, and having their walls folded in such a manner as to give rise to a variable number (generally twelve) of septal laminæ. The cœnenchyma, so called, is composed of slender tubes, of smaller size than the true corallites, packed closely side by side, crossed, like the corallites, by regular transverse tabulæ, but destitute of septa. The soft parts occupy only the parts of the corallum above the uppermost tabulæ, and therefore only a surface-layer of the colony is actually alive. The polypes are completely retractile, with eight pinnately-fringed tentacles, and eight mesenteries. The mesenteries, however, have no correspondence with the septa, which are twelve in number as a rule. The septa are thus seen to be pseudo-septa, and they cannot be regarded as being homologous with the septa of the *Zoantharia sclerodermata.* The so-called cœnenchymal tubes are occupied by sacs lined by the endoderm, which are closed externally, but communicate freely with the body-cavities of the polypes by means of transverse canals; and Mr Moseley suggests, with great probability, that these are really of the nature of rudimentary sexless polypes.

The genus *Heliopora* is not known as occurring in the fossil condition, but it is represented by various extinct types, dating from the Lower Silurian period.

Of the extinct types, the Silurian and Devonian genus *Heliolites* (fig. 116) is the most important. It has a well-

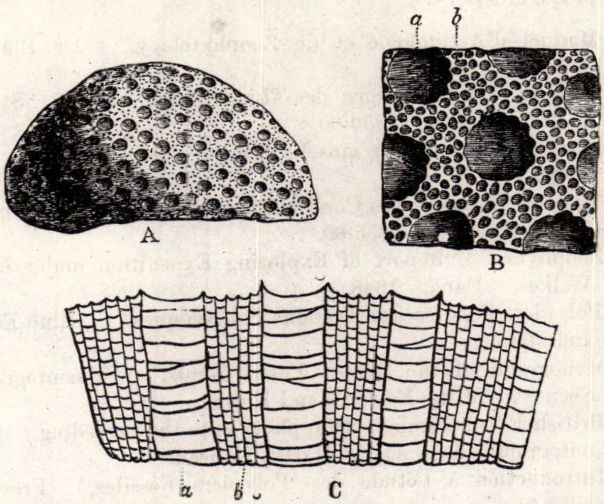

Fig. 116.—A, Small colony of *Heliolites megastoma*, of the natural size; B, Small portion of the surface of the same, magnified, showing the calices (*a*) and the mouths of the cœnenchymal tubes (*b*); C, Vertical section of the same, enlarged, showing the tabulate corallites (*a*), and the tabulate tubes of the cœnenchyma (*b*). (Original.)

developed sclerodermic corallum, with comparatively large-sized, tubular, regularly tabulate corallites, usually possessing distinct but rudimentary septa, intermingled with a copious cœnenchyma formed of tabulate geometric tubuli, much smaller than the corallites, and destitute of septa. The so-called "cœnenchymal" tubules were probably occupied in the living state by rudimentary or imperfect polypes.

In the Silurian *Plasmopora*, the corallum is very similar to that of *Heliolites*, but the cœnenchyma is more vesicular, and its tubules are not so distinct. The genus *Propora*, also Silurian, is very like the preceding, but the calices are exsert. Lastly, we may mention the genus *Polytremacis*, which hardly differs from *Heliolites*, save in its granular surface, and which is remarkable in being found in rocks as modern as the Cretaceous.

LITERATURE.

[In the subjoined list of some of the more important works treating of the fossil corals, no distinction has been made as regards the different orders, since most works of the kind quoted deal with members of two or more of the orders.]

1. "Manuel d'Actinologie et de Zoophytology." De Blainville. 1834-37.
2. "Klassen und Ordnungen des Thier-Reichs," vol. ii. 'Strahlenthiere.' Bronn. 1859-60.
3. "Histoire des Animaux sans Vertèbres," vol. ii. Lamarck. (Ed. 2, 1836.)
4. "Histoire Naturelle des Coralliaires ou Polypes proprement dits." Milne-Edwards. 1855-60.
5. "Zoophytes." 'Report of Exploring Expedition under Captain Wilkes.' Dana. 1848.
6. "Polypiers Fossiles des Terrains Paléozoiques." Milne-Edwards and Haime.
7. "Monograph of the British Fossil Corals." 'Palæontographical Society.' Milne-Edwards and Haime.
8. "British Fossil Corals." (Supplement to the preceding.) 'Palæontographical Society.' Martin Duncan.
9. "Introduction à l'étude des Polypiers Fossiles." Fromentel. 1858-61.
10. "Reports on the British Fossil Corals." 'Rep. Brit. Assoc.,' 1869-71. Martin Duncan.
11. "On the Affinities of the Palæozoic Tabulate Corals with Existing Species." 'Amer. Journ. Sci. and Art,' 1872. Verrill.
12. "Affinities of the Anthozoa Tabulata." 'Ann. of Nat. Hist,' ser. 4, vol. xviii. 1876. Lindström.
13. "Monographie der Sclerodermata Rugosa aus der Silur-formation Estlands," &c. Dybowski. 1873.
14. "Beiträge zur Kenntniss Fossiler Korallen." 'Deutsch. Geol. Ges.' 1870. Kunth.
15. "Guynia and Haplophyllia." 'Phil. Trans.,' 1872. Martin Duncan.
16. "Corals." 'Encyclopædia Brit.' 9th ed., vol. vi. 1877. Nicholson.

17. "Contributions to the Study of the Chief Generic Types of Palæozoic Corals." 'Ann. Nat. Hist.' 1875-76. Thomson and Nicholson.
18. "First Report on the Palæontology of the Province of Ontario." 1874. Nicholson.
19. "Fossil Corals of the State of Michigan." 1876. Rominger.
20. "British Palæozoic Fossils." 1851. M'Coy.
21. "Petrefakten Deutschlands." 1826-33. Goldfuss.
22. "Nouvelles Recherches sur les Animaux Fossiles du Terrain Carbonifère de la Belgique." Premiere Partie. 1872. De Koninck.
23. "On the Structure and Relations of the Alcyonarian Heliopora cærulea," &c. 'Phil. Trans.,' vol. clxvi. 1876. Moseley.

CHAPTER XII.

SUB-KINGDOM III.—ANNULOIDA.

ECHINODERMATA.

SUB-KINGDOM III. ANNULOIDA.—*Animals in which the alimentary canal is completely shut off from the general cavity of the body. Nervous system distinct. A peculiar system of canals, usually communicating with the exterior and containing water derived from the outside, and termed the "water-vascular" or "aquiferous" system, is present in all. In none is the body of the adult composed of definite segments, or provided with "bilaterally disposed successive pairs of appendages."*

This sub-kingdom was proposed by Huxley, as a provisional arrangement, to include the two groups of the *Echinodermata* (Sea-urchins, Star-fishes, &c.) and the *Scolecida* (Tape-worms, Round-worms, Wheel-animalcules, &c.) Whether this arrangement be ultimately retained or not matters not at all to the palæontologist, as no member of the *Scolecida* is known in the fossil condition. The palæontologist, therefore, has simply to deal with the *Echinodermata*, the complete distinctness of which, as a group, is beyond question.

CLASS ECHINODERMATA.

The class *Echinodermata* comprises the animals known commonly as Sea-urchins, Star-fishes, Brittle-stars, Sea-lilies, and Sea-cucumbers, and is distinguished by the fact that *the external envelope of the body* ("perisome") *has the power of secreting calcareous matter to a greater or less extent. The integument is, therefore, either composed of calcareous plates articulated together, or is coriaceous, and has granules or spicules of lime developed in it. The water-vascular system usually communicates with the exterior, and generally subserves locomotion. The adult animal exhibits more or less distinctly a "radial symmetry," or star-like arrangement of its parts, but the young animal is more or less bilaterally symmetrical.*

The *Echinodermata* are divided into the following seven orders:—

1. *Echinoidea.—Ex.* Heart-urchin (Spatangus).
2. *Asteroidea.—Ex.* Star-fish (Uraster).
3. *Ophiuroidea.—Ex.* Brittle-star (Ophiura).
4. *Crinoidea.—Ex.* Stone-lily (Encrinus).
5. *Cystoidea.—Ex.* Hemicosmites.
6. *Blastoidea.—Ex.* Pentremites.
7. *Holothuroidea.—Ex.* Trepang (Holothuria).

The above is not a true or natural arrangement of the orders of the *Echinodermata*, but it is convenient for many reasons to consider them in this sequence. As regards the general distribution of the class, the *Echinodermata* are represented more or less abundantly in all the great formations from the Upper Cambrian to the present day. The orders *Cystoidea* and *Blastoidea* are not only extinct, but are exclusively Palæozoic; while in the *Crinoidea* we have an order which has passed its prime, and appears to be verging on extinction. On the other hand, the orders *Echinoidea, Asteroidea, Ophiuroidea,* and *Holothuroidea* appear to have attained their maximum of development at the present day. The *Asteroidea* and *Ophiuroidea* commence in the Silurian period. The *Echinoids* commence in the Upper Silurian, but reach no marked development till we enter upon Mesozoic deposits. Lastly, the *Holothurians,* as might be expected from the soft nature of their integuments, are hardly known as fossils, though they seem to have existed at any rate as early as the Carboniferous period.

Order I.—Echinoidea.

The members of this order—commonly known as Sea-urchins—are characterised by the possession of *a more or less globular, heart-shaped, discoidal or depressed body, encased in a "test" or shell, which is composed of numerous calcareous plates, immovably connected together. The intestine is convoluted, and there is a distinct anus. The mouth is usually armed with calcareous teeth, and is always situated on the inferior aspect of the body, but the position of the vent varies.*

As a matter of course, the palæontological student has to deal with nothing except the test of the Echinoids and its

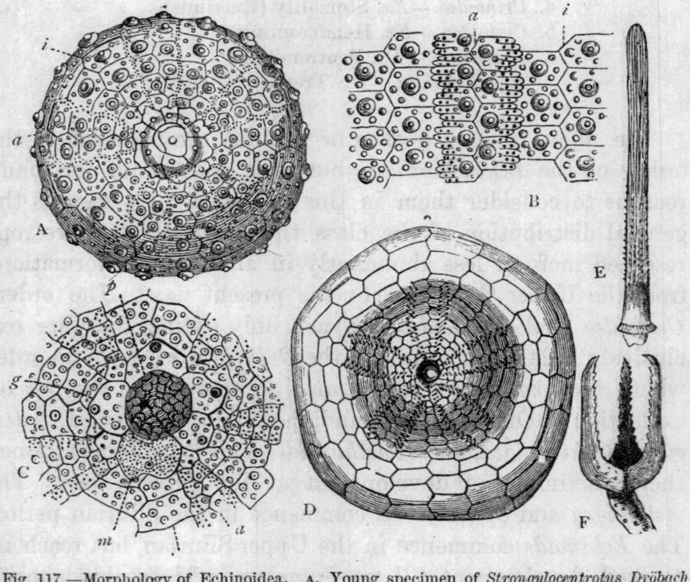

Fig. 117.—Morphology of Echinoidea. A, Young specimen of *Strongylocentrotus Dröbachiensis*, viewed from above. B, Small portion of the test of the same, magnified. C, Summit of the test of *Echinus sphæra*, magnified. D, *Clypeaster subdepressus*, viewed from above, showing the petaloid ambulacra. E, Spine of *Porocidaris purpurata*. F, Pedicellaria of *Toxopneustes lividus*. *a, a,* Ambulacral areas; *i, i,* Interambulacral areas; *g,* Genital plate; *o,* Ocular plate; *m,* Madreporiform tubercle; *p,* Membrane surrounding the anus. (Figs. A, B, and D are after A. Agassiz.)

appendages, and these must be described in some detail. The "test" of the *Echinoidea* may be regarded as essentially composed of the so-called "corona" and of the "apical disc," though other less important elements are present as well. The "corona" is the main element of the test, and includes all the calcareous covering of the animal except the scattered plates round the mouth and anus and the "apical disc." The test is composed of numerous calcareous plates, firmly united to one another by their edges, arranged in rows (fig. 117, A), and bearing different names, according to their position and function. In the curious Urchins which form the family of the *Echinothuridæ*, and in some of the Palæozoic Echinoids, the plates of the test overlap one another in an imbricating manner, so that the shell becomes flexible. As a rule, however, the corona forms an immovable case or box, within which the animal is contained; and its growth is carried on by means of additions made to the edge of each individual plate, by means of an organised membrane which

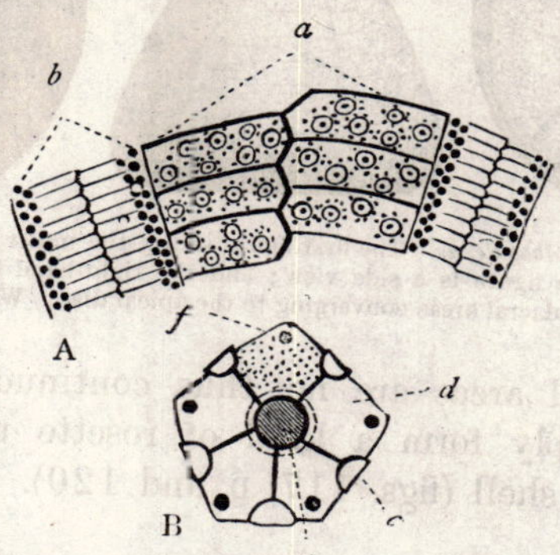

Fig. 118.—Morphology of Echinoidea. A, Portion of the test of *Galerites hemisphericus*, enlarged, showing an inter-ambulacral area (*a*), and an ambulacral area (*b*). B, Genital and ocular disc of *Hemicidaris intermedia*, enlarged: *c*, Ocular plate; *d*, Genital plate; *e*, Anal aperture; *f*, Madreporiform tubercle. (After Forbes.)

passes between the sutures, or the lines where the plates come in contact with one another.

In all recent and most fossil Echinoids, the test is composed of twenty meridional rows of calcareous plates, which are arranged in ten alternating zones or areas (fig. 117, A). Each zone, therefore, is composed of two rows of plates. In five of these zones (figs. 117, B, and 118) the plates are of large size, and are not perforated by any apertures. These zones are called the "interambulacral areas." The remaining five zones alternate with the former, and are composed of very much smaller plates, which are perforated by minute apertures or pores. Through these apertures are emitted the little suctorial tubes of the water-vascular system—the so-called "ambulacral tubes" or "tube-feet"—by means of which the animal moves. Hence these zones of perforated plates are termed the "ambulacral areas" or "poriferous zones."

In one great group of the Echinoids the ambulacral areas pass from the centre of the base of the shell to its summit (figs. 117, A, and 119), when they are said to be "perfect" (*ambulacra perfecta*) or "simple." In another great group

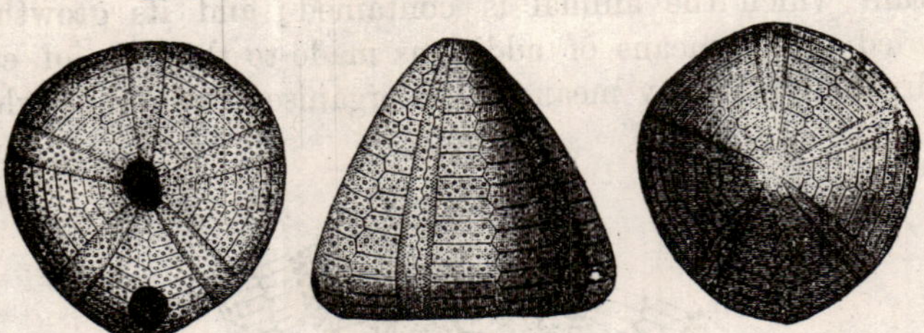

Fig. 119.—*Galerites albogalerus.* The first figure shows the under surface with the mouth and anus; the middle figure is a side view; and the right-hand figure shows the upper surface, with the ambulacral areas converging to the apical disc. White Chalk.

the ambulacral areas are not thus continuous from pole to pole, but simply form a kind of rosette upon the upper surface of the shell (figs. 117, D, and 120). In these cases

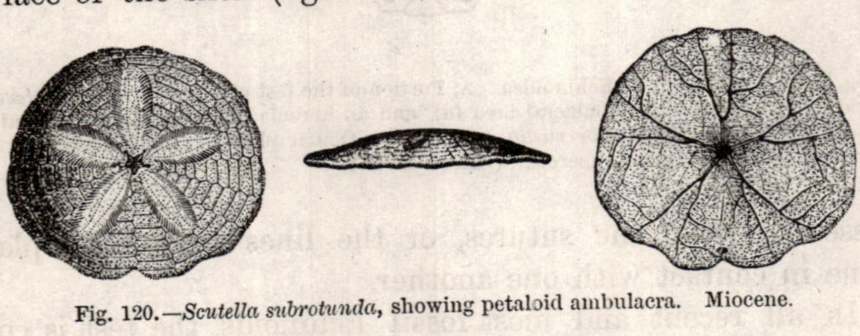

Fig. 120.—*Scutella subrotunda*, showing petaloid ambulacra. Miocene.

—as in the common Heart-urchins—the ambulacral zones are said to be "circumscript" (*ambulacra circumscripta*) or "petaloid."

The most important external structures of the corona are

the tubercles and spines. The tubercles are rounded elevations upon which the spines are carried (figs. 117, A, and 121). They vary much in their dimensions, and receive special names, according to their size or position on the test. Ordinarily the tubercle consists of a rounded ball or hemisphere (the "mamelon") supported upon a conical process (the "boss") which arises from the plate. The ball of the tubercle may or may not be perforated for the insertion of a ligament which is attached to the articular surface of the spine. In many cases (as in fig. 121) the base of the tubercle is surrounded by a round or oval, smooth and excavated space which is termed the "areola" or "scrobicule."

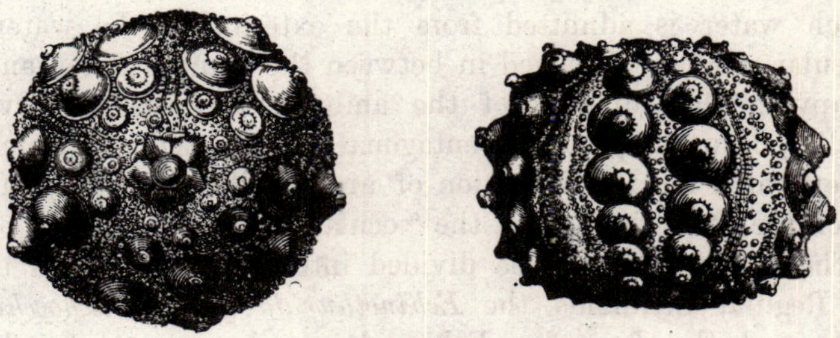

Fig. 121.—*Hemicidaris crenularis*, showing tubercles, the larger of which are perforated, and are surrounded by an areola. Oolite.

The spines are movable appendages which are jointed to the tubercles by a sort of "ball-and-socket" or "universal" joint. They are used defensively and in locomotion, and vary much in length and shape. Sometimes they are very minute; at other times they attain a length considerably exceeding the diameter of the test. Sometimes they are slender, tapering, and truly spine-like (fig. 117, E); at other times they are thickened, ovate, or almost globular (fig. 122). The spine fits on to the rounded head of the tubercle by a concave articular surface ("acetabulum"), and there may or may not be a pit at the bottom of this, for the attachment of the ligament before spoken of. Above the acetabulum or socket of the spine there is a prominent ridge or ring, more or less "milled,"

Fig. 122.—Spine of *Cidaris glandiferus*.

for the attachment of the muscular fibres which move the spine.

The "apical disc" or "genital disc" occupies the summit of the test, and is generally composed of ten plates (figs. 117, C, and 118, B). Five of these plates are of comparatively large size, and are termed the "genital plates," each being perforated by the duct of an ovary or testis. Each genital plate occupies the summit of one of the interambulacral areas. One of the genital plates (the right antero-lateral plate) is larger than the others, and supports a spongy tubercle, perforated with many apertures, like the rose of a watering-pot, and termed the "madreporiform tubercle" (fig. 118, B). This structure protects the mouth of the canal by which water is admitted from the exterior to the water-vascular system. Wedged in between the genital plates, and occupying the summits of the ambulacral areas, are five smaller, heart-shaped, or pentagonal plates, each of which is perforated for the reception of an "ocellus" or eye, and which are therefore termed the "ocular plates."

The *Echinoidea* may be divided into the four groups of the Regular Echinoids, the *Echinothuridæ*, the *Perischoechinidæ*, and the Irregular Echinoids; and we may briefly consider the more important characters, the geological range and the leading type of these divisions.

I. REGULAR ECHINOIDS (*Echinoidea endocyclica*).—In this group of the Echinoids the mouth is situated in the centre of the base (fig. 123), and the vent is placed at the summit

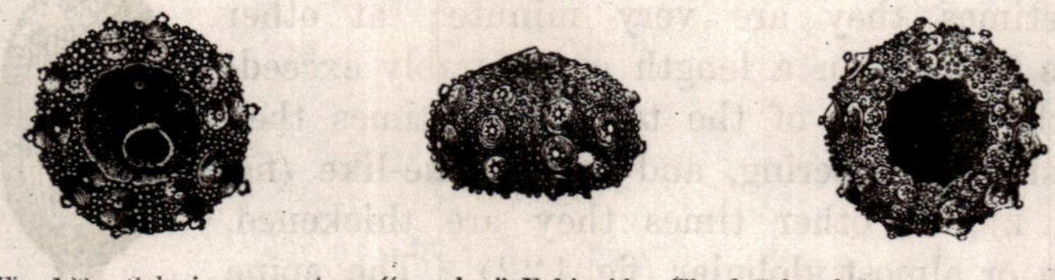

Fig. 123.—*Salenia personata*, a "regular" Echinoid. The left-hand figure represents the upper surface of the shell, and shows the anus surrounded by the apical disc. The right-hand figure shows the mouth in the centre of the base.

of the test, surrounded by the genital disc. They have a test which is almost always circular, or spheroidal, or, it may be, depressed; and the mouth is armed with a complicated masticatory apparatus. The Regular Echinoids comprise the families of the *Cidaridæ, Hemicidaridæ, Diademadæ, Echinidæ*, and *Saleniadæ.*

In the *Cidaridæ,* the test is spheroidal and more or less flattened at the oral and anal poles. The ambulacral areas (fig. 124), are very narrow, often flexuous, and never provided with large tubercles. The interambulacral areas are wide, and carry large perforated tubercles, surrounded by an areola, and supporting the primary spines. The spines are of two sizes, the primary ones usually more or less cylindrical, clavate, or fusiform, and generally longitudinally ridged or tuberculate (fig. 122). Of the genera of this family, *Cidaris* itself (fig. 124) is the most important, ranging from the Trias to the present day. *Rhabdocidaris* is Jurassic and Cretaceous; *Diplocidaris* is Jurassic; and the *Porocidaris* of the Secondary and Tertiary periods has now been detected by Sir Wyville Thomson in a living condition.

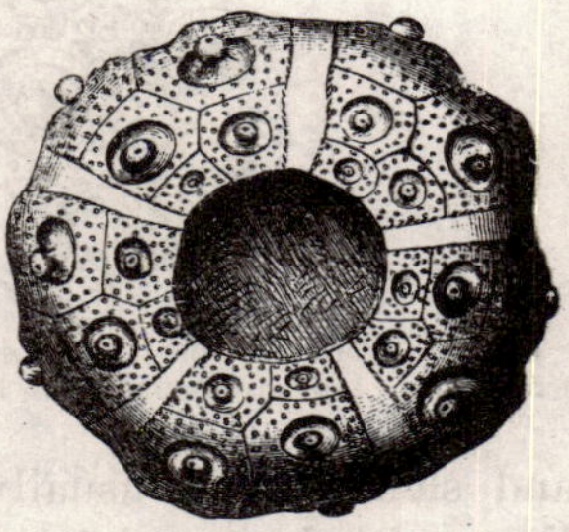

Fig. 124.—Under surface of the test of *Cidaris vesiculosa*, of the natural size. Chalk. (After Wright.)

In the *Hemicidaridæ* (fig. 121) the test is spheroidal, more or less depressed, and with a close general resemblance to that of the *Cidaridæ.* The ambulacral areas, however, though narrow and mostly undulated, are wider than in the *Cidaridæ,* and are provided with comparatively large tubercles, which may be developed inferiorly only (*Hemicidaris*), or which may extend along the entire length of the area (*Acrocidaris*). The interambulacral areas are wide, and carry very large perforated tubercles with crenulated bosses. The spines are usually long, cylindrical, and tapering. The type-genus of this family is *Hemicidaris* (fig. 121), which ranges from the Upper Trias to the Lower Cretaceous inclusive. In *Acrocidaris,* of the Jurassic period, the ambulacral areas are comparatively wide.

In the family of the *Diademadæ*—sometimes made to include that of the *Hemicidaridæ*—the test is circular or pentagonal, more or less depressed and flat below. The ambulacral areas are wide, and carry two rows of large primary tubercles (fig. 125), equal in size to the two or more rows of tubercles upon the interambulacra. The tubercles are sometimes perforated, sometimes imperforate, and they may or may not be crenulated. The spines are cylindrical

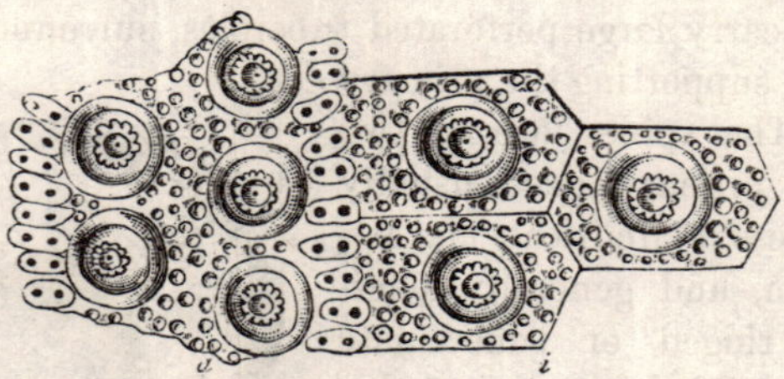

Fig. 125.—Portion of the test of *Pseudodiadema Fittonii*, enlarged four times. *a*, Ambulacral area; *i*, Interambulacral area. Lower Greensand (Cretaceous). (After Wright.)

and slender, and usually of considerable length. In the living *Diadema* and in *Astropyga* the spines are long, tubular, and covered with imbricated scales arranged in oblique rings. As the type of the family we may take *Pseudodiadema*, which ranges from the Liassic to the Tertiary inclusive, and in which the spines are solid and microscopically striated, while the tubercles are perforated (fig. 125). In the Cretaceous *Cyphosoma* the tubercles are solid. *Hemipedina*, again, of the Oolites, is very like *Pseudodiadema*, but the tubercles are

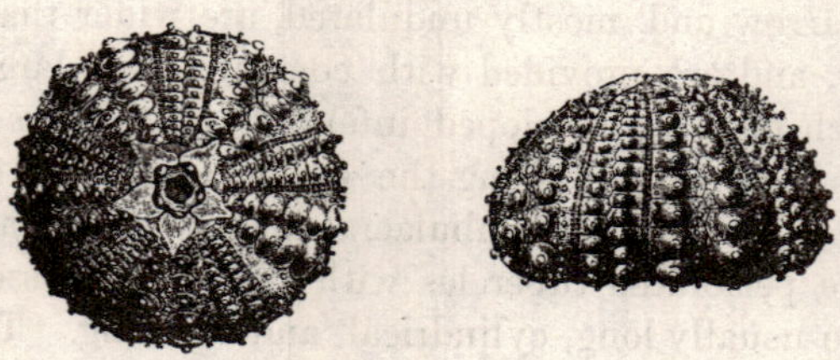

Fig. 126.—*Goniopygus major*, viewed from above and sideways, of the natural size. Cretaceous.

not crenulated. We may also place here the Cretaceous genus *Goniopygus* (fig. 126), which presents us with a type intermediate between the *Diademadæ* and *Saleniadæ*. It approaches the latter family more particularly in the great size of the apical disc; but this structure is only composed of the normal ten plates, and wants the supernumerary or "suranal" plate of the *Saleniadæ*.

In the family of the *Saleniadæ* the test is generally spheroidal, hemispherical, or depressed, and the ambulacral areas are always narrow, sometimes straight, sometimes flexuous, and without large primary tubercles. The interambulacral areas are always provided with two rows of large tubercles, with crenulated bosses, which may or may not be perforated. The leading character of the family, however, is to be found in the apical disc (figs. 123 and 127), which is of unusually large size, and possesses a supernumerary or "suranal" plate in addition to the ten normal plates. This suranal plate (fig. 127, *s*) is placed in front of the anus, and it may be single, or it may be broken up into several (not more than eight) elements. Of the genera of the *Saleniadæ*, *Acrosalenia* is essentially Jurassic; *Peltastes* and *Goniophorus* are Cretaceous; and *Salenia* itself (fig. 123), commencing in the Cretaceous, is not only found in the Tertiary rocks, but has now been detected in a living condition.

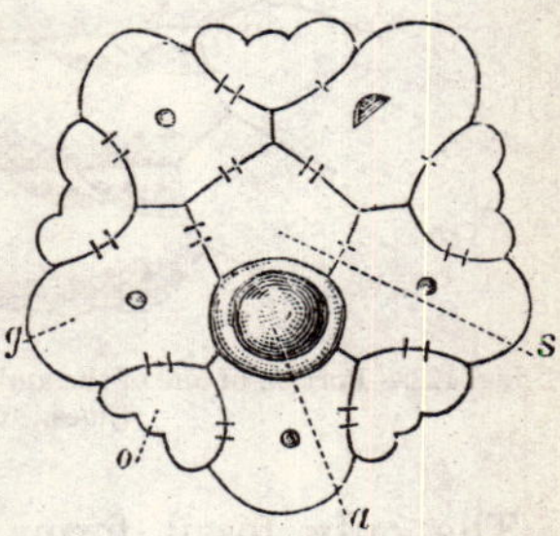

Fig. 127.—Apical disc of *Peltastes Wrightii*, one of the *Saleniadæ*. *a*, Anus; *g*, One of the genital plates; *o*, One of the ocular plates; *s*, Suranal plate. Twice the natural size. Cretaceous (Lower Greensand). (After Wright.)

Lastly, in the great family of the *Echinidæ* the test is usually globular or hemispherical; and the ambulacral areas are comparatively wide, and always carry two or more rows of tubercles. The interambulacral areas are wide, and carry primary tubercles, which are always imperforate, and are never of very large size. The spines are short and awl-shaped, and their surface is marked with fine longitudinal lines. Of the genera of this family, *Glypticus*, *Magnotia*, *Polycyphus*, the great group termed *Stomechinus*, and others, occur in the Jurassic; *Codiopsis* is Cretaceous; and *Temnechinus* is Tertiary and Recent.

II. Echinothuridæ.—In this small but highly remarkable division of the *Echinoidea* the test is "regular," the anus being placed in the centre of the apical disc, and the ambulacral areas being continuous; but the plates of both the ambulacral and interambulacral areas are imbricated and overlap one another (fig. 128), the test thus becoming flexible. In this abnormal character, the *Echinothuridæ* agree with some of the Palæozoic Urchins, but they differ from these, and agree with the ordinary Regular Echinoids in having the test composed of no more than twenty rows of plates.

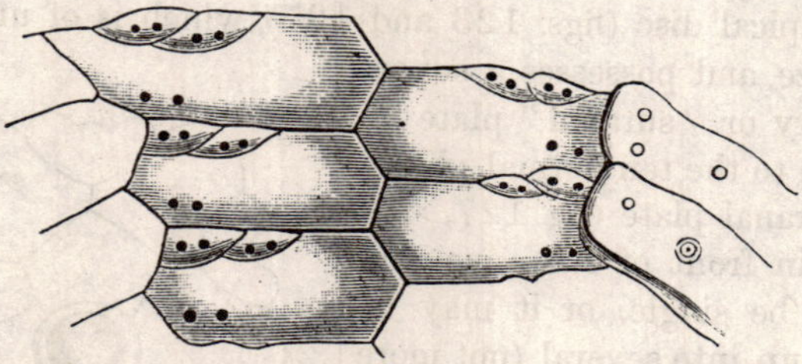

Fig. 128.—Portion of one of the ambulacral areas of *Echinothuria floris*, enlarged four times. Chalk. (After Wright.)

The only fossil forms of this group, as yet discovered, are referable to the Cretaceous genus *Echinothuria*, the true affinities of which have now been elucidated by the discovery of the extraordinary living types referred to the genera *Calveria* (or *Asthenosoma*) and *Phormosoma*.

III. PERISCHOECHINIDÆ.—In this group we have a series of singular Palæozoic Sea-urchins, which agree with the two preceding sections in having a "regular" test, but which differ from all known Echinoids, living or extinct, in having the test composed of more than twenty rows of plates. The test is still divided into five ambulacral and five interambulacral *areas*, but there is a multiplication of the rows of plates in either the ambulacra or interambulacra, or in both. The apical disc consists of no more than the normal ten plates, and the anus is placed in its centre. The ambulacral areas are continuous from pole to pole.

In the genus *Palæchinus* (fig. 129) the test is spheroidal, and its plates abut against one another without any overlapping. The ambulacral areas are comparatively narrow, of two rows only, each plate perforated by two pores. The interambulacral areas are broad, and are composed of from five

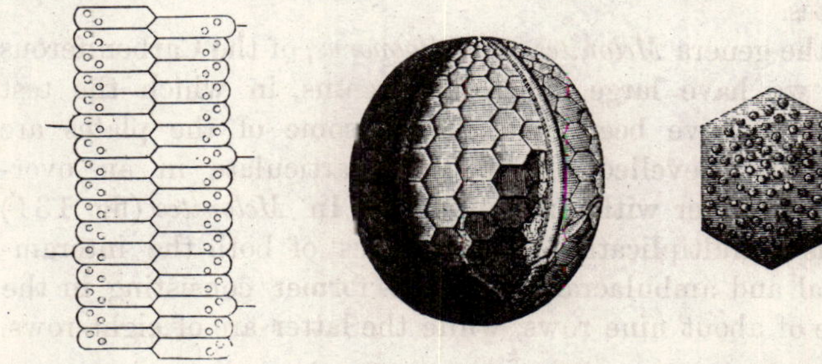

Fig. 129.—*Palæchinus ellipticus.* The left-hand figure shows a portion of an ambulacral area enlarged. The right-hand figure exhibits a single interambulacral plate.

to eight rows of plates. The apical disc (fig. 130, A) has five triply-perforated genital plates, and an equal number of doubly-perforated ocular plates (Baily), but the latter are not recognised by De Koninck. The genus is said to occur in the Silurian, but is principally known as occurring in the Carboniferous rocks.

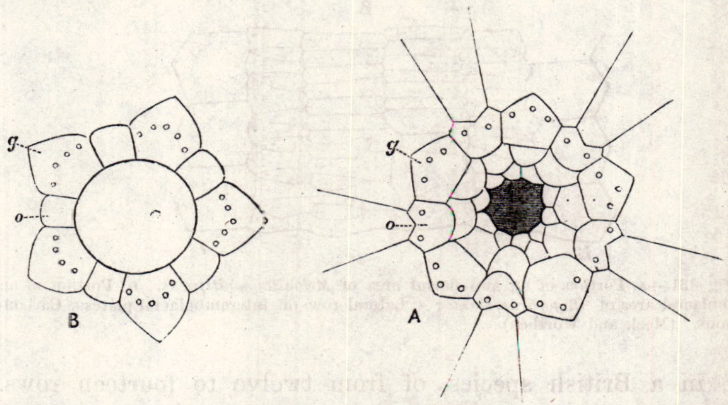

Fig. 130.—A, Apical disc of *Palæchinus*, enlarged (after Baily). B, Apical disc of *Melonites* (after Meek and Worthen): *g*, One of the genital plates; *o*, One of the ocular plates.

The genus *Eocidaris*, of the Devonian, Carboniferous, and Permian formations, also appears to have rigidly articulated plates. The interambulacra are of five or more rows, some of which disappear before reaching the poles. Each interambulacral plate carries a primary tubercle, not surrounded by a ring. The ambulacral areas are composed of two rows of plates.

In the genera *Melonites* and *Oligoporus*, of the Carboniferous rocks, we have large spherical Urchins, in which the test appears to have been rigid, though some of the plates are occasionally bevelled off, so as to articulate in an overlapping manner with one another. In *Melonites* (fig. 131) there is a multiplication of the plates of both the interambulacral and ambulacral areas, the former consisting in the middle of about nine rows, while the latter are of eight rows,

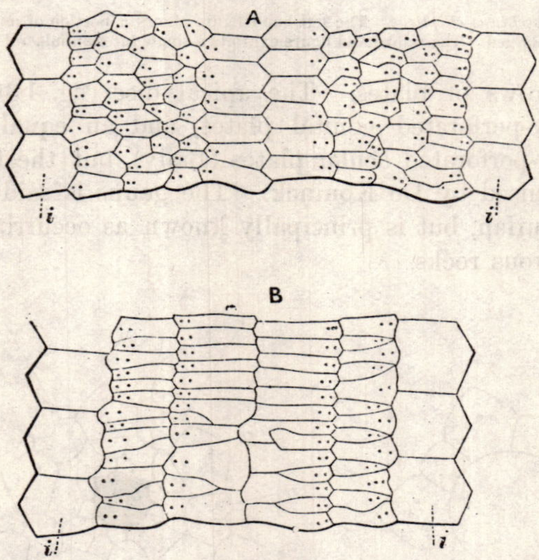

Fig. 131.—A, Portion of an ambulacral area of *Melonites multiporus*. B, Portion of an ambulacral area of *Oligoporus Danæ*: *i*, Lateral row of interambulacral plates. Carboniferous. (Meek and Worthen).

or, in a British species, of from twelve to fourteen rows. The central two rows of ambulacral plates are larger than the rest and elevated above them, and each plate of these areas is doubly perforated. The apical disc (fig. 130, B) is composed of the normal ten plates, but the ocular plates are sometimes imperforate, and the genital plates are furnished with from three to five pores.

Oligoporus (fig. 131, B) is very similar to *Melonites*, but the ambulacral areas consist each of only four rows of plates.

In the genus *Archæocidaris*, well known as a Carboniferous type, the test is spheroidal, the ambulacra are only two-rowed, and the interambulacra are wide, and are composed of three or more rows of plates. The interambulacral plates carry each a large perforated primary tubercle, and have the curious character, as shown by Mr John Young, that some of their edges are bevelled off, this clearly appearing to indicate a certain amount of flexibility in the test, similar to what we have seen in the group of the *Echinothuridæ*.

A much nearer approach to the type of the *Echinothuridæ*, however, is made by some other genera of the *Perischoechinidæ*. In the forms in question (*Perischodomus*, *Lepidechinus*, &c.) the plates of the test overlap one another in an imbricating manner, as in the recent *Calveria*, and the test must have been quite flexible. In the *Echinothuridæ*, however, the imbrication of the interambulacral plates is from above downwards, and that of the ambulacral plates from below upwards, exactly the reverse of this arrangement taking place among the flexible *Perischoechinids*. Moreover, the former have only twenty rows of plates in the test, whereas this number is exceeded in the latter. In the Carboniferous *Perischodomus* there are two rows of ambulacral and five rows of interambulacral plates, and in the nearly allied *Lepidechinus*, of the Devonian and Carboniferous, there are as many as eleven rows of interambulacral plates. The Carboniferous genus *Rhoechinus* is very nearly allied to the two preceding, but the interambulacral plates have no primary tubercles. Lastly, in the Carboniferous *Lepidesthes* we have another flexible Perischoechinid, with imbricated plates, but the ambulacra were composed of no less than ten rows of plates, and the interambulacral areas are comparatively narrow, and are composed of several rows of plates.

IV. Irregular Echinoids (*Echinoidea exocyclica*).—In this section of the *Echinoidea* the test is generally of an oblong, pentagonal, heart-shaped, or discoidal figure (as in the common "Heart-urchins" and "Cake-urchins"); *the anus is situated outside the apical disc,* usually marginal or submarginal in position; there are mostly only four genital plates in the apical disc; and the mouth is mostly destitute of

Fig. 132.—*Discoidea cylindrica,* an "irregular" Echinoid. The right-hand figure shows the summit of the shell, with the genital disc. The left-hand figure shows the base of the shell, on which are situated both the mouth and anus. Cretaceous.

a masticatory apparatus. The Irregular Echinoids may be divided into the families of the *Echinoconidæ, Collyritidæ, Echinonidæ, Echinobrissidæ, Echinolampadæ, Clypeastridæ, Ananchytidæ,* and *Spatangidæ.*

In the *Echinoconidæ* the test is usually circular or pentagonal, the ambulacral areas narrow, and the plates of both

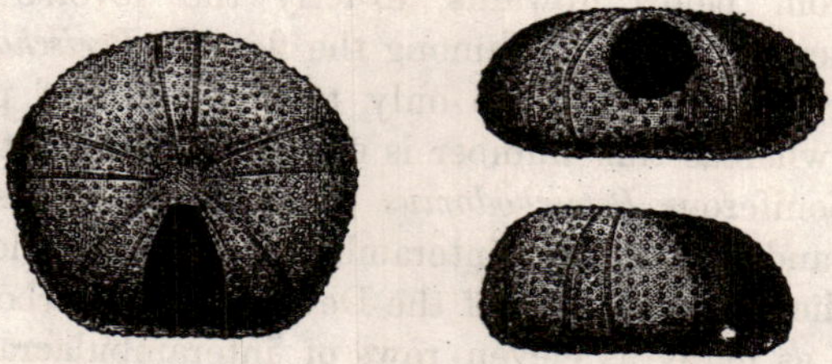

Fig. 133.—*Pygaster truncatus,* viewed from above, from behind, and from one side. Cretaceous.

areas carrying small, crenulated, and perforated tubercles. The mouth is inferior and central in position, toothed; the vent is on the upper surface, marginal, or inferior; the apical disc central, superior, and composed of the normal ten plates. The spines are short and awl-shaped.

All the members of this family are found in the Oolitic and Cretaceous rocks. The genus *Pygaster* (fig. 133) commences in the lowest Jurassic deposits, and extends upwards into the

Cretaceous. It is easily recognised by the great size of the anal aperture, which is oblong or pyriform in shape, and is placed upon the upper surface of the test. The genus *Hyboclypus* (fig. 134), again, is an aberrant type, since the mouth,

Fig. 134.—*Hyboclypus gibberulus*, viewed from above, from one side, and from below. Jurassic.

though inferior, is not central; the opening of the anus is placed in a longitudinal dorsal valley; and the posterior two ambulacral areas are disjoined from the anterior three, converging (as in the *Collyritidæ*) to two distinct points upon the upper surface of the test. Of the remaining genera of this family the two most important are *Galerites* (fig. 119) and *Discoidea* (fig. 132), constituting two nearly related types which are widely distributed in the Cretaceous rocks. *Holectypus* (fig. 136, A), lastly, is closely allied to *Discoidea*, and is principally confined to the Jurassic period.

In the small and singular family of the *Collyritidæ* (fig. 135), the test is usually ovate or heart-shaped; the ambulacral areas are narrow, and are "disjunct," meeting at two

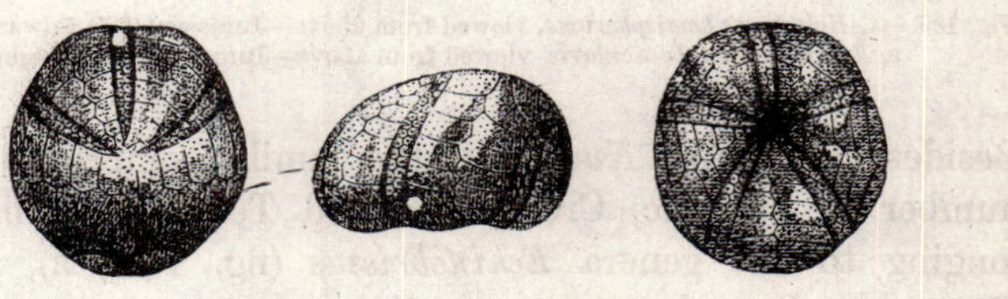

Fig. 135.—*Collyrites (Disaster) Eudesi*, viewed from above, from one side, and from below. Jurassic.

more or less remote points upon the upper surface; and the tubercles are small, perforated, and crenulated. The mouth is inferior and excentric; the anus is supra-marginal; and the elements of the apical disc are detached. This family is wholly confined to the Jurassic and Cretaceous rocks, and by far the most important genus contained in it is *Collyrites* (*Disaster* or *Dysaster*) itself.

In the small family of the *Echinonidæ* the test is oval, the ambulacral zones meeting at the apical disc, and the tubercles neither perforated nor crenulated; this last character separating the family from the *Echinoconidæ*. There are only four genital plates in the apical disc; the mouth is inferior, central, and toothless; and the anus is usually supra-marginal. The only fossil genus is the *Pyrina* of the Cretaceous rocks.

In the family of the *Echinobrissidæ* (*Cassidulidæ* in part) the shape of the test is variable; the tubercles are small, imperforate, uncrenulated, and surrounded by sunken *areolæ;* the spines are short and slender; and the ambulacral areas are petaloidal, lanceolate above, or sub-petaloidal. The mouth is inferior, sub-central, and edentulous; and the anus (fig. 136, B) is dorsal, opening in a valley, or supra-marginal.

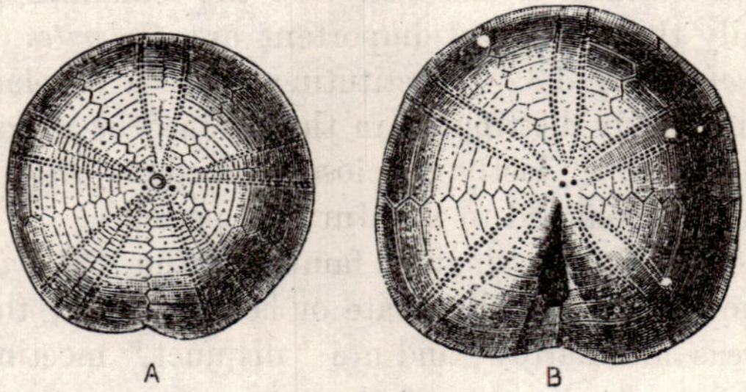

Fig. 136.—A, *Holectypus hemisphericus*, viewed from above—Jurassic (after Edward Forbes); B, *Echinobrissus clunicularis*, viewed from above—Jurassic (after Wright).

Besides the living *Nucleolites*, the family comprises a large number of Jurassic, Cretaceous, and Tertiary Urchins, belonging to the genera *Echinobrissus* (fig. 136, B), *Clypeus*, *Catopygus*, *Pygaulus*, &c.

In the family of the *Echinolampadæ* (*Cassidulidæ* in part) we have Urchins in many respects agreeing with the *Echinobrissidæ*, especially in their petaloidal or sub-petaloidal ambulacra, but having the anus marginal, infra-marginal, or supra-marginal, in no case opening into a dorsal sulcus.

Moreover, the ambulacral zones are expanded and petaloid in the immediate neighbourhood of the mouth, and so form an "oral rosette" in the centre of the base, while the tubercles are often perforated. Of the more important genera of this family, *Pygurus* is Oolitic and Cretaceous, *Conoclypus* and *Echinanthus* are Cretaceous and Tertiary, and *Echinolampas* is Tertiary and Recent.

In the family of the *Clypeastridæ* the test is usually circular or elliptical, generally depressed, the surface covered with small tubercles surrounded by sunken, ring-like *areolæ*, and carrying hair-like spines. The dorsal portions of the ambulacral zones are wide and petaloid, and the ambulacral pores are confined to the apical "rosette" thus formed. The mouth is inferior, central, and armed with teeth; and the anus is marginal or infra-marginal. The numerous types included in this family range from the Cretaceous to the present period, the most important genera being *Clypeaster*, *Scutella* (fig. 120), *Echinocyamus*, and *Echinarachnius*.

In the family of the *Ananchytidæ* (= *Echinocoridæ*, Wright) the test is usually ovate or heart-shaped, the mouth is toothless, excentric, and advanced forward, and the anus is marginal, infra-marginal, or supra-marginal (fig. 137, B). The ambulacral areas are not petaloidal, and the ambulacral pores are not confined to a dorsal rosette. There are four genital pores in the apical disc; the tubercles are small, perforated, and crenulated; and the spines are minute.

The chief genus of this family is *Ananchytes* itself (fig. 137, B), represented by many forms in the Cretaceous rocks. *Holaster*, abundantly represented in the Cretaceous, is very nearly allied to *Ananchytes*, but has the anus supra-marginal. In *Cardiaster*, again, which is also Cretaceous, there is the additional character of the existence of a "fasciole," which passes beneath the anus and is continued on the sides of the test. As will be seen immediately, the presence of "fascioles"—that is to say, of circumscribed bands of microscopic granules, occupying definite areas and positions on the test—is highly characteristic of the *Spatangidæ*, towards which *Cardiaster* thus makes an approach.

Lastly, in the family of the *Spatangidæ* or *Brissidæ*

("Heart-urchins") the test is oval, oblong, or commonly heart-shaped; the ambulacra (fig. 137, A) are petaloid, the anterior one unpaired, usually lodged in a groove or "sulcus," and thus rendering the skeleton bilaterally symmetrical.

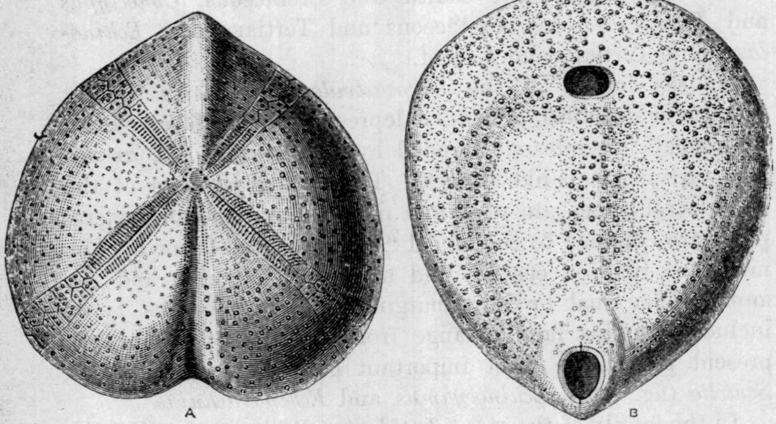

Fig. 137.—A, Upper surface of *Micraster coranguinum*, of the natural size—Cretaceous; B, Under surface of *Ananchytes ovata*, of the natural size—Cretaceous. (After Edward Forbes.)

The anus is posterior and supra-marginal. The mouth is inferior, excentric, and toothless. The tubercles are small, and support hair-like spines, but there are larger, crenulated, and perforated tubercles for supporting larger spines. As a rule, bands of microscopic tubercles known as "fascioles" (fig. 138) are present, and occupy different positions in different genera. Sometimes the "fasciole" surrounds the ambulacral rosette, when it is said to be "peripetalous;" sometimes it is "internal," surrounding the unpaired ambulacrum; sometimes it surrounds the sides, and is said to be "lateral;" at other times it runs round the test, and is termed "marginal;" and lastly, it may be limited to the base of the anal aperture, when it is termed "sub-anal." Of the

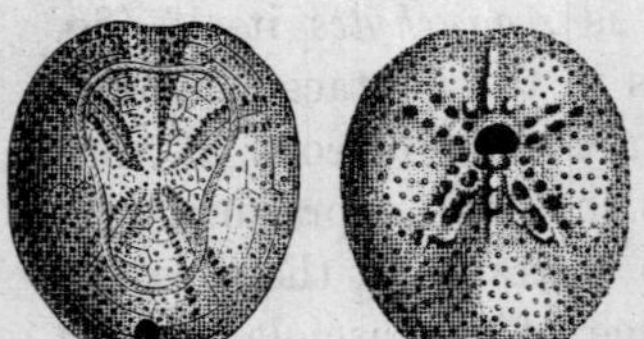

Fig. 138.—*Gualtieria Orbignyana*, viewed both from above and below. The left-hand figure shows the "fasciole" cutting the ambulacral rosette.

genera of the *Spatangidæ*, *Micraster* (fig. 137, A), *Echinospatagus*, *Epiaster*, and others are Cretaceous; *Hemiaster* and *Periaster* are both Cretaceous and Tertiary; *Gualtieria* (fig. 138), *Macropneustes* and *Pericosmus* are Tertiary; and *Spatangus*, *Schizaster*, *Eupatagus*, *Brissus*, and *Brissopsis* are represented both in Tertiary deposits and in our recent seas.

CHAPTER XIII.

ASTEROIDEA AND OPHIUROIDEA.

ORDER II.—ASTEROIDEA.

THE order *Asteroidea* or *Stellerida* comprises the ordinary "star-fishes," and is defined by the fact that *the body* (fig. 139) *is star-shaped or pentagonal, and consists of a central "disc," surrounded by five or more lobes or "arms." The arms are truly prolongations of the body, are hollow, and contain prolongations of the stomach in their interior. The arms are, further, grooved on their under surface for the reception of the ambulacral or water-vascular vessels. From these grooves the tube-feet are protruded in two or four rows. The integument (perisome) is leathery, but is more or less calcified by the development in it of plates, granules, and spines of carbonate of lime. The mouth is inferior in position, and is toothless. An anus is usually present, but may be absent.*

The two most striking features which distinguish the Star-fishes from the Sea-urchins are the star-like figure of the former, and the fact that the body is not enclosed in an immovable calcareous box or "test," as it is in the latter. The integument of the *Asteroidea* is, however, so richly provided with calcareous matter, that though more or less soft and flexible during life, it is quite capable of being preserved in a fossil condition. It is, of course, wholly with the calcareous secretions of the animal that the palæontologist has to deal; and we may therefore dispense with any further account of the soft parts, beyond what is contained in the above definition.

In their form the Star-fishes differ considerably, though in most the figure is markedly stellate. The animal consists of a central body or "disc," which gives off radiating processes or "arms," but the size of the disc is very different in different species, and the arms vary greatly in length and in number. In many living and extinct forms the "disc" is inconspicuous, and appears to be formed simply by the junction of the bases of the arms, which in this case are normally five in number. The living *Urasters* and *Cribellæ*, and the extinct *Palæasters* (fig. 139), may be taken as examples of this state of parts. In other forms, as in the Sun-stars (*Solaster*) and the extinct *Lepidasters* and *Plumasters*, the disc is very broad, exceeding or equalling the length of the arms in its diameter; whilst the rays vary in number, from eight or ten up to thirty or more. In the Cushion-stars (*Goniaster* and *Goniodiscus*), again, the body is pentagonal, disc-shaped, and flattened on the two sides, and the arms can only be recognised by the ambulacral grooves on the lower surface (fig. 140).

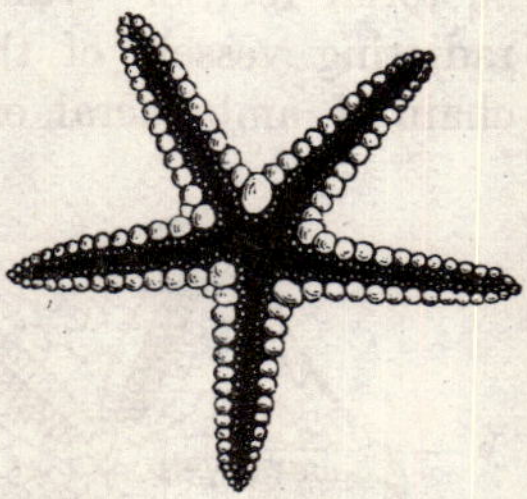

Fig. 139.—*Palæaster Niagarensis*, Hall. Lower Silurian.

On the upper surface of the body, placed nearly in the centre of the disc, is the aperture of the anus, when this is present; but the genera *Astropecten*, *Ctenodiscus*, and *Luidia* are destitute of a vent. Also on the upper surface is the "madreporiform tubercle," in the form of a spongy or striated disc placed at the angle of junction of two rays. It has the same function as in the Echinoids, serving to protect the entrance to the water-vascular system. Ordinarily there is a single madreporiform tubercle, but in some genera there are two, three, or more tubercles; and there seems in some cases to be a correspondence between the number of the arms and the number of madreporic plates.

Placed in the centre of the lower surface is the mouth, at the angles of which are the so-called "oral plates" (fig. 140). Radiating from the mouth are a series of furrows, varying in number with the arms, and termed the "ambulacral grooves." Each ambulacral groove is continued along the lower surface of one of the arms, tapering gradually towards the extremity of the latter. The floor of each groove is constituted by a double row of minute calcareous pieces—the "ambulacral ossicles"—which are movably articulated to one another at their inner ends. At the bottom of each groove is lodged one of the radiating canals of the water-vascular system or ambulacral system, from which are given off the rows of suctorial feet, or "tube-feet." It follows from this that the radiating vessels of the ambulacral system are *outside* the chain of ambulacral ossicles, so that these latter are to be regarded as an *internal* skeleton, and they do not correspond with any part of the skeleton of Echinoids[1]—at least they do not correspond with the perforated ambulacral plates of the Sea-urchins. The ambulacral ossicles, however, of the Star-fishes are of such a form that by their apposition an aperture or pore is formed between each pair. By means of these pores (fig. 140, *a*) the tube-feet communicate with a series of little bladders, or "ampullæ," placed *above* the chain of ossicles. These perforations, however, do not correspond with the perforated plates of the Echinoid test, and the tube-feet of the Star-fishes pass through no "poriferous" plates on their way to the exterior.

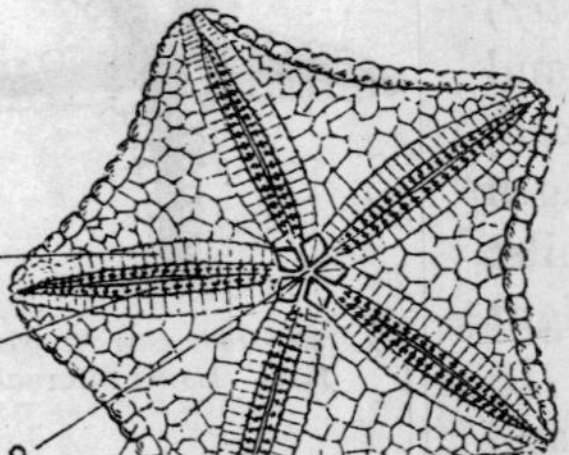

Fig. 140.—Diagram of a Star-fish (*Goniaster*), showing the under surface, with the mouth and ambulacral grooves. *a*, Ambulacral ossicles, with the ambulacral pores between them; *b*, Adambulacral plates, bounding the ambulacral grooves; *m*, Marginal plates (wanting in many species); *o*, Oral plates, placed at the angles of the mouth.

This may be rendered more intelligible by examining a section of the arm of a Star-fish from which the soft parts have been removed (fig. 141). In such a section the ambulacral ossicles (*a*, *a*) are seen in the centre of the lower

[1] The structures in the *Echinus*, which are truly homologous with the ambulacral ossicles of the *Asteroidea* and *Ophiuroidea*, are the so-called "auriculæ."

surface, united along the middle line by their inner extremities. They are so placed as to form a kind of elongated pent-house, and immediately beneath the line where the ossicles of one side are articulated with those of the other side is placed the ambulacral vessel (*b*). Superficial to this, again, is a nerve-cord; so that the whole chain of ambulacral ossicles is placed in the midst of the soft parts of the animal, and is thus clearly an internal skeleton. At their outer extremities the ambulacral ossicles are articulated by the intervention of the "adambulacral plates" (fig. 140, *b*), with plates belonging to the external or integumentary skeleton, to be immediately described. As before said, the shape of the ambulacral ossicles is such that a pore is formed by the

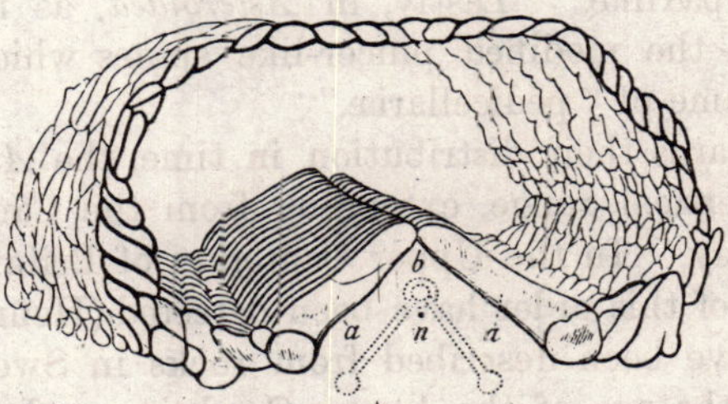

Fig. 141.—Section of the ray of *Uraster rubens*. *a*, *a*, Ambulacral ossicles; *b*, Position of the ambulacral vessel; *c*, *c*, Plates of the external skeleton; *n*, Nerve-cord. The dotted lines show the tube-feet proceeding from the ambulacral vessel.

apposition of each pair; and by these apertures each tube-foot communicates with a vesicle placed internal to the chain of ossicles. It will be seen, however, that the tube-feet (indicated by the dotted lines in the figure) do not pass through these apertures, or through any other pores of the skeleton, on their way to the surface. The "poriferous zones" of the Sea-urchins are part of the external skeleton, and are not represented in the Star-fishes. On the other hand, the integumentary skeleton in the Star-fishes *is absent* along the ambulacral areas, or along the areas occupied by the ambulacral grooves.

Leaving the ambulacral ossicles or internal skeleton of the *Asteroidea*, we come now to the integumentary skeleton. This consists of a vast number of small calcareous pieces, or "ossicles," united together so as to form a species of chain-armour. The ossicles are generally united with one another in a reticulated manner, and the interspaces between them are filled by the coriaceous integument. In some genera there is a single or double row of large plates round the borders of the disc and arms (fig. 140, *m*). These are termed the "marginal plates." The integument in the Star-fishes is also furnished with spines, tubercles, and granules of calcareous matter. The spines vary in their development and in their position in different Star-fishes; but there is commonly a row of spines along each side of each of the ambulacral grooves. In some genera (as in *Solaster, Luidia, Ctenodiscus,* &c.) there are spines the summits of which carry bunches or tufts of minute calcareous processes. These are termed "paxillæ." Lastly, in *Asteroidea,* as in *Echinoidea,* there are the modified pincer-like spines which are known by the name of "pedicellariæ."

As regards their distribution in time, the *Asteroidea* have a long vertical range, extending from the Cambrian to the present day. In the Upper Cambrian of Britain, remains of members of this order have been detected (Henry Hicks), and others have been described from rocks in Sweden, believed to be of the age of the Lower Cambrian. In the Silurian seas Star-fishes were comparatively abundant, and their remains are found, though rarely, in the subsequent Devonian and Carboniferous formations. In the Secondary deposits, and more especially in the Jurassic and Cretaceous, Star-fishes, often belonging to existing genera, are far from uncommon; and other types, closely related to living forms, are found in the Tertiary.

As regards living forms, the order *Asteroidea* may be divided into the following five families:—

Family 1. *Asteriadæ* or *Asterocanthiidæ.*—Four rows of ambulacral feet.

Fam. 2. *Astropectinidæ.*—Two rows of ambulacral feet; back flattish, netted with tubercles, which carry radiating spines at the tip ("paxillæ").

Fam. 3. *Oreastridæ.*—Two rows of ambulacral feet; skin granular, pierced by minute pores.

Fam. 4. *Asterinidæ.*—Two rows of ambulacral feet; body discoidal or pyramidal, sharp-edged : skeleton of imbricate plates ; dorsal wart single, rarely double.

Fam. 5. *Brisingidæ.*—Arms long and rounded, sharply marked off from the disc. Ambulacral grooves not reaching the mouth ; two rows of ambulacral feet.

Owing to the imperfect state of preservation in which the remains of fossil Star-fishes are usually found, it is difficult or impossible to speak definitely as to the precise affinities of many of the extinct species. The *Brisingidæ* are not certainly known to occur as fossils, and the *Asterinidæ* are but poorly represented ; the great majority of fossil Star-fishes being thus referable to the families of the *Asteriadæ*, the *Astropectinidæ*, and the *Oreastridæ*. It should be borne in mind, however, that the singular family of the *Brisingidæ*, forming an intermediate group between the *Asteroidea* and the *Ophiuroidea*, may possibly prove to be of very ancient origin, and to be represented by such Silurian and Devonian types as *Protaster*, *Eugaster*, *Tæniaster*, &c., which we may provisionally consider among the *Ophiuroids*. Lastly, as Star-fishes are always of rare occurrence as fossils, and as they are therefore of little importance to the general student, we shall content ourselves here with simply glancing at the more important types which have made their appearance in the successive geological periods, briefly noting some of the special characters of the more interesting ancient forms.

In the Silurian period the genus *Palæaster* (figs. 142 and 143, B) is the most important. In this genus we have Star-fishes in which the body is five-armed, the disc being very small ; the ambulacral grooves are placed on the lower surface of the arms, and are furnished with two rows of ambulacral ossicles and pores, bordered on each side by a row of "adambulacral plates," which are, in turn, bordered by a series of "marginal plates." On the dorsal surface are three or more rows of plates, which are stated to fit closely together, instead of forming pores by their junction. The genus *Palæaster* comprises some species of considerable size, and ranges from the Lower Silurian to the Devonian. The Silurian genus *Urasterella* (= *Stenaster*) is in many respects like *Palæaster*, but the ambulacral grooves are margined by a row of adambulacral plates only, without a second row of marginal plates. *Petraster*, also Silurian, has an incomplete series of disc-plates between the adambulacral and marginal rows of plates, but is otherwise almost identical with *Palæ-*

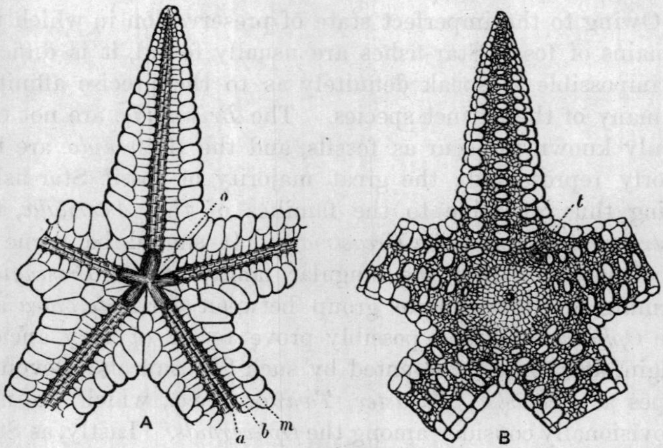

Fig. 142.—*Palæaster eucharis*, Devonian (after Hall). A, Under side of a specimen, four of the arms being cut short; B, Upper side of the same. *a*, Ambulacral ossicles, lying in the ambulacral grooves; *b*, Adambulacral plates; *m*, Marginal plates; *o*, One of the oral plates; *t*, Madreporiform tubercle.

aster; whilst the Silurian and Upper Cambrian *Palasterina* (fig. 143, A) has the disc still more extensively developed, and is further distinguished by the fact that the plates of the adambulacral series, which are placed at the angles of the

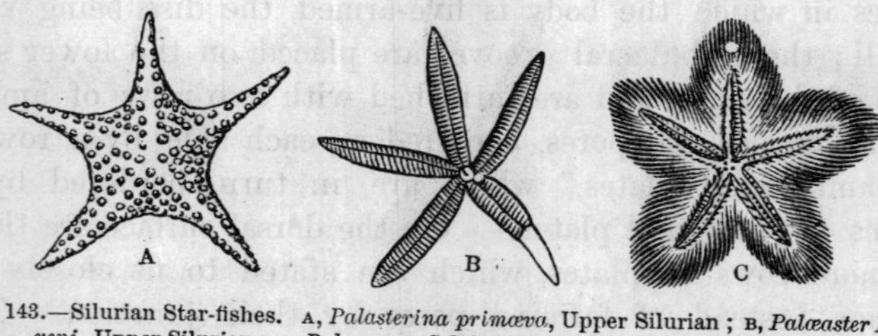

Fig. 143.—Silurian Star-fishes. A, *Palasterina primæva*, Upper Silurian; B, *Palæaster Ruthveni*, Upper Silurian; C, *Palæocoma Colvini*, Upper Silurian. (After Salter.)

oral aperture, are large and triangular. The genera above mentioned can hardly be grouped with any existing family of the *Asteroidea*, as they possess several striking peculiarities of their own, more especially the presence of a row of "adambulacral plates," exceeding in size the ordinary plates of the skeleton. They might thus with propriety form a distinct family, which might be termed *Palasteriadæ*, and which would to some extent fill up the interval between the families of the *Asteriadæ* and *Astropectinidæ*. *Urasterella* may also be best referred to the same group, especially as the plates upon the ventral surface are closely fitted together, though it wants "marginal plates," and in the characters of the upper surface presents points of resemblance to the *Urasters*.

We are not, however, without ancient representatives of existing families of Star-fishes. Thus, the Upper Silurian *Lepidaster* and *Trichotaster*, and the Devonian *Helianthaster* represent the living Sun-stars (*Solaster*), with their wide disc

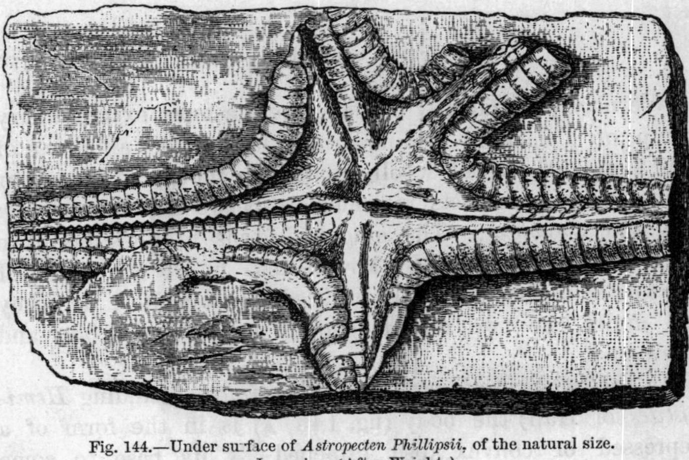

Fig. 144.—Under surface of *Astropecten Phillipsii*, of the natural size. Jurassic. (After Wright.)

and numerous arms; Devonian and Carboniferous species of the living genus *Astropecten* (or *Asterias*) have been detected, and the Upper Silurian *Palæocoma* of Salter (fig. 143, c) may possibly be an old form of the living "Bird's-foot Star-fishes" (*Palmipes*).

Leaving the Palæozoic period, the Secondary deposits are known to contain a large number of Star-fishes, which need not detain us here, as they are almost wholly referable to living genera, though the species are distinct. In the Trias (Muschelkalk), we have the genus *Pleuraster*, doubtfully separable from *Astropecten*. In the Jurassic period we have the earliest representatives of the living genera *Uraster*, *Solaster*, *Luidia*, *Astrogonium*, and *Goniaster*. The existing genus *Astropecten* (fig. 144), easily recognised by the great spine-bearing marginal plates, is largely represented; while the extinct *Tropidaster* forms a link between this and *Uraster*; and *Plumaster*, also extinct, is a near ally of the recent *Luidia*.

In the Cretaceous rocks almost all the known forms belong to existing genera (such as *Goniaster*, *Stellaster*, *Astrogonium*, *Palmipes*, and *Oreaster*); while in the Tertiary deposits we meet only with the generic types of the present day.

Agelacrinidæ.

We may provisionally consider here a most extraordinary group of Palæozoic Echinoderms, the precise affinities of which are at present wholly uncertain, though they appear to be in some respects intermediate between the *Asteroidea* and the *Cystoidea*. The singular forms in question have been grouped together by Mr Billings under the name of *Edrioasteridæ*, but are better entitled *Agelacrinidæ*; and they fall under the two related generic types *Agelacrinus* and *Edrioaster*.

In the genus *Agelacrinus* or *Agelacrinites* (including *Hemicystites* of Hall) the body (fig. 145, A) is in the form of a depressed or convex disc, attached by its base to some foreign body. The upper surface of the disc (which is really the *ventral* surface) is covered with numerous small calcareous plates, which may or may not overlap in an imbricating manner, and exhibits five curved "arms," which radiate from the centre. The rows of plates forming the arms are so disposed, in some instances at any rate, as to leave between them distinct "pores," penetrating the thickness of the test; so that the arms clearly correspond with the "ambulacral grooves" of the Star-fishes. The opening of the mouth appears to be placed in the centre of the five arms; and in one of the spaces between the arms is situated a little pyramid of from five to nine calcareous plates, which forms the valvular "ovarian aperture" (fig. 145, A, *o*). The genus is wholly Silurian, Devonian, and Carboniferous, and it has been generally placed under the order *Cystoidea*. It differs, however, from these in its total absence of a stalk of attachment, and in the possession of ambulacral pores, perforating the test. On the other hand, it recalls the *Asteroidea* in the structure of the ambulacral grooves; while the sometimes articulated, sometimes imbricated interradial (interambulacral) plates recall to mind the *Perischoechinidæ*; though we know of no parasitic or sessile examples of the Star-fishes or Sea-urchins.

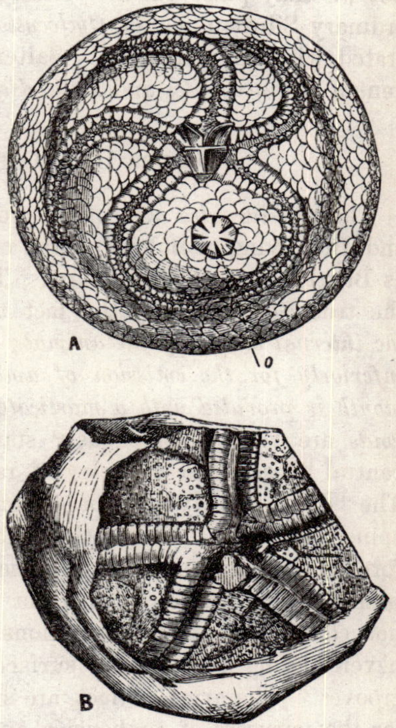

Fig. 145.—A, Upper surface of *Agelacrinus Cincinnatiensis*, enlarged two and a half diameters. Lower Silurian (after Hall). B, Upper surface of an imperfect specimen of *Edrioaster Bigsbyi*, of the natural size. Lower Silurian (after Billings). *o*, Ovarian pyramid.

In the allied genus *Edrioaster* (fig. 145, B) the body is sessile, circular, discoid, and convex, and covered with irregular polygonal plates. The mouth is large and somewhat pentagonal. Radiating from the mouth are five ambulacral grooves, each formed of a double series of plates ("ambulacral ossicles"), with two pores between each pair of plates. There are thus four rows of ambulacral tube-feet, as shown by the pores, which penetrate the entire thickness of the test. An ovarian aperture appears to be present in one of the interradial spaces. This genus is only known as occurring in the Lower Silurian, and it is clearly very closely allied to *Agelacrinus*, though it has not yet been demonstrated to have been attached to foreign bodies; and its other characters perhaps entitle it to generic distinction. Whether or not we may place in the vicinity of *Agelacrinus* the extraordinary Silurian genus *Cyclocystoides*, cannot at present be stated with any certainty. Salter would also refer here the genera *Echinocystites* and *Palæodiscus*.

ORDER III.—OPHIUROIDEA.

The *Ophiuroidea* are often grouped with the *Asteroidea*, and the living members of the order are known commonly as Brittle-stars and Sand-stars. They are distinguished from the true Star-fishes by the fact that *the "disc" contains all the internal organs of the animal; the "arms" are not grooved inferiorly for the emission of ambulacral tube-feet; and the mouth is provided with a masticatory apparatus.* The *Ophiuroids* are very conspicuously star-shaped, and consist of a central "disc" and a series of radiating "arms" (fig. 146). The "disc" is truly disc-shaped, and is covered with granules, spines, or scales. From the disc proceed the arms, in the form of long and slender processes, which may be simple or branched, but which differ from the arms of Star-fishes in not containing any prolongations from the stomach, and in never having their under surfaces furrowed by ambulacral grooves. The arms, in fact, are special processes superadded for the purposes of prehension and locomotion, and rendered necessary by the fact that the ambulacral system takes no part in the function of locomotion, as it does in the Star-fishes. A madreporiform tubercle, however, is present, and is placed on the inferior surface of the body, being commonly concealed by one of the plates surrounding the mouth. The mouth, as in the Star-fishes, is placed in the centre of the lower surface of the disc; but the stomach terminates blindly; and there is, therefore, no anal aperture.

Each arm is furnished with an internal and an external

skeleton. The internal skeleton consists, as in the Starfishes, of a chain of "ambulacral ossicles," placed along the centre of the arm. The ossicles are, however, now amalgamated in pairs, each coalescing with its fellow on the opposite side; so that in place of a double row of movably articulated ossicles, we have a single row of bilaterally-symmetrical pieces.

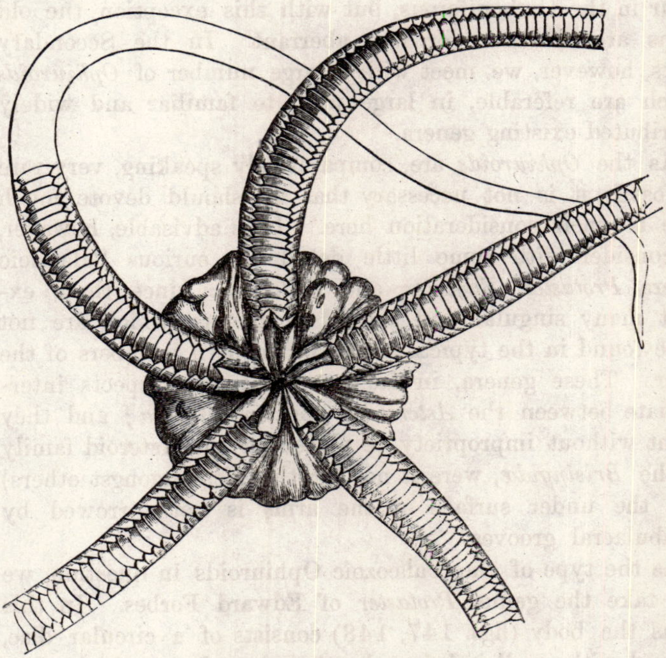

Fig. 146.—Under surface of *Ophioderma* (*Ophioglypha?*) *Gaveyi*, of the natural size. Jurassic (Lias). (After Wright.)

The external skeleton of the arms is composed of four rows of plates, one on the dorsal surface, one on the ventral, and one on each lateral surface. The lateral plates generally carry more or less developed spines. In the extinct genera *Eugaster* and *Protaster* (fig. 148) the plates of the ventral row are double; and in *Ptilonaster* (fig. 148) there are four ventral rows of plates. The disc, as before said, has a well-developed external skeleton of scales, granules, and spines; but there are never any of those modified spines which are known as "pedicellariæ," and which occur in the Asteroids and Echinoids.

As regards their distribution in time, the *Ophiuroids* make their first appearance in the Lower Silurian, and they are represented by various ancient types in the Upper Silurian and Carboniferous. The living genus *Ophiura* is said to occur in the Carboniferous, but with this exception the old forms are all more or less aberrant. In the Secondary rocks, however, we meet with a large number of *Ophiuroids* which are referable, in large part, to familiar and widely distributed existing genera.

As the *Ophiuroids* are, comparatively speaking, very rare as fossils, it is not necessary that we should devote much time to their consideration here. It is advisable, however, to consider with some little detail the curious Palæozoic genera *Protaster*, *Eugaster*, and *Ptilonaster*, since these exhibit many singular and special characters, which are not to be found in the typical and more modern members of the order. These genera, in fact, are in many respects intermediate between the *Asteroidea* and *Ophiuroidea*; and they might without impropriety be placed in the Asteroid family of the *Brisingidæ*, were it not for the fact (amongst others) that the under surface of the arms is not furrowed by "ambulacral grooves."

As the type of the Palæozoic Ophiuroids in question, we may take the genus *Protaster* of Edward Forbes. In this genus the body (figs. 147, 148) consists of a circular disc, covered with small imbricated calcareous plates, which gives origin to five long and flexuous arms. The chief peculiarity of the genus is to be found in the structure of the arms, which show the peculiarity that they possess two rows of ventral plates, instead of one, and that these plates are opposite to one another (Salter), or very slightly alternating (Hall). These plates, moreover, are so disposed as to give origin to a series of distinct *pores* (fig. 148, E). If, as would appear to be the case, *Tæniaster* of Billings be really identical with *Protaster*, then the genus dates from the Lower Silurian and ranges into the Devonian. In the Devonian genus *Eugaster* (fig. 148, A and B) the structure is essentially the same as in *Protaster*, but the disc is prolonged along the bases of the arms, and the plates of the disc are articulated by their edges, and do not overlap. In the Devonian *Ptilonaster* (fig. 148, E), again, we have a form fundamentally similar to the preceding, but having *four* rows of perforated plates on the lower surface of the arms. The chief question as to the affinities of the types just alluded to turns upon the true nature of the perforated plates, in double or quadruple series, placed along the lower surface of the arms These have generally been regarded as identical with the "ventral plates" of the arms of the ordinary Ophiuroids — structures which belong wholly to the integumentary skeleton, and which are not represented in the Asteroids. It does not seem certain, however, that this is the true nature of the perforated plates of the under surface of the arms of the genera above alluded to; and it is not impossible that they are really the *internal* "ambulacral ossicles" of the arms, exposed to view by the destruction of the superficial "ventral plates" during fossilisation. Should this turn out to be the case, the structure of *Protaster* and its allies is not so abnormal. If, on the other hand, these perforated plates are really the "ventral plates," then these genera exhibit the peculiarity that the under surfaces of the arms are pierced by "ambulacral pores."

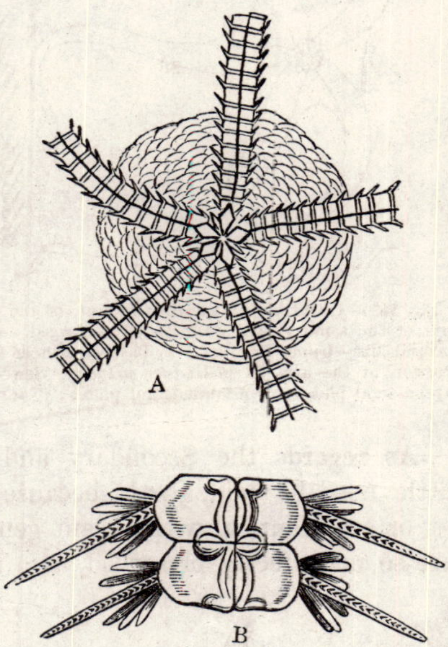

Fig. 147.—*Protaster Sedgwickii*. Upper Silurian. A, Disc and bases of the arms, magnified; B, Portion of an arm greatly enlarged.

The only other Palæozoic type which needs notice here is

Eucladia, described by Dr Henry Woodward from the Upper Silurian, which apparently belongs to the section of the *Ophiuroids* represented at the present day by *Euryale*, in which the arms, instead of being simple, are bifurcated.

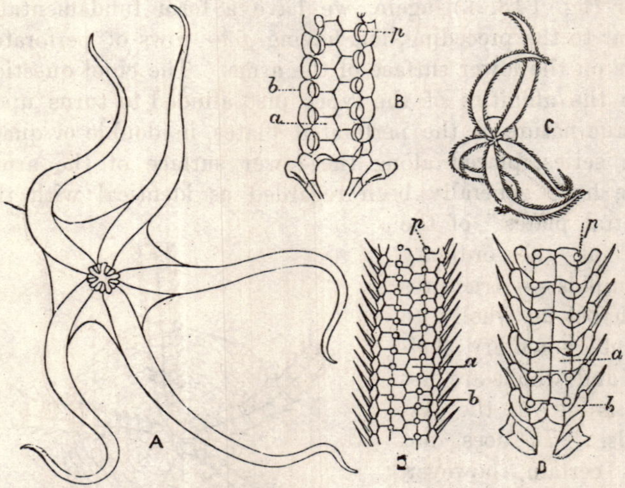

Fig. 148.—A, Outline of *Eugaster Logani*, of the natural size—Devonian. B, Base of an arm of the same viewed from below, enlarged. C, Outline of *Protaster Forbesi*, of the natural size—Upper Silurian. D, Base of arm of same, viewed from below, enlarged. E, Portion of the arm of *Ptilonaster princeps*, viewed from below, enlarged—Devonian: *a*, Ambulacral plates; *b*, Adambulacral plates; *p*, Pore. (After Hall.)

As regards the Secondary and Tertiary *Ophiuroids* very little need be said, partly because they approximate closely to, or are identical with, recent generic types, or because they are so imperfectly preserved, as a rule, that their determina-

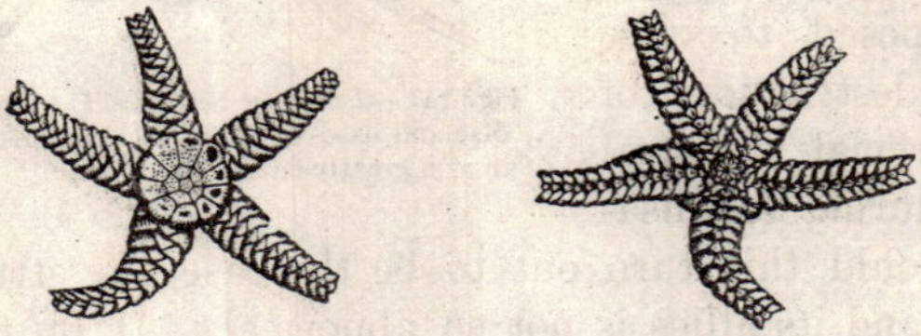

Fig. 149.—*Aspidura loricata*. Muschelkalk.

tion is exceptionally difficult. In the Trias appear the genera *Acroura* and *Aspidura* (fig. 149), the latter being supposed to be confined to this period and to be characteristic of the Muschelkalk. In the Jurassic and Cretaceous periods, as in the Tertiary, remains of Ophiuroids are by no means very uncommon, and they are generally referred partly to extinct types (such as *Acroura*). Most of them, however, are apparently referable to well-known existing genera, such as *Ophioderma* (fig. 146), *Ophiocoma* (= *Ophiurella*), *Ophioglypha* (= *Ophiura*, Forbes), *Ophiolepis*, and *Amphiura*.

CHAPTER XIV.

ORDER IV.—CRINOIDEA.

THE Crinoids or Sea-lilies are *Echinodermata*, in which *the body is fixed, during the whole or a portion of the existence of the animal, to the sea-bottom by means of a longer or shorter, jointed, and flexible stalk.* The body is distinct, composed of articulated calcareous plates, bursiform, or cup-shaped, and provided with branched arms, which are typically from five to ten in number, are independent of the visceral cavity, and are grooved on their upper surfaces. (The position of the body being reversed, the *upper* surface is *ventral*; whilst the *dorsal* surface is *inferior*, and gives origin to the pedicle.) The tubular processes, however, which are given off from the radiating ambulacral canals of the *Crinoidea*, unlike those of the *Echinoidea* and *Asteroidea*, are not used in locomotion, but have probably a respiratory function. The mouth is central, and looks upwards, an anal aperture being sometimes present, sometimes absent. The ovaries are situated beneath the skin in the grooves on the ventral surfaces of the arms or pinnules, as are also the ambulacral or respiratory tubes. The arms are furnished with numerous lateral branches or "pinnulæ." The embryo is "free and ciliated, and develops within itself a second larval form, which becomes fixed by a peduncle" (Huxley).

If we take such a living Crinoid as *Rhizocrinus* (fig. 150), we shall be able to arrive at a comprehension of the leading characters of this order. *Rhizocrinus* is one of those Crinoids which is permanently rooted to some foreign object by the base of a stalk which is composed of a number of calcareous pieces or articulations. In some cases (as in *Apiocrinus*) the base of the stem or "column" is considerably expanded. In other cases the column is simply "rooted by a whorl of terminal cirri in soft mud" (Wyville Thomson). The joints of the column are movably articulated to one another, the joint-surfaces often having a very elaborate structure, so that the entire stem possesses in the living state a greater or lesser amount of flexibility. Each joint is perforated centrally by a canal, which has been very inappropriately termed the "alimentary canal," but which in truth has nothing to do with the digestive system of the animal. At the summit of the stem is placed the body, which is termed the "calyx," and which is usually more or less cup-shaped, pyriform, bursiform, or discoidal. The calyx exhibits two surfaces, a dorsal and a ventral, of which the dorsal is composed, wholly or in part, of calcareous plates articulated by their margins, whilst the former is, in the living forms, composed of a more or less leathery integument, strengthened by the deposition in it of numerous small plates of carbonate of lime; whereas in many extinct forms it too is composed of articulated cal-

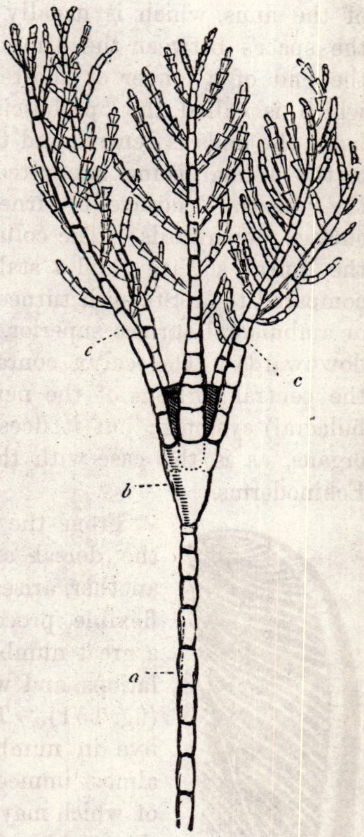

Fig. 150.—Crinoidea. *Rhizocrinus Lofotensis*, a living Crinoid (after Wyville Thomson), four times the natural size. *a*, Stem; *b*, Calyx; *c*, *c*, Arms.

careous plates. The ventral surface exhibits the aperture of the mouth, which may be subcentral or may be very excentric, and which in many extinct forms is wholly concealed from view. The ventral surface also exhibits the aperture of the anus, which is usually placed excentrically in one of the spaces between the arms, and which is often carried at the end of a longer or shorter tubular eminence or process, which is called the "proboscis." Sometimes, on the other hand, the anus is central, and the mouth is excentric. Owing to the animal being supported on a stalk, it is evident that the "ventral" surface is turned upwards, and the "dorsal" surface downwards. The column springs from the centre of the dorsal surface; and a stalked Crinoid may therefore be compared to a Star-fish turned upside down, with its lower or ambulacral surface superior, and its dorsal surface looking downwards. The calyx contains the digestive canal, and the central portions of the nervous and water-vascular (ambulacral) systems; but it does not contain the reproductive organs, as is the case with the visceral cavity of the other Echinoderms.

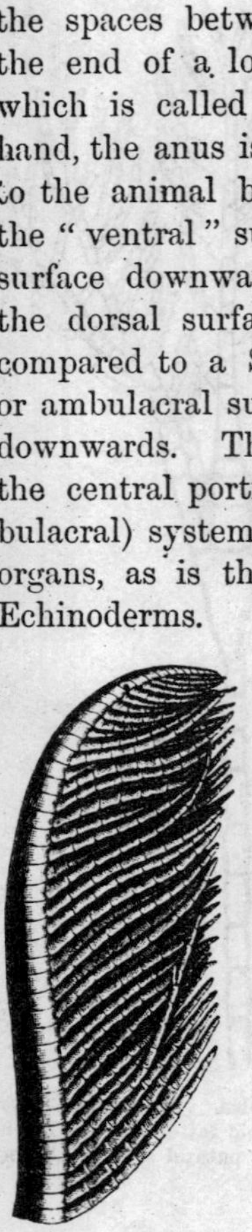

Fig. 151.—Portion of an arm of *Platycrinus*, showing the lateral pinnulæ.

From the margins of the calyx, where the dorsal and ventral surfaces join one another, arises a series of longer or shorter flexible processes, which are composed of a great number of small calcareous articulations, and which are termed the "arms" (fig. 151). The arms are usually primarily five in number, but they generally divide almost immediately into two branches, each of which may again subdivide; the branches thus produced perhaps again dividing, until a crown of delicate graceful filaments is formed. The arms carry smaller lateral branches or "pinnulæ" on both sides; and they are not hollow like the arms of the Star-fishes, nor do they contain any prolongations of the stomach. The upper surface of the arms and pinnulæ is covered with a soft membrane, and below this are placed the reproductive organs.

The generative organs are therefore not placed within the calyx, and it follows of necessity that there is no generative opening or "ovarian aperture" in the walls of the calyx. The ventral surfaces of the arms and pinnulæ are furnished with grooves, which in the living species are seen to be covered with vibratile cilia. The brachial grooves coalesce till they constitute five (sometimes fewer) primary grooves, which are continued from the bases of the arms to the mouth. The action of the cilia gives rise to a constant current of sea-water, bearing organic matter in suspension; and this current proceeds from the brachial grooves to the mouth. In this way the animal obtains its food. As the bases of the arms are separated from the mouth by an intervening space, it follows that the brachial grooves are continued over the ventral surface of the calyx, till they reach the oral opening.

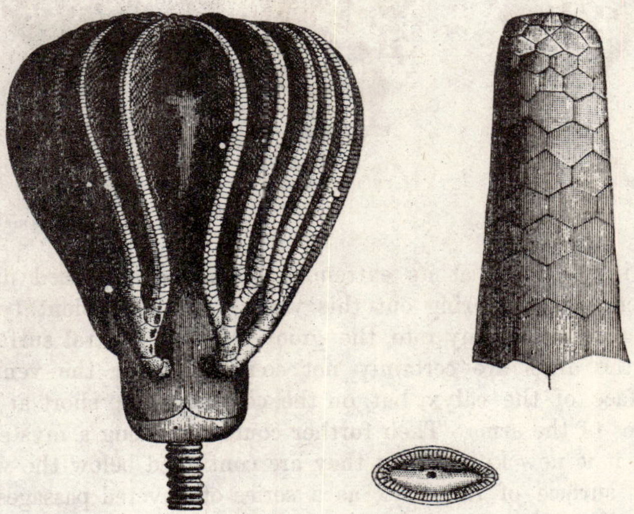

Fig. 152.—*Platycrinus tricontadactylus.* Carboniferous. The left-hand figure shows the calyx, arms, and upper part of the stem, and the figure next this shows the surface of one of the joints of the column. The right-hand figure shows the proboscis.

There is no doubt that it is by the above arrangement that the living Crinoids obtain their food, and the mechanism seems to have been essentially the same in many extinct species. In the Palæozoic Crinoids, however, there seems to have been a modification of this arrangement. In these forms, as in *Actinocrinus* (fig. 153), the arms have much the structure of those of the recent Crinoids, and are deeply grooved on their ventral surfaces. The ventral surface of the calyx, however, exhibits no central aperture, but only a proboscidiform tube, which arises from one of the inter-radial spaces (*i.e.*, one of the intervals between two of the arms). This tube is often of great length, and a good deal of controversy has taken place as to its nature. Without entering into the conflicting views upon this subject, it may be stated that the preponderance of authority is overwhelmingly in favour of the view that this "proboscis" is an anal tube,

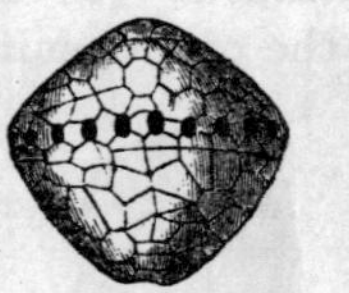

Fig 153.—Calyx of *Actinocrinus rotundus*.

Fig. 154.—Calyx of *Actinocrinus Konincki*.

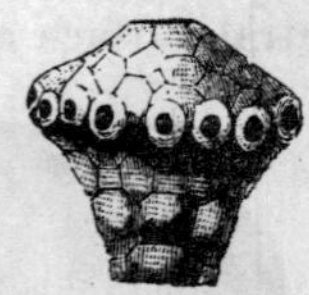

Fig. 155. — Calyx of *A. Verneuillanus*. The arms are wanting, and the apertures at their bases are seen.

having the vent at its extremity, all analogies based upon recent forms bearing out this view. In the ancient types in question, at any rate, the grooves on the ventral surfaces of the arms are certainly not continued over the ventral surface of the calyx, but, on the contrary, stop short at the bases of the arms. Their further course was long a mystery; but it is now known that they are continued below the ventral surface of the calyx as a series of covered passages or tunnels, the external apertures of which are placed at the points where the arms spring from the disc (see figs. 153-155). These covered channels are simply roofed over by the calcareous integument of the calyx; and they converge to a central point in the middle of the ventral surface of the disc. Here is placed the mouth, concealed by the calcareous plates of the perisome. In this point of their structure, therefore, there is an extremely important distinction between the older types of Crinoids and the later ones, though the process by which the microscopic organisms which serve as food are collected from the surrounding water and conveyed to the mouth, seems to have been in both cases essentially the same. Moreover, even in the living *Antedon*, as shown by Dr Carpenter, the true mouth is situated at a little distance below the apparent mouth, as formed by the point of convergence of the brachial furrows; so that if we imagine these furrows to be roofed over by calcareous plates, where they cross the ventral surface of the disc, we should have a condition of parts closely resembling what we find in the Palæocrinoids.

The stalked or "pedunculate" Crinoids of the present day are few in number, and are mostly inhabitants of the deep sea. We find, however, various and widely distributed representatives of another group of Crinoids — namely, the "sessile" Crinoids, all of which are generally known as "Feather-stars." In all these, such as the living *Comatula*, *Antedon*, *Actinometra*, &c., and the extinct *Saccosoma* (fig. 156) and *Solanocrinus*, the animal is only *stalked* when young, and in its adult condition leads a free life. The young form in the members of this group is supported upon a jointed calcareous column, by which it is fixed to some foreign object; and at this stage it in no respect differs from the ordinary stalked Crinoids. At a certain period of its existence, however, the calyx drops off its column, and becomes a locomotive animal. It now has a near resemblance to one of the Brittle-stars (*Ophiuroidea*); but is distinguished, not only by its developmental history, but also by its internal and skeletal structure, by the possession of lateral pinnæ to the arms, and in having the reproductive organs situated external to the body proper. In the Feather-stars, moreover, the dorsal surface of the disc, at the point where the column was originally inserted, carries a series of jointed filaments or "cirri," by which the animal can moor itself to any foreign object. These may be regarded as homologous with the "side-arms" of the column of certain Crinoids. When the animal is thus temporarily moored by its dorsal cirri, it is placed in the ordinary position held by the Crinoids—namely, with the mouth and ventral surface of the disc looking upwards. When creeping about, on the other hand, by means of the long and flex-

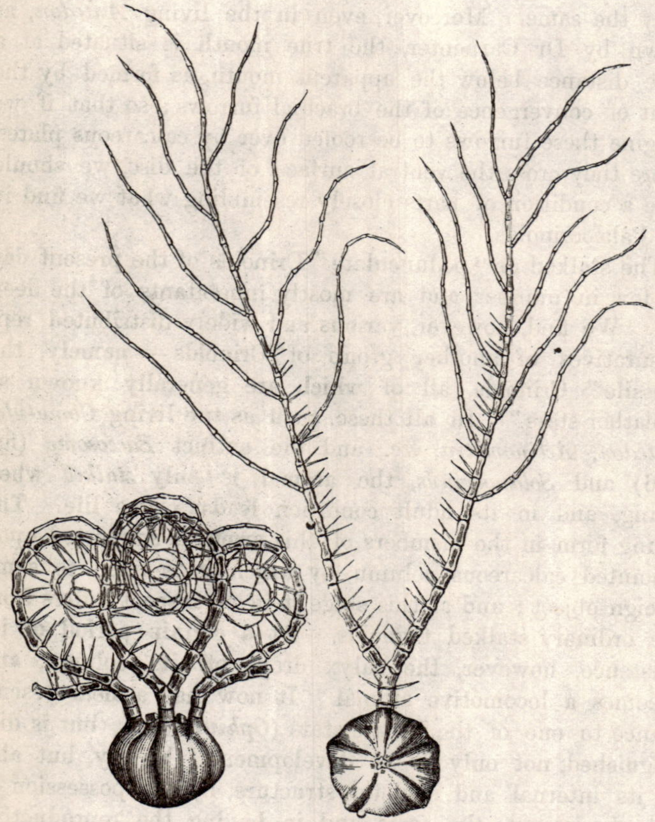

Fig. 156.—*Saccosoma pectinata*, a free Crinoid. Jurassic.

ible arms, the animal occupies the position held by the Star-fishes and Ophiuroids—namely, with the mouth and ventral surface of the disc directed downwards, or towards the ground.

Having now given a general account of the structure of the Crinoids, it remains to consider some of their parts in greater detail. In the first place, as regards the "column," we find that the stem of attachment is composed of a great number of separate plates or "articulations" placed one above the other, and so jointed to one another that whilst

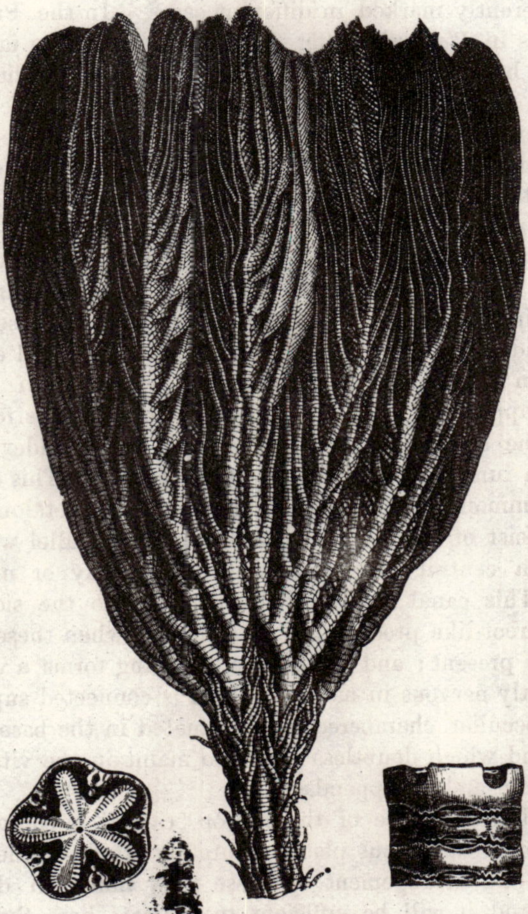

Fig. 157.—*Pentacrinus fascicul[illegible]*, showing the form of the column, and one of the facets of a joint. The central figure shows the arms, and the summit of the column with side-arms.

the amount of movement between any two pieces must be very limited, the entire column acquires more or less flexibility. In the Palæozoic Crinoids, with few exceptions, the column was round; but in *Platycrinus* it is oval or elliptical (fig. 152). In the genera *Pentacrinus* (fig. 157) and *Extracrinus* the column is pentagonal in outline; but much less markedly so in the former than in the latter genus. The joints articulate with one another by surfaces or facets which are differently marked in different cases. In the Palæozoic forms, as in *Platycrinus* (fig. 152), the articulating facets are marked by more or less numerous striæ which radiate from near the centre of the joint. In most of the Mesozoic genera, on the other hand, as in *Pentacrinus* (fig. 157), the articulating facets are united by crenated ridges arranged in a pentapetalous figure. In many cases, as in *Extracrinus* and *Pentacrinus*, the column is furnished with more or less numerous "auxiliary" arms, or "side-arms," the function of which is not altogether clear. The column increases in height by the interpolation of new joints between the base of the calyx and the highest articulation of the stem; and each articulation is pierced by a variously shaped perforation. Hence, by the apposition of the successive joints there is formed a tube—the so-called "alimentary canal" of the older writers—which runs the entire length of the column. This canal is most commonly round, but it may be pentapetalous, or it may consist of four or five canals running parallel with and around a central tube, into which they may or may not open. This canal sends off diverticula into the side-arms and the root-like processes of attachment, when these structures are present; and it contains, in living forms, a vascular axis (partly nervous in nature?) which is connected superiorly with a peculiar chambered organ situated in the base of the calyx, and which doubtless serves to maintain the vitality of the column and its appendages.

The dorsal surface of the cup or "calyx" is composed of a number of calcareous plates accurately fitted together. The number and arrangement of these vary much in different genera, and it will be sufficient to indicate here their general disposition.[1] Reposing directly upon the summit of the

[1] Different authorities have employed the various terms connected with the plates in the calyx of the *Crinoidea* in such different senses that the subject of the anatomy of the Crinoidal skeleton has been rendered very difficult to

column is a series of plates which are termed "basal" from their position, and which constitute the "pelvis" of Miller. The "basals" may be five, four, three, or rarely two in number, and they form the lowest portion of the cup. In general the basals are free, and are simply articulated by their edges, but in some cases it is believed that they are more or less undistinguishably fused with one another into a single mass. In many cases the "basals" are succeeded by a

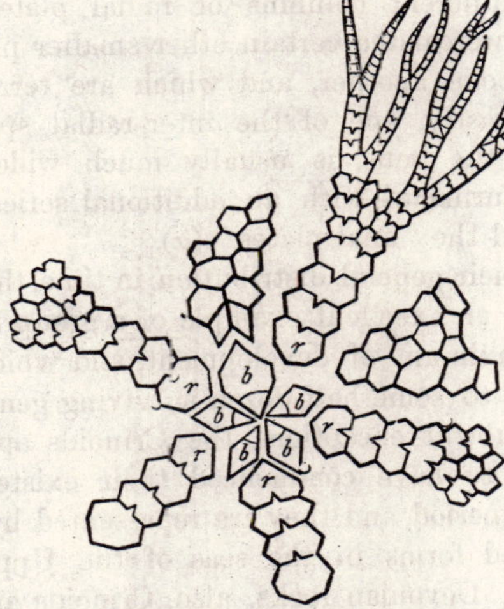

Fig. 158.—Diagram to show the structure of the calyx in the fossil Crinoids. *b*, Basals; *r*, Radials; *i*, Inter-radials; *a*, Anal plates. Calyx of *Forbesiocrinus*. After De Koninck and Le Hon. (As it has now been discovered that this genus really possesses three small basals, below the plates here represented, the pieces lettered *b* are truly "parabasals.")

second row or cycle of plates, which may be regarded, with Professor Beyrich, as a second series of basals, but which are properly regarded as something special, and are termed the "parabasals" or "sub-radials." The basals (fig. 158, *b*) are

comprehend. To this cause of confusion must be added the fact that different observers have employed the same terms for parts which do not appear to be truly homologous. In the present state of uncertainty upon the whole of this subject, it has seemed sufficient to give here merely a general account of the nomenclature which has hitherto been most generally employed in works descriptive of the fossil Crinoids, though this nomenclature is probably often not strictly correct.

in other cases succeeded directly by a series of two or three rows of plates, which are directly superimposed upon one another, and which form the foundations of the arms (*r*, *r*). These are termed the "radials" (the "costæ" of Miller), and are termed "primary," "secondary," and "tertiary," according to their distance from the basals. The last radial plates, or those furthest from the column (sometimes called the "axillary radials"), give origin to the arms. The radial plates are arranged in a series of vertical columns, which extend from the summit of the basals to the bases of the arms. Between the different columns of radial plates, however, there may be intercalated certain other smaller plates, which alternate with one another, and which are termed "inter-radials" (*i*). Lastly, one of the inter-radial spaces, corresponding with the anus, is usually much wider than the others, and is furnished with an additional series of plates, which are called the "anal plates" (*a*).

As regards their general distribution in time, the *Crinoidea* present us with an excellent example of a group which early attained its maximum of development, and which has now dwindled down to some half-dozen surviving genera. With one or two doubtful exceptions, the Crinoids appear, so far as yet known, to have commenced their existence in the Lower Silurian period, and they are represented by numerous and very varied forms in the seas of the Upper Silurian period. In the Devonian rocks, also, Crinoids are plentiful, and many generic types are peculiar to this period. It is in the earlier portion, however, of the Carboniferous period that the Crinoids attain their highest development. The Carboniferous Limestone is in many places, over wide areas, and for a thickness of many yards, almost entirely made up of the *débris* of Crinoids; and in many places it is so charged with the remains of these organisms as to deserve and acquire the name of "crinoidal limestone" or "entrochal marble." It is in the Palæozoic period, then, that the Crinoids attain their maximum, both numerically and as regards the number of genera and species. Taken as a whole, the Palæozoic Crinoids are distinguished by the characters already mentioned—namely, by having the brachial grooves conveyed to the mouth in concealed channels or tunnels, in having mostly a rounded column, and in the fact that the articular facets of the column-joints are usually marked by numerous simple radiating striæ. As a rule, also, the Palæocrinoids have a calyx, in which the dorsal surface has a preponderating development as compared with the ventral surface.

Coming to Mesozoic times, Crinoids are still abundant, though certainly reduced in numbers as compared with their development in the Palæozoic period. Taken as a whole, the Mesozoic Crinoids are characterised by having the brachial grooves continued over the ventral surface of the disc, and by having the ventral portion of the calyx more extensively developed than the dorsal; whilst the articular surfaces of the column-joints are usually morticed to one another by means of crenated ridges which have a flower-like arrangement. Neither of these latter characters, however, is constant. In the Mesozoic rocks, also, appear for the first time remains of the *free* Crinoids allied to the living *Comatulæ*.

In the Tertiary period the Crinoids are reduced to a very small number of forms, the most important of which is the genus *Pentacrinus*, which is represented at the present day by the living *Pentacrinus caput-Medusæ* of the Antilles and various other species. The best-known species is the *P. subbasaltiformis* of the Eocene. At the present day, Crinoids are moderately abundant, if we look only to the free forms like *Comatula*, which are largely represented, and attain their maximum in recent seas. On the other hand, the stalked Crinoids are reduced to a comparatively very limited number of generic types represented by a comparatively small number of species.

A good classification of the Crinoids is still a desideratum. Commonly they have been divided into two groups termed respectively *Articulata* and *Tesselata*, according as the radial plates of the calyx are freely articulated to one another, or are immovably joined together without articulation. This arrangement is a far from satisfactory one; but if accepted, we should have all the Mesozoic and Kainozoic Crinoids (except *Marsupites*) in the division of the *Crinoidea articulata*. On the other hand, the division of the *Crinoidea tesselata* would correspond with the Palæocrinoids, comprising all the species known from the Palæozoic formations.

The division of the Crinoidea into stalked and free forms is in many respects inapplicable as a basis of zoological classification. There can, however, be no doubt but that the free Crinoids are structurally an advance upon the fixed forms. It is therefore of interest to note that the stalked Crinoids had attained their maximum in the Palæozoic rocks, and had even commenced to decline before the free Crinoids first made their appearance in the Mesozoic Series.

In the absence of any satisfactory classification of the Crinoids, it will be sufficient here to briefly consider the

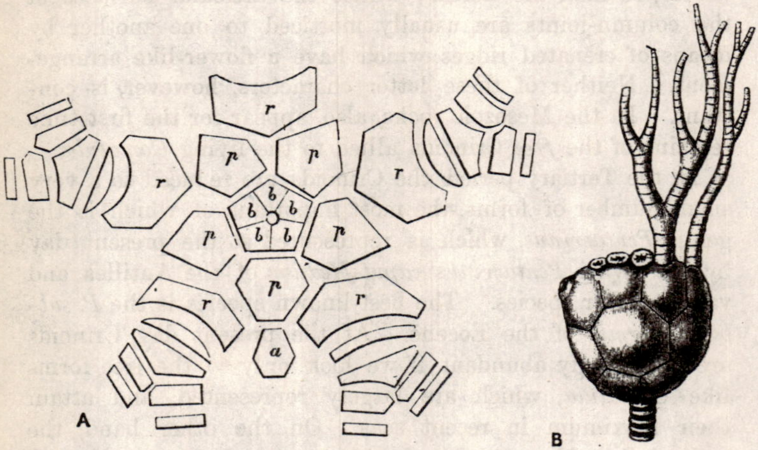

Fig. 159.—A, Diagram, showing the dissected calyx of *Cyathocrinus* (after Hall). *b*, Basals; *p*, Parabasals; *r*, Lowest of the three radials ("primary radials"); *a*, Anal plate. B, Calyx and part of the arms of *Cyathocrinus planus*, of the natural size. Carboniferous.

leading types and geological range of the more important families of the order. It should be borne in mind, however, that though some of these families are undoubtedly natural assemblages, the same cannot be affirmed of all of them, while there often exist considerable and legitimate doubts as to the true position of many of the genera.

The family of the *Cyathocrinidæ*, as represented by *Cyathocrinus* itself, presents us with one of the simplest types of the Pedunculate Crinoids. In *Cyathocrinus* (fig. 159) the calyx is somewhat globular, consisting of five "basals," alternating with an equal number of "parabasals" or "subradials," these in turn being followed by the "radials." There are generally three "radials" to each arm, the primary radials being comparatively large, while the other ones are small. There are no "inter-radials." The structure of the upper surface of the calyx in this genus has not yet been completely made out. The vault of the calyx is always slightly arched, or comparatively flat and depressed, and there appears to be a small anal proboscis on one side. A central mouth has been described as present; but specimens examined by Meek and Worthen would go to show that the apparent mouth is due to breakage, and that the summit of the calyx is really vaulted over by calcareous plates, the mouth, as in the Palæocrinoids generally, being thus hidden from view. The genus *Cyathocrinus* is mainly confined to the Carboniferous and Permian rocks, though examples have been described from both the Silurian and Devonian. The Carboniferous genus *Zeacrinus* is also related to *Cyathocrinus*, but the basals are very small, and there are four, six, or more anal plates. We may place here also the curious *Ichthyocrinus* of the Silurian and Carboniferous (fig. 162, *b*), the arms of which are frequently bifurcated.

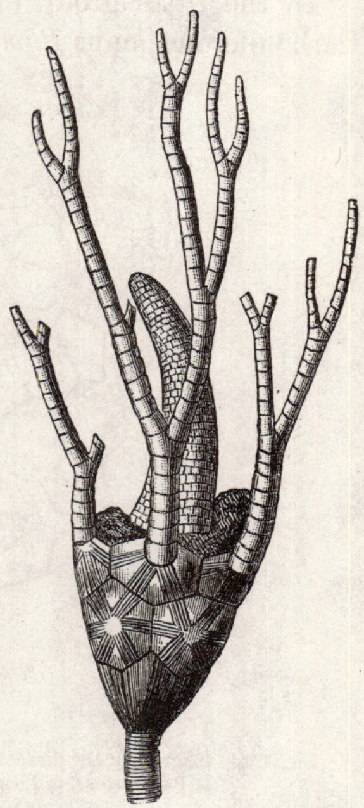

Fig. 160.—Calyx and part of the arms of *Poteriocrinus radiatus*, showing the proboscis. Carboniferous. (After De Koninck and Le Hon.)

Allied to the preceding is the family of the *Poteriocrinidæ*, of which *Poteriocrinus* is the type. In this genus (fig. 160) the cup consists, as in *Cyathocrinus*, of five basals, five parabasals, and a variable number of radials, the primary radials being the largest; and there are no inter-radials. There are, however, four to six anal plates; and the upper surface of the calyx, instead of being depressed, is always swollen and convex, and furnished with a very large anal tube or "proboscis." The genus (with various sub-genera) commences in the Silurian, is present in the Devonian, and abounds in the Carboniferous period, after which it disappears. If the Carboniferous genus *Zeacrinus* be proved to have a large proboscis, it will have to be removed from the *Cyathocrinidæ* and placed here. The Silurian genus *Dendrocrinus* also stands very close to *Poteriocrinus*, and possesses a greatly-developed proboscis, in some species of extraordinary length.

In the little group of the *Rhodocrinidæ*, typified by the Carboniferous genus *Rhodocrinus* (fig. 161) there are five basals

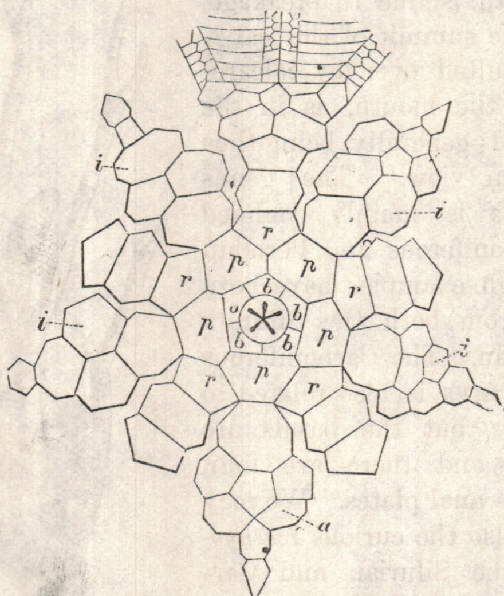

Fig. 161.—Diagram of the dissected calyx of *Rhodocrinus* (after Schultze). *b*, Basals; *p*, Parabasals; *r*, First radials; *i*, Inter-radials; *a*, Anal plates.

and five parabasals or sub-radials; there are three cycles of radial plates; there are from six to eight "inter-radials" in each inter-radial space; the anal plates are eight to twelve in number; and the arms, varying in number from ten to twenty, are bifurcated two or three times during their course. Very nearly allied to *Rhodocrinus*, though apparently separable from it, is the Carboniferous genus *Gilbertsocrinus*.

All the preceding families are provided with a zone of "parabasal" or "sub-radial" plates, and if we regard this character as being of high classificatory value, we must consider here the groups of the *Taxocrinidæ* and *Anthocrinidæ*, both of which in some respects differ widely from the typical Palæocrinoids. In the *Taxocrinidæ*, typified by the genus *Taxocrinus* (fig. 162, *c*) of the Silurian, Devonian, and Carboniferous deposits, there are three very small basals, succeeded by five large sub-radial or parabasal pieces, which support from three to seven cycles of radials. The inter-radials may be wanting, and are always few in number. The discovery that *Forbesiocrinus*, of the Devonian and Carboniferous, really possesses three minute basals (not represented in fig. 158, after De Koninck and Le Hon, in which the plates lettered *b* thus really are the "parabasals"), places this genus in the immediate neighbourhood of *Taxocrinus*, from which it is only separated by the possession of numerous inter-radial and anal plates. *Taxocrinus* is remarkable, as compared with almost all the Palæocrinoids, in apparently having the upper surface of the calyx covered by soft integument only, instead of being vaulted over with calcareous plates.

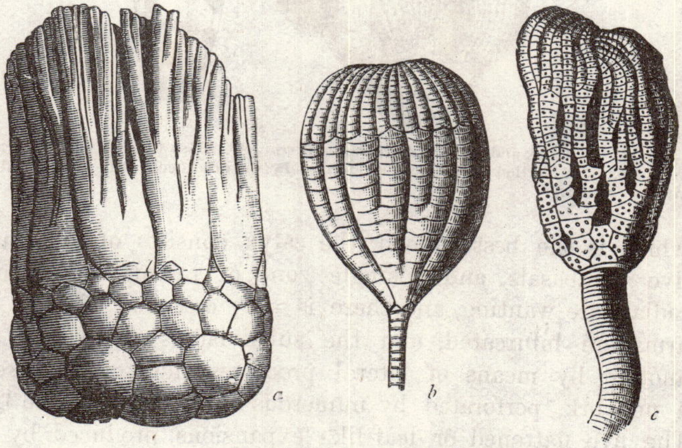

Fig. 162.—Upper Silurian Crinoids. *a*, Calyx and arms of *Eucalyptocrinus polydactylus*, Wenlock Limestone; *b*, *Ichthyocrinus lævis*, Niagara Limestone, America; *c*, *Taxocrinus tuberculatus*, Wenlock Limestone. (After M'Coy and Hall.)

In the family of the *Anthocrinidæ* we have only the extraordinary Silurian genera which have been described as *Anthocrinus* and *Crotalocrinus.* In the former of these (fig. 163),

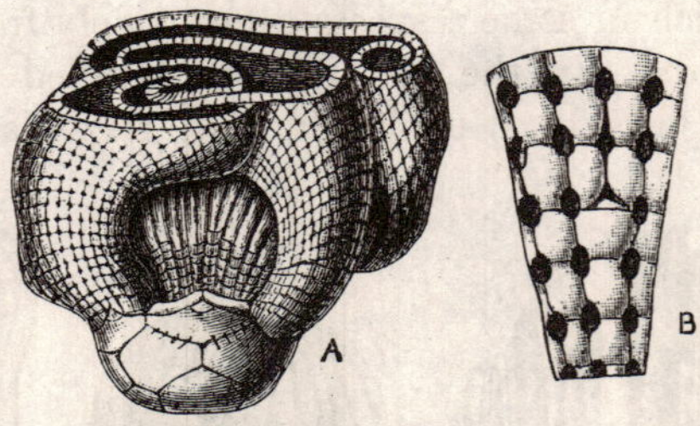

Fig. 163.—A, Calyx and arms of *Anthocrinus Loveni*, cut across to show how the arms are rolled up; B, A portion of the network formed by the arms, enlarged. Upper Silurian. (After J. Müller.)

which is the best known, the calyx consists of five basals, five parabasals, and a single zone of radials, while interradials are wanting, and there is only one anal plate. The arms are bifurcated, and the subdivisions unite with one another by means of lateral processes, thus giving rise to a network, perforated by numerous apertures (fig. 163, B). The five flattened or leaf-like expansions, produced by this curious metamorphosis of the arms, are rolled up like the petals of a flower. *Crotalocrinus* in the conformation of its arms appears to resemble *Anthocrinus,* except that the edges of the arms appear to be united, but the structure of the calyx is somewhat different.

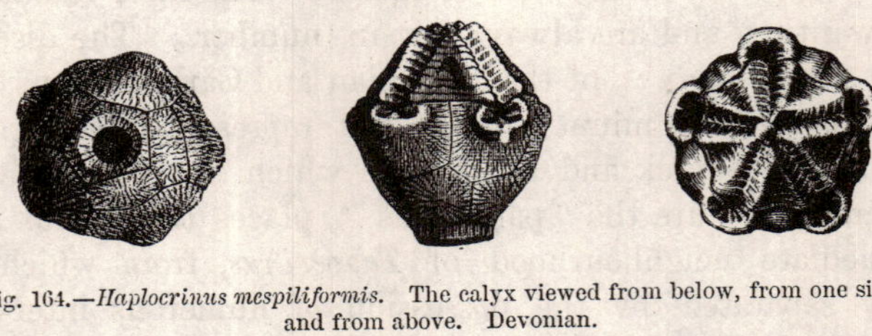

Fig. 164.—*Haplocrinus mespiliformis.* The calyx viewed from below, from one side, and from above. Devonian.

Lastly, we must place here the little family of the *Haplocrinidæ.* In the Devonian genus *Haplocrinus* (fig. 164) the calyx is small and globular, with five small basals, succeeded by a cycle of three parabasals. There are five radials, two of which stand directly upon the basals, while the others rest on the parabasals. The upper surface is covered by five inter-radials, so disposed as to form a pyramid. In the related genus *Triacrinus*, of the Silurian and Devonian, there are only three basals, and a single parabasal.

We come now to a series of the stalked Crinoids in which there are no parabasal plates. Foremost among these is the great family of the *Actinocrinidæ*, of which *Actinocrinus* itself is the type. In this genus the calyx (figs. 153-155), though very variable in shape, consists of three basals (fig. 165, *b*), which articulate upwards with three cycles of radials, the "sub-radials" being wanting. The axillary radials carry each a double series of brachial plates, which support the variously divided arms. There are three or more anal plates, of which the lowest (fig. 165, *a*) always rests upon the basals directly. There is a variable number of inter-radials, and the column is round. The upper surface of the cup is vaulted over with calcareous plates, and the brachial grooves are continued beneath the vault thus formed, as so many tunnels, to the central and concealed mouth. The anus may or may not be extended into a proboscis, and it is sometimes very excentric, sometimes subcentral. It has been shown that in many of the *Actinocrinidæ* there exists in the interior of the calyx a singular convoluted calcareous plate, of a reticulated texture, shaped somewhat like an ordinary Bubble-shell (*Bulla*), occupying the vertical axis of the body, and often of large size. This has been compared with the calcareous structures present in the "sand-canal" of various Echinoderms; but it is probably rather an extreme

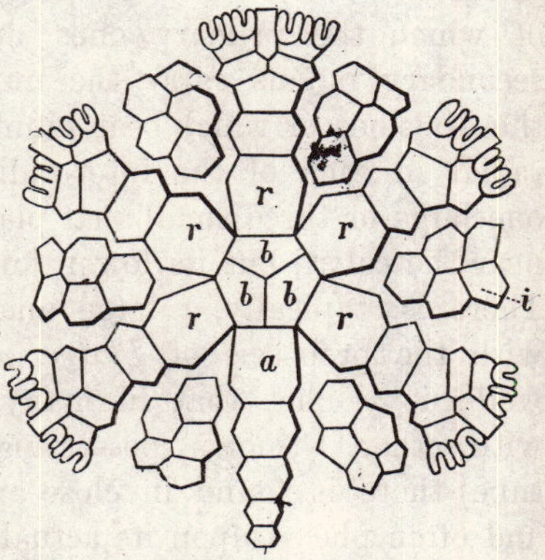

Fig. 165.—Diagram of the dissected calyx of *Actinocrinus* (after Schultze). *b*, Basals; *r*, Radials; *i*, Inter-radials; *a*, The lowest of the anal plates.

development of the discoidal calcareous plates, which have been described as strengthening the double wall of the spirally-twisted alimentary canal in the living *Comatula.* The genus *Actinocrinus* appears to commence in the Upper Silurian, and is also represented in the Devonian; but it attains its maximum in the Carboniferous, and is wholly unknown in later deposits. *Megistocrinus, Agaricocrinus,* and *Batocrinus* are Carboniferous forms very closely allied to *Actinocrinus,* and perhaps of but sub-generic value. The Silurian *Periechocrinus* is another close ally of *Actinocrinus,* and the Carboniferous *Amphoracrinus* and *Dorycrinus* only differ from it in comparatively trifling particulars.

In the Devonian genus *Melocrinus,* though the structure is in many respects similar to that of *Actinocrinus,* there are *four* basals, and the lowest anal plate is separated from the basals by the primary radials.

In the family of the *Platycrinidæ,* as typified by *Platycrinus* itself (fig. 152), the cup consists of three basals (as in *Actinocrinus*), which support two cycles of radial plates, of which the primary ones are much the largest. The secondary radials carry the numerous bifurcated arms, all the divisions of which bear pinnulæ. There is a single inter-radial in each of the inter-radial spaces, and there may be one large or three small anal plates. The column is rounded near the calyx, but its lower joints are oval and compressed. There is, typically, a large anal proboscis. In connection with the proboscis of *Platycrinus,* we may just notice the well-known fact that in many specimens (as is the case with other Crinoids possessing a similar elongated anal tube) there is found in close apposition with the proboscis, and often placed upon its actual summit, the shell of a fossil Univalve (apparently almost always, or always, a species of *Platyceras*). It was originally supposed that the Crinoid had been fossilised in the act of eating the Mollusc—the anal tube being regarded as the mouth—but all the living Crinoids feed upon microscopic animalcules, and this supposition is therefore, *prima facie,* an improbable one. It has also been shown by Meek and Worthen that the *Platyceras* must have lived for a long time attached to the proboscis of the Crinoid, since the lip of its shell has closely adapted its form to that of the surface to which it is attached. We may therefore safely accept the conclusions reached by these observers, that the *Platyceras* was in the habit of attaching itself parasitically to the side or summit of the proboscis of *Platycrinus* and other Crinoids, thus obtaining a share of the minute animalcules upon which its host lived.

Platycrinus is essentially a Carboniferous genus, though forms certainly not more than sub-generically differing from it occur in the Upper Silurian and Devonian. The Devonian genus *Hexacrinus* is very closely allied to *Platycrinus*, but it possesses one large anal plate, which rests directly upon the basals (as in *Actinocrinus*). In *Coccocrinus*, of the Upper Silurian and Devonian, we have an aberrant type of this family, in which the summit of the calyx is covered by five large pentagonal plates, forming a depressed pyramid.

In the extraordinary family of the *Eucalyptocrinidæ* is only the abnormal *Eucalyptocrinus* (*Hypanthocrinus*) of the

A 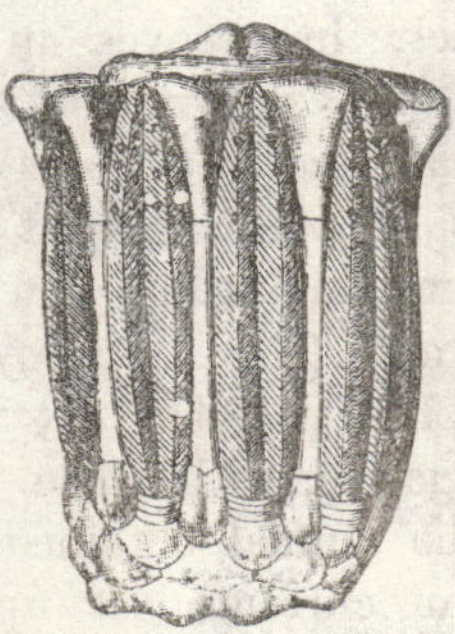 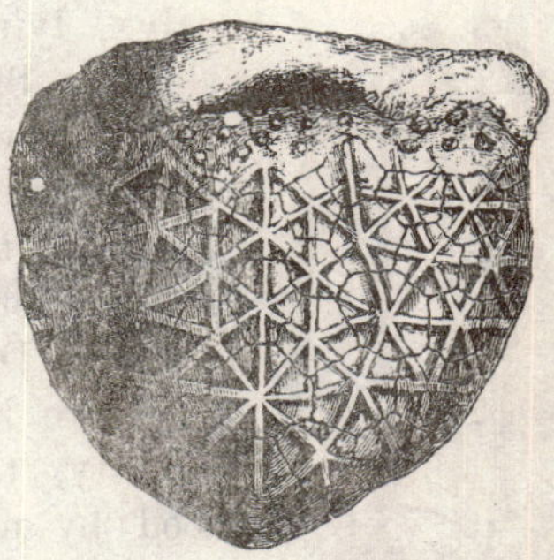 B

Fig. 166.—A, Calyx and arms of *Eucalyptocrinus rosaceus*, viewed from one side, of the natural size—Devonian (after Schultze); B, Calyx of *Glyptocrinus basalis*, viewed from one side, of the natural size—Lower Silurian (after M'Coy).

Silurian and Devonian (figs. 162, *a*, and 166, A). In this genus the calyx is inverted upon itself, the calicine cup being deeply concave at its base, so as to look like the bottom of a wine-bottle. As now known, there are *four* small basals (as in *Melocrinus*), which carry five large primary radials. These are strongly bent, one half of each passing up into the basal funnel, while the other half appears on the lower and lateral aspects of the cup. Two other rows of radials succeed these, the tertiary radials being unusually large, and each supporting the bases of two arms. The inter-radial plates are developed in a most singular manner, so as to form a series of five linear, clavate processes, which separate and support the arms; five other precisely similar processes being borne by the axillary radials. The arms thus come to lie in deep grooves or niches in the sides of the calice, the upper surface of which they do not reach. The upper surface is completely vaulted over, and is mainly formed by the upper ends of the ten inter-brachial processes just spoken of, in the centre of which is a small circular (oral?) aperture, surrounded by four plates. Without entering further into the structure of this genus, it is clear that it presents us with an entirely unique type, in which the mode of existence must have been very unlike that of the ordinary Crinoids. This is shown by the fact that the arms are shorter than the calyx, lying freely in grooves in its sides. Billings has shown that the brachial grooves in *Eucalyptocrinus* perforate the walls of the calyx at the bases of the arms, and thus gain the interior of the cup (Geol. Survey of Canada, Decade III. p. 24). It may therefore be suggested, with some probability, that the animal obtained its food by means of currents set up in the surrounding water by the cilia lining the brachial grooves (as in the ordinary Crinoids), but that the real oral aperture was placed at some unknown point, deep within the cup, and that the opening at the top of the calyx, usually supposed to be the mouth, is really the *anus*.

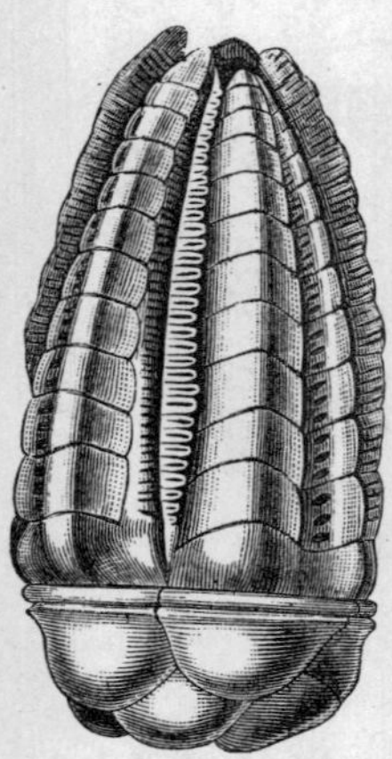

Fig. 167.—Side view of the cup of *Cupressocrinus crassus*, with the arms folded up, of the natural size. Devonian. (After Schultze.)

The family of the *Glyptocrinidæ*, represented by the Silurian *Glyptocrinus*, agrees with the preceding families in having no parabasals, but there are *five* basals, succeeded by three rows of radials. There are six inter-radials in each inter-radial space, and seven or more anal plates. The arms are well developed, the column is annulated or moniliform, and the plates are usually ornamented with radiating ridges (fig. 166).

In the little Devonian family of the *Gasterocomidæ*, as represented by *Gasterocoma* (= *Epactocrinus*), there are five basals, surrounding the uppermost joint of the column, and succeeded by a single cycle of five radials, the upper edges of which are hollowed out for the reception of the brachial grooves on their way into the interior of the calyx. The anus is quite excentric, and is placed just above one of the five basals, surrounded by a ring of five small plates, surmounted by a sixth larger one. There are five arms; and the column is obtusely quadrangular, and is perforated by four canals, surrounding, or confluent with, a larger central tube.

In the *Cupressocrinidæ*, as represented by the singular *Cupressocrinus* of the Devonian, the calyx (fig. 167) is cup-shaped, the centre of its base being formed by the expanded uppermost joint of the column, which is surrounded by five basals, which carry a single cycle of five large radials. These support five much smaller plates, which may be secondary radials, and these give origin to the five arms. The arms are undivided, diminish gradually towards their apices, and by the accurate apposition of their edges form together a pentagonal pyramid, when the animal is at rest. The arms are deeply grooved on their ventral surfaces, but these grooves are roofed over by four parallel rows of small calcareous plates, which leave centrally a small ambulacral sulcus. The upper surface of the calyx appears to have been covered with a leathery perisome; the mouth is central, and the anus ex-

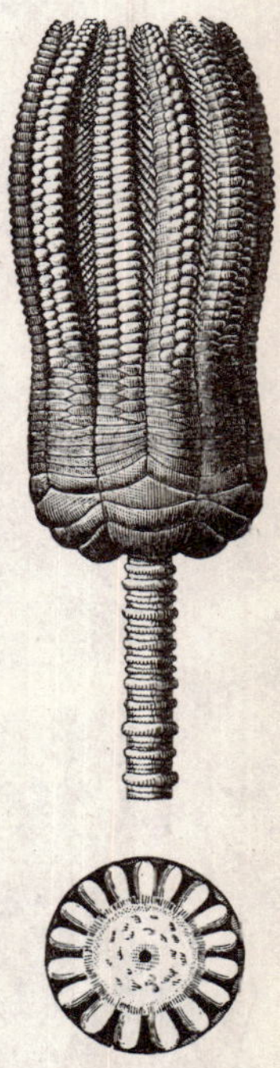

Fig. 168. — *Encrinus liliiformis.* Muschelkalk. The lower figure shows the articulating surface of one of the column-joints.

centric and not prolonged into a proboscis; and there is a peculiar internal skeleton (as in various Palæocrinoids) for the support of the viscera. The column is obtusely quadrangular, annulated, and traversed by a large central canal surrounded by four smaller tubes. The Carboniferous and Devonian *Symbathocrinus*, and the Devonian *Phimocrinus* are nearly related to *Cupressocrinus*.

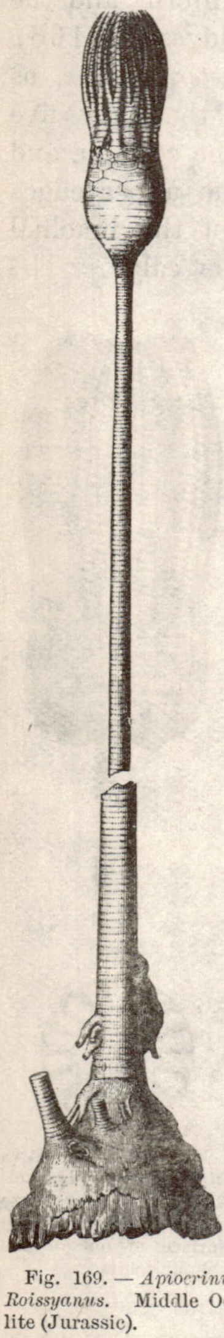

Fig. 169. — *Apiocrinus Roissyanus*. Middle Oolite (Jurassic).

In the family of the *Encrinidæ*, we have Crinoids with freely-articulated radial plates (*Crinoidea articulata*). The calyx consists of five basals, five parabasals, and three cycles of radials (fig. 168), but there are no inter-radials. The arms are composed of a double series of alternating pieces, and carry pinnules on their inner faces. The column is long, composed of round joints, with articulating surfaces grooved in a radiating manner, and pierced by a small round central canal. The chief genus is *Encrinus* itself, well known by the beautiful *E. liliiformis* (fig. 168) of the Muschelkalk. All the genera of the family are Triassic.

In the *Eugeniacrinidæ* are imperfectly known and comparatively small Crinoids, principally from the Jurassic, but extending upwards into the Cretaceous. In *Eugeniacrinus* the calyx consists of five basals, and three cycles of radials; but there are no parabasals or inter-radials.

In the *Pentacrinidæ*, typified by the well-known genus *Pentacrinus* (fig. 157), the calyx is small, composed of five basals and three cycles of radials, without either parabasals or inter-radials; the arms are long, abundantly and irregularly ramified; and the column is pentagonal, the articulating surfaces being formed by crenated ridges arranged in a flower-like manner. The principal genus is *Pentacrinus* (including *Extracrinus*), which made its first appearance in the Trias, is abundantly represented in the Jurassic, is found in less numbers in the Cretaceous and Tertiary, and still survives at the present day.

In the *Apiocrinidæ*, or "Pear-Encrinites," the column is long and rounded, and is rooted inferiorly to the sea-bottom. The upper column-joints gradually increase in diameter till they attain the width of the calyx; or, if this does not occur, the basal plates of the calyx are fused, so as to appear like the highest joint of the stem. In the typical members of the family the calyx consists of five basals and three cycles of radials, but there are no inter-radials or parabasals. The type genus of the family is the Jurassic *Apiocrinus* (fig. 169); *Guettardicrinus*, and *Millericrinus* are also Jurassic; the curious genus *Bourgueticrinus* is Cretaceous and Tertiary; and *Rhizocrinus* (fig. 150) is found living at the present day.

Intermediate between the ordinary stalked Crinoids and the typical free sessile forms (such as *Comatula*) is the family of the *Holopidæ*, in which the animal is permanently fixed, though there is no stem. In the living *Holopus*, the type of the group, the basals, with the first, and probably also the second, radials, are fused together, "forming the wall of a tube-like body-chamber, which is cemented beneath to the foreign body to which the Crinoid is attached, by an irregularly expanded calcareous base" (Sir Wyville Thomson). There are five arms, and there does not appear to be any anal opening. The genus *Cyathidium*, from the Chalk of Faxöe, appears to be hardly separable generically from *Holopus*.

We have, lastly, to consider very briefly the ordinary sessile Crinoids, none of which are known to have appeared in deposits older than the Secondary, and which have played a comparatively unimportant part in geological history. The first family of these is that of the *Marsupitidæ*, in which the calyx in many points of its structure resembles that of the stalked Palæocrinoids, but in which there is no peduncle. In *Marsupites* (fig. 170), the "Tortoise-Encrinite," which we may take as the type of the group, the calyx is of large size, and the centre of its base is formed by a single plate, which may be regarded as the uppermost segment of the stem, or as an azygous basal, or as formed by the anchylosis of several basals. Round this are two cycles of parabasals (or one cycle of basals and one of parabasals), surmounted by a single cycle of radials. There are five arms, and the upper

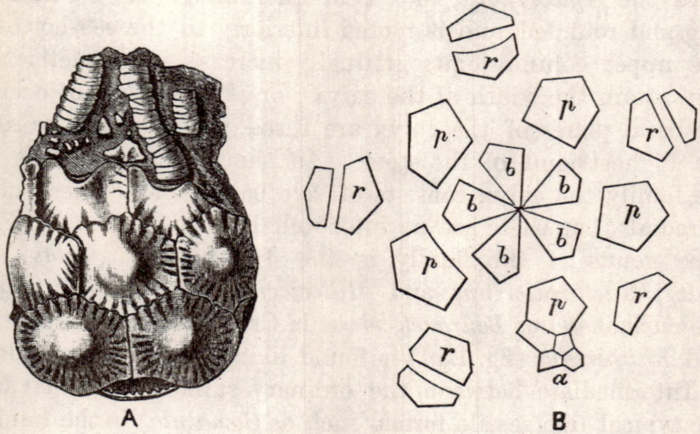

Fig. 170.—A, Side view of the calyx and bases of the arms of *Marsupites ornatus*, from the Chalk. B, Diagram (after Hall) of the dissected calyx of *Agassizocrinus*: *b*, Basals; *p*, Parabasals; *r*, Radials; *a*, Anal plates.

surface of the calyx appears to have been covered by a leathery membrane, containing calcareous ossicles. *Marsupites* is wholly confined to the Cretaceous; and the allied genus *Uintacrinus*, with ten arms, is from rocks of the same age in North America. In the curious *Astylocrinus* and *Agassizocrinus*, of the Carboniferous, we have two closely allied, or identical, Palæozoic types of sessile Crinoids. In these forms (fig. 170, B) the base of the calyx is formed of five anchylosed basals, succeeded by five parabasals, which in turn are surmounted by two cycles of radials. *Agassizocrinus* appears also to possess a small number of anal plates.

Finally, in the family of the *Comatulidæ*, represented by our living Feather-stars, we have Crinoids, which are pedunculate in the early stages of their existence, but which in the adult state lead an entirely free life. In their free condition, the "peduncle" of the stalked Crinoids is represented only by the uppermost joint of the column, which is dilated to form the so-called "centro-dorsal basin," and which carries one or more whorls of short jointed pinnules or "cirri"—the homologues of the "side-arms" of *Pentacrinus.* In the calyx itself the basals are generally comparatively large, and are anchylosed with one another, and often with the primary radials, to form the so-called "rosette." There are usually three cycles of radials; and the arms may retain their primitive number of five, or may be once or twice bifurcated. The most important fossil genus of the family is the Jurassic *Solanocrinus;* but the living *Antedon* has been detected in the Tertiaries, and has been quoted, along with other genera, from older deposits. As shown by Mr P. Herbert Carpenter, the fossils of the Chalk known as *Glenotremites* consist really of the united centro-dorsal piece and radial pentagon of a Comatulid. The Jurassic genus *Saccosoma* (fig. 156), though a free Crinoid, apparently allied to *Comatula,* is at present but imperfectly understood.

CHAPTER XV.

CYSTOIDEA, BLASTOIDEA, AND HOLOTHUROIDEA.

ORDER V.—CYSTOIDEA.

THE *Cystoidea* are *Echinodermata* in which *the body was generally spheroidal, pedunculate or sessile, enclosed by calcareous articulated plates, some of which are porous and are connected with respiration. Arms rudimentary, mostly reduced to the pinnulæ only. Reproductive organs contained within the interior of the calyx.*

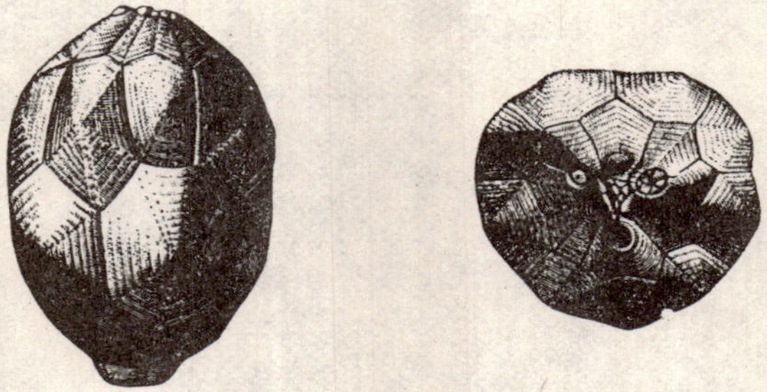

Fig. 171.—*Hemicosmites pyriformis*, one of the Cystideans. The right-hand figure shows the upper surface of the calyx. Lower Silurian.

In general form the Cystideans are globular, oval, pear-shaped, conical, or sub-cylindrical, and they resemble the Crinoids in consisting of a stem or "column" and a body or "calyx." The column is composed of a succession of calcareous joints, and in no respect differs from the column of the Crinoids. In *Lepadocrinus* (fig. 172, D), however, it is doubtful if the column was affixed to any foreign body, for its lower extremity is composed of a single, long, spindle-shaped piece. There are also forms in which the mode of attachment of the body is still unknown. The calyx consists of a a number of polygonal calcareous plates accurately fitted together, and enclosing all the viscera of the animal. Sometimes the plates are indefinite in number and arrangement; but in other cases the number of plates is limited, and they are arranged according to a definite plan. In other cases the number and arrangement of the plates may be definite on one side of the calyx and indefinite on the other.

As regards the structure of the arms of the Cystideans, there is a good deal of diversity. In some forms there were

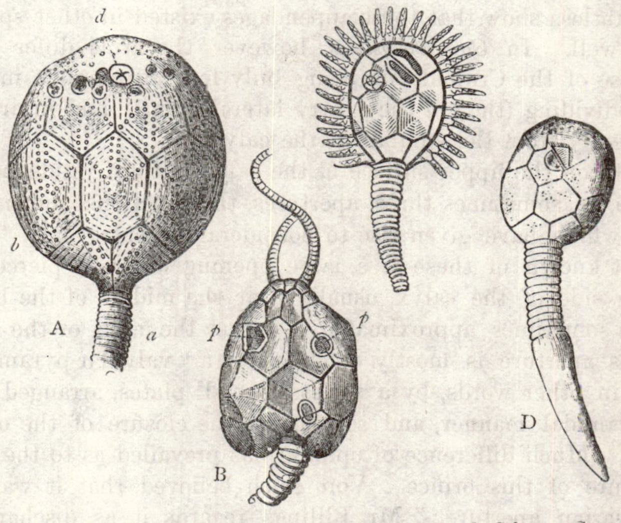

Fig. 172.—Cystideans. A, *Caryocrinus ornatus*: *a*, Column; *b*, Calyx; *c*, Scars where pinnulæ were attached; *d*, Valvular pyramid. B, *Pleurocystites squamosus* (dorsal side): *p*, *p*, Two of the pectinated rhombs. C, *Pseudocrinus bifasciatus*. D, *Lepadocrinus Gebhardi*.

no true arms, comparable with the arms of the Crinoids; but there were small jointed processes, which were attached to the calyx, and which represent the "pinnulæ" or lateral branches of the arms of the Crinoids. In some cases, grooves —corresponding with the brachial grooves of the Crinoids— are seen to extend from the point of attachment of each pinnula to the summit of the calyx. In other cases, as in

Apiocystites, *Codaster*, and *Callocystites*, arms were present, but they were bent backwards and immovably soldered down to the surface of the calyx. The arms spring in these cases from the apex of the calyx, and are anchylosed by their dorsal surfaces to the body. On their ventral surfaces the arms are grooved by furrows which clearly correspond with the brachial grooves of the Crinoids, and on each side of these grooves is a row of pinnulæ. In one or two cases there is only a single row of pinnules, and the arm seems to have been fastened to the calyx by one of the lateral surfaces, instead of by the dorsal surface. In one Cystidean only (viz., *Comarocystites punctatus*, Billings), are there free arms as in the true Crinoids; but further researches will doubtless show that these appendages existed in other species as well. In *Comarocystites*, however, the arms differ from those of the Crinoids in being only four in number, in not subdividing (though they carry lateral pinnæ), and in arising directly from the summit of the calyx.

Upon the upper surface of the calyx in the Cystideans are two, or sometimes three, apertures, the functions and nature of which have given rise to considerable controversy. The best known of these is a large opening which is pierced in one side of the calyx, usually near the middle of the body, but sometimes approximated to either the apex or the base This aperture is mostly defended by a "valvular pyramid;" or, in other words, by a series of small plates, arranged in a pyramidal manner, and serving for the closure of the opening. Much difference of opinion has prevailed as to the true nature of this orifice. Von Buch believed that it was an "ovarian aperture;" Mr Billings regards it as discharging the functions of both the mouth and anus; whilst Sir Wyville Thomson, Mr Salter, and other high authorities, regard it as being the anus, and as corresponding with the proboscis of the Crinoids. That it is not an "ovarian orifice" may be regarded as certain, so that the question is narrowed to its being the anus alone, or an "oro-anal" orifice. In the living *Leskia mirabilis*, one of the Sea-urchins, both the mouth and vent are closed by converging triangular valves, which doubtless correspond with the "pyramid" of the

Cystideans. This recent form, however, is not sufficient to decide the present question since in it *both* the mouth and the anus exhibit this valvular apparatus. Upon the whole, therefore, this question must be regarded as undecided, though the analogies of recent forms would lead to the belief that the "pyramid" of the Cystideans is truly the anus, and that the mouth must be sought for between the bases of the arms, when these are present.[1]

A second aperture is placed near the centre of the summit of the calyx, between the bases of the arms, when these exist. This opening has not been universally detected, and its true nature is doubtful. By Mr Billings it is believed to be what he terms an "ambulacral orifice"—*i.e.*, an aperture by which the ambulacral vessels passed from the interior of the calyx to the exterior. The analogies of recent forms, however, would support the view that this aperture is the mouth, in which case the "pyramid" must be the anus. There is, however, in some cases a third aperture of small size, always placed near the apex, and this has been regarded as being truly the anus.

Many Cystideans were further provided with a system of minute pores or fissures, penetrating the plates of the calyx, and often arranged in definite groups. These groups are known as "pectinated rhombs" or "hydrospires," and their exact function is not absolutely certain; though there can be little hesitation in accepting the views of Billings and Rofe that they acted as respiratory organs. Their form differs considerably in different genera. In many cases the porous surfaces or "pectinated rhombs" have the figure of diamond-shaped areas, divided into two halves, of which one belongs to one plate of the perisome, and the other to the contiguous plate (fig. 173). Generally, if not invariably, the pores upon the one half of the rhomb are connected each

[1] Sir Wyville Thomson has recently shown that the mouth in the living *Hyocrinus* and *Rhizocrinus* is protected by a pyramid of five calcareous valves; but this aperture is central, and thus differs from the laterally-placed "pyramid" of the Cystideans. It seems not improbable that the mouth of the Cystideans was not only central in position, but that it resembled the mouth of the Palæocrinoids in being hidden from view by the calcareous plates covering the vault of the calyx.

with its corresponding pore on the other half by means of canals or grooves, which may or may not be visible externally. The pores in the pectinated rhombs appear to be the external openings of canals which perforate the test vertically, and open into a series of flat tubes or parallel reduplications of a thin shelly membrane; and this structure leads to the belief that the sea-water was admitted to their interior, and that they thus were enabled to serve in the

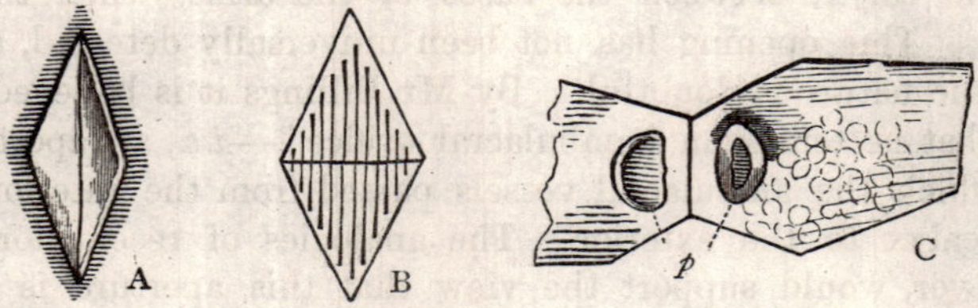

Fig. 173.—A, Pectinated rhomb of *Glyptocystites multiporus* (Billings). B, Pectinated rhomb of *Pleurocystites* (Billings). C, Two plates of *Callocystites Jewetti* (Hall), showing the pectinated rhombs (*p*). All enlarged.

aeration of the perivisceral fluid. While many Cystideans possess regular pectinated rhombs, of the structure above noted, others appear to have a completely imperforate test (*Cryptocrinus, Malocystites, Trochocystites*); while in other cases there are pores or fissures distributed without any very definite arrangement over most or all the plates (*Echinosphærites, Caryocystites, Palæocystites,* &c.)

As regards the distribution of the Cystideans in time, they are not only wholly extinct, but they are exclusively confined to the earlier portion of the Palæozoic period. They appear to have commenced their existence in the Upper Cambrian period, the earliest known examples being two extremely simple forms (*Trochocystites* and *Eocystites*) from the "primordial zone" of North America. Other forms have been described as occurring in the "primordial zone" of Bohemia. In the Chazy and Trenton Limestones of North America, of Llandeilo-Caradoc age (Lower Silurian), and on the same horizon in Russia, Scandinavia, and Bohemia, Cystideans are found, often in vast numbers, though in a very fragmentary condition. In the Bala Limestone (Lower

Silurian) of Britain, Cystideans are abundant fossils. In the Upper Silurian (Niagara and Lower Helderberg) of North America, and on the same horizon (Wenlock and Ludlow) of Britain, Cystideans still occur, though their remains are not so plentiful. Lastly, Cystideans certainly existed in the Devonian period, the genus *Strobilocystites* being an undoubted representative of this order in strata of this age in Iowa. Some, however, of the so-called "Cystideans" of the Devonian rocks appear to be probably referable to the *Protozoa*, and others belong to the *Blastoidea*. With the close of the Upper Silurian age, however, the order must have unquestionably greatly diminished, and we have no evidence of its having survived the Devonian period, except in the aberrant *Codaster* and *Codonites* of the Carboniferous.

As regards their zoological affinities, the *Cystoidea* are related to both the *Crinoidea* and the *Blastoidea*. From the Crinoids they are separated by the, generally speaking, less definite or unsymmetrical development of the plates of the perisome; by the absence or rudimentary condition of the arms; by the usual possession of definitely arranged pores or fissures in the test, which subserve a respiratory function; and by the fact that the reproductive organs appear to have been always lodged in the interior of the calice, instead of being situated under the integument of the arms or pinnulæ. To the Blastoids the Cystideans are more nearly related; but in the former the internal cavities of the hydrospires and pinnulæ are directly united through the intervention of the ambulacral pores (Billings), whereas in the latter they have no connection with one another. Moreover, the recumbent arms are directly imbedded in the substance of the test, whereas in the Cystoids such structures, if present at all, lie above and upon the plates of the test.

The classification of the Cystoids is still in a very unsatisfactory condition, largely in consequence of the fact that the true structure of many types is yet very imperfectly understood. There are, however, some fairly recognisable groups, which may be briefly noticed, along with some of the more prominent types belonging to each. In the first of these groups, as typified by *Cryptocrinus* (fig. 174, B), the plates of the test are all solid and imperforate. In the genus just mentioned the calyx is globular, composed of comparatively few plates, which have a tolerably definite arrangement, there being three basals, succeeded by two ranges of pentagonal plates. There is a "valvular pyramid" of five small plates, but the arms are unknown. The genus *Malocystites*, from a very low horizon in the Silurian series (Chazy Limestone), has also an imperforate test, and differs from *Cryptocrinus* chiefly in having the calyx composed of an indefinite series of plates. Another form of the same group is the *Trochocystites* of Barrande, from the "Primordial"

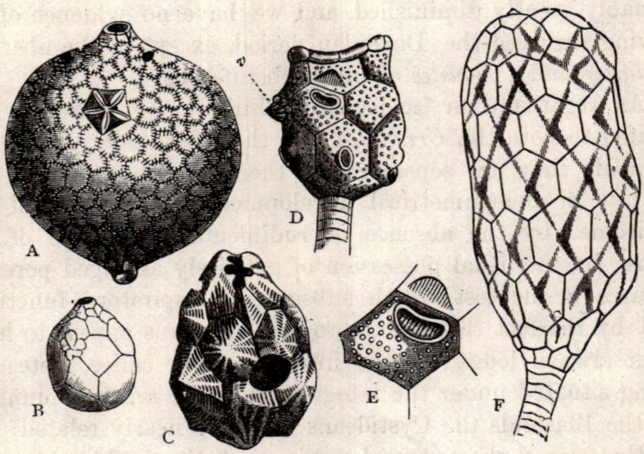

Fig. 174.—A, *Echinosphærites aurantium.* B, *Cryptocrinus lævis.* C, *Echinoencrinus Senkenbergi.* D, *Echinoencrinus armatus.* E, One of the "pectinated rhombs" of the last, enlarged. F, *Holocystites cylindricus.* *v*, Valvular pyramid—Silurian. All the specimens are viewed from one side. (A, B, and C are after Von Buch; L is after Edward Forbes; and F is after Hall.)

(Upper Cambrian) of Bohemia. This last-named type is of special interest from its high antiquity, the only other Cystoids of like age being the *Eocystites* of the "Acadian group" of St John's and the *Protocystites* of the "Menevian group" of St David's, both of which are only known by detached fragments, and are ill understood. The aberrant genus *Lichenocrinus*, of the Lower Silurian, may also possibly belong to this group of the Cystideans, but if so, it forms the type of a quite peculiar section. We may temporarily place here likewise a group of forms, in which no pores or fissures have yet been detected in the test, though it is possible that such really exist. Among these, two of the most remarkable types are *Holocystites* and *Gomphocystites*, both from the Upper Silurian (Niagara group) of North America. In *Holocystites* (fig. 174, F) the calyx is long and sub-cylindrical, and is composed of six or more ranges of hexagonal or polygonal plates, which have a tuberculated surface, but are not known to be perforated by any respiratory pores. A short jointed stalk is present, but there is no evidence as to the existence of arms or pinnulæ. In *Gomphocystites* (fig. 175, F) the calyx has a very peculiar shape, being pyriform, very narrow below, and inflated at its summit. The integumentary covering consisted of numerous ranges of plates, which were granulated superficially, but show no signs of pores. There is, further, the character that the summit is grooved by five sessile arms, which surround a central opening, and have the same spiral arrangement as in *Agelacrinus*.

In another group of *Cystideans* the plates of the test are indefinite in number, and are not arranged in a symmetrically radial manner, but most or all of them are perforated by respiratory pores. The type-genus of this group is *Echinosphærites* (fig. 174, A), in which the respiratory pores are numerous, and are situated along the margins of the plates, those of any one plate being united in pairs with the pores of contiguous plates by canals which are visible externally. The genus has been quoted from the Devonian, but appears to be truly confined to the Silurian period. *Caryocystites*, of the Lower Silurian, is another close ally of the preceding, but has an elongated and sub-cylindrical calyx. In *Sphæronites*, on the other hand, though there are some general resemblances to *Echinosphærites*, there is the distinction that the pairs of connected pores are upon the same, not upon different plates, the canals which unite them not running across the sutures between contiguous plates. Both *Sphæronites* and the allied *Glyptosphærites* seem to be essentially Silurian, and the Devonian fossils which have been referred to the former will probably find a place elsewhere.

In another great group of *Cystidea*, lastly, we have genera in which the plates of the test are definite in number, and more or less clearly arranged in a quinary manner, while the respiratory pores are usually confined to special and limited areas of the perisome ("pectinated rhombs"). The number of types included in this group is very large, and only some of the most important can be noticed here. One of these—namely, the *Pleurocystites* of Billings (fig. 172, B)—is a transitional link between the present and the preceding groups; since the perisomatic plates of the dorsal side of the calyx are large and definitely arranged, while those of the ventral side are small, numerous, and indefinite. There are two free, jointed pinnulæ, or unbranched arms, and there are three "pectinated rhombs." The genus is wholly Lower Silurian. A more typical member of this group is *Echinoencrinus* (fig. 174, C and D), in which, as in various allied genera, the calyx is composed of four series of plates, there being four basals and five plates in every range above this. There are three "pectinated rhombs," and, so far as known, no arms. The genus is found in the Lower and Upper Silurian. *Glyptocystites*, of the Lower Silurian, is a near ally of the preceding, but possesses numerous "pectinated rhombs" (ten to twelve in number), no other known genus having more than three of these organs. *Apiocystites*, *Pseudocrinus*, and *Callocystites*, form a closely-related sub-group, differing from *Glyptocystites* in various points, but agreeing with it in possessing five recumbent arms, which are soldered to the calyx, on the surface, in the form of so many grooves, bordered on each side (when perfect) with a row of short jointed pinnulæ. In *Pseudocrinus* (fig. 172, C) the calyx is orbicular, and the arms, though recumbent, project considerably above the surface; in *Apiocystites* (fig. 175, E) the body is oblong and tetragonal; and in *Callocystites* the body is ovoid (fig. 175, D), and the arms are deeply furrowed. These three genera are characteristic of the Upper Silurian period, as are *Lepadocrinus* (fig. 172, D) and *Prunocystites*; the former of these being peculiar in having the last joint of the column very much elongated and pointed, while the latter possesses long, free brachial processes in place of recumbent arms.

In another small section of the preceding great group are the two well-known Silurian genera *Hemicosmites* and *Caryo-*

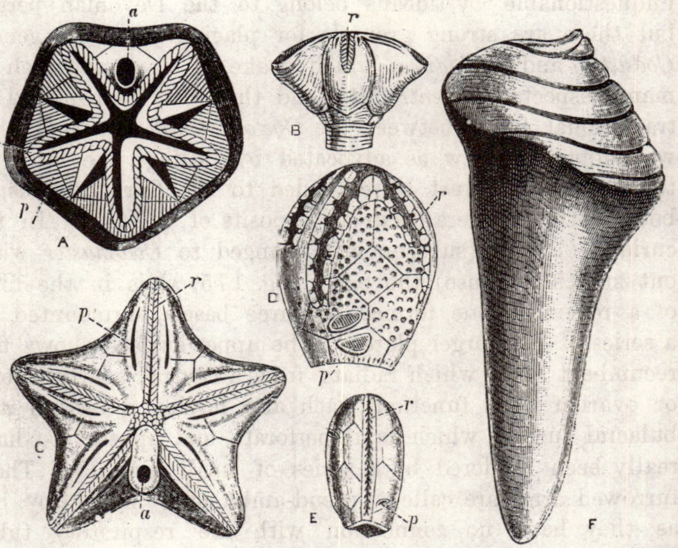

Fig. 175.—Cystideans. A, *Codaster acutus*, showing the upper surface of the calyx, enlarged, from the Carboniferous Limestone (after M'Coy). B, *Codonites stelliformis*, side view of the calyx, of the natural size (after Meek and Worthen), from the Carboniferous Limestone. C, Summit of the calyx of the same, enlarged. D, Side view of the calyx of *Callocystites Jewettii*, of the natural size, showing the brachial furrows, from the Upper Silurian (after Hall). E, Side view of the calyx of *Apiocystites pentrematoides*, showing one of the brachial grooves, of the natural size (after E. Forbes). F, Side view of the cast of the calyx of *Gomphocystites glans*, showing the inflated summit, and spiral brachial grooves, of the natural size—Upper Silurian (after Hall). *a*, Anal aperture; *p*, Respiratory fissures, or rhombs; *r*, Recumbent arm, or pseud-ambulacrum.

crinus, which agree with one another in having numerous pores on each plate, these pores being separate externally, but united in pairs by canals which run on the internal faces of the plates. *Hemicosmites* (fig. 171) has an ovoid calyx, the respiratory pores being disposed on each plate in regular lines, and no arms are known. In *Caryocrinus* (fig. 172, A), on the other hand, the ovoid calyx is surmounted by 12-20 free jointed pinnulæ arranged in three groups. There is a long and cylindrical column, and each plate carries a series of respiratory pores or spongy tubercles, in lines radiating from its centre. The genus is a link between the *Cystoidea* and the *Crinoidea.* As before remarked, the last unquestionable Cystideans belong to the Devonian period, but there are strong grounds for placing here the genera *Codaster* and *Codonites,* which make a close approach in many respects to *Pentremites,* and thus may be regarded as transitional forms between the *Cystoidea* and *Blastoidea.* If we adopt this view, as advocated by Billings, the range of the Cystideans must be extended to the Carboniferous, as both of these genera occur in deposits of this age. In the curious *Codaster* (subsequently changed to *Codonaster* without sufficient cause) the calyx (fig. 175, A) is in the form of a reversed cone, formed by three basals, surmounted by a series of five larger plates. The upper surface shows five recumbent arms, which radiate from a central aperture (oral or ovarian (?) in function), each arm having a central ambulacral furrow, which is imperforate, and appears to have really been bordered by a series of small pinnulæ. These furrowed arms are called "pseud-ambulacra" by M'Coy, but as they have no connection with the respiratory tubes ("hydrospires"), they do not correspond precisely with the structures so named in the Blastoids. On the other hand, there exist five large fissured areas, which correspond in structure and function with the "pectinated rhombs" or "hydrospires" of the true Cystoids, of which two are incomplete, and are placed on each side of a large lateral (oro-anal (?) or anal) aperture, while the other three are bisected by the recumbent arms.

In *Codonites* (fig. 175, B and C) the general form of the calyx is much the same as in *Codaster,* and the summit carries five recumbent arms (or "pseud-ambulacra"), which have a central furrow, and probably carried pinnulæ, and which agree with the corresponding structures in *Codaster,* and differ from those of *Pentremites,* in not being perforated by any apertures. There is, further, no central orifice at the point where the brachial furrows coalesce; but there is a large lateral aperture (supposed to be oro-anal) in one of the inter-radial spaces. The respiratory organs or "hydrospires," lastly, are reduced to ten fissures situated on the sides of, and parallel with, the brachial grooves.

Fossils of Uncertain Affinities.

We may very briefly consider here four genera which have been at one time or other regarded as belonging to the *Cystoidea*, but which probably have really affinities of a very different nature, though their exact zoological position is still open to doubt The genera in question are *Pasceolus*, *Sphærospongia*, *Nidulites*, and *Cyclocrinus*, all of them found in the Silurian or Devonian rocks, all obviously related to one another in form and general structure, and all as yet partially understood. If they are Cystideans, they can only be regarded as a peculiar and very aberrant group of this order; but it would seem more likely that they will find their true place in the vicinity of *Receptaculites* and *Ischadites*.

The genus *Pasceolus* (fig. 176, *a* and *b*) was created by Mr Billings for the reception of some curious Silurian fossils of an ovate or globular form, having an integument composed of hexagonal or pentagonal plates, closely united with one another, apparently deeply concave towards the interior, and furnished with a lateral aperture, which is said to be surrounded by six plates differing in form from the rest. A peduncle is also supposed to have been present. Mr Billings first supposed that *Pasceolus* might possibly belong to the *Tunicata*, but he subsequently came to the conclusion that its zoological position was entirely uncertain; while Messrs Verrill and Niles seem to regard it as a Cystidean.

The genus *Sphærospongia* is in a hardly more satisfactory position. It was founded originally by Salter for fossils such as the *Sphæronites tesselatus* of the Devonian, believed generally to be of a Cystidean nature, and in general structure nearly resembling *Pasceolus*. The body (fig. 176, *c*) is ovate or pyriform, hollow within, and covered by hexagonal or pentagonal plates (fig. 176, *d*), but there is no evidence of any aperture in the test, nor do we know anything as to its mode of attachment. Salter regards the genus as referable to the Sponges.

Cyclocrinus, again (fig. 176, *e*), has a globular body, also covered with hexagonal or pentagonal plates, the apex being

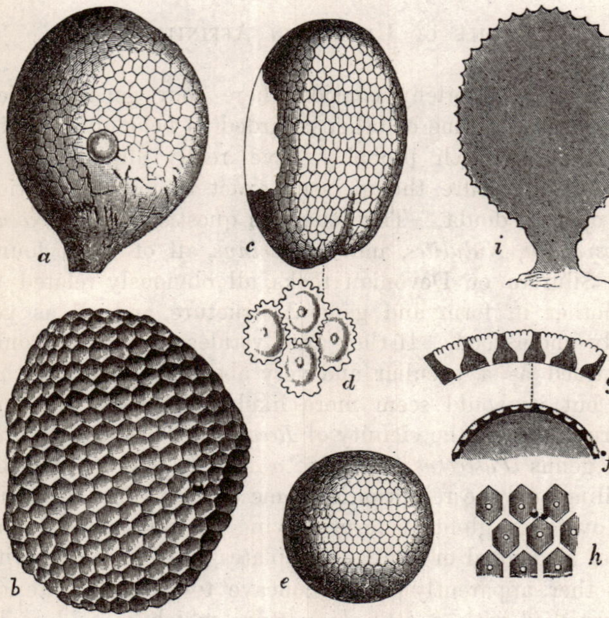

Fig. 176.—*a*, *Pasceolus Halli*, of the natural size (after Billings); *b*, *Pasceolus globosus*, of the natural size (after Billings); *c*, *Sphærospongia melliflua*, of the natural size (after Salter); *d*, Four of the integumentary plates of the same, enlarged; *e*, *Cyclocrinus Spaskii*, of the natural size (after Ferd. Roemer); *f*, Part of a vertical section of the same; *g*, Part of the last, showing the structure of the integument, enlarged; *h*, Part of the mould of the inner surface, enlarged; *i*, Diagram of a vertical section of *Nidulites favus*, showing the form of the body and integumentary plates, and the supposed peduncle (original). All the specimens are from the Silurian.

provided with an aperture and the opposite extremity with a small peduncle. The plates themselves are cup-like, of considerable thickness (fig. 176, *f* and *g*), with an inward central projection which leaves a median pit in casts of the interior (fig. 176, *h*).

Lastly, in *Nidulites* (fig. 176, *i*) we have ovate, globular, or pyriform, hollow bodies, which were probably attached by a pedicle, and which have an integument composed of closely approximated hexagonal plates. No aperture in the test has been observed, and the plates are deeply concave, each with an inward projection, which leaves a corresponding pit in the cast.

Various views have been held as to the affinities of all these bodies, but the most probable conjecture, as before remarked, is that they are allied to *Receptaculites,* and are therefore large and aberrant forms of *Rhizopoda.* They are nearly related to one another, and may prove all to belong to a single generic type. All are found in the Silurian rocks, and the only one which survives into the Devonian is *Sphærospongia.*

ORDER VI.—BLASTOIDEA.

The *Blastoidea* or "Pentremites" are *Echinodermata* in which *the body is enclosed in an armour of closely-fitting calcareous plates, and is fixed to some foreign object by a slender column. From the summit of the calyx radiate five areas which are grooved longitudinally, and striated across, and which carry a row of jointed pinnulæ on each side. No free arms; reproductive organs contained within the calyx; respiratory apparatus of the type of the hydrospires of the Cystideans.*

The *Blastoidea* in many respects closely resemble the Crinoids, but are at once separated from them by the total want of free arms, as well as by the structure of the calyx. They are also nearly allied to those Cystideans which have a definitely-constructed calyx and recumbent arms; but they are separated by the peculiar arrangement of the calycine plates, and by the constant association of the respiratory pores and tubes ("hydrospires") with the arms.

Taking *Pentremites* as the type of the order, we find the body or "calyx" to be of a more or less globose, ovoid, or pyriform shape, supported upon a short, jointed column, and inclosed in an integument of articulated calcareous plates, which are few in number, and have a definite form and arrangement. The lowest plates of the cup (figs. 177, 178, A and B) are three "basals," one small and quadrangular, the other two larger and pentagonal. According to Lyon and Billings, each of these three basals is really composed of an upper and lower piece closely anchylosed together. Above the basals (*b*) come five large plates (figs. 177, 178, *d*), which are usually called "radials," though it is not certain that they correspond with the plates so called in the true Crinoids. Each of these "radial" plates is rendered forked or deeply cleft by the presence of a deep sinus or excavation in its upper margin. This sinus encloses the lower portion

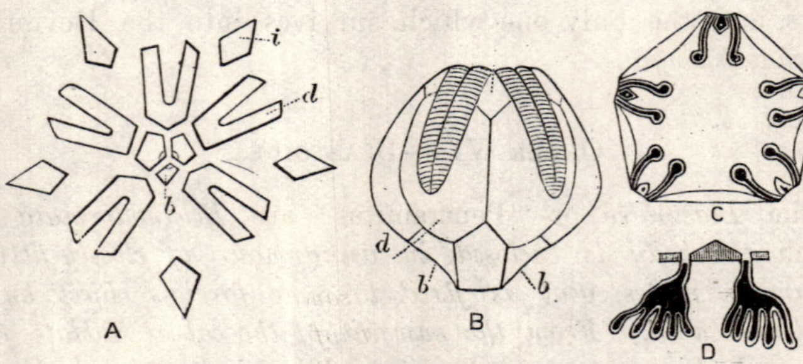

Fig. 177.—Structure of Blastoids. A, Calyx of *Pentremites*, dissected, showing the basals (*b*), the forked "radials" (*d*), and the "deltoid" plates (*i*). B, Side view of the calyx of *Pentremites cervinus*—the letters as before. C, Section across the calyx of *Pentremites ellipticus*, showing the respiratory tubes cut transversely in their course below the pseud-ambulacra. D, Section across one of the pseud-ambulacral areas of *Pentremites florealis*, showing the compound nature of the respiratory tubes, enlarged. (After Hall and Rofe.)

of a pseud-ambulacrum, and these plates, from their shape, are often spoken of simply as the "forked plates." Towards the summit of the calyx the adjacent arms of the fork of each contiguous pair of "radial" plates are separated to receive one of a series of much smaller rhomboidal pieces, which are thus five in number, and are known as the "deltoid" plates (figs. 177, 178, *i*). Between the "deltoid" plates and the forked ends of the "radials" are situated the "pseud-ambulacra." These (figs. 177, 178, *p*) are broad petaloid areas which pass in a radiating manner from the centre of the summit of the calyx to its margins. It is these which give to the summit of the body its resemblance to a flower-bud, upon which the name of the order is founded (Gr. *blastos*, a bud; and *eidos*, form). The petaloid pseud-ambulacra are not particularly well named, as it is certain that they represent a series of five recumbent *arms* —essentially similar to the arms of the Crinoids, but imbedded in the calycine integument, and modified for their connection with the respiratory tubes. Each pseud-ambu-

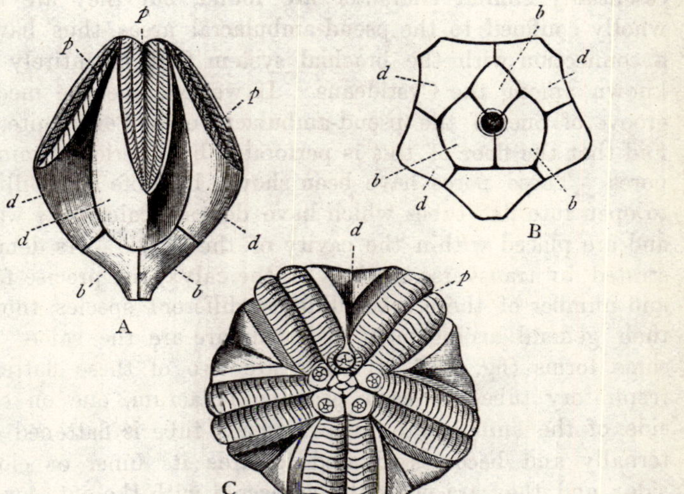

Fig. 178.—Blastoidea. A, *Pentremites pyriformis*, side view. B, Base of the same. C, Summit of *Pentremites conoideus: b, b*, Basals; *d, d*, Radials; *p, p*, Pseud-ambulacra. C, Shows the central pentagonal aperture, surrounded by the five openings at the summit of the deltoid plates. Carboniferous.

lacrum is grooved by a median furrow, or "ambulacral groove," is deeply striated transversely, and, when perfect, carried on each side a series of short jointed filaments, corresponding with the "pinnulæ" of the arms of the Crinoids.

There are two other points as to the calyx of the *Blastoidea* which require notice—namely, the respiratory tubes and the apical apertures. As to the first of these, we have seen that, in the majority of the Cystideans, the test is more or less freely pierced by pores, which admit the sea-water to a series of delicate internal tubes, or sacs, and which may thus reasonably be supposed to have exercised a respiratory function. In some Cystoids these pores and their internal tubes are connected with almost all the plates of the test; in other cases they are restricted to a few (generally three) definite and limited areas—the so-called "pectinated rhombs." In the *Blastoidea* structures of an essentially similar character are found, but they are now wholly confined to the pseud-ambulacral areas, thus having a connection with the brachial system that is entirely unknown among the Cystideans. If we examine the median groove of one of the pseud-ambulacra of a Pentremite, we find that the floor of this is perforated by a series of minute pores. These pores have been shown by Rofe and Billings to open into flat tubes which have delicate calcareous walls, and are placed within the cavity of the calyx. As demonstrated by transverse sections of the calyx, the precise form and number of these tubes differ in different species, though their general arrangement and structure are the same. In some forms (fig. 177, C) there are two of these flattened respiratory tubes to each pseud-ambulacrum, one on each side of the ambulacral furrow. Each tube is flattened externally and becomes dilated towards its inner or closed side; and they are not only connected with the exterior by the minute pores in the ambulacral furrow, but they also open in pairs by apical apertures, which will be immediately described. In other species, though there are really only two tubes to each ambulacral furrow, each of these two becomes compound by a reduplication of its delicate wall, and thus splits up into secondary tubes (fig. 177, D). That the system of pores and internal tubes here described is respiratory in function, and strictly homologous with the system of "hydrospires" or "pectinated rhombs" in the Cystideans cannot be doubted, though there is here the peculiarity of their being directly connected with the recumbent arms.

The pseud-ambulacral areas have been treated above as if composed of a single piece each; but in reality each is compound in its nature. Thus the floor of the ambulacral furrows is formed by the so-called "lancet-plates" of Roemer, which generally, or always, are very narrow bifurcate plates, which start at the centre of the calyx and send a long arm down contiguous sides of any given pair of pseud-ambulacra. Thus each ambulacral furrow is formed by the junction of the arms of two "lancet-plates."

At the summit of the calyx of the *Blastoidea* are usually six apertures. One of these is larger than the others, and is placed in the centre of the calyx, at the point towards which the grooves of the pseud-ambulacra converge. This central aperture has generally, and doubtless rightly, been regarded as the mouth, but Mr Billings dissents from this view. The remaining five apertures (fig. 178, c) are placed at the summit of each deltoid plate, between two pseud-ambulacra, and embraced each by the fork of one of the lancet-plates. Four of the apertures are equal in size, and have generally been regarded as "ovarian;" but they are looked upon by the above-mentioned eminent authority as being connected with the respiratory system, and it is known that the internal respiratory tubes reach the surface by all the five smaller apertures. The fifth is of larger size than the other four, and is usually regarded as partly ovarian and partly anal; but it is looked upon by Mr Billings as "oro-anal."

Like the Cystideans, the *Blastoidea* are not only extinct, but are exclusively Palæozoic. The Cystideans, however, attained their maximum at a very early period of Palæozoic time; whereas the Pentremites flourished most abundantly in the Carboniferous seas, towards the close of the Palæozoic period. The Cystideans had nearly died out before the close of the Upper Silurian period; and it is a noteworthy fact that it is just at that point that the *Blastoidea* seem to make their first appearance. Very sparingly represented in the Upper Silurian rocks, the *Blastoidea* increase largely in numbers during the Devonian period; but they attain their maximum of development in the seas of the earlier portion of the Carboniferous period. With one problematical exception, no member of the order is known to have survived the close of the Carboniferous epoch.

The number of genera included in the order *Blastoidea* is very small, though there is a considerable number of species, and individuals are often very abundant in particular strata.

The genus *Pentremites*[1] itself (fig. 179, *a* and *b*) is much the most important, but as its structure has been taken as typical of that of the entire order, nothing further need be said on this head here. The species of the genus range from the Upper Silurian (*P. Rheinwardtii*, Troost) to the Carboniferous, but they attain their maximum in the latter period, and are not known in any higher deposits. Very nearly allied to *Pentremites* is the genus *Nucleocrinus* (= *Eleacrinus*, Roemer), which differs from the former chiefly in the fact that one of the deltoid plates is split into two by the intercalation of a supplementary anal plate, and by the comparatively small size of the forked "radials" and the correspond-

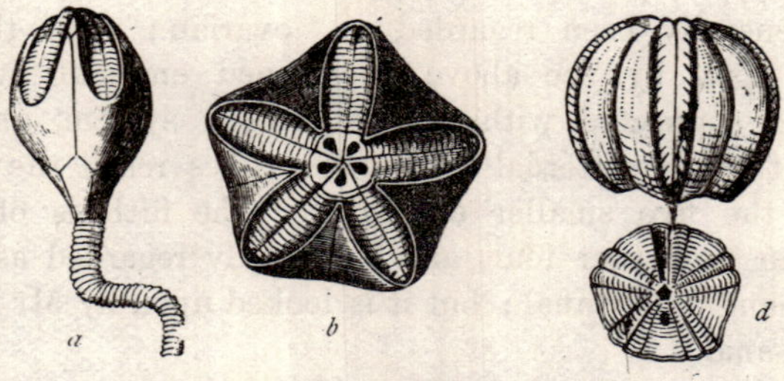

Fig. 179.—Morphology of Blastoidea. *a*, *Pentremites pyriformis*, viewed sideways, showing a portion of the column; *b*, Summit of the calyx of *Pentremites cervinus*, showing the pseud-ambulacral areas and the apical apertures; *c*, Side view of *Granatocrinus melonoides*; *d*, Summit of *Granatocrinus neglectus.* (Figs. *a* and *b* are of the natural size; *c* and *d* are slightly enlarged.) (After Hall, and Meek and Worthen.)

ingly increased length of the "deltoids." The genus is characteristically Devonian, but extends into the Carboniferous. *Granatocrinus*, again (fig. 179, and *d*), is intermediate as regards the relative size of its "radials" and "deltoids" between *Pentremites* and *Nucleocrinus*, and it is peculiar in having the pseud-ambulacra very narrow, and so greatly elongated as almost to divide each radial plate into two pieces. The genus is confined to the Carboniferous rocks. Lastly, we may provisionally place in the *Blastoidea* the extraordinary genus *Astrocrinites*, of the Carboniferous rocks, which agrees with *Pentremites* in general form and structure,

[1] The generic name has been often, but unnecessarily, changed, upon philological grounds, to *Pentatremites* or *Pentatrematites*.

but differs from all known genera of this order in having only *four* pseud-ambulacral areas, in having been free and not provided with any peduncle, and in having the calycine plates ornamented with prominent tubercles.

ORDER VII.—HOLOTHUROIDEA.

The last order of the *Echinodermata* is that of the Holothurians or "Sea-cucumbers," in which *the body is vermiform or slug-shaped, and the calcareous matter secreted by the integument is reduced to scattered spicules* (Synapta), *or rarely is present in the form of imbricated scales* (Psolus).

As might have been expected from the generally soft nature of their integuments, the Holothuroids are hardly known as fossils, and they merely require to be mentioned here. The only remains referred with any probability to this order are certain calcareous spicula which have been found in deposits as old as the Carboniferous, and also in strata of both Mesozoic and Tertiary age, and which have been regarded as belonging to forms related to *Synapta*, while the shield of a species of *Psolus* has been found in Post-Tertiary deposits in Bute.

It seems very probable, however, that the more universal use of the microscope by palæontological observers will result in the discovery of the anchor-shaped spicules and wheels of the *Synaptidæ* in all the more modern, at any rate, of the formations which compose the crust of the earth.

LITERATURE.

GENERAL.

1. "Die Klassen und Ordnungen des Thierreichs." Vol. ii., Strahlenthiere. Bronn. 1860.
2. "Ueber den Bau der Echinodermen." J. Müller. Abhandl. der Berl. Akad. 1853.
3. "Lethæa Geognostica." Bronn. Third edition. 1851-56.
4. "Index Palæontologicus." Bronn. 1850.
5. "Catalogue of British Fossils." Morris.

ECHINOIDEA.

6. "Monographies d'Echinodermes vivants et fossiles." Louis Agassiz. 1838-41.

7. "Catalogus systematicus ectyporum Echinodermatum fossilium, &c. Louis Agassiz. 1840.
8. "Catalogue raisonné des familles, des genres, et des espèces de la classe des Echinodermes." Louis Agassiz and Desor. 1847.
9. "Monograph of the British Fossil Echinodermata." Thomas Wright. 'Palæontographical Society.'
10. "Monograph of the Echinodermata of the British Tertiaries." Edward Forbes. 'Palæontographical Society.'
11. "Paléontologie Francaise, Terrains Crétacés," tom. vi. and vii. 'Echinodermes.' A. D'Orbigny and G. Cotteau. 1856-57 and 1861-67. (Still in process of continuation by Cotteau.)
12. "Synopsis des Echinides fossiles." E. Desor. 1855-59.
13. "Palæozoic Fossils." M'Coy. 1851. (*Perischoechinidæ.*) Also "Carboniferous Fossils of Ireland." 1844.
14. "Echinothuridæ and Perischoechinidæ." R. Etheridge, jun 'Quart. Journ. Geol. Soc.,' vol. xxx. pp. 307-316. 1874.
15. "Illinois Geological Report." Messrs Meek and Worthen. Vol. ii. 1866. (Perischoechinidæ.)
16. "Notes on Palæozoic Echini." W. Keeping. 'Quart. Journ. Geol. Soc.,' vol. xxxii. pp. 35-41. 1876.
17. "Petref. Germaniæ." Goldfuss. 1826.
18. "Echinoides tertiaires de la Suisse." De Loriol. 1876.

ASTEROIDEA AND OPHIUROIDEA.

19. "System der Asteriden." Müller and Troschel. 1843.
20. "British Fossil Echinodermata." 'The Asteroidea.' Thomas Wright. 'Palæontographical Society.'
21. "British Fossil Asteriadæ." 'Mem. Geol. Survey,' vol. ii., Part ii., and Decade iii. Edward Forbes. 1848 and 1850.
22. "New Palæozoic Star-fishes." Salter. 'Ann. and Mag. Nat. Hist.,' ser. 1, vol. xx. 1857.
23. "Asteriadæ of the Lower Silurian Rocks of Canada." Billings 'Geol. Surv. of Canada,' Decade iii. 1858.
24. "The Genus Palæaster," &c. James Hall. 'Twentieth Ann. Rep on the State Cabinet.' 1868.

CRINOIDEA, CYSTOIDEA, AND BLASTOIDEA.

25. "A Natural History of the Crinoidea," &c. J. S. Miller. 1821.
26. "Memoir on the Pentacrinus Europæus." J. V. Thomson. 1827
27. "Histoire naturelle des Crinoides vivants et fossiles." A. D'Orbigny 1840. (Unfinished.)
28. "Ueber den Pentacrinus Caput-Medusæ." J. Müller. 1843.
29. "Recherches sur les Crinoides du terrain carbonifère de la Belgique." De Koninck and Le Hon. 1854.
30. "The Crinoidea of the Lower Silurian Rocks of Canada." E Billings. 'Figures and Descriptions of Canadian Organic Remains,' Decade iv. 1859.

31. "Monographie der Echinodermen der Eifler Kalk." Schultze. 'Denkschr. der K. Akad. der Wiss.' 1867.
32. "Monograph of Recent and Fossil Crinoids." T. Austin and T. Austin, jun. 1843. (Unfinished.)
33. "On the Cystideæ of the Lower Silurian Rocks of Canada." E. Billings. 'Figures and Descriptions of Canadian Organic Remains,' Decade iii. 1858.
34. "Notes on the Structure of the Crinoidea, Cystidea, and Blastoidea." E. Billings. 'Palæozoic Fossils,' vol. ii. 1874.
35. "Ueber Cystideen." Von Buch. 1845.
36. "Russische Sphæroniten." Von Volborth. 1846.
37. "British Fossil Cystideæ." Edward Forbes. 'Mem. Geol. Survey,' vol. ii. part ii. 1848.
38. "Monographie der fossilen Krinoideen-Familie der Blastoideen." Ferdinand Roemer. 1851.
39. "Notes on Echinodermata." Rofe. 'Geol. Mag.,' Dec. i., vol. ii. 1865.
40. "Palæontology of New York." James Hall. Vols. i.-iii. 1843-1859. (Descriptions of Silurian Crinoids and Cystideans.)
41. "Geology of Iowa.," vol. i. part ii. James Hall. 1858. (Descriptions of Carboniferous Crinoids and Blastoids.)
42. "Twentieth Report on the State Cabinet." James Hall. 1868. (Silurian Cystideans.)
43. "Palæontology of Illinois," vol. v. Meek and Worthen. 1873. (Carboniferous Crinoids and Blastoids.)
44. "Palæontology of Ohio," vols. i. and ii. Meek. 1873 and 1875. (Descriptions of Silurian and Carboniferous Crinoids.)
45. "Lethæa Rossica." Eichwald. 1860.
46. "Petref. Germaniæ." Goldfuss. 1826.
47. "Holopus and Cyathidium." Sir C. Wyville Thomson. 'Proc. Roy. Soc., Edin,' vol. ix. 1877.
48. "Structure of Palæozoic Crinoids." Wachsmuth. 'Amer. Journ. Sci. and Arts.' 1877.
49. "The Palæozoic Crinoids." Sir C. Wyville Thomson. 'Trans. Roy. Soc., Edin.' 1871.
50. "Structure and Classification of Crinoids." Lütken. 'Geol. Mag.,' Dec. i., vol. iv. 1867.
51. "Structure of Crinoidea." Beyrich. 'Ann. and Mag. Nat. Hist.,' ser. iv., vol. viii. 1871.

CHAPTER XVI.

SUB-KINGDOM IV.—ANNULOSA.

FOSSIL ANNELIDA.

SUB-KINGDOM ANNULOSA.—The Annulose animals are distinguished by the possession of *a body which is composed of numerous segments arranged longitudinally one behind the other. A nervous system is always present, and consists of a double chain of ganglia running along the ventral surface of the body and traversed anteriorly by the gullet. The limbs (when present) are turned towards that side of the body upon which the chief masses of the nervous system are situated.*

The sub-kingdom *Annulosa* may be divided into two primary sections, according as the body is provided with articulated appendages or not; these divisions being known respectively as the *Arthropoda* (or *Articulata*) and *Anarthropoda*. The first of these comprises the Crabs, Lobsters, and the like (*Crustacea*), the Spiders and Scorpions (*Arachnida*), the Centipedes and Millipedes (*Myriapoda*), and the true Insects (*Insecta*). The latter comprises the Spoon-worms (*Gephyrea*), the Arrow-worms (*Chætognatha*), the Leeches (*Hirudinea*), the Earth-worms (*Oligochæta*), the Tube-worms (*Tubicola*), and the Sand-worms or Sea-worms (*Errantia*); the last four groups constituting the class of the Ringed Worms or *Annelida*.

Regarded as a whole, the great Annulose sub-kingdom seems to have commenced at least as early as the *Echinodermata* and the *Cœlenterata*. Both the Anarthropodous and

Arthropodous divisions of the sub-kingdom are represented in the Upper Cambrian; and the former, at any rate, is represented in the Lower Cambrian period, along with the earliest traces of life known to us, except the *Eozoön* of the Laurentian Series.

ANNELIDA.

In the Anarthropodous division of the *Annulosa* the locomotive appendages are never distinctly jointed or articulated to the body; and the integument, though usually capable of secreting chitine or horny matter, is almost always quite soft and flexible. The Spoon-worms (*Gephyrea*), the Arrow-worms (*Chætognatha*), and two orders of the Annelides (viz., the Leeches and the Earth-worms), are wholly unknown in the fossil condition, and need not be considered here. There remain only two orders of the Annelides (viz., the Tube-worms or *Tubicola*, and the Sand-worms or *Errantia*) which come under the observation of the palæontologist, and neither of these requires much notice. In both orders, as throughout the division, the integument is more or less soft, and there are no internal hard structures, except, occasionally, horny jaws; hence it is doubtful if we have any example of the fossilised body of these creatures, though such have been alleged to occur. The Tubicolous Annelides, however, protect themselves by a tube of lime, sand, or adventitious particles, and these investing tubes are often preserved in the fossil condition. The Errant Annelides, again, besides their fossilised jaws, have left numerous traces of their past existence in the form of filled-up burrows or meandering trails upon the soft sand and mud of the sea-bottom; and from these we know that the Annelides commenced their existence at least as early as the Lower Cambrian period, obscure traces of their presence having been even detected in the Laurentian Series.

ORDER TUBICOLA.—The Tubicolous Annelides are distinguished by the fact that *the body is protected by a tube, within which the animal can withdraw itself by means of tufts of bristles carried on the sides of the body. The gills are placed on or near the head, generally in two lateral tufts; hence the name of "Cephalobranchiate Annelides," applied to this order* (fig. 180).

The protecting tube of the Tubicolous Annelides may be composed of carbonate of lime (*Serpula*), of grains of sand (*Sabellaria*), or of sand, pieces of shell, and other adventitious particles cemented together by a glutinous secretion from the body (*Terebella*); or it may be simply membranaceous or leathery (*Sabella*). Sometimes the tube is free and non-adherent (*Pectinaria*); more commonly it is attached to some submarine object by its apex or by one side (*Serpula* and *Spirorbis*). Sometimes the tube is single (*Spirorbis*); sometimes the animal is social, and the tubes are clustered together in larger or smaller masses (*Sabellaria*).

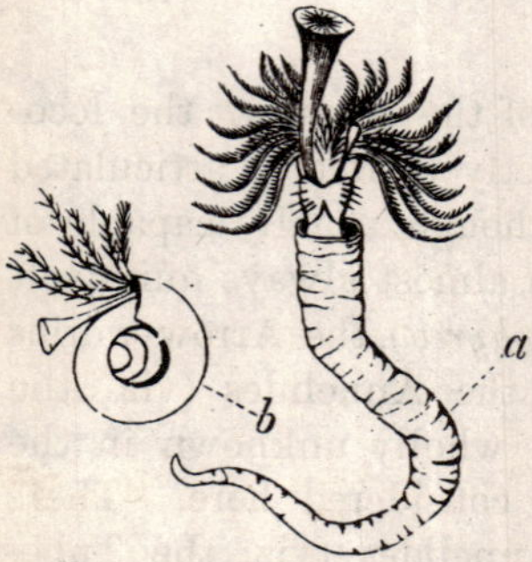

Fig. 180.—Tubicola. *a*, *Serpula contortuplicata*, showing the branchiæ and operculum; *b*, *Spirorbis communis*.

When the tube is calcareous, it presents certain resemblances to the shells of some of the Molluscs, such as *Vermetus* and *Dentalium*. In the living state it is easy to make a distinction between these, for the Tubicolar Annelides are in no way organically attached to their tubes, whereas the Molluscs are always attached to their shell by proper muscles. In the fossil condition, however, it may be very difficult to refer a given calcareous tube to its proper place. As a general rule, however, the calcareous tubes of Annelides, such as *Serpula*, are less regular and symmetrical than those of *Vermetus*, whilst the latter is partitioned by shelly septa, which do not exist in the former. Again, the tube of *Dentalium* is open at both ends, whereas it is closed at one extremity in the *Serpulæ*. In the Annelidous genus *Ditrupa*, however, the tube is open at both ends, so that this distinction is one not universally applicable.

Tubicolar Annelides are known from the Silurian rocks upwards, almost every great period having representatives of the order, though many of the fossils referred to this group are of a more or less problematical nature. The genus *Spirorbis* has survived from the Upper Silurian period to the present day; and forms very nearly allied to, if not actually identical with, the recent *Serpulæ*, are found in almost all formations, beginning with the Silurian.

The chief Palæozoic genera of *Tubicola* are *Cornulites*, *Conchicolites*, *Ortonia*, *Serpulites*, *Trachyderma*, *Spirorbis* (*Microconchus*), and *Serpula*. The genus *Cornulites* (fig. 181) is Silurian, and the best known species is *C. serpularius*. In this singular form the tube is of considerable length—often three or four inches—with a wide aperture at one end, and tapering gradually to its opposite extremity, which is often curved, and seems to have been attached to some solid body. The tube is calcareous, with very thick walls, the substance of which is composed of a number of cellular cavities. Externally the tube is ringed with transverse annulations, and marked with fine longitudinal striæ. The internal cast of the tube has the form of a series of inverted conical rings, of small width, arranged in an imbricated manner. The tube appears to have been solitary, and is rarely found attached. The microscopic structure of the thick wall of *Cornulites* is very peculiar, and exhibits a thin outer laminated layer and a delicate internal wall, the space between these being divided into small lenticular compartments or cells by means of curved calcareous septa. So peculiar is this structure, that it raises some doubt as to the true nature and affinities of *Cornulites*.

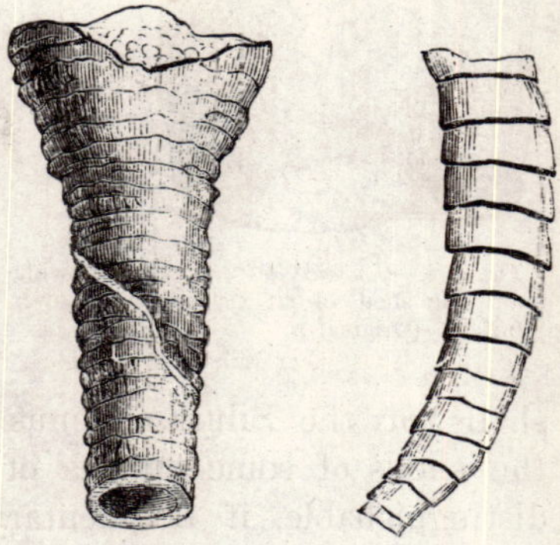

Fig. 181. — *Cornulites serpularius*. Upper Silurian. The right-hand figure represents a cast of the tube. (Original.)

In the genus *Conchicolites* (fig. 182) we have a much smaller Annelide, living socially, and forming masses of clustered tubes growing attached to dead shells in Lower Silurian strata. As in *Cornulites*, the tube of *Conchicolites* appears to have been calcareous, but it is comparatively thin, and has none of the vesicular structure so characteristic of the former. The tube of *Conchicolites* is made up of a series of short conical rings, inserted into one another in an imbricated manner, with their broader ends turned away from the mouth of the tube. It is worthy of notice that the casts of *Conchicolites*, from their possession of the above structure, exhibit a close resemblance to the shells of the Silurian genus *Tentaculites;* whilst *casts* of the shells of some species of the latter are absolutely undistinguishable, if fragmentary, from casts of the tubes of the former. This is a remarkable fact, since *Tentaculites* has often been regarded as a genus of Tubicolar Annelides; but there are strong reasons for believing that it is truly referable to the *Mollusca*, and belongs to the order of the Pteropods.

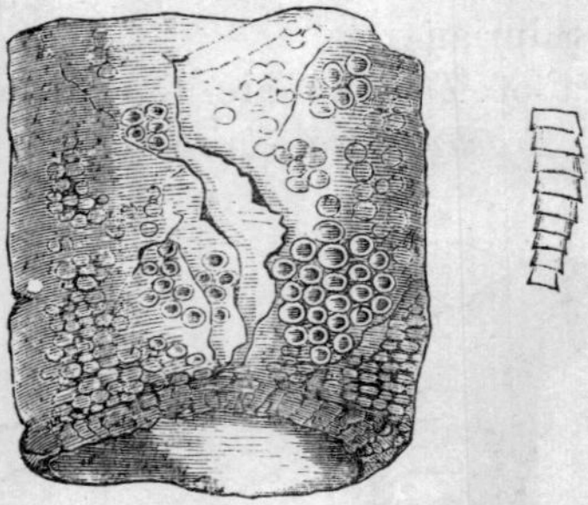

Fig. 182.—*Conchicolites gregarius*, growing upon the shell of an *Orthoceras*. Lower Silurian. (Original.)

Another remarkable genus is *Ortonia* (fig. 183), which ranges from the Silurian to the Carboniferous. In this genus the tube is calcareous, and is adherent along one side to a shell, coral, or other foreign body. By this fact, it is distinguished from *Conchicolites*, which is only attached by the *base* of the tube, and is otherwise free. Moreover, in the latter the tubes are clustered and gregarious, whereas those of *Ortonia*, though frequently found in numbers attached to the same shell, are always individually quite separate. The tubes of *Ortonia* (fig. 183, B) are conical in shape, straight or slightly curved, ringed with imbricating annulations, and either compact or exhibiting an apparent development of a cellular structure on the free surface, similar to that present in *Cornulites*. Though with a wide geological range, the genus seems to be most abundant in the Lower Silurian (Cincinnati group) of North America, and in the Wenlock Limestone (Upper Silurian) of Britain.

The genus *Serpulites* was instituted by Murchison for certain smooth semi-calcareous tubes, often of great length,

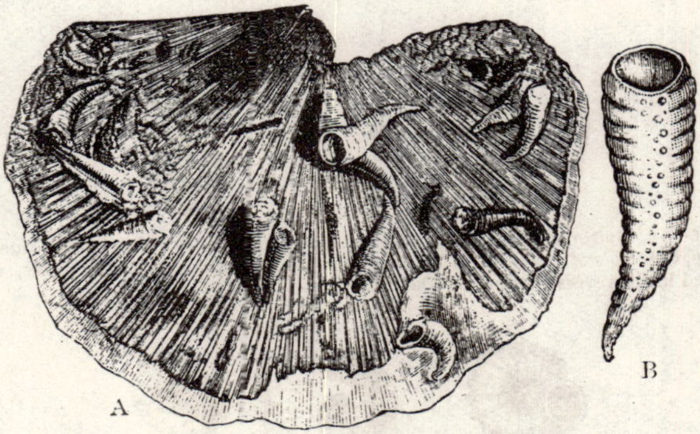

Fig. 183.—A, Tubes of *Ortonia conica* (Nich.) growing upon the valve of *Strophomena alternata*, natural size; B, A single tube of the same, enlarged.

and apparently unattached, which occur in the Silurian series. These tubes in some species reach a length of over a foot, with a diameter of an inch, and their true nature is very doubtful. The genus *Trachyderma*, again, was proposed by Phillips for the casts of membranous flexible tubes which are found in the Silurian rocks. These are transversely wrinkled or plaited, and though the tube itself has disappeared, there can be little doubt entertained as to their Annelidous nature.

The genus *Spirorbis* (figs. 184-186) is characterised by the possession of a shelly calcareous tube, which is coiled into a flat spiral, one side of which is cemented to some foreign body. The spiral may be either right-handed or left-handed, and the shell generally occurs in numbers together, attached to dead shells or to the remains of plants. The genus commences to be represented in the Upper Silurian rocks, in which *S. Lewisii* is an abundant fossil. Other species occur in the Devonian, often in considerable abundance, attached to the shells of Molluscs or the exterior of corals (figs. 184,

185); and similar forms are found in the Carboniferous. The *Spirorbis carbonarius* (fig. 186) of the latter formation is an abundant and well-known type, and is remarkable for being not uncommonly found attached to the exterior of fossil land-plants, which leads to the belief that it must have lived plentifully in the salt-marshes of the Carboniferous coasts.

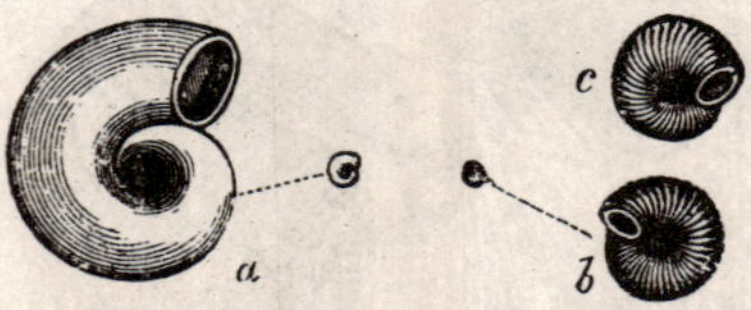

Fig. 184. —*a*, *Spirorbis omphalodes*, natural size and enlarged—Devonian, Europe and America; *b*, *Spirorbis Arkonensis*, of the natural size and enlarged; *c*, The same, with the tube twisted in the reverse direction—Devonian, America. (Original.)

Fig. 185.—*a*, *b*, *Spirorbis laxus*, enlarged—Upper Silurian, America; *c*, *Spirorbis spinuliferus*, of the natural size and enlarged—Devonian, Canada. (After Hall and the Author.)

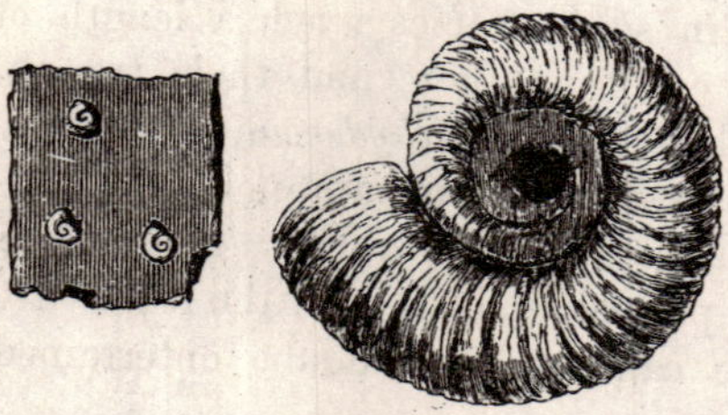

Fig. 186.—*Spirorbis* (*Microconchus*) *carbonarius*, natural size, attached to a fossil plant, and magnified—Carboniferous. (After Dawson.)

Other species have been described from the Permian, and the genus continues to be well represented in both Mesozoic and Tertiary deposits; while living forms, apparently little different from the fossil ones, abound in recent seas.

The genus *Serpula* (fig. 187) possesses a long shelly tube, usually more or less tortuous, sometimes solitary, sometimes aggregated, and fixed to some foreign body by part of its surface. Species of this genus have been described from the Upper Silurian, Devonian, and Carboniferous[1] formations. In the Trias some species occur, and many forms are known from the Oolitic and Cretaceous formations, whilst they are equally numerous in the Tertiary series.

Very nearly allied to *Serpula*, and probably hardly, if at all distinguishable in the fossil condition, are the genera *Filograna* and *Vermilia*. Taking these three genera together, the occurrence of Tubicolar Annelides of this particular type in past time must be said to be fairly frequent, though remains of this nature are naturally more numerous in Mesozoic and Kainozoic strata than in deposits of Palæozoic age.

Fig. 187.—*Serpula flagellum*. Oxford Clay (Jurassic).

Of the other fossils referred to the Tubicolar Annelides, the only one which needs notice is the genus *Ditrupa*. The tube in this genus is unattached, open at both extremities, and very closely resembling the shell of a *Dentalium*. This genus does not seem to have come into existence till the close of the Cretaceous period; but it is found in great abundance in the London Clay (Eocene) and the Crag (Pliocene).

ORDER ERRANTIA.—The Errant Annelides are characterised by the fact that *the body is covered with lateral tubercles carrying tufts of bristles. The animal* (fig. 188) *leads a free life, and is not confined to a tube. The gills are placed along the back or sides of the body; hence the name of "Dorsibranchiate Annelides," often applied to this order.*

[1] The *Serpula parallela* of M'Coy, of the Carboniferous rocks, is now known to be really the siliceous "rope" of a Sponge (*Hyalonema*).

The only fossils which can be asserted positively to be actually portions of the bodies of Errant Annelides are the minute petrifactions representing the horny jaws of Sea-worms, resembling in general form and structure the jaws

Fig. 188.—*Cirrhatulus grandis*, an "Errant Annelide," in its living condition. (After Verrill.)

of the living Nereids. Fossil jaws of this nature have been described from the Lower Silurian under the name of *Nereidavus* (Grinnell), and Mr George Jennings Hinde has detected various forms of these bodies in the Silurian, Devonian, and Carboniferous formations.

Apart from the jaws, the integument of the Errant Annelides always secretes chitinous matter to a certain extent, but in no case does this go so far as to give rise to a regular and resistant exoskeleton. There is, therefore, no ground for surprise if we should be unable to point to any fossils which can be properly regarded as the petrified *bodies* of these animals. It is true that palæontologists have often described fossil remains (such as *Nereites* and *Phyllodocites*) as being of this nature; but no structure has ever been detected in the bodies in question; and till some adequate explanation can be given of the process by which it is possible for the soft body of an Annelide, lying on the shore or in shallow water, to be replaced by an accurate mould in mud or sand, it is best to believe that these fossils are really of the nature of "tracks." Very numerous remains which have been referred to Errant Annelides have been noted as occurring throughout the entire geological series, wherever we meet with muddy or sandy deposits; but the true nature of many of these is still in the highest degree uncertain. Some are, perhaps, really referable to the vegetable kingdom; others are almost certainly formed by Molluscs, or by Crustaceans; others are of entirely dubious affinities; while others are, doubtless, really due to the operation of Errant Annelides. It may be added that the fossil remains which have been referred to Nemertean Worms cannot at present be separated, in any satisfactory manner, from those formed by Errant Annelides. Thus the so-called *Nemertites* of the Silurian is just as likely to be Annelidan as Nemertean, and the nature of the *Legnodesmus* of the Solenhofen Slates is wholly problematical. In fact, the entire subject of the remains of fossil Errant Annelides is one of the most obscure and difficult with which the palæontologist is called upon to deal; and all that can be done here is to glance at some of the leading points of interest connected with it, under the following heads:—

I. *Burrows of Habitation.*—Various living Annelides live buried in the sand or mud, between tide-marks or in shallow water, and communicate with the surface by means of a perpendicular shaft or burrow. Such shafts may, for convenience' sake, be termed "burrows of habitation," though the animal forms a fresh one at will, as it moves from one spot to another; and, as a matter of course, they run in a direction more or less opposed to the surfaces of the laminæ of the rock, being often quite vertical. Sometimes such burrows are hollow, but they are more commonly filled up by the matrix of the rock. The most important genera which have been founded upon remains of this kind are

Scolithus, *Histioderma*, and *Arenicolites*, all of which occur in rocks of Cambrian or Silurian age. *Scolithus* is founded upon long burrows, which are nearly straight, and descend vertically through the rock (fig. 189). They often become somewhat widened out superiorly, and are generally found in great numbers together. They occur abundantly in the Potsdam Sandstone (Upper Cambrian), and Clinton formation (Upper Silurian) of North America, and also in the hard sandstones of the Stiper Stones in Shropshire (Upper Cambrian). They have been supposed to have been formed

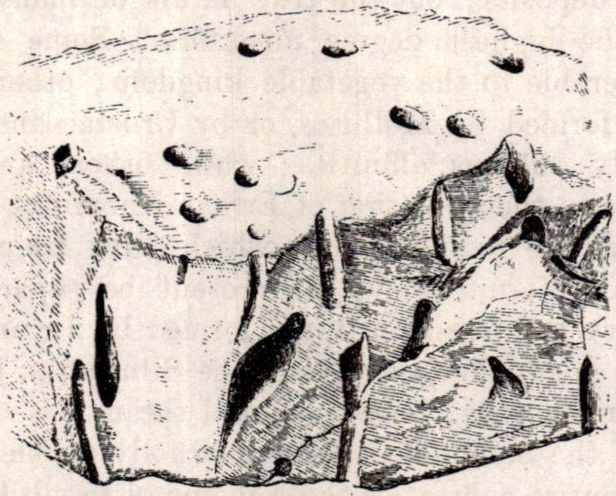

Fig. 189.—Annelide-burrows (*Scolithus Canadensis*), from the Potsdam Sandstone (Upper Cambrian). (After Billings.)

by sea-weeds, but there is little doubt that they are truly the burrows of Annelides. The somewhat problematical fossil upon which the genus *Histioderma* is founded is described as a curved burrow, from one to nearly four inches in length, terminating in a trumpet-shaped opening, which is placed in the centre of a small mound. The genus *Arenicolites*, again (fig. 190), includes small double burrows, which form loops, shaped like the letter U, opening on the surface by two apertures placed close to one another.

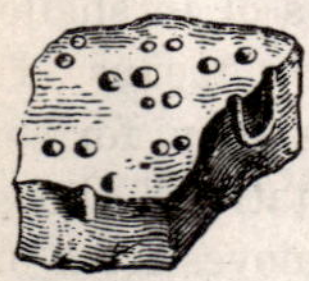

Fig. 190.—Burrows of *Arenicolites didymus*. From the Longmynd (Lower Cambrian).

The mouths of these burrows are thus placed in pairs, one orifice being supposed to be an aperture of entrance for the worm, and the other one of exit. Burrows of this nature occur abundantly in the Lower Cambrian strata of the Longmynd, and are also far from uncommon in deposits of Silurian age.

II. *Wandering Burrows.*—Various recent Annelides, among which the common Lug-worm (*Arenicola piscatorum*) is a notable example, form long, wandering, irregular, and tortuous tunnels in the sand of the sea-shore, at a little distance below the surface. In these cases the worm subsists upon particles of organic matter disseminated through the sand or mud, through which, therefore, it literally eats its way. The burrows thus formed are, consequently, very irregular; they principally have a horizontal direction; if formed by many individuals, they may cross or intersect one another in various ways; and as the worm proceeds on its course, they become filled up in the rear of the advancing animal by the sand which has been passed through the alimentary canal, some of this sand being often voided at some point at the surface to form the tortuous "worm-casts," with which every wanderer on the sea-shore is so familiar. Bodies which we may reasonably assume to be of essentially the same nature as the filled up "wandering burrows" of worms like the living Lug-worm, are well known to all workers amongst the more ancient, muddy, and sandy strata of the earth's crust, and they have received various names, and have had very various origins ascribed to them. They usually present themselves as irregularly cylindrical, worm-like elevations of the surfaces of the strata (fig. 191), which usually are more or less parallel with the laminæ of deposition, but often run somewhat obliquely to these, so as to thread successive laminæ to one another. Generally, they differ slightly in texture and colour from the surrounding rock—as can well be supposed, if their origin be as above described—and from their somewhat superior hardness, they resist disintegration by weathering, and thus come to stand out prominently above the surface. Though they may be much matted together, and may thus appear to branch, it is not probable that they really subdivide. It would appear that many of the fossils of the Palæozoic rocks which have been referred to the Fucoids, under the generic titles *Palæochorda*, *Palæo-*

phycus, &c., are really the filled-up burrows of wandering marine worms, and all such remains may at present be grouped together under the common name of *Planolites*. At the same time, some fossils of a similar general appearance and occurring in similar strata (*Cruziana*, for example) may really be of vegetable origin, and these will be briefly noticed later on.

The fossils known as *Lumbricaria*, so abundant in the Solenhofen slates (Jurassic), have, again, been generally re-

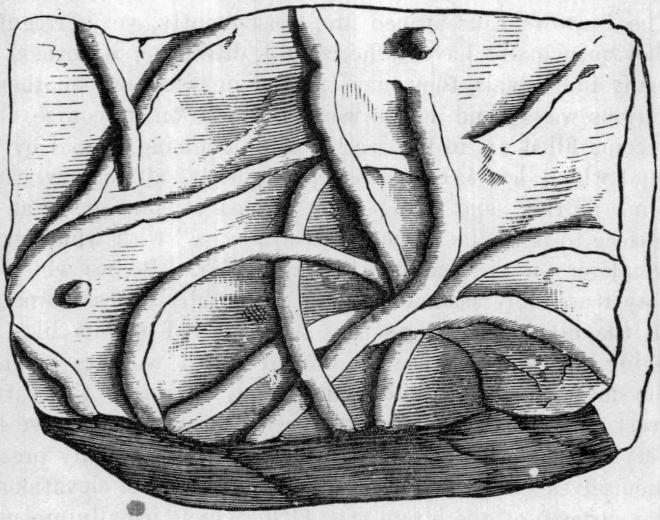

Fig. 191.—*Planolites vulgaris*, the filled-up burrows of a marine worm. Upper Silurian (Clinton Group), Canada. (Original.)

garded as casts of the alimentary canal—true "worm-casts" in fact—of marine worms; and certainly they present every appearance of these. They are stated, however, to consist commonly of *crystalline* carbonate of lime, and this would render their true nature doubtful.

III. *Trails and Tracks.*—Lastly, we have to deal with a great group of fossils which have been supposed to be of the nature of the "trails" of Errant Annelides—that is to say, markings formed by the animal dragging its soft body over the surface of wet sand or mud, between tide-marks or in shallow water. Markings of this nature are extremely abundant in many of the older rocks, and in many cases no doubt can be entertained as to their being really the tracks of some marine animal. Even in these cases, however, it is at present impossible, in the majority of instances, to discriminate between the trails produced by Annelides and those formed by Univalve Molluscs. There are, nevertheless, certain tracks which we may fairly assert to be Annelidan. This is especially true of the Silurian fossils upon which the genera *Nereites* (fig. 192, B) and *Phyllodocites* (fig. 192, A) have been founded. In these cases we have long, sinuous, and often sharply-bent impressions on the surfaces of the strata, which consist of a central, broader or narrower axis, representing the body of the worm, and of a series of lateral, more or less leaf-like markings, representing the foot-tubercles. These tracks, and others like them, have commonly been supposed to represent the actual *body* of the Annelide, now replaced by mud; but, as before remarked, it is very difficult to conceive of such a replacement, and it is more likely that we have simply the trail of the animal formed by its serpentine wandering over the surface of soft mud.

Another fossil, which is extremely abundant in the Silurian rocks of some localities, and which has generally been supposed to be the track of an Annelide, is *Myrianites.* In ordinary specimens of this genus (fig. 192, C) all that is seen is that the surfaces of the strata are marked by winding and tortuous linear impressions, of extremely small comparative width, and easily recognisable from the matrix by their darker colour and slightly different texture. These meandering markings wind over the surface of the stone in indefinite undulations, often appearing to cross one another; and no one, looking at such a specimen, would be inclined to doubt that he had to deal with the trails left upon the mud of the sea-shore by some soft-bodied marine animals, though he might question if these could be Annelides. Other specimens, however, of the same fossil, which have been carefully examined by the author, prove conclusively that, in spite of appearances, *Myrianites* is not only not Annelidan in its nature, but that it cannot possibly be the *track* of any animal whatever. It can be shown, in fact, that the narrow serpentine markings upon the surface of the stone,

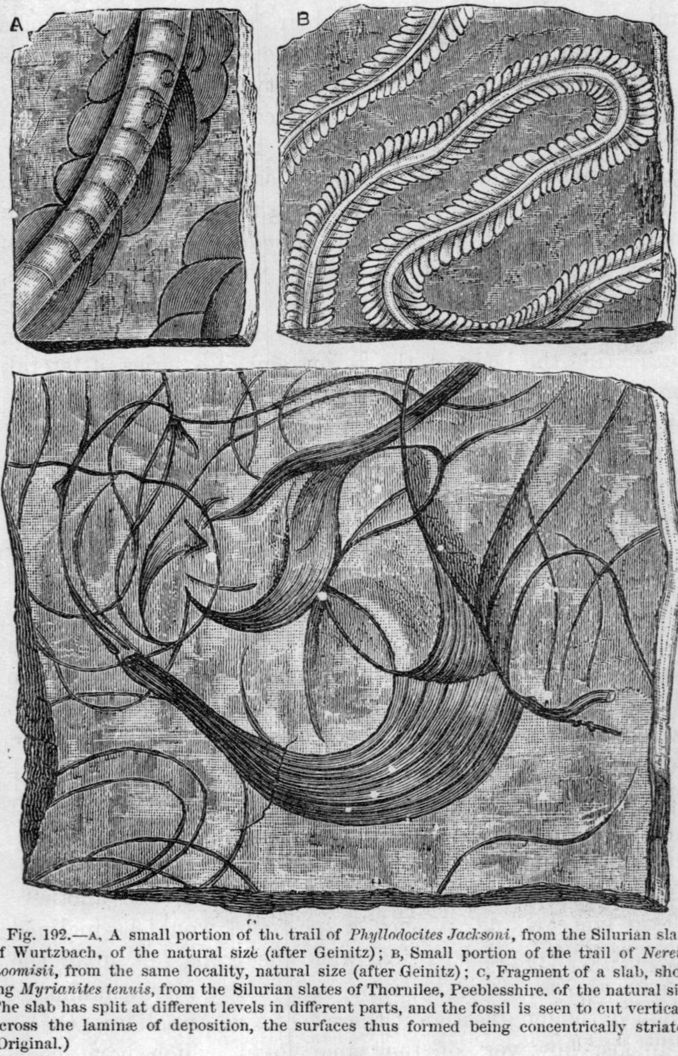

Fig. 192.—A, A small portion of the trail of *Phyllodocites Jacksoni*, from the Silurian slates of Wurtzbach, of the natural size (after Geinitz); B, Small portion of the trail of *Nereites Loomisii*, from the same locality, natural size (after Geinitz); C, Fragment of a slab, showing *Myrianites tenuis*, from the Silurian slates of Thornilee, Peeblesshire, of the natural size. The slab has split at different levels in different parts, and the fossil is seen to cut vertically across the laminæ of deposition, the surfaces thus formed being concentrically striated. (Original.)

which are universally understood under the name *Myrianites*, are really *the cut edges of thin vertical laminar expansions, sinuously folded, and seen in horizontal section.* In fig. 192, C, a portion of one of the specimens referred to is figured, from which it will be seen that the fossil *cuts directly across the laminæ of deposition,* its actual surface (where exposed by exfoliation of a part of the slab) being marked with concentric striæ. It is therefore quite clear that *Myrianites* was really a thin erect, folded, leaf-like expansion, of some kind or another, and that what palæontologists have described under this name is only the horizontally-cut edge of this expansion as seen on the surface of the stratum. What *Myrianites* really is, is quite an open question. It is, perhaps, a peculiar form of Fucoid. That it is not Annelidan seems perfectly certain.

Another fossil which has generally been regarded as referable to the Errant Annelides is the *Crossopodia* of M'Coy (fig. 193), also very abundant in certain Silurian strata. In

Fig. 193.—*Crossopodia Scotica,* a supposed Annelide track. Silurian. (After M'Coy.)

this fossil there is a central narrow groove, which winds in serpentine bends over the surface of the stone, and is supposed to represent the body of the animal, bounded on each side by a broader and generally ill-defined space, supposed to represent the foot-tubercles. That *Crossopodia,* however, should be the petrified *body* of an Errant Annelide seems almost incredible, and that it is even the *track* of one of these creatures is extremely improbable. In well-preserved specimens of any size, the impressions known under this name are seen to wind backwards and forwards over the stone in a succession of long loops which are placed quite close together, and which could hardly have been produced by any animal in a movement of forward progression. There is, indeed, some evidence that the impressions of *Crossôpodia* really cut directly across the laminæ of deposition to some depth, and that they have some direct, though at present not understood, connection with *Myrianites*.

As might have been expected, any fossils which can be supposed with any probability to be the tracks of Annelides, or of other marine animals, present themselves as *depressed* or *concave* markings on the upper surfaces of the strata. The *casts* of these markings, however, are often to be observed on the *under* surfaces of the beds, and these, as a matter of course, present themselves as *convex* or *elevated* impressions. When the beds are vertical, or when the specimens are not found actually *in situ*, it is impossible to distinguish between these two classes of specimens; especially as *some* elevated impressions, supposed to be tracks, do really occur on the *upper* surfaces of the strata. Such impressions, in the opinion of Principal Dawson, "have been left by denudation of the surrounding material, just as footprints on dry snow sometimes remain in relief after the surrounding loose snow has been drifted away by the wind, the portion consolidated by pressure being better able to resist the denuding agency."

Before leaving this obscure subject finally, it may be well to notice briefly one or two considerations which bear upon the question of the origin and real nature of markings such as we have been considering. In the first place, the late Mr Albany Hancock, in an extremely able memoir, advocated the view that the vermiform fossils of the older rocks may have been, in general at any rate, produced by *Crustaceans*. He showed that similar markings are produced, at the present day, by small Amphipod Crustaceans (*Sulcator arenarius* and *Kröyera arenaria*), which burrow immediately below the sand on the sea-shore, and give rise to the following appearances: (1.) Large tracks, about 3-8ths of an inch wide, slightly raised, ribbon-like in shape, with a median groove, often intricate and convoluted, sometimes knotted, and several feet in length; (2.) Narrow wedge-shaped furrows, 2-10ths of an inch wide, winding capriciously and often abruptly over the surface; (3.) Nodulated or articulated tracks, consisting of a small furrow, with a rounded ridge on one side. Mr Hancock showed that tracks of these three kinds are actually produced by the above-named small Crustaceans, which burrow beneath the sand, but a short way below the surface, "the arch or tunnel thus formed partially subsiding, as the creature moves forwards, and breaking along the centre," thus giving rise to a median groove. There is no doubt that the phenomena so carefully observed by Mr Hancock throw considerable light upon the subject of the supposed Annelide tracks of muddy and sandy sediments; but there is room for much hesitation before concluding that any of these tracks, in the older rocks at any rate, were really formed by Crustaceans like the living *Sulcator arenarius.* One ground for such hesitation need alone be brought forward here—namely, that the so-called "Annelide-tracks" of the older Palæozoic rocks often occur in vast numbers, in finely-levigated deposits, and throughout a thickness of sometimes hundreds of feet of strata, and that it is almost inconceivable that traces of the makers should not have been detected in the same beds, supposing them to have been formed by animals, which, like Crustaceans, have a skeleton highly susceptible of preservation in the fossil condition.

Principal Dawson, again, suggests "that Algæ and also land-plants, drifting with tides and currents, often make the most remarkable and fantastic trails," which might easily be mistaken for the tracks of Annelides. This suggestion is a very valuable one, but certainly will not explain the origin of the majority of the so-called "Annelide-tracks" of the Palæozoic rocks, the regular serpentine form of which is one of their most remarkable features. The same distinguished authority remarks that "Lingulæ, when dislodged from their burrows, trail themselves over the bottom like worms, by means of their cirri," and that "colonies of these creatures, so abundant in the Primordial, may, when obliged to remove, have covered the surfaces of beds of mud with vermicular markings." Little weight, however, can be attached to this suggestion, since there is only one existing *Lingula*—namely, the very aberrant *Lingula pyramidata* — which has any power of locomotion in its adult condition; and we have no right to assume that any of the extinct forms were endowed with this quite exceptional faculty, unless distinct evidence to this effect can be brought forward, the occurrence of markings which might possibly have been formed by Palæozoic representatives of *Lingula pyramidata* not constituting evidence of the required kind.

LITERATURE.

1. "Palæontology of New York," vols. i. and ii. 1843 and 1852. James Hall.
2. "Fossils of the Longmynd." Salter. 'Quart. Journ. Geol. Soc.,' vol. xiii. 1857. Also 'Mem. Geol. Survey,' vol. iii., Appendix. 1866.
3. "Vermiform Fossils from the Mountain Limestone." Albany Hancock. 'Ann. and Mag. Nat. Hist.,' ser. 3, vol. ii. 1858.
4. "The Taconic System." Emmons. 1844.
5. "The Silurian System." Murchison. 3d ed. 1859. (Description of Annelides by Macleay.)
6. "Petrefacta Germaniæ." Goldfuss. 1826.
7. "Ueber ein Æquivalent der takonischen Schiefer Nord-Amerikas in Deutschland." Geinitz and Liebe. 'Verhandl. d. K. Leopold. Carolin. deutsch. Akad.' 1867.
8. "Ueber fossile Würmer aus dem lithographischen Schiefer in Bayern." Ehlers. 1869.
9. "Impressions and Footprints of Aquatic Animals on Carboniferous Rocks." Principal Dawson. 'Amer. Journ. Sci. and Arts.' 1873.
10. "Cornulites, Tentaculites, and Conchicolites." Nicholson. 'Amer. Journ. Sci. and Arts.' 1872.
11. "On the Genus Ortonia." Nicholson. 'Geol. Mag.,' Dec. 1, vol. ix. 1872.

CHAPTER XVII.

ARTHROPODA.

CRUSTACEA.

DIVISION ARTHROPODA, OR ARTICULATA.—The members of this division of the sub-kingdom *Annulosa* are distinguished from the preceding division of the *Anarthropoda* by the possession of *jointed appendages, articulated to the body. The body is composed of a series of segments or "somites," arranged along a longitudinal axis; each segment occasionally, and some always, being provided, at some period of life or other, with articulated appendages. Both the segmented body and the articulated limbs are more or less completely protected by an external skeleton formed by the deposition of horny* (*chitinous*) *matter, sometimes along with variable amounts of carbonate and phosphate of lime, in the integument.*

The division *Arthropoda* includes four great classes of animals which are very generally spoken of as the "Articulate Animals." These classes are the *Crustacea* (Lobsters, Crabs, &c.), the *Arachnida* (Spiders, Scorpions, &c.), the *Myriapoda* (Centipedes and Millipedes), and the *Insecta* (Insects). All these classes came into existence in the Palæozoic period, the division being represented by Crustaceans as early as the Upper Cambrian at any rate, and doubtfully in the Lower Cambrian. Owing to the fact that the Crustaceans alone lead an habitually aquatic life, the remains of this class, as might be expected, preponderate largely over those of the other three. The air-breathing classes of the

Arachnida, *Myriapoda*, and *Insecta*, naturally, have not left abundant traces of their existence in past time, a state of things which is assisted by the nature of their integuments, which are rarely as hard and resisting as those of the Crustaceans.

CLASS I.—CRUSTACEA.

The *Crustaceans* are *Articulate animals in which the breathing organs (when distinct) are in the form of gills, and the mode of existence is almost always more or less aquatic. The body is protected by a chitinous or sub-calcareous exoskeleton or "crust," and the number of pairs of articulated limbs is generally from five to seven. Some of the locomotive appendages are often carried upon the segments of the abdomen, and there are two pairs of jointed feelers or "antennæ."*

The body of a typical Crustacean, such as a Lobster (fig. 194), consists of a definite number of somites placed one behind the other, and divisible into three regions—a *head*, *thorax*, and *abdomen*. Most authorities regard the body as being typically composed of twenty-one somites, of which seven go to the head, seven to the thorax, and seven to the abdomen. All these somites, except the last, may be provided with a pair of appendages each. The last segment of the abdomen, however, never carries any appendages. This segment is known as the "telson" (fig. 194, 1, *t*), and it is variously regarded as a somite without appendages, or as an *unpaired appendage* placed in the middle line of the body. If this latter view be adopted, the body of a typical Crustacean will consist of only twenty segments, instead of twenty-one. The telson is very greatly developed in some Crustaceans, such as the King-crabs, and less so in the extinct *Eurypterida*.

Generally speaking, a greater or less number of the somites are amalgamated together, rendering it difficult to recognise their existence unless they bear appendages—each pair of appendages indicating a separate somite. Very commonly the segments of the head and thorax are welded together into a single mass, which is termed the "cephalothorax," and which really consists of fourteen coalescent segments. The cephalothorax is generally covered by a great shield or buckler, which is termed the "carapace" (fig. 194, 1, *ca*), and which is produced by an enormous development of the dorsal walls of one or two of the cephalic somites.

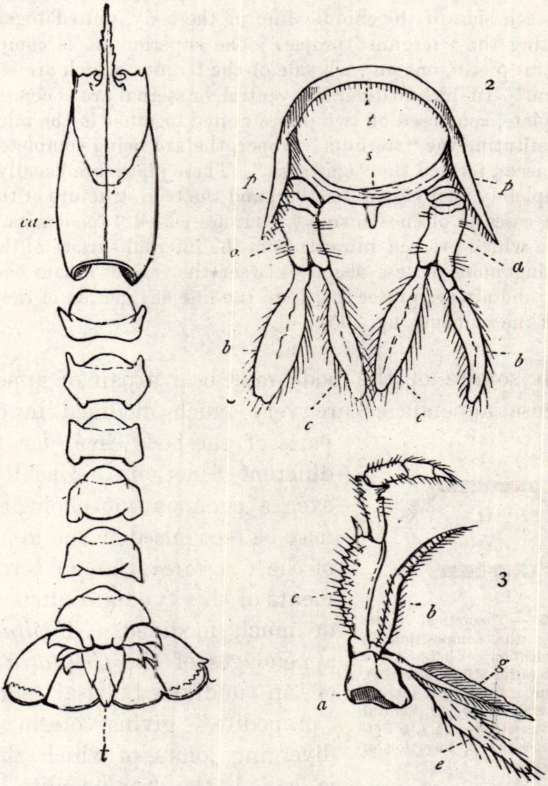

Fig. 194.—Morphology of Lobster. 1. Lobster with all the appendages, except the terminal swimmerets, removed, and the abdominal somites separated from one another: *ca*, Carapace; *t*, Telson. 2. The third abdominal somite separated: *t*, Tergum: *s*, Sternum; *p*, Pleuron; *a*, Propodite; *b*, Exopodite; *c*, Endopodite. 3. One of the last pair of foot-jaws or maxillipedes: *e*, Epipodite; *g*, Gill; the other letters as before.

Each segment of the body may be regarded as essentially composed of a convex upper plate, termed the "tergum," which is closed below by a flatter plate, called the "sternum," the line where the two unite being produced downwards and outwards into a plate which is called the "pleuron," or "pleura" (fig. 194, 2).

Strictly speaking, the composition of the typical somite is considerably more complex, each of the primary arcs of the somite being really composed of four pieces. The tergal arc is composed of two central pieces, one on each side of the middle line of the body, united together, and constituting the "tergum" proper. The superior arc is completed by two lateral pieces, one on each side of the tergum, which are termed the "epimera." In like manner the ventral or sternal arc is composed of a central plate, composed of two pieces united together in the middle line, and constituting the "sternum" proper, the arc being completed by two lateral pieces, termed the "episterna." These plates are usually more or less completely anchylosed together, and the true structure of the somite in these cases is often shown by what are called "apodemata." These are septa which proceed inwards from the internal surface of the somite, penetrating more or less deeply between the various organs enclosed by the ring, and always proceeding from the line of junction of the different pieces of the segment (fig. 195).

Each somite of the body may bear a pair of appendages, and these appendages are very much modified in different parts of the body, in order to fulfil different functions. Usually, however, a common morphological type may be recognised in the appendages of the *Crustacea*, though certain elements of this type are often wanting or much modified. *Typically*, the appendages of the *Crustacea* consist of an undivided basal portion or "propodite," giving origin to two diverging joints, of which the inner is called the "endopodite," whilst the outer is known as the "exopodite." In such an appendage as the "swimmeret" of a Lobster (fig. 194, 2), these fundamental parts are readily recognisable; but either the exopodite or endopodite, or both, may be wanting, or they may be very much modified in shape and form.

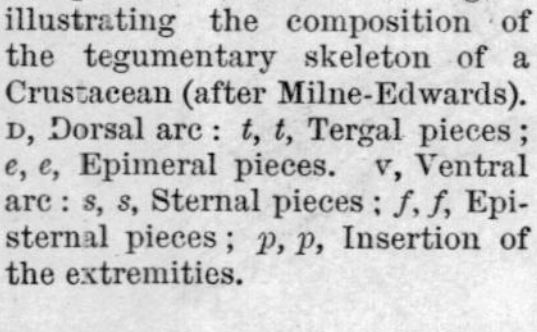

Fig. 195. — Theoretical figure illustrating the composition of the tegumentary skeleton of a Crustacean (after Milne-Edwards). D, Dorsal arc: *t*, *t*, Tergal pieces; *e*, *e*, Epimeral pieces. V, Ventral arc: *s*, *s*, Sternal pieces; *f*, *f*, Episternal pieces; *p*, *p*, Insertion of the extremities.

It is impossible to give any general view of the appendages of a Crustacean; but it may be as well to name the appendages which are present in one of the higher forms, such as the Lobster, in which all the somites, except the telson, carry a pair of appendages each. The somites of the head and thorax are amalgamated into a single mass, termed the "cephalothorax," which is protected above by the "carapace," and carries the appendages on its lower surface. The *first* segment of the head carries a pair of eyes, which are "compound," and are borne upon long stalks, formed by the propodite of the appendage. The *second* segment of the head carries a pair of small jointed feelers, which are known as the "lesser antennæ" or "antennules." Each consists of a short propodite, and a much-segmented endopodite and exopodite, which are nearly of equal length. The *third* segment of the head carries a pair of very long feelers, which are known as the "great antennæ." Each consists of a short propodite and a long and jointed endopodite, with a rudimentary exopodite. The *fourth* segment of the head carries a pair of jaws, which are known as the "mandibles." Each mandible consists of a large propodite, with no exopodite, but with a small endopodite, which is known as the "mandibular palp." Between the bases of the mandibles also is placed the aperture of the mouth, which is bounded *in front* by a plate, known as the "labrum" (upper lip) or "hypostoma," and *behind* by a forked plate, known as the "labium" (lower lip) or "metastoma." The *fifth* segment carries another pair of jaws, which are known as the first pair of "maxillæ;" whilst the *sixth* segment carries another pair of the same, known as the second pair of "maxillæ." The *seventh* and last segment of the head carries the first of three pairs of what are generally known as "foot-jaws" or "maxillipedes." Each foot-jaw is merely an ordinary limb, consisting of propodite, exopodite, and endopodite, but modified to assist in mastication. The *eighth* segment (the first of the thorax) carries a second pair of foot-jaws, and the *ninth* segment (the second of the thorax) bears a third pair of the same. The *tenth* segment (the third of the thorax) carries a pair of jointed limbs, consisting of propodite and endopodite alone, without any exopodite. These limbs are greatly developed, and their extremities form a pair of pincers or "chelæ," so that they constitute the "nipping-claws" of the Lobster.

The *eleventh* segment (the fourth of the thorax) carries a second pair of limbs, also "chelate," but much smaller than the preceding; and the *twelfth* segment (the fifth of the thorax) carries another pair of the same. The *thirteenth* segment (the sixth of the thorax) carries a pair of limbs like the preceding, but with simply-pointed extremities; and the *fourteenth* segment (the last of the thorax) carries another pair of the same; so that there are altogether five pairs of ambulatory limbs, carried respectively by the 10th, 11th, 12th, 13th, and 14th somites of the body; or, in other words, by the last five segments of the thorax. Of the seven segments of the abdomen—completing the total of twenty-one—the first six carry each a pair of appendages, which are used as swimming organs, and which are termed the "swimmerets." Each swimmeret (fig. 194, 2) consists of a propodite and a flattened exopodite and endopodite; and the last pair is greatly widened out and expanded, forming with the telson a powerful swimming-tail. The telson or last abdominal segment carries no appendages, and is simply placed between the last pair of swimmerets.

As regards the general distribution of the *Crustacea* in time, remains of the class are comparatively abundant in all formations except the very oldest; as might have been expected from the generally chitinous or sub-calcareous nature of their integuments and their aquatic habits. Owing also to their habit of periodically casting their shell, a single individual may leave repeated traces of himself, and the number of fossils may considerably exceed that of the individuals which actually underwent fossilisation. The Crustaceans appear to have commenced their existence in the Cambrian period, remains of members of this class being tolerably abundant in the higher portion of this formation. The Palæozoic formations, taken as a whole, are characterised by the predominance of the orders *Trilobita*, *Eurypterida*, *Ostracoda*, and *Phyllopoda*, of which the two former are exclusively confined to this period. All the other orders of *Crustacea*, which have left any traces of their past existence at all, appear to have come into existence before the close of the Palæozoic period. Upon the whole, however, there has been a marked progression in proceeding from the older formations to the present day. The Trilobites and Eurypterids of the older Palæozoic rocks, though highly organised so far as their type is concerned, are in many respects inferior to later forms, whilst they present some striking points of resemblance to the larval forms of the higher groups. The great group of the Stalk-eyed Crustaceans—undoubtedly the highest of the entire class—is not represented at all till we reach the Carboniferous rocks: and it is not till we come into the Secondary period that we find any great development of this group, whilst its abundance increases to a marked extent in the Tertiary period, and it attains its maximum at the present day. Similarly, of the two sub-orders of the *Merostomata*, the *Eurypterida* are confined to the earlier portion of the Palæozoic period, whilst the more highly organised and less larval King-crabs (*Xiphosura*), with few exceptions, did not make their appearance till the Eurypterids had disappeared, at the close of the Carboniferous period.

The following table shows the orders of the *Crustacea*, and a short account will be given of the distribution in time of those which are known to occur as fossils. The structure also of the extinct groups will be shortly described. The orders marked with an asterisk do not occur as fossils, or only doubtfully so, and will not be considered here.

TABULAR VIEW OF THE DIVISIONS OF THE CRUSTACEA.

Sub-class I. EPIZOA (*Haustellata*).
Order 1. *Ichthyophthira.**
„ 2. *Rhizocephala.**

Sub-class II. CIRRIPEDIA.
Order 3. *Thoracica.* { Balanidæ. Verrucidæ. Lepadidæ.
„ 4. *Abdominalia.**
„ 5. *Apoda.**

Sub-class III. ENTOMOSTRACA.
Order 6. *Ostracoda.* } *Legion*, Lophyropoda.
„ 7. *Copepoda.** }

Order 8. *Clădocera.**
,, 9. *Phyllopoda.*
,, 10. *Trilobita.*
} *Legion*, Branchiopoda.
,, 11. *Merostomata.*

Sub-class IV. MALACOSTRACA.
Division A. EDRIOPHTHALMATA.
Order 12. *Læmodipoda.**
,, 13. *Isopoda.*
,, 14. *Amphipoda.*
Division B. PODOPHTHALMATA.
Order 15. *Stomapoda.*
,, 16. *Decapoda.*
Tribe *a. Macrura.*
,, *b. Anomura.*
,, *c. Brachyura.*

SUB-CLASS CIRRIPEDIA.

Animal free when young, but permanently attached in the adult condition to some foreign body by the anterior extremity of the metamorphosed head. The visceral cavity of the adult

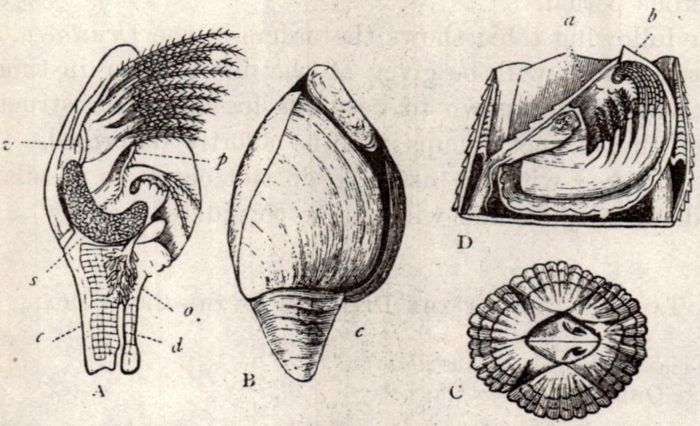

Fig. 196.—Morphology of Cirripedia. A, *Lepas pectinata*, one of the Barnacles, one side of the shell being removed, enlarged four times: *c*, Peduncle; *d*, Cement-duct; *o*, Ovary; *s*, Ovisac; *v*, Vas deferens; *p*, Penis. B, *Pœcilasma fissa*, enlarged five times: *c*, Peduncle. C, *Balanus balanoides*, viewed from above, of the natural size. D, *Balanus tintinnabulum*, with the shell on one side removed to show the animal: *a*, One of the valves ("scutum") of the operculum; *b*, Another valve ("tergum") of the operculum. (After Darwin and Pagenstecher.

protected by a calcareous shell of several pieces, or by a coriaceous envelope. Abdomen free and rudimentary. Thoracic segments usually carrying six pairs of forked ciliated limbs.

The *Cirripedia* include three orders, of which only the order *Thoracica* has ever been found in a fossil condition, or is ever likely to be so. In this order are the common Acorn-shells (*Balanidæ*) and Barnacles (*Lepadidæ*), in which the body is protected by a more or less complete calcareous shell. The Acorn-shells are generally known as the "Sessile Cirripedes," because the shell is directly attached by its base to some foreign body, whereas the Barnacles are commonly known as the "Pedunculated Cirripedes," because the shell is supported upon a stalk or "peduncle." Besides these, the order *Thoracica* comprises a third family, that of the *Verrucidæ*, in which the shell resembles that of the *Balanidæ* in being sessile, but differs in being unsymmetrical, and in some other particulars.

It is with the *shell* of the Cirripedes that the palæontologist has to deal; and we may therefore consider briefly the chief parts of the shell in the Sessile and Pedunculated Cirripedes respectively. It will not be necessary, however, to enter into minute details on this complicated subject, and it will be sufficient to indicate the leading facts of importance.

In the symmetrical Sessile Cirripedes or *Balanidæ*, commonly known as Acorn-shells, the animal is protected by a calcareous shell formed by calcifications within the walls of the first three cephalic segments. The animal is placed within the shell, head downwards, and is fixed to the centre of a shelly or membranous plate, which closes the lower aperture of the shell, and which is termed the "basis" (fig. 197, A, *l*). The "basis" is fixed by its outer surface to some foreign object, and is sometimes compact, sometimes porous. Above the basis rises a limpet-shaped, conical, or cylindrical shell, which is open at the top, but is capable of being completely closed by a pyramidal lid or "operculum." Leaving the operculum out of consideration at present, the sides of the shell are seen to be composed of from four to eight separate pieces or valves, or, as they are technically called, *compartments*. These compartments are usually closely contiguous by their lateral margins, and are separated by lines of division or "sutures;" but they are sometimes anchylosed together. Each compartment consists of a central portion, which is termed the "paries" (fig. 197, B, *p*), which is attached by its base to the "basis" of the shell. The "paries" grows downwards, so that the whole shell increases by additions made round the base. The paries of each compartment is flanked by wing-like portions, which differ from the paries in appearance, and are called "radii" and "alæ," according to their shape (fig. 197, B, C). Sometimes the paries has a "radius" on both sides, sometimes "alæ" on both sides, and sometimes an ala on one side and a radius on the other.

The separate compartments of the shell receive special names according to their position. The compartment at the end of the shell where the animal thrusts out its cirrated

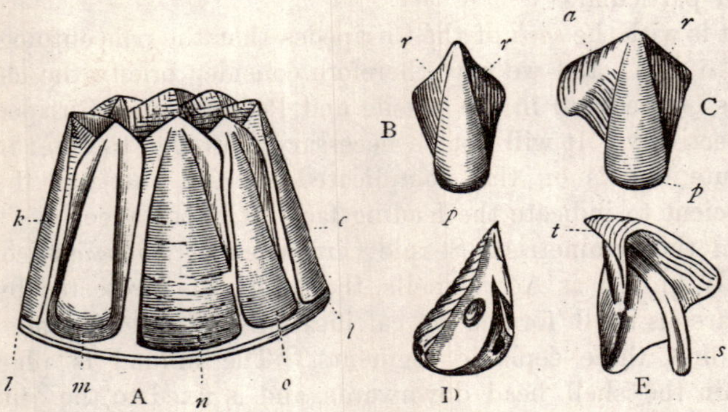

Fig. 197.—Shell of Balanidæ. A, Diagram of the shell of *Balanus*: *l*, *l*, Basis; *c*, Carina; *k*, Rostrum; *m*, Rostro-lateral compartment; *n*, Lateral compartment; *o*, Carino-lateral compartment. B, Compartment with two radii (*r*, *r*), flanking the paries (*p*). C, Compartment with a radius (*r*) on one side, and an *ala* (*a*) on the other side of the paries. D, Internal view of the scutum. E, Internal view of the tergum, showing the spur (*s*) and the beak (*t*). (After Darwin.)

limbs, is called the "carina" (fig. 197, A); and the compartment immediately opposite to this "rostrum." The remaining compartments are "lateral," the one nearest the carina "carino-lateral," the one nearest the rostrum "rostro-lateral," and the middle one simply "lateral" (fig. 197, A); but the three rarely coexist.

The "operculum" or lid of the shell consists of two pairs of valves, known as the "scuta" and "terga," forming a little pyramid or cone, attached within the orifice of the shell by a membrane. Each scutum opens and shuts against its fellow along one margin (the "occludent" margin), and articulates with one of the terga along the opposite margin. Similarly, each tergum opens and shuts against its fellow along one margin (the "carinal" margin), and articulates with one of the scuta along the opposite margin. The apex of the terga (fig. 197, E) often forms a prominent beak, and the basal margin is furnished with a process or "spur." The scuta and terga are not only movable, but are furnished with proper depressor muscles.

As regards the distribution of the *Balanidæ* in time, the oldest known representative of the family, so far as is certainly known, has been indicated by Mr Seeley as occurring in the Lias, and has been made the type of a new genus under the name of *Zoocapsa*. So far as is known, no member of the group occurs in any Palæozoic deposit; and negative evidence is in this case of considerable value, as the *Balani* possess a shell which is readily preserved, whilst they adhere to all sorts of marine bodies. With the above-mentioned exception (which may, perhaps, be referred to the *Verrucidæ*), no fossil Balanoid has hitherto been discovered in sediments older than the commencement of the Tertiary period. The genus *Balanus* is the earliest of the group, and appears under several specific forms in the Eocene rocks. In the Miocene and Pliocene deposits, the *Balanidæ* are abundantly represented by *Balanus* itself, and in the latter by the genera *Acasta*, *Pyrgoma*, and *Coronula*.

The remaining family of the Sessile Cirripedes is that of the *Verrucidæ*, comprising only the single genus *Verruca*. In many respects the *Verrucidæ* approach the *Balanidæ*, but the shell is composed of six valves only, and is unsymmetrical, whilst the scuta and terga (forming the operculum), though movable, are not furnished with a depressor muscle. The *Verrucidæ* appear, so far as is known, to have commenced their existence towards the close of the Secondary period, the Chalk having yielded one species. *Verruca Strömia* is found in the Coralline and Red Crags (Pliocene), in Glacial deposits, and in existing seas.

The third family of the *Cirripedia Thoracica* is that of the *Lepadidæ* or Pedunculated Cirripedes, commonly known as

Barnacles." In these (fig. 198) the animal differs from the Sessile Cirripedes in having its anterior extremity greatly elongated, forming a stalk or "peduncle" by which it is fixed to some foreign object. At its free extremity the peduncle bears the "capitulum," which corresponds to the shell of the Balanoids, and is composed of various calcareous pieces, united by a membrane, moved upon one another by appropriate muscles, and protecting in their interior the body of the animal with its various appendages. The peduncle is cylindrical, of varying length, flexible, and furnished with proper muscles. In some species the peduncle is naked, and cannot be preserved in the fossil condition; but in other cases the peduncle is furnished with calcareous scales (*Loricula* and *Turrilepas*, fig. 200), in which case it is readily preserved. The "capitulum" (fig. 199), as before said, corresponds with the shell of the *Balani*, and is generally much flattened. It consists ordinarily of five or more valves united to one another by membrane, usually with marked interspaces; but the valves may be rudimentary or wanting, and the entire capitulum may be membranous. The parts of the capitulum correspond ideally with the parts of the shell in the Balanoids. In the latter, however, the shell is for the most part composed of the "compartments," and the "operculum" is comparatively small and insignificant. In the Lepadoids, on the other hand, the valves which correspond with the operculum of the Balanoids are disproportionately developed, and the valves which correspond with the compartments of the Balanoids are much less conspicuous, and are often partially absent. The most important and per-

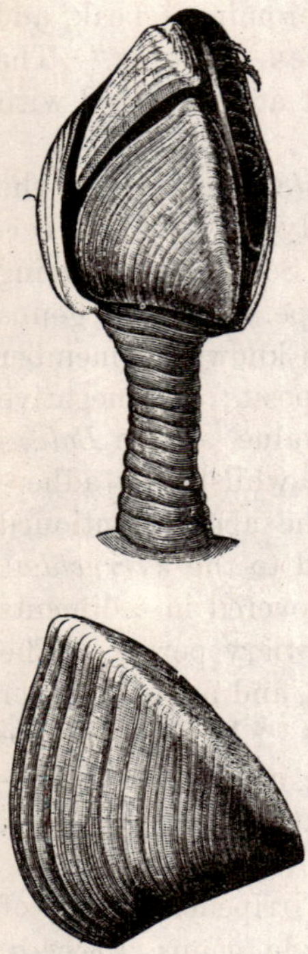

Fig. 198.—*Anatifa lepas*, a recent Pedunculated Cirripede. The lower figure shows the scutum detached.

sistent of the valves are the "scuta" (fig. 199, *b*), which protect the front part of the body, and correspond with the valves bearing the same name in the operculum of the Balanoids. The next most important are the "terga" (fig. 199, *a*), which protect the dorso-lateral surface. A pair of scuta and a pair of terga are present, and these are the largest of all the valves. The "carina" and "rostrum" are placed along the edges of the capitulum, the former being much the most important, and there may be a "sub-carina" and "sub-rostrum." The remaining valves, with the carina and rostrum, correspond with the proper shell of the Balanoids; but they are often wanting or rudimentary, and they require no further consideration here.

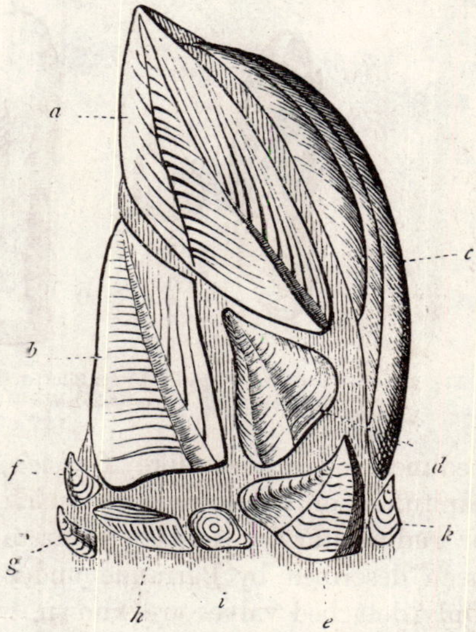

Fig. 199.—Capitulum of a Pedunculated Cirripede. *a*, Tergum; *b*, Scutum; *c*, Carina; *d*, Upper latus; *e*, Carino-latus; *f*, Rostrum; *g*, Sub-rostrum; *h*, Rostral latus; *i*, Infra-median latus; *k*, Sub-carina. (After Darwin.)

As regards the distribution of the Pedunculated Cirripedes in time, until recently no member of the family was certainly known to have existed in the Palæozoic period. Dr Henry Woodward, however, has described a very interesting form from the Upper Silurian rocks, under the name of *Turrilepas* (fig. 200, A). In this singular fossil the peduncle was furnished with intersecting rows of plates, as in *Loricula*. These plates, when detached and occurring in an isolated condition, are not unlike the shells of certain of the Pteropods, and fragmentary specimens may be, and have been, taken for Chitons. The genus *Turrilepas* is identical with the *Plumulites* of Barrande, and species have been now detected in the Silurian formation of Bohemia, Britain, and North America. The distinguished Bohemian palæontologist just mentioned regards *Turrilepas* not as being the scaly

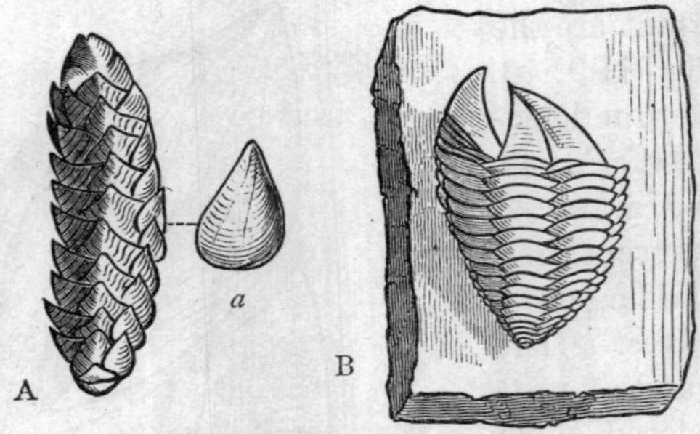

Fig. 200.—A, *Turrilepas Wrightii*—Upper Silurian (after Woodward): *a*, A plate of the same magnified. B, *Loricula pulchella*—Chalk (after Darwin).

peduncle of a form like *Loricula*, but as being truly the capitulum of a Lepadoid, in which the peduncle is wanting or rudimentary. Another Silurian genus of Lepadoids has been described by Barrande under the name of *Anatifopsis.* Only detached valves are known, which have a general likeness to those of *Lepas*, each valve being somewhat quadrilateral, and having the lower part of its base marked out into one or two horizontal segments, which are more or less separated from the body of the valve. No undoubted pedunculated Cirripedes have as yet been detected in the Devonian, Carboniferous, or Permian rocks; and, with the exception of the ancient types just mentioned, the oldest known representatives of the Lepadoids belong to the genus *Pollicipes*, species of which have been discovered in the Rhætic beds (Upper Trias), and in the Stonesfield Slate (Lower Oolites). In the Cretaceous period, the *Lepadidæ* reached their maximum of development, about 80 per cent of the known fossil forms having been recognised in strata of this age. The Tertiary forms are few in number, and belong to the genera *Scalpellum*, *Pollicipes*, and *Pœcilasma*, the last being only a sub-genus of the living *Anatifa*.

CHAPTER XVIII.

CRUSTACEA (Continued).

SUB-CLASS ENTOMOSTRACA.

THE Entomostracous Crustaceans are defined by Professor Rupert Jones as follows: "*Animal aquatic, covered with a shell, or carapace, of a horny consistency, formed of one or more pieces, in some genera resembling a cuirass or buckler, and in others a bivalve shell, which completely or in great part envelops the body and limbs of the animal; in other genera the animal is invested with a multivalve carapace, like jointed plate-armour; the branchiæ are attached either to the feet or to the organs of mastication; the limbs are jointed, and more or less setiferous. The animals, for the most part, undergo a regular moulting or change of shell as they grow; in some cases this amounts to a species of transformation.*"

The orders commonly included in the sub-class *Entomostraca* are the *Ostracoda*, *Copepoda*, *Cladocera*, *Phyllopoda*, *Trilobita*, and *Merostomata* (comprising the sub-orders *Xiphosura* and *Eurypterida*). Of these, the *Copepoda* and *Cladocera* may be left out of consideration, as they are not certainly known to occur in the fossil condition.

ORDER OSTRACODA.

Minute Crustaceans having the entire body enclosed in a shell or carapace, which is composed of two valves united along the back by a membrane. The valves are capable of being closed by an adductor muscle, the insertion of which is marked in the interior of each valve by a tubercle, pit, or group of spots, or by both spots and a pit. The branchiæ are attached to the posterior jaws, and there are only two or three pairs of feet, which subserve locomotion, but are not adapted for swimming.

Of the living Ostracode Crustaceans, a great many inhabit fresh water (*Cypris*); others live in fresh or in brackish waters (*Candona*); lastly, others are exclusively confined to the sea (*Cythere* and *Cypridina*). They generally swarm in the localities in which they occur, and from their habit of periodically shedding their valves, considerable accumulations of their shells may be formed under favouring circumstances.

It is only the carapace-valves of the Ostracode Crustaceans that are preserved in the fossil condition, with the rarest exceptions; and the general form of the carapace is often very similar in different genera. Hence the palæontologist has to rely, in the discrimination of these minute fossils, upon small variations of shape, differences in the thickness of the valves, the characters of the edges of the valves, or the manner in which they are hinged to one another, or, lastly, the surface-ornamentation. Besides the difficulty attaching to the study of the fossil *Ostracoda* from their small size and general similarity of appearance, it is often by no means easy to distinguish between the cephalic and the posterior extremity of the body. When not alike, the most contracted extremity is to be regarded as the head, and the widest as the hinder end of the carapace. The former, as a rule, carries grooves or tubercles when such structures are present at all. The tubercles of the test, where developed, appear to represent the eye; and the grooves and intervening lobes, which are found in many forms, have been aptly compared by Barrande to the furrows and lobes of the glabella of Trilobites. There are many types, however, in which there are no conspicuous external markings, and in which the two ends of the carapace are similar. The Mesozoic and Tertiary *Ostracoda* are very small, and the same is true of a large number of Palæozoic species; but among the latter we find some comparatively colossal types, which range from one to three inches in length. As a rule, also, the Palæozoic *Ostracoda* are plain, or are simply striated or granular; whereas the Mesozoic and Tertiary forms are commonly ornamented with projecting tubercles, or in various other ways. As regards their general distribution in time, the *Ostracoda* certainly commence in the Upper Cambrian, and are even doubtfully represented in the Lower division of this formation. They existed under many and varied types in the Silurian, and are abundant in all the succeeding formations, till the recent period is reached. As matter of course, the remains of this group of Crustaceans with which we have chiefly to do, are principally those of the *marine* members of the order, and this is especially true as regards the Palæozoic species. In later deposits, however, the occurrence of Ostracodes of types now found in fresh or brackish water is by no means uncommon.

The total number of fossil *Ostracoda* is very large, the difficulties attending their study, for reasons already stated, are quite exceptionally great; and it is impossible, as a rule, for the student to discriminate species, or often, even genera, unless he should have made the group a subject of special investigation. Here, therefore, no attempt will be made to give even the briefest analysis of the families or genera of the order; but it may be well to shortly characterise some of the common or more remarkable types, with special reference to the Palæozoic forms, with which, upon many grounds, it is desirable that the student should have some acquaintance.

Among the Palæozoic Ostracodes, the first group that may be noticed is that of which *Leperditia* (fig. 201, C and D) is the type. In this genus the two valves are unequal in size, smooth, nearly oblong, bean-shaped, with the posterior end wider than the anterior. There is a small tubercular eye-spot, placed on the head, near the hinge, and underneath and behind this is a slightly inflated area, corresponding with an excavation of the shell interiorly, and exhibiting reticulated or areolar muscular markings. Behind the eye-spot is generally a vertical groove, which begins at the dorsal margin, and extends a short way across the valves. The genus ranges from the Silurian to the Carboniferous. *Isochilina* (fig. 201, B) nearly resembles the preceding, but the valves are equal. It is not uncommon in the Silurian. *Aristozoë*,

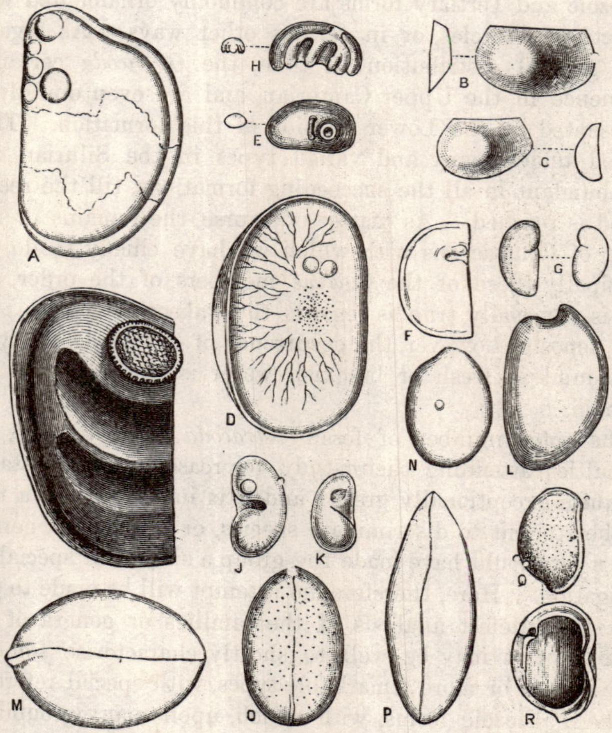

Fig. 201.—Types of Ostracoda. A, *Aristozoë memoranda*, of the natural size—Silurian; B, *Isochilina Ottawa*, enlarged four times (left valve and ventral view)—Silurian; C, *Leperditia Josephiana*, of the natural size (right valve and anterior view)—Silurian; D, *Leperditia solitaria*, enlarged, showing the eye-spot and muscular impression—Silurian; E, *Primitia strangulata*—Silurian; F, Outline of the same, enlarged; G, Right and left valves of *Primitia tarda*—Silurian; H, *Beyrichia complicata*—Silurian; I, *Beyrichia oculifera*, showing the elevated eye-spot, greatly enlarged—Silurian; J, *Entomis pelagica*, right valve, enlarged twice—Silurian; K, *Entomis impendens*, enlarged—Silurian; L, *Cypridina Wrightiana*, left valve, enlarged four times—Carboniferous; M, Dorsal view of a small example of *Entomoconchus Scouleri*, enlarged four times—Carboniferous; N, Left valve of *Polycope simplex*, enlarged eight times—Carboniferous; O, *Cypris Browniana*, viewed dorsally, enlarged twenty-five times—Pleistocene; P, *Candona candida*, viewed ventrally, and similarly enlarged—Pleistocene; Q, *Cythere punctata*, right valve, similarly enlarged—Pliocene; R, Interior of left valve of the same. (After Barrande, Rupert Jones, M'Coy, Hall, and G. S. Brady.)

(fig. 201, A), *Orozoë*, and *Callizoë* are Silurian genera allied to *Leperditia*, but they have tubercular eminences developed upon the head, and they are exceptionally large in point of size. *Aristozoë* reaches sometimes the extraordinary length of three inches. *Primitia* (fig. 201, E-G) is another ancient type, essentially Silurian in its distribution, and characterised by its equivalve, convex, oblong carapace, indented with a vertical dorsal groove of variable depth. Related to *Primitia* is the genus *Beyrichia* (fig. 201, H and I), which ranges from the Silurian to the Carboniferous, and is more particularly distinguished by the possession of two or three transverse grooves, which start at the hinge, and pass partially or wholly across the valves. In the curious *Beyrichia oculifera*, the eye-spot forms a prominent and faceted tubercle, unlike that of any other Ostracode (fig. 201, I). Two other members of the ancient family of the *Leperditiadæ* may be just mentioned—namely, *Kirkbya* and *Moorea*, both commencing in the Upper Silurian, and ranging, the former to the Permian, and the latter to the Carboniferous.

Another great group of *Ostracoda* is that of the *Cypridinadæ*, with numerous living forms, and represented in past time principally by the three generic types, *Cypridina*, *Entomis*, and *Entomoconchus*, the last two of these being entirely extinct. In *Cypridina* (fig. 201, L) the carapace is produced in front into a beak-like projection, below which is a hollow or notch facing the ventral margin. Many of the so-called *Cypridinæ* of the older Palæozoic rocks are now known to be referable to other types, but the genus is well represented in the Carboniferous, and exists at the present day. *Entomis* (fig. 201, J and K) is a Silurian and Devonian genus, which resembles some of the *Leperditiadæ* in having a dorsal groove, indenting the valves transversely, and sometimes reaching the ventral margin, and having a rounded tubercle placed below its lower end. One species of this genus, formerly known as a *Cypridina*, is so abundant in certain of the Devonian strata of Germany as to have gained for these the name of "Cypridinen-Schiefer." *Entomoconchus* (fig. 201, M), again, is a large form, with a thick and globose carapace, having a much less developed notch in front than in *Cypridina*. It is confined to the Carboniferous rocks. Among other Carboniferous *Cypridinadæ* may be mentioned the genera *Cyprella*, *Cypridella*, *Cypridellina*, *Sulcana*, and *Rhombina*.

Another group of the Ostracodes is that characterised by the genus *Polycope* (fig. 201, N), in which the carapace-valves are subequal and thin, not markedly notched in front, and having no beak. Though represented by living species, the only undoubted members of this genus which have been detected in a fossil state are from the Carboniferous rocks.

The genus *Cytherella* is the type of another group, in which the valves are very thick and calcareous, and are not notched in front. In *Cytherella* itself the right valve is much larger than the left, overlapping throughout the whole circumference, and "presenting round the entire inner margin a distinct groove, into which the valve of the opposite side is received" (G. S. Brady). The genus ranges from the Carboniferous to the present day; and we may provisionally place with it the *Cytherellina* and *Æchmina* of the Silurian.

Lastly, we have the great group of Ostracodes represented by the families of the *Cypridæ* and *Cytheridæ*, "including all the fresh-water and a vast majority of the marine *Ostracoda*, and embracing all the forms classed by the earlier writers under the two great genera *Cypris* and *Cythere*" (G. S. Brady). In the *Cypridæ*, as typified by *Cypris* (fig. 201, O) itself, the valves are thin and smooth, and more or less sinuate below. The family seems to have attained its maximum at the present day, but it is represented in deposits as old as the Carboniferous by forms which are believed to be referable to the living genera *Candona* and *Bairdia*. In the *Cytheridæ* the carapace is hard and calcareous, generally with a rough or uneven surface. This family, also, is mainly Recent, Tertiary, and Secondary; but it is represented in the Palæozoic period by the living genus *Cythere*, and by the extinct *Thlipsura*.

ORDER PHYLLOPODA.

Crustacea, mostly of small size, the carapace protecting the head and thorax, or the body entirely naked. Feet numerous, never less than eight pairs, mostly foliaceous or leaf-like, branchial in function.

Most of the living Phyllopods are inhabitants of fresh waters, but some live in the sea (*Nebalia*), and others affect waters which are abnormally salt (*Artemia*). The two most interesting recent forms, as bearing on fossil examples of the order, are *Limnadia* and *Apus*, both of which live in fresh water. In *Limnadia* the body is enclosed in an oval bivalve carapace, and there are from eighteen to thirty pairs of membranous leaf-like feet. In *Apus* the carapace is clypeiform, and protects a considerable portion of the abdomen; and there are sixty pairs of feet, of which all but the first pair are foliaceous.

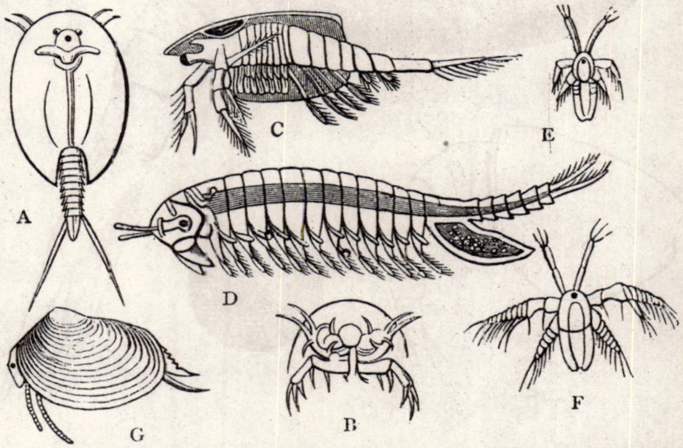

Fig. 202.—Morphology of recent Phyllopoda. A, *Lepidurus Angassi*, viewed dorsally; B, Under side of head of the same; C, *Nebalia bipes*, one side of the carapace being removed, so as to show the branchial feet; D, *Branchipus stagnalis*, female; E and F, Young stages of the same; G, A magnified specimen of *Estheria*, in its living state.

Leaving out of sight the genera *Estheria* and *Leaia*, the Phyllopods are almost exclusively Palæozoic in their distribution, and are chiefly, though not exclusively, known by their carapace-valves. The best-known genera are the *Hymenocaris* of the Lingula flags; the *Caryocaris* of the Skiddaw Slates; the *Peltocaris* and *Discinocaris* of the Silurian, with their allies *Aptychopsis*, *Pterocaris*, and *Cryptocaris*; the *Ceratiocaris* of the Silurian, Devonian, and Carboniferous; and the *Dithyrocaris* of the Carboniferous Limestone. These forms have a general resemblance to one another, and are believed to be most nearly allied to the recent *Apus*, whilst they are exclusively Palæozoic. The genus *Aspidocaris*, however, is allied to *Discinocaris*, and is found in the Triassic period.

In *Hymenocaris* (fig. 203, *b*) the carapace is comparatively large, sub-triangular, apparently not bivalved; there are nine free abdominal segments, and the last carries three pairs

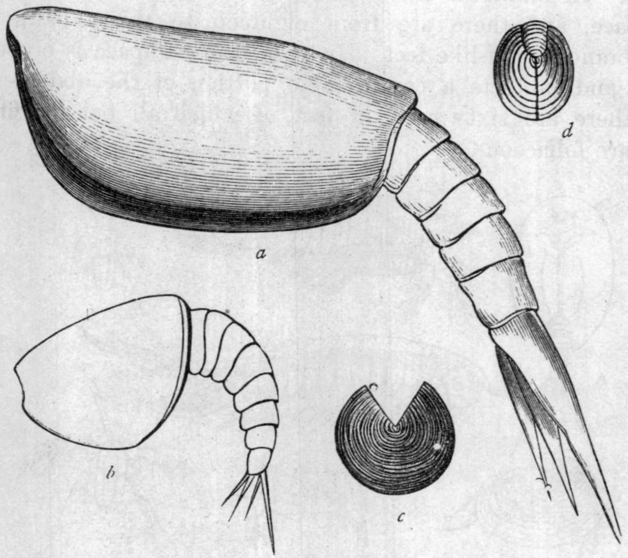

Fig. 203.—Palæozoic Phyllopods. *a*, *Ceratiocaris papilio*—Upper Silurian (Salter); *b*, *Hymenocaris vermicauda*—Upper Cambrian (Salter); *c*, *Discinocaris Browniana*, without the "rostrum"—Lower Silurian (Original); *d*, *Peltocaris aptychoides*—Lower Silurian (Woodward).

of unequal lanceolate appendages. In *Caryocaris* the carapace is bivalved, pod-shaped, and truncated behind, and the last abdominal segment carries three spines. Allied to the preceding, but much more important, and with a much wider range in time, is the genus *Ceratiocaris* (fig. 203, *a*). In this type the carapace is bivalve, hinged along the back, the valves being semi-oval or pod-like in shape, contracted in front, with an abrupt posterior truncation. The surface of the carapace is marked with fine linear striæ, and there are fourteen or more body-rings, of which the last five or six were free. The telson is composed of three long spinose appendages; and the jaws have been recognised in a few instances, either detached or in connection with the body of their former possessor. Some species of the genus reached a great size, *C. Ludensis* growing to a length of two feet. It has also been shown that the caudal segments of some specimens bear gill-like plates freely depending from the under surface; and these organs, which have been compared with the supplementary abdominal gill-feet in *Nebalia*, by Dr Henry Woodward, were probably present in all the species of this genus. *Dithyrocaris* (fig. 204), of the Carboniferous, is allied to the preceding in general form, but it has only one free body-ring. The telson consists of three spines, and the semi-oval valves are marked by conspicuous longitudinal ridges.

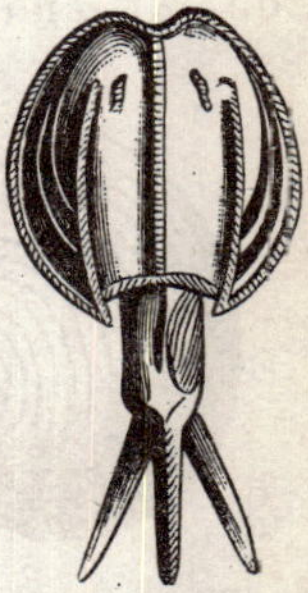

Fig. 204.—*Dithyrocaris Scouleri*, viewed from above, reduced slightly in size. Carboniferous Limestone. (After M'Coy.)

Another remarkable group of Phyllopods, of which all the known forms except the Triassic *Aspidocaris* are Palæozoic, may be considered to be typified by the Silurian genus *Peltocaris*. In this genus the carapace (fig. 203, *a*) is approximately circular, and consists of two valves of a semicircular form, which are united along the back by a straight median hinge. The valves, however, are so notched in front as to leave a space which is completed by the addition of a third parabolic valve or "rostrum." The entire carapace is concentrically striated, but the body-rings are unknown. The *Discinocaris* of the Silurian (fig. 203, *c*) is very like the preceding, but the two lateral valves are anchylosed, and the rostrum is triangular. One species of this genus attains a diameter of seven inches across the carapace. *Aptychopsis*, also Silurian, has the straight "medial suture" and triangular rostrum of *Discinocaris*, but the valves are united by a dorsal suture, as in *Peltocaris*. *Pterocaris*, from the same formation, resembles the genus just mentioned, but the valves are only united by suture for a short space anteriorly, while they diverge behind, so as to form a wide posterior notch. Lastly, in the Silurian *Cryptocaris* the carapace is semicircular, and its two halves are united in front at an umbonal point, and are only imperfectly divided behind by a straight ridge or fold.

More widely distributed than any of the preceding is the genus *Estheria* (fig. 202, G), which appears to be related to the living *Limnadia*, and which in some respects constitutes a connecting-link between the Phyllopods and the Ostracodes. The body in *Estheria* is enclosed in a bivalve carapace (fig. 205, A), and the feet are foliaceous. The valves of the

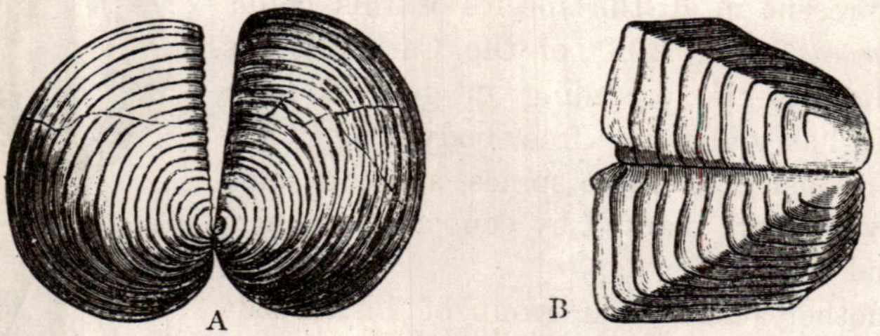

Fig. 205.—A, Carapace of *Estheria ovata*, magnified six diameters—Trias; B, Carapace of *Leaia Leidyi*, magnified five diameters—Lower Carboniferous. (After Rupert Jones.)

carapace have a well-marked beak or "umbo," and are hinged to one another along a dorsal line. From these circumstances, and from their being marked with numerous concentric lines of growth, the carapace-valves of *Estheria* very closely resemble the shells of certain Bivalve Molluscs, for which they have often been mistaken. The valves are usually sub-triangular, ovate, or sub-quadrate in form, and they possess a horny texture.

The living *Estheriæ* (fig. 202, G) are, without exception, inhabitants of fresh or, rarely, brackish water; and no one of the recent twenty-four species has been detected in the sea. This would afford a strong presumption that the deposits in which fossil *Estheriæ* occur were deposited in fresh or brackish water; but they not uncommonly occur in conjunction with undoubted marine remains. They appear, on the whole, to occur most frequently in those accumulations that "have been decidedly the result of brackish-water inundations, and of more permanent lagoons" (Jones). Fossil *Estheriæ* occur in the Devonian, Carboniferous, Permian,

Triassic, Jurassic, Cretaceous, and some Tertiary deposits; but they appear to have attained their maximum development towards the close of the Triassic period.

The genus *Leaia* (fig. 205, B) is very nearly allied to *Estheria*, and comprises small Bivalved Crustaceans, with "dark, horny, sub-quadrate valves, obliquely ridged from umbo to angles, and ornamented with distinct lines of growth parallel with the border" (Jones). *Leaia* is a very widely distributed genus, but all the known species belong to either the Carboniferous or Permian rocks. It has been suggested, however, that the obscure Silurian genera *Myocaris* and *Ribeiria* are allied to *Leaia*.

Lastly, we may notice that a Phyllopod nearly allied to the living fresh-water genus *Branchipus* (fig. 202, D) has been detected by Dr Henry Woodward in the Eocene formation, and has been described by him under the name *Branchipodites vectensis*. The much older *Branchipusites* (rightly *Branchipodites*) *anthracinus*, of the Coal-measures, has been supposed to have similar relationships, but this cannot be said to be certainly established.

ORDER TRILOBITA.

Crustaceans in which the body is usually more or less distinctly trilobed; there is a cephalic shield, usually bearing a pair of sessile compound eyes; the thoracic somites are movable upon one another, and are very variable in number; the abdominal segments are coalescent, and form a caudal shield; there is a well-developed upper lip or "hypostome."

As regards the general structure of the Trilobites, the body was protected by a well-developed chitinous shell or "crust," which covered the whole dorsal surface of the body, and which usually exhibits more or less markedly a division into three longitudinal lobes (fig. 206), from which the name of the order is derived. In some cases, however, as in the genera *Homalonotus* and *Illænus*, this trilobation is only obscurely marked. Though closely agreeing in their fundamental features, Barrande observes that it is not uncommon for the same species of Trilobite to present itself under two forms, one broad and the other long, and he regards the broad forms as the females, and the long forms as the males, of the species.

The crust exhibits a well-marked division into three regions, which are commonly found detached and separate from one another. These three regions are—1, a cephalic shield; 2, a variable number of movable "body-rings" or thoracic segments; and 3, a caudal shield or "pygidium."

The cephalic shield or buckler (fig. 206) is generally more

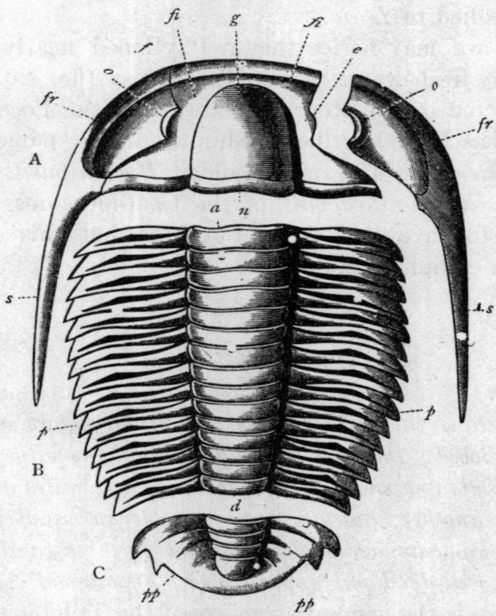

Fig. 206.—The skeleton of a Trilobite (*Angelina Sedgwickii*), partially dissected A, Head-shield. B, Movable rings of the thorax. C, Tail or abdomen. *g*, Glabella, in this species without furrows; *fi*, Fixed cheeks; *e*, Eye-lobe; *o*, Eye; *f*, Facial suture; *fr*, Free cheeks; *s*, Head-spines; *p*, Pleuræ; *pp*, Anchylosed pleuræ of pygidium.

or less semicircular in shape, and is composed of a central and two lateral pieces, of which the two latter may or may not be united in front of the former. The central portion of the cephalic shield is usually elevated above the remainder. It is termed the "glabella" (fig. 206, *g*), and it protected the region of the stomach. The form of the glabella varies a good deal. Usually it is widest in front (fig. 207), but its width may be nearly uniform, or it may be widest posteriorly and contracted in front, as in *Calymene.* The glabella is bounded at the sides by two grooves, which are known as the "axal furrows," and is marked off behind by a third groove, which is termed the "neck-furrow." The surface of the glabella may be quite smooth, but it is ordinarily divided into "lobes" by "grooves," which originate in the axal furrows, and pass inwards towards the middle line (fig. 207). These furrows mark the position of the segments which compose the glabella, and they are sometimes continuous from side to side. Usually there are three pairs of these furrows, a lower or basal, a middle or ocular, and an upper or frontal furrow; but there may be an additional pair of furrows in front of these. In some cases, as in *Illænus* (fig. 212), the glabella is very indistinctly marked off from the rest of the shield.

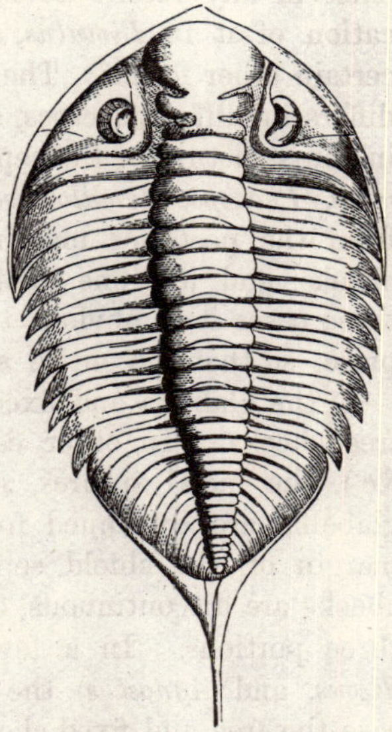

Fig. 207.—*Phacops* (*Dalmanites*) *limulurus.* Upper Silurian.

The grooves of the glabella probably mark off so many *segments,* to which organs of prehension and mastication were attached inferiorly, and they are marked internally by corresponding ridges, to which muscles must have been attached. Sometimes (*Illænus, Ellipsocephalus, Encrinurus,* &c.) they are obsolete, as also occurs in particular species of other genera (*Trinucleus* and *Æglina*). Typically, three pairs of grooves are present, but *Phacops* has four, and so have some other types.

At each side of the glabella, and continuous with it, is a small semicircular area, which is termed the "fixed cheek" (fig. 206, *fi*). The glabella, with the "fixed cheeks," is separated from the lateral portions of the cephalic shield, termed the "movable" or "free cheeks," by a peculiar suture or line of division, which is known as the "facial suture" (fig. 206, *f*). No such peculiar line of division is known to exist in any recent Crustacean; but there is a faint indication of it in *Limulus*, and some doubtful traces of it in certain other forms. The course taken by the facial sutures differs in different cases, and causes an important difference in the structure of the cephalic shield. In some cases (*Asaphus*, *Phacops*, *Homalonotus*, &c.) the facial sutures, starting from the posterior margins of the buckler, skirt the fixed cheeks, and join one another in front of the glabella. In these cases it is obvious that the free cheeks form a single piece, so that the entire shield consists of but two portions —1, the glabella and fixed cheeks; and 2, the amalgamated free cheeks. In other cases (*Paradoxides*, *Illænus*, *Proetus*, &c.), the facial sutures, instead of joining in front of the glabella, are continued forward, till they cut the anterior margin of the shield separately. In these cases the free cheeks are discontinuous, and the cephalic shield consists of three portions. In a few genera (as in *Trinucleus*, *Microdiscus*, and *Agnostus*) the facial suture is absent, in which case the free and fixed cheeks are fused with one another.

The posterior angles ("genal angles") of the free cheeks are very commonly prolonged into longer or shorter spines, and the free cheeks also bear the eyes. The eyes are compound, and consist of an aggregation of facets, covered by a thin cornea. They are generally crescentic or reniform in shape, and are invariably *sessile*, in the sense that they are never supported upon movable stalks. In some cases, however, they are carried upon longer or shorter prominences. The eyes differ much in size, and they are wanting in a few forms, such as *Agnostus*, *Ampyx*, some of the *Trinuclei*, and certain forms of *Conocephalus*. Though usually facets are easily detected in such as have eyes of any size, there are some (*Bronteus*, *Arethusina*, *Proetus*, &c.) in which the eyes are smooth. In any case, the number of lenses varies greatly, there being as few as fourteen facets, or as many as fifteen thousand in each eye in different types (Barrande).

Behind the cephalic shield comes the thorax, composed

of a variable number of segments which are not soldered together, but are capable of more or less movement upon each other. The amount of movement thus allowed varies, but in several genera (*e. g.*, *Calymene* and *Illænus*) it was sufficiently great to allow of the animal completely rolling itself up after the manner of a hedgehog. The number of body-rings or segments in the thorax varies from no more than two (*Agnostus*) to as many as twenty-six (*Harpes ungula*). Ordinarily the thorax (fig. 211) is strongly trilobed, and each body-ring exhibits the same trilobation, being composed of a

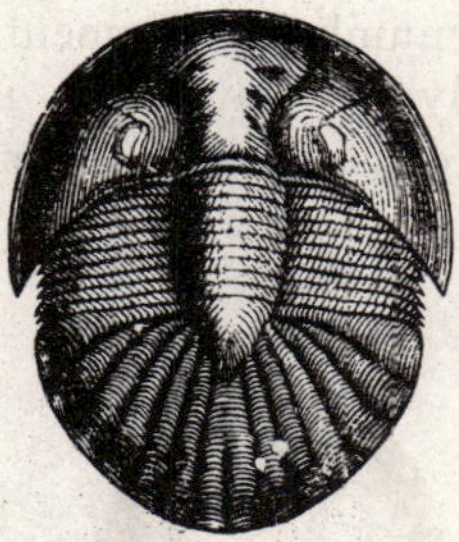

Fig. 208.—*Bronteus lunatus.*

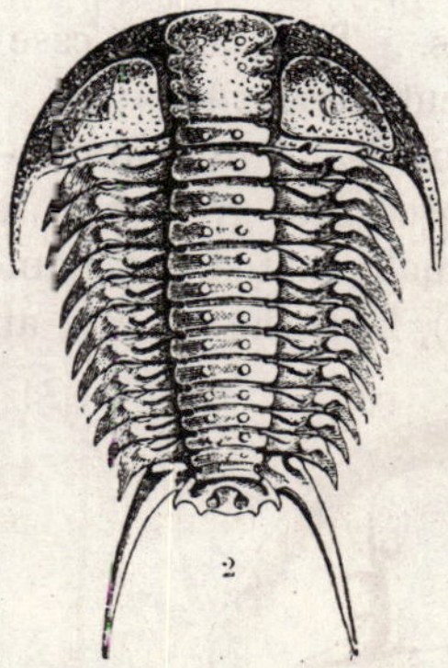

Fig. 209.—*Cheirurus pleurexanthemus.*

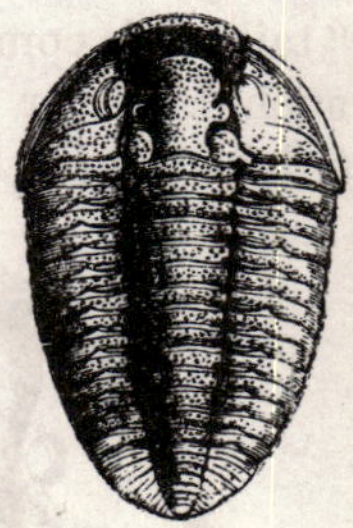

Fig. 210.—*Calymene Blumenbachii.*

central, more or less convex portion, called the "axis," and of two flatter side-lobes, termed the "pleuræ." The pleuræ are in one piece with the axis, but are separated from it by a more or less pronounced groove, the "axal furrow." In one type of pleuræ, each of these structures carries a deep longitudinal groove or sulcus upon its upper surface (as in *Asaphus, Ogygia, Phacops, Calymene*, &c.) In another type, on the other hand (as in *Cheirurus, Bronteus, Acidaspis*, &c.), the place of the sulcus is taken by a similarly situated oblique *ridge*. The pleuræ are always bent downwards towards their ends, and also commonly bent backwards as well, the point where the backward curvature begins being the "fulcrum" of Salter. In the Trilobites with grooved pleuræ, more especially, the pleuræ imbricate and overlap; and the "fulcrum" of each is often bevelled off, so as to form a facet upon which the pleura immediately in front plays, thus allowing the animal to *roll up*. In the state of complete enrollment, the under surface of the pygidium is closely applied to the corresponding surface of the head-shield, thus entirely concealing the ventral aspect of the animal. Some forms (such as *Homalonotus, Lichas, Triarthrus, Olenus, Paradoxides, Æglina*, &c.) are not known to have been endowed with the power of rolling up. Though the trilobation of the thorax is usually very well marked (figs. 208-211), at other times the axis is very broad, and the axal furrows more or less inconspicuous. This is the case in *Illænus* (fig. 212), and to a less extent in *Homalonotus.*

The caudal shield or "pygidium"—commonly called the "tail"—is composed of a greater or less number of segments anchylosed or amalgamated. Commonly, the pygidium is trilobed (fig. 211), like the thorax, and consists of a central

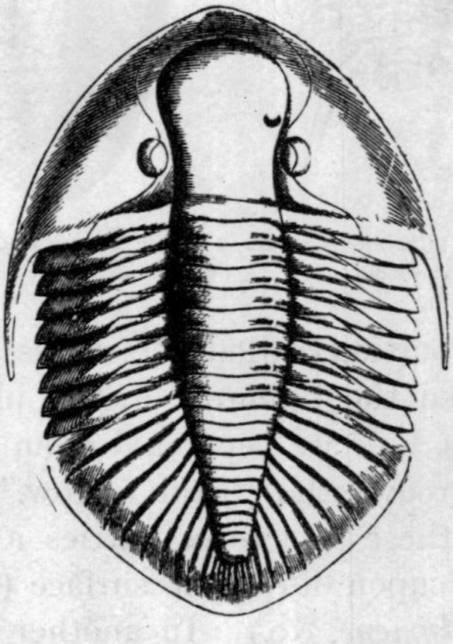

Fig. 211.—*Asaphus Canadensis* (Chapman). Lower Silurian.

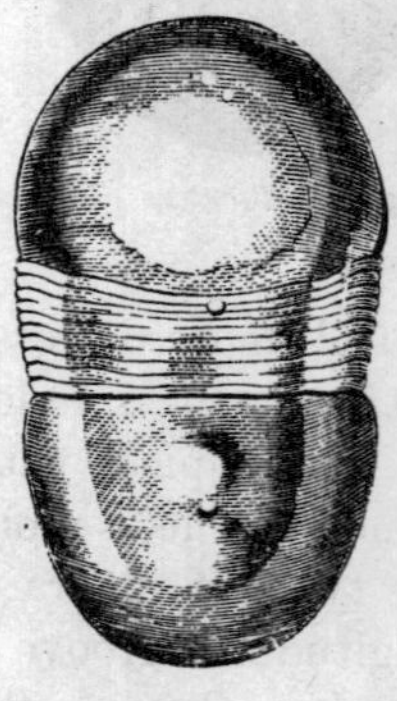

Fig. 212.—*Illænus Barriensis* (Murchison). Lower Silurian.

elevated "axis" and of a marginal "limb." The limb is separated from the axis by axal furrows, and usually exhibits on its surface the lines which indicate the component pleuræ, as well as the longitudinal furrows on the faces of these. The extremity of the pygidium is sometimes simply rounded, with an "entire" margin; but it may be prolonged into a shorter or longer spine or "mucro" (fig. 207), and the ends of the pleuræ may also be extended into spine-like projections (fig. 213). The number of rings in the tail varies from two (*Sao hirsuta*) to twenty-eight (species of *Amphion*).

With regard to the condition of the *under* surface of the Trilobites, the progress of our knowledge has been slow, and is still far from complete. They appear to have lived on muddy bottoms, in shallow water, and they probably swam on their backs, as do the modern *Apus* and the larvæ of *Limulus;* but in spite of the innumerable specimens which

Fig. 213.—Glabella and pygidium of *Dikellocephalus magnificus*, Quebec Group (Lower Silurian). (After Billings.)

have been exhumed from the rocks, the only structure which had, until within the last few years, been detected on the lower surface was the upper lip. The margin of the head-shield (as that of the pygidium also) is turned under in the form of a downward and inward inflection, or "doublure;" and to the centre of this, in many forms, is attached the "lip-plate." The shape of this varies, but it was usually a broad forked plate situated in front of the mouth, and doubtless corresponding with the upper lip—"labrum" or "hypostome"—of living Crustaceans, its form closely resembling that of the lip-plate of the recent *Apus*, one of the Phyllopods. The next advance in our knowledge was the description by Dr Henry Woodward of a specimen of *Asaphus platycephalus*, in which, in addition to the lip-plate, there is a jointed filament (fig. 214, *p*), apparently springing from a maxilla, and being the remains of a maxillary "palpus."

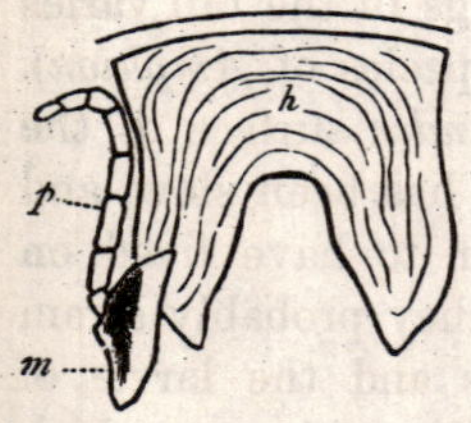

Fig. 214. — Buccal organs of *Asaphus platycephalus* (after Woodward). *h*, Hypostome; *p*, Palpus; *m*, Maxilla (?).

Still more recently, a specimen of the large Trilobite, *Asaphus platycephalus* (fig. 215), was described by Mr Billings as possessing organs which this distinguished palæontologist regards as being the remains of legs. The structures in question are in the form of eight pairs of apparently jointed appendages, which correspond with the eight segments of the thorax, arising near the middle line, and curving forwards. Dr Henry Woodward corroborates the view propounded by Mr Billings, that these structures are of the nature of ambulatory legs. Professors Dana and Verrill, on the other hand, regard these remains as being "the semi-calcified arches in the membrane of the ventral surface, to which the foliaceous appendages or legs were attached."

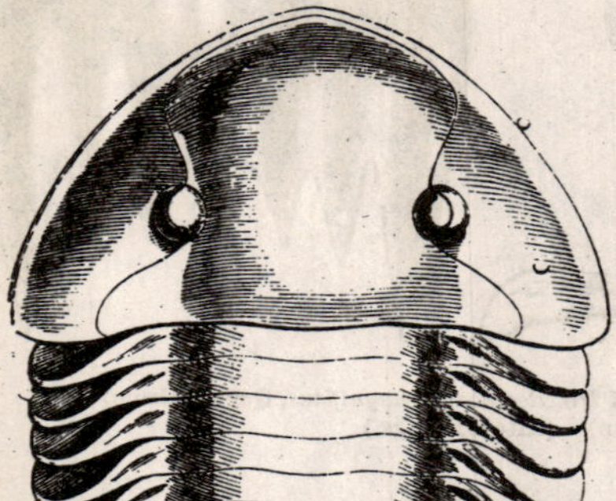

Fig. 215.—*Asaphus platycephalus*. Lower Silurian.

The last additions to our acquaintance with this difficult subject have been made by Mr C. D. Walcott, whose researches have been carried on by the method of making thin transverse and longitudinal sections of rolled-up specimens. This observer has shown that the visceral cavity of the Trilobites (fig. 216, *b*) was bounded inferiorly by a thin membrane, which is attached to the lower margin of the dorsal crust all round. This ventral membrane was strengthened by calcified arches, which in turn supported the appendages beneath. As to these latter our knowledge is not yet complete, but we know that there existed a row of articulated appendages on each side of the middle line below. The thoracic appendages seem to have been slender five-jointed legs, in which the terminal segment forms a pointed claw,

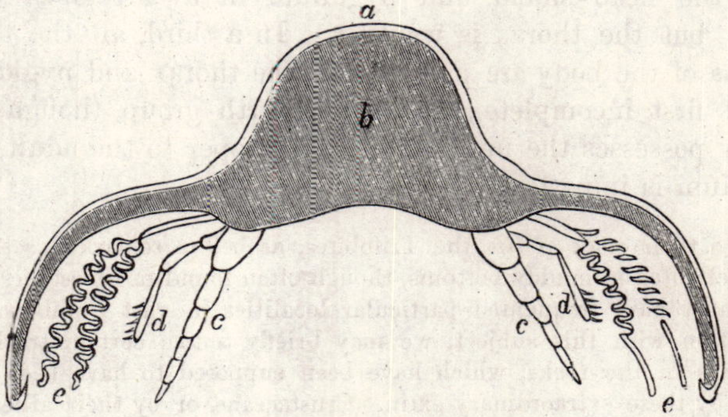

Fig. 216.—Transverse section of the thorax of *Calymene senaria*, partially restored (after C. D. Walcott). *a*, Dorsal crust; *b*, Visceral cavity, continued laterally to the pleural margins of the dorsal crust; *c*, Legs, restored; *d*, Epipodite; *e*, Spiral gills. Enlarged six times.

and the basal segment carries a jointed appendage, regarded by Mr Walcott as homologous with the "epipodite" of many recent Crustaceans. On each side of the thoracic cavity there is, also, attached a row of bifid spiral appendages (fig. 216, *e*), of the nature of gills; and branchial appendages were probably attached to the bases of the thoracic limbs as well. With regard to the appendages of the head, the mouth is situated behind the hypostome, and is bounded by four pairs of jointed manducatory appendages, the basal joints of which are, partly or wholly, modified to act as jaws.

With regard to the *development* of the Trilobites, the eggs have been noticed by both Barrande and Walcott. They are spheroidal or cylindroidal in shape, and mostly about a twenty-fifth to a fiftieth of an inch in diameter, and they seem to have been deposited in clusters. The larval condition of the Trilobites is only known in certain cases, and it is possible that the young may often have been naked. This subject has been chiefly worked out by Barrande, who has shown, that, so far as our present knowledge goes, the development of the Trilobites follows one or other of four principal lines. In one group of forms (*e.g.*, *Sao hirsuta*), the most minute larvæ observed possess a head-shield, but have no pygidium, and the thorax is either wanting or rudimentary. In another type (*Agnostus*), the larva has both the head-shield and pygidium in a developed condition, but the thorax is wanting. In a third, all the three regions of the body are present, but the thorax and pygidium are at first incomplete; and in a fourth group, though the thorax possesses the number of rings proper to the adult, the pygidium is imperfect.

As to their *mode of life*, the Trilobites, as before remarked, seem to have delighted in muddy bottoms, though often found in limestones, and they must have frequented particular localities in vast numbers. In connection with this subject, we may briefly notice certain tracks or markings in the rocks, which have been supposed to have been produced by these extraordinary extinct Crustaceans, or by their allies the King-crabs and Eurypterids. The most interesting of the tracks in question are those which have been described by Professor Owen from the Potsdam Sandstone (Upper Cambrian) of Canada, under the name of *Protichnites*. The tracks upon which this genus is founded (fig. 217, A and B) consist essentially of a median groove, with a number of depressions or footprints on each side in corresponding pairs, these being often arranged in answering groups, of seven or eight pairs each. Sometimes the pits or footprints are replaced by shallow grooves, on each side of the main median groove (fig. 217, B). The tracks of *Protichnites* indicate the existence in the Upper Cambrian of some animal of very considerable size, since they are sometimes half a foot or more in width. That the animal producing these tracks was a Crustacean can hardly be doubted; the median groove being made by the tail-spine, and the lateral markings by the feet; and as we know that large Trilobites actually lived during this period, it seems most reasonable to suppose that we have in these the real makers of the tracks. Principal Dawson, however, has shown that tracks of a closely similar nature are formed by the living King-crabs (*Limulus*), and he would therefore ascribe *Protichnites* rather to the Eurypterids. The same eminent observer has also shown that smaller forms of *Protichnites* occur in the Carboniferous; and he ascribes these to the Limuloid genus *Belinurus*. The curious track known as *Climactichnites* (fig. 217, C) is likewise found in the Potsdam formation, and consists of a band about six inches wide, crossed by straight ridges, and bounded by beaded margins. These were probably formed by the same animal as that which produced *Protichnites*, and Dawson has shown that the living *Limulus*, when it uses its swimming feet, gives rise to a ladder-like track of the same kind. Prof. Chapman believes that both *Protichnites* and *Climactichnites* are really of vegetable origin. The only other fossil which need be mentioned in this con-

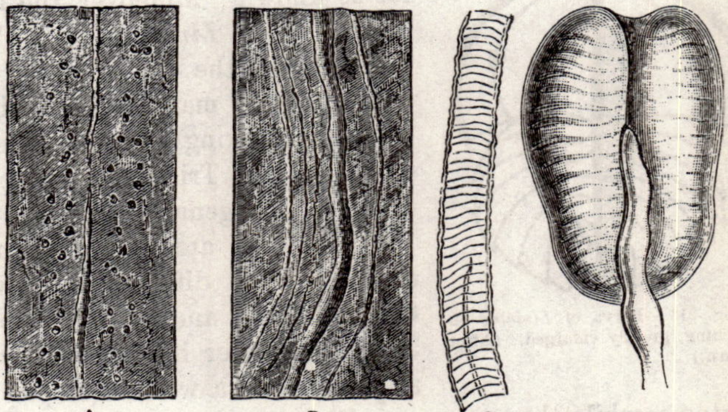

A B C D

Fig. 217.—Supposed Crustacean tracks and burrows. A, Portion of the track of *Protichnites alternans*, from the Potsdam Sandstone, reduced to one-tenth of the natural size; B, Portion of the track of *Protichnites lineatus*, from the same formation, similarly reduced; C, Portion of *Climactichnites Wilsoni*, from the Potsdam group, reduced to one-thirtieth of the natural size; D, *Rusichnites* (*Rusophycus*) *bilobatus*, from the Clinton formation (Middle Silurian), reduced one-half. (After Owen and Hall.)

nection is the curious *Rusichnites*, which is of common occurrence in the Silurian of North America, and is also found in the Carboniferous. Originally described as a plant under the name of *Rusophycus*, its name was changed by Dawson to *Rusichnites*, in accordance with his belief that it really represents the *casts* of the burrows of Trilobites, and that it can be shown to be sometimes connected with footprints consisting of a double series of transverse markings. In form, *Rusichnites* (fig. 217, D) presents itself as an oval, cylindroidal body, deeply furrowed or bilobed, by means of a longitudinal sulcus, the lateral halves being transversely ridged or grooved. The body thus constituted may be independent, or may stand in apparent connection with a cylindrical and slender appendage.

As regards their zoological *affinities*, the Trilobites are related most closely to the *Phyllopoda* and the *Xiphosura*, though they also exhibit points of agreement with the *Iso-*

poda. They resemble the Phyllopods principally in the indefinite segmentation of the body, and in the likeness between their hypostome and the lip-plate of the recent *Apus*. On the other hand, they agree with the *Xiphosura* in the possession of compound and sessile eyes, in the possession of thoracic branchiæ, and to some extent in the nature of their mouth organs, so far as known. Moreover, the larva of the recent *Limulus* (fig. 218) is destitute of the tail-spine of the adult, and in many respects shows a most striking resemblance to certain of the Trilobites, and especially to the genus *Trinucleus* (fig. 230). There are, however, many and important differences between the Trilobites and the *Xiphosura*. Thus the former have free thoracic segments, and a well-developed hypostome, while there are no ocelli; the latter, on the contrary, having the thoracic segments anchylosed, the limbs entirely converted into organs of prehension and mastication, the hypostome rudimentary, and the head-shield furnished with ocelli as well as compound eyes. Lastly, the Trilobites agree with the *Isopoda* in possessing compound sessile eyes, in having the abdominal somites amalgamated to form a caudal shield, in the occasional power of rolling into a ball, the thoracic segments being free and movable, and in the absence of ocelli. They differ from the Isopods in the fact that the thoracic segments of the latter are always definite and almost always seven in number, and in their apparent possession of a much larger number of locomotive appendages.

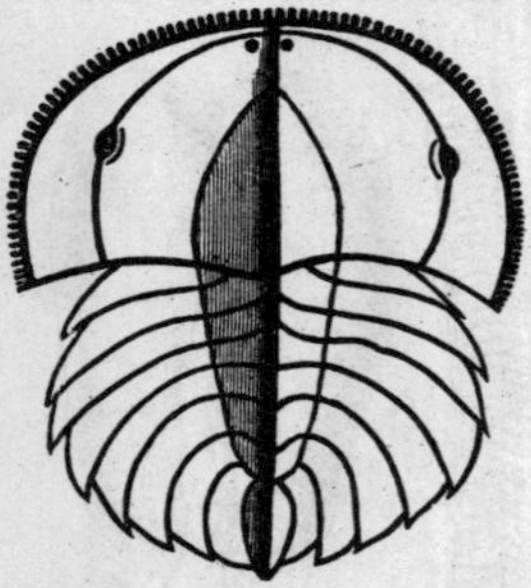

Fig. 218.—Larva of *Limulus* on hatching, greatly enlarged. (After Dohrn.)

The general facts as to the *distribution* of the *Trilobita* in past time are soon told. The Trilobites are exclusively Palæozoic, their range extending from the Upper Cambrian to the Lower Carboniferous. If the *Palæopyge Ramsayi* of the Longmynd beds be a Trilobite, then the order has its commencement in the Lower Cambrian. In the Upper Cambrian rocks the order attained a wonderful development, the number of generic and specific types already known from this formation being very large, while individuals are sometimes extremely abundant. Some of these so-called "Primordial" Trilobites attain a very great size, being only surpassed amongst later forms by some species of *Asaphus.* They mostly belong to the families of the *Agnostidæ, Conocephalidæ,* and *Paradoxidæ,* and to the genera *Paradoxides, Conocoryphe, Sao, Ellipsocephalus, Hydrocephalus, Arionellus, Dikellocephalus,* &c. Along with these, however, are genera such as *Agnostus* and *Olenus,* which pass up into the Silurian. Some of the "Primordial" Trilobites are not so highly organised as their successors, as shown by the occasional absence of eyes, or want of the facial suture, by the imperfect segmentation of the glabella, or by the multiplication or diminution of the number of the body-rings; but others do not exhibit any inferiority to those of later periods.

In the Lower and Upper Silurian rocks the Trilobites attain their maximum of development, the leading families being the *Asaphidæ, Phacopidæ, Trinucleidæ, Cheiruridæ,* and *Calymenidæ,* and the chief genera being *Asaphus, Ogygia, Phacops, Trinucleus, Ampyx, Cheirurus, Encrinurus, Calymene,* and *Homalonotus.* In the Devonian rocks, again, Trilobites are tolerably abundant, though less so than in the preceding series. The commonest Devonian genera are *Phacops, Homalonotus,* and *Bronteus.* Lastly, the order seems to die out before the close of the Carboniferous period, being represented in the Carboniferous period solely by the four genera, *Phillipsia, Brachymetopus, Griffithides,* and *Proetus.*

In the following, we shall give a brief summary of the families of the *Trilobita,* indicating the principal genera of each, and their distribution in time. No strictly zoological arrangement of these families is as yet possible, except in a general sense, but we have, in the main, adhered to the classification proposed by M. Barrande, one of the most illustrious palæontologists of the age.

1. HARPEDIDÆ.—Cephalic shield large, and horse-shoe-shaped, its posterior angles greatly prolonged, and its mar-

gin or "limb" perforated by pores (fig. 219). The glabella is conical, the eyes small and of few facets. There are twenty-five or twenty-six thoracic segments, the pleuræ are grooved, and the pygidium is extremely small. The only genus of the family is *Harpes* itself, which is characteristic of the Lower and Upper Silurian.

Fig. 219.—*Harpes ungula*, from the Upper Silurian of Bohemia. (After Barrande.)

2. Remopleuridæ.—In this family the head is greatly developed, semicircular in shape, the genal angles produced into spines. The glabella shows the lateral grooves, and the facial sutures unite in front of it. The eyes are very long and are reticulated. There are eleven body-rings, with grooved pleuræ, and the pygidium is very small, and is often reduced to two segments. This family contains only the single genus *Remopleurides* (fig. 220), which is confined to the Silurian period, occurring in both the Lower and Upper divisions of this formation.

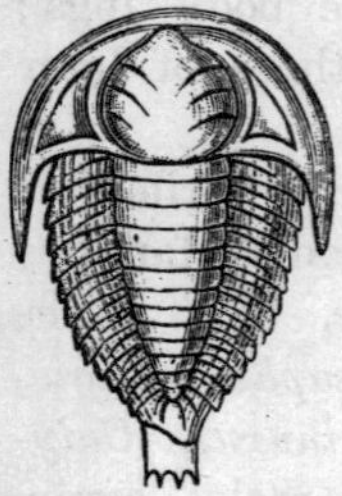

Fig. 220.—*Remopleurides radians*. Upper Silurian. (After Barrande.)

3. Paradoxidæ or Olenidæ.—Head-shield well developed, crescentic, the genal angles produced. The glabella typically widest anteriorly, with well-marked grooves. The facial sutures nearly parallel, cutting the head-shield separately; the eyes large. The body is very long (fig. 221, *a*); the thorax has from twelve to twenty segments, with grooved pleuræ; the pygidium being usually small and of few segments. The family is essentially characteristic of the Upper Cambrian.

The principal genus of this group is *Paradoxides* itself (fig. 221, *a*, and fig. 222), with its long and trilobed body, sometimes reaching a length of two feet or more. The thorax in this genus is greatly elongated, and consists of sixteen or twenty rings, while the axis of the pygidium often contains two segments only. The eyes are long, reniform, and smooth. The genus is characteristic of the "Primordial" or Upper Cambrian. *Plutonia* and *Anopolenus*, with a similar geological range, are closely related to *Paradoxides*; but the former of these two genera has a narrow glabella and a tuberculated surface, while the latter has the last two pleuræ of the thorax dilated and bent backwards, and the pygidium has wide lateral lobes. In *Hydrocephalus*, again, the glabella is immensely inflated, with a median longitudinal groove; and the facial sutures cut the outer margins of the head-shield, so that the free cheeks become much reduced in size, and

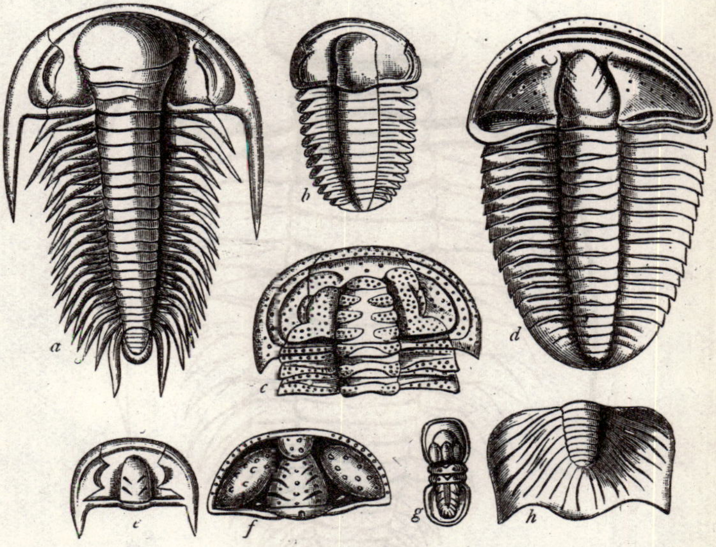

Fig. 221.—*Paradoxidæ* and *Conocephalidæ*. *a*, *Paradoxides Bohemicus*, reduced in size; *b*, *Ellipsocephalus Hoffi*; *c*, *Sao hirsuta*; *d*, *Conocoryphe Sultzeri*—(all the above, together with fig. *g*, are from the Upper Cambrian or "Primordial Zone" of Bohemia); *e*, Head-shield of *Dikellocephalus Celticus*, from the Lingula Flags of Wales; *f*, Head-shield of *Conocoryphe Matthewi*, from the Upper Cambrian (Acadian Group) of New Brunswick; *g*, *Agnostus rex*, Bohemia; *h*, Tail-shield of *Dikellocephalus Minnesotensis*, from the Upper Cambrian (Potsdam Sandstone) of Minnesota. (After Barrande, Dawson, Salter, and Dale Owen.)

the genal spines are attached to the fixed cheeks. Eyes are wanting, and there are only twelve body-rings. A more important and widely distributed genus of this family is *Olenus* (fig. 223), in which the general characters are very similar to those of *Paradoxides*, but the glabella is contracted in front so as to become conical, and its furrows are often extended completely across it: there are only fourteen body-rings; and the pygidium is well developed. *Parabolina* includes *Oleni* with only twelve body-rings, and is wholly Upper Cambrian, while *Olenus* appears to survive in much diminished numbers in the Lower Silurian. Several other genera belong to this family, but the only one which needs mention is *Dikellocephalus* (figs. 213 and 221, *h*), in which

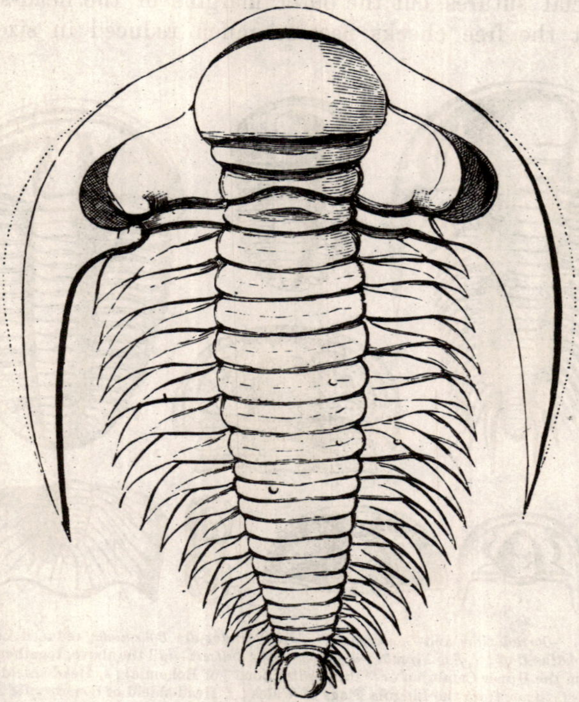

Fig. 222.—*Paradoxides Micmac.* Upper Cambrian. (After Dawson.)

the most striking feature is the broad, fan-like, often spined tail, with its short many-ringed axis. The facial sutures are united in front of the glabella (Billings), and the transverse grooves of the glabella are like those of many *Oleni* in joining from side to side. The genus is Upper Cambrian, but appears to occur also in the lowest Silurian deposits.

4. CONOCEPHALIDÆ.—This family is a convenient one to retain, though it does not seem at present possible to separate

it from the preceding by any rigid definition. Its members resemble the *Paradoxidæ* in general characters, but the glabella is narrow in front, or, at any rate, not dilated in this region, and the tail is usually moderately developed, while the thoracic rings are not so numerous as in the typical forms of the latter. The type-genus is *Conocoryphe* or *Conocephalites* (fig. 221, *d* and *f*), which has resemblances to both *Olenus* and *Calymene*, its glabella approaching that of the latter in its comparatively great posterior width and its contraction in front. Eyes are usually, but not always present; the facial sutures are discontinuous; the fixed cheeks are large and the free cheeks small; and there are fourteen or fifteen body-rings, while the tail

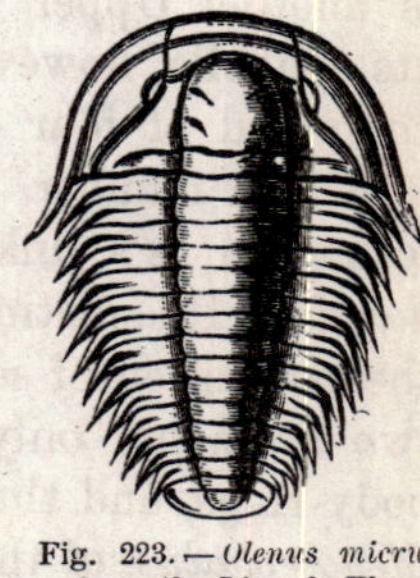

Fig. 223.—*Olenus micrurus*, from the Lingula Flags. (After Salter.)

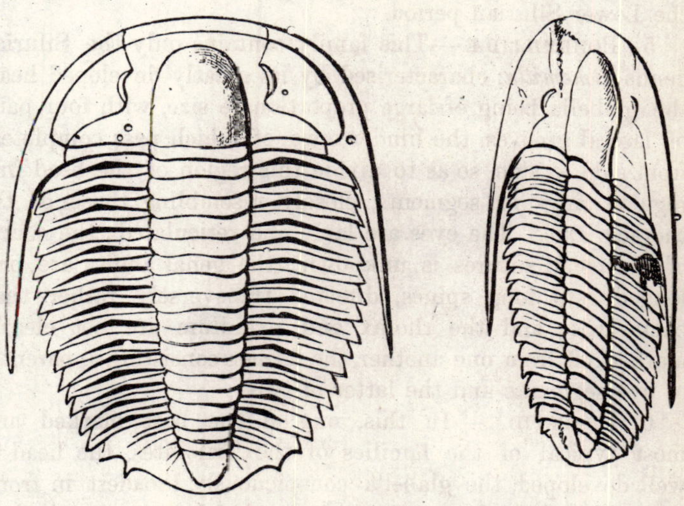

Fig. 224.—*Angelina Sedgwickii*, in its natural condition, and distorted by cleavage. Upper Cambrian. (After Salter.)

has from two to eight rings. The genus is represented by very numerous species in the Upper Cambrian. *Ellipsocephalus* (fig. 221, *b*) is related to the preceding, but the glabella is subquadrate, smooth, and convex, and there are twelve to fourteen body-rings. *Angelina* (fig. 224) is another Upper Cambrian genus, with affinities to *Olenus*. Its glabella, however, is destitute of grooves, and the tail is composed of four or five rings, while the body-segments are fifteen in number. The genus *Sao* (fig. 221, *c*), of the same formation, is a link between the present family and the *Paradoxidæ*. It is distinguished by its prominent furrowed glabella, the possession of seventeen body-rings, and the minute tail of two segments only. *Arionellus*, also Cambrian, has sixteen body-rings, and three caudal segments; and the allied genus *Menocephalus*, of the Lower Silurian, is chiefly separated from the former by its much more convex glabella. Lastly, *Bathyurus* (Upper Cambrian and Lower Silurian) and *Bathyurellus* (Lower Silurian) form in some respects intermediate types between the *Conocephalidæ* and the *Asaphidæ;* and the Olenoid genus *Triarthrus* survives to near the middle of the Lower Silurian period.

5. BOHEMILLIDÆ.—This family contains only the Silurian genus *Bohemilla*, characterised by its greatly developed head, the glabella being of large proportionate size, with four pairs of lateral grooves, the hinder ones of which pass completely from side to side, so as to divide this region of the head into regular rings or segments, closely resembling those of the thoracic axis. The eyes are large and reticulated; the course of the facial sutures is unknown; the genal angles are prolonged into long spines, directed transversely rather than backwards; and the thorax and pygidium are not clearly marked off from one another, the former consisting apparently of five segments and the latter of two.

6. PHACOPIDÆ.—In this, one of the best marked and most typical of the families of the Trilobites, the head is well developed, the glabella conspicuously broadest in front, with three lateral grooves, and the facial sutures united in front of the glabella, and cutting the outer margins of the cephalic buckler behind. The eyes are large and faceted (fig. 225); there are eleven body-rings, with grooved pleuræ; and the condition of the pygidium is variable. The genus *Phacops* itself (with the sub-genera *Trimerocephalus*, *Phacops*

proper, *Acaste, Chasmops, Dalmanites,* and *Cryphæus*) forms the entire family, and ranges from the base of the Lower Silurian to the Devonian, inclusive. Some of the above-mentioned sub-genera may perhaps be of generic value; but they are none of them separated by conspicuous characters from *Phacops* proper.

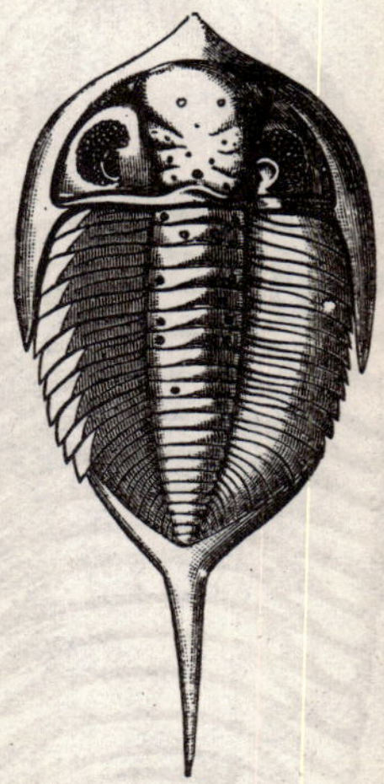

Fig. 225.—*Phacops longicaudatus.* Upper Silurian (Wenlock) of Britain and North America.

7. Calymenidæ.—In this family the crust is often tuberculated or granulated; the head is well developed; the glabella is widest behind; and the facial sutures are discontinuous, and terminate at the posterior angles of the cephalic shield. There are thirteen segments in the thorax, with grooved pleuræ; the tail is of moderate size; and the condition of the eyes is variable. In *Calymene* itself (fig. 210) the head is usually crescentic, with rounded genal angles; and the glabella is conical, strongly convex, with deep axal furrows, and divided by three deep lateral grooves on each side, all the "lobes" thus formed being globose, and the hindmost being the largest. The eyes are minute and reticulated, but are rarely recognisable. The tail is convex, with a well-marked axis; and the hypostome is forked. The genus is found both in the Lower and Upper Silurian, and the three best known species are the nearly-allied *C. Blumenbachii, C. senaria,* and *C. brevicapitata.* The genus *Homalonotus* is the only other member of this family, and agrees with *Calymene* in most respects. The glabella, however (fig. 226), is almost destitute of any trace of lobation, while the body is greatly elongated and is but faintly trilobed. The best known species is the *Homalonotus delphinocephalus* of the Upper Silurian.

8. Proetidæ.—In this family the head is of variable size, semicircular, sometimes with rounded, sometimes with spinose genal angles. The facial sutures are not continuous;

there are eight to twenty-two body-rings, with grooved pleuræ; and the eyes are smooth.

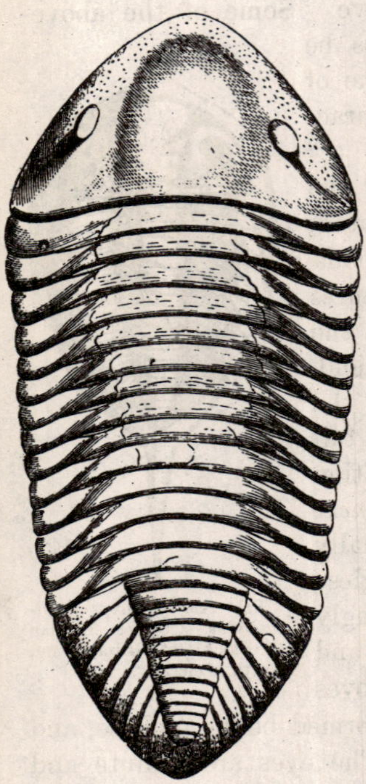

Fig. 226.—*Homalonotus delphinocephalus.* Upper Silurian.

In *Proetus* itself (fig. 228, C) the head-shield is semicircular; the glabella has three pairs of lateral furrows; the eyes are large, semicircular, of numerous facets, covered by a thin cornea; there are eight or ten body-rings, and the tail has an "entire" border. The genus ranges from the Lower Silurian to the Carboniferous. *Cyphaspis* (fig. 228, B), of the Silurian, differs from the preceding chiefly in its more convex glabella, with circumscribed basal lobes, its ovoid and remote eyes, and the generally greater number (fifteen to seventeen) of the body-rings. *Arethusina* (fig. 228, A), also Silurian, has its glabella much shortened, while the body-rings are as many as twenty-two in number. *Carmon* is another Silurian genus allied to *Proetus*, but it has neither eyes nor facial sutures, and it possesses eleven body-rings. The genus *Harpides* (apparently = *Erinnys* of Salter) is an interesting type, which appears to be intermediate between the *Proetidæ* and *Olenidæ*, and which carries back the range of the former into the Upper Cambrian. It has the "limb" of the head-shield very wide, and covered with a network of radiating and bifurcated nervures. On the other hand, the *Proetidæ* are represented in the Carboniferous rocks, not only by *Proetus* itself, but also by the genera *Phillipsia*, *Griffithides*, and *Brachymetopus*, some of which have been detected in North America in deposits of the age of the Coal-measures, and which are the most modern examples of the order *Trilo-*

bita at present known to us In *Phillipsia* (fig. 227) the general form is like that of *Proetus,* but there are only nine body-rings. The tail is semi-oval, distinctly segmented, and with an "entire" margin; and the eyes are large, reticulated, and reniform. *Griffithides* resembles *Phillipsia* in having nine body-rings, but the glabella is pyriform, and is destitute of lateral furrows, while the eyes are small.

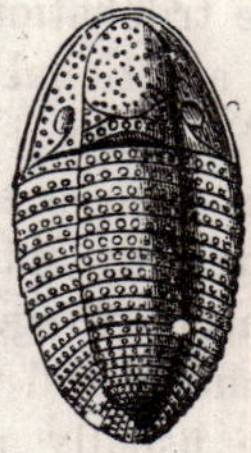

Fig. 227.—*Phillipsia seminifera,* of the natural size. Carboniferous.

9. LICHADÆ.—In this family we have only the Silurian genus *Lichas* (fig. 228, D) in which the head is very convex, and the frontal grooves of the glabella are extended backwards, so as to enclose a central lobe. The facial sutures are discontinuous, and the eyes are small and smooth. There are eleven segments to the thorax, with grooved pleuræ; and the pygidium is larger than the head,

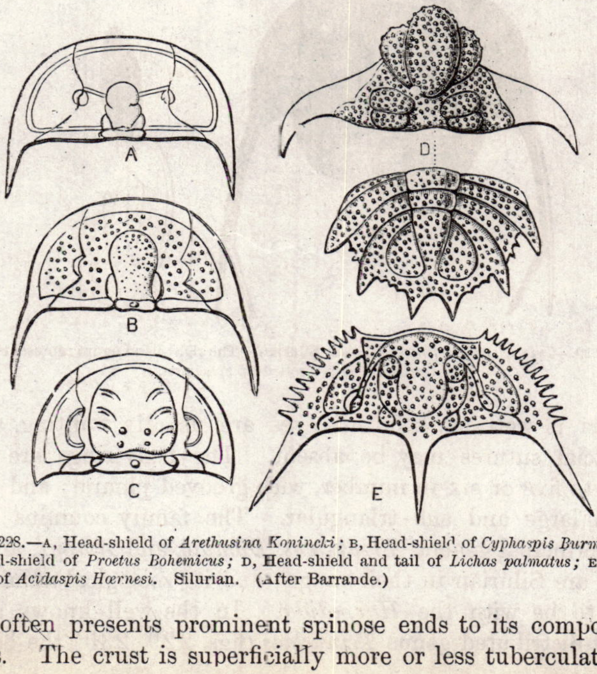

Fig. 228.—A, Head-shield of *Arethusina Konincki;* B, Head-shield of *Cyphaspis Burmeisteri;* C, Head-shield of *Proetus Bohemicus;* D, Head-shield and tail of *Lichas palmatus;* E, Head-shield of *Acidaspis Hœrnesi.* Silurian. (After Barrande.)

and often presents prominent spinose ends to its component rings. The crust is superficially more or less tuberculated.

10. ACIDASPIDÆ.—Like the preceding, this family contains only a single genus—viz., *Acidaspis* itself. In this characteristically Upper Silurian type (fig. 228, E), the usual form of the head-shield in the Trilobites is somewhat masked, the trilobation being rendered indistinct by the presence of an additional and secondary pair of axal furrows, which mark off a central inflated portion of the glabella. The thorax has nine or ten rings, with ridged pleuræ, which are terminated by spines; while the tail is very small, and has its margin fringed with spines. The facial sutures are continuous, and the eyes are smooth. The species of *Acidaspis* are usually readily recognised by their highly ornamented and spinose crust.

11. TRINUCLEIDÆ.—In this singular family (fig. 229) the head-shield is enormously developed, with a wide margin or "limb," which is usually perforated by rounded pores. The

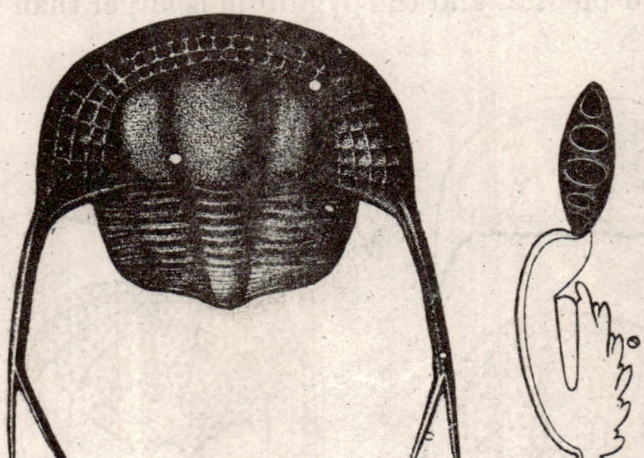

Fig. 229.—*Trinucleus Pongerardi*—Lower Silurian. The right-hand figure represents a vertical section of a rolled-up specimen.

glabella is well marked, but eyes are usually wanting, and the facial sutures may be absent. The body-rings are reduced to five or six in number, with grooved pleuræ; and the tail is large and sub-triangular. The family contains the three principal genera, *Trinucleus, Dionide,* and *Ampyx,* all of which are Silurian in their range; and its zoological affinities seem to be with the *Harpedidæ*. In the well-known and widely-distributed genus *Trinucleus* (figs. 229, 230) the body is distinctly trilobed; the "limb" of the head-shield is composed of two lamellæ, and is perforated by numerous larger or smaller foramina; and the "genal angles" are prolonged into conspicuous spines which are usually single, but are forked in *T. Pongerardi.* The glabella is prominent and pyriform, with mere traces of lateral grooves, the facial sutures being rudimentary, and the cheeks being tumid, and generally furnished on each side with a small tubercle seemingly representing the eyes. There are six body-rings; and the tail is triangular, with a distinct axis, and having its margin entire and striated. The genus *Dionide* has a sub-quadrangular glabella; and the "limb" of the head-shield, though perforated, is rudimentary, and by absence of the facial sutures becomes continuous with the cheeks. The eyes are wanting, and there are six body-rings. In the curious Lower Silurian genus *Ampyx* (fig. 231) the head is sub-triangular, with spinose genal angles, and without a

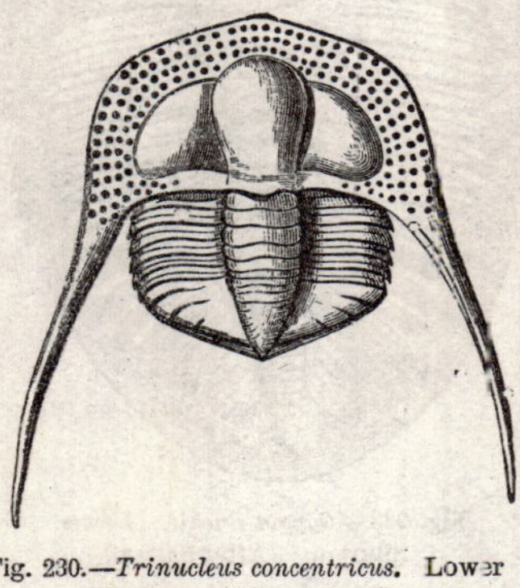

Fig. 230.—*Trinucleus concentricus.* Lower Silurian. (After Salter.)

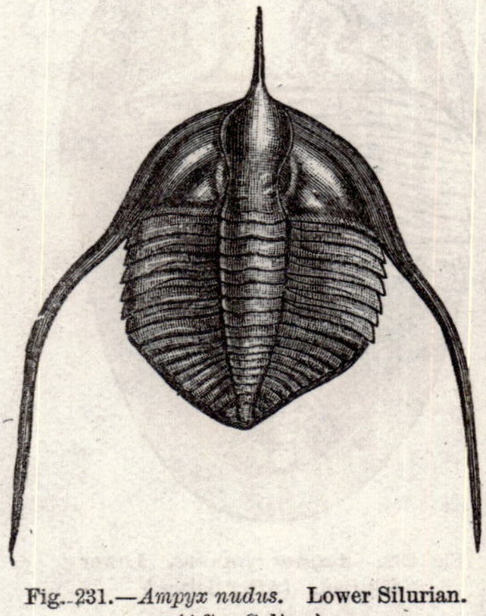

Fig. 231.—*Ampyx nudus.* Lower Silurian. (After Salter.)

perforated "limb." The glabella is prolonged forwards in front of the head-shield as a triangular process, which is often extended into a long spine, while its lateral grooves are obsolete, and eyes are wholly wanting. The facial sutures are present and are discontinuous. There are five or six body-rings, and the tail is sub-triangular. *Endymion* is allied to both *Trinucleus* and *Ampyx*, but wants the perforated border of the former, and the prolonged glabella and genal spines of the latter.

12. Asaphidæ.—Large Trilobites, generally oval, and never furnished with spines or tubercles on their surface. The eyes smooth, and the facial sutures terminating on the posterior margin. The cephalic and caudal shields generally of large size, the glabella of the former often obscure, and the latter sometimes exhibiting no indication of its component segments. The body-rings usually eight in number, with grooved pleuræ. The family is characteristically Lower Silurian, and the two principal genera are *Asaphus* and *Ogygia*. In the genus *Asaphus* (figs. 211, 215, 232) the general trilobation is somewhat indistinct, and the caudal

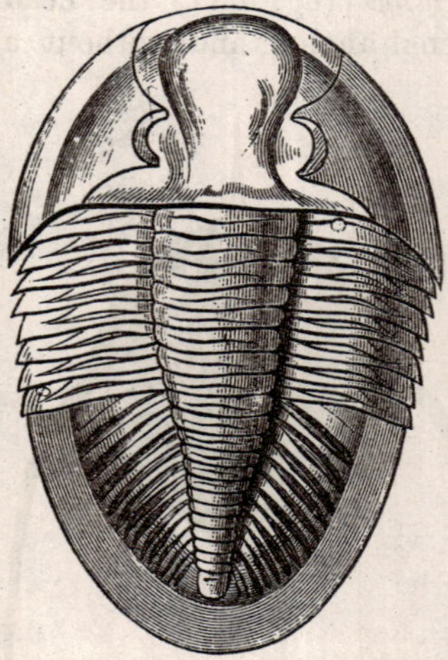

Fig. 232.—*Asaphus tyrannus*. Lower Silurian. (After Salter.)

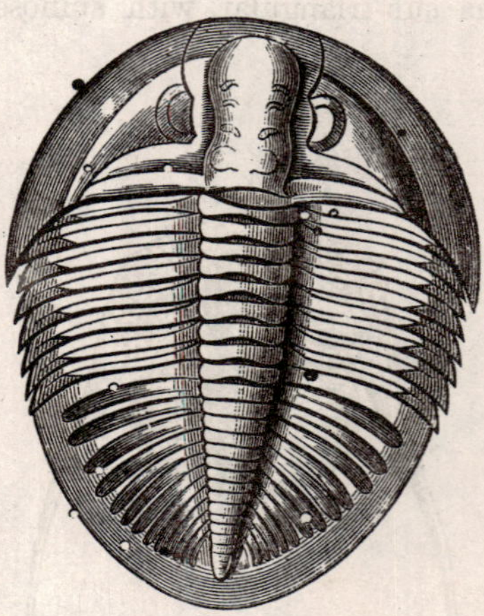

Fig. 233.—*Ogygia Buchii*. Lower Silurian. (After Salter.)

shield is at least equal to the head in size. The genal angles of the head-shield may be rounded or spinose, and the glabella is not marked off by conspicuous axal furrows. The facial sutures are discontinuous, the eyes crescentic, the hypostome deeply forked, and the pygidium may or may not show a conspicuous axis, its hinder extremity being usually rounded, and its margin always entire. This important genus, though not unknown in the Upper Cambrian, is characteristic of the Lower Silurian, in which it is represented by numerous species, some of which attain the extraordinary size of two feet in length. The genus *Ogygia* (also Lower Silurian) is closely allied to *Asaphus* in general form and proportions. The axis of the tail is, however, more conspicuously marked than in most *Asaphi* (fig. 233), the hypostome is *rounded* and not cleft, the glabella is distinctly furrowed, and the pleuræ of the thorax have only rudimentary "fulcra." *Niobe,* of the Upper Cambrian, is an early form of this family, which is intermediate between *Asaphus* and *Ogygia,* having the round hypostome and lobed glabella of the latter, while it approaches the former in its wide glabella and its obtuse and faceted pleuræ. *Barrandia* and *Stygina* are both related to *Ogygia;* and the Upper Cambrian genus *Psilocephalus* forms a link between *Asaphus* and *Illænus;* while the genus *Nileus* may be almost indifferently placed with either the *Asaphidæ* or the *Illænidæ.*

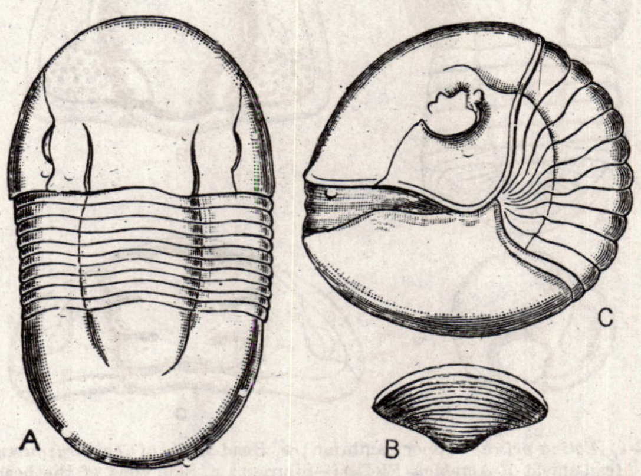

Fig. 234.—A, A complete example of *Illænus Davisii*, in its unrolled state; B, Hypostome of the same; C, *Illænus (Bumastus) Barriensis*, rolled up. Lower Silurian. (After Salter.)

13. ILLÆNIDÆ.—In this family the head and tail are greatly developed, and, as in *Asaphidæ,* are approximately equal in size (fig. 234). The glabella is broad and rounded, destitute of lobes, and with the axal furrows hardly at all

marked. There are eight to ten body-rings; and the axis of the tail is truncated or wanting, and in no case exhibits definite segmentation. The free cheeks are extremely small; the facial sutures are separate; and the eyes are crescentic.

If *Nileus* be regarded as belonging to the *Asaphidæ,* the only well-defined genus in this family is *Illænus* itself (fig. 234). In this genus the head is large, convex, and tumid, as is also the tail, the glabella and pygidial axis being hardly marked out, and being wholly unsegmented. In the typical forms of this group (*Illænus* proper), the axis of the thorax is not disproportionately wide, and the axal furrows are distinct; but in others (*Bumastus*), the thoracic axis is extremely wide, and is hardly separated from the pleuræ by recognisable axal furrows.

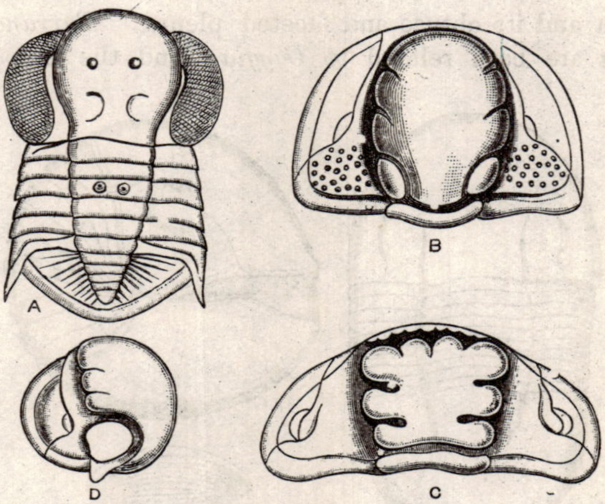

Fig. 235.—A, *Æglina prisca*—Upper Cambrian; B, Head-shield of *Cheirurus juvenis*—Lower Silurian; C, Head-shield of *Amphion Fischeri*—Silurian; D, Side view of the head-shield of *Sphærexochus mirus*—Lower Silurian. (After Barrande and Salter.)

14. ÆGLINIDÆ. — This family contains only the single genus *Æglina* (fig. 235, A), of the Upper Cambrian and Lower Silurian, and is chiefly distinguished from the preceding by the much larger size of the eyes, and the smaller number of body-rings. The head and tail are both of large size, the latter with a truncated axis; the glabella is not conspicuously marked out; the facial sutures are discontinuous;

and the eyes are extremely large and reticulate; while the segments of the thorax are reduced to five or six in number.

15. CHEIRURIDÆ.—In this large and important family the head-shield is well developed, with discontinuous facial sutures, which terminate on its outer margin. The glabella is usually highly convex or tumid, with well-marked axal furrows and lateral grooves. There are ten to twelve, generally eleven, body-rings, and the pygidium is small, of from three to five segments, the pleuræ terminating in free ends. The family ranges from the Upper Cambrian to the Devonian, but is principally characteristic of the Silurian rocks. In *Cheirurus* itself (fig. 235, B) the head is semicircular, with rounded or pointed genal angles, and with a strongly-arched glabella, which is deeply grooved by the lateral furrows. There are generally eleven body-rings, with ridged or slightly grooved pleuræ; and the tail has a well-marked axis of four rings, its pleuræ being prolonged into points or spines. *Amphion* (fig. 235, C) is nearly related to *Cheirurus*, but has from fifteen to eighteen body-rings, and exhibits minor differences as well. *Placoparia*, again, agrees with *Cheirurus* in having eleven body-rings, and also in the form and lobation of the glabella, but it is destitute of both eyes and facial sutures, as is also the genus *Areia*. *Sphærexochus*, lastly, (fig. 235, D), while agreeing with *Cheirurus* in its eleven body-rings and in the form of its tail, is distinguished by the extreme inflation of the glabella, and the presence of no more than three segments in the pygidium. We may also place here the very singular and aberrant Silurian genera *Staurocephalus* and *Deiphon*. In the former of these the general characters correspond with those of *Cheirurus*, but the anterior or "frontal" portion of the glabella is enormously swollen, and forms a great globular projection in advance of the line of the cheeks. In the still more curious *Deiphon* the fixed cheeks are rudimentary, and are reduced to two curved spines, which spring from the sides of the swollen glabella, and carry the faceted eyes, while the free cheeks are obsolete. The axis of the tail is formed of four or five rings, and the pleuræ are prolonged into two spines on each side, one of these being formed by the first segment only,

while the other and much larger one is made up of the amalgamated extremities of the remaining segments.

16. ENCRINURIDÆ.—This family is principally or exclusively characteristic of the Upper and Middle Silurian periods, and is related to the preceding through the intervention of *Amphion*. The head is fairly developed, the genal angles rounded or pointed, and the facial sutures discontinuous, and cutting the outer angles of the cephalic buckler. Eyes are present, though not of large size, and the glabella may or may not exhibit distinct lateral grooves. The surface is tuberculated, and some or all of the body-rings may bear spines. The thorax consists of eleven segments; and the tail, though moderate in size, has a well-marked axis, which is composed of very numerous rings.

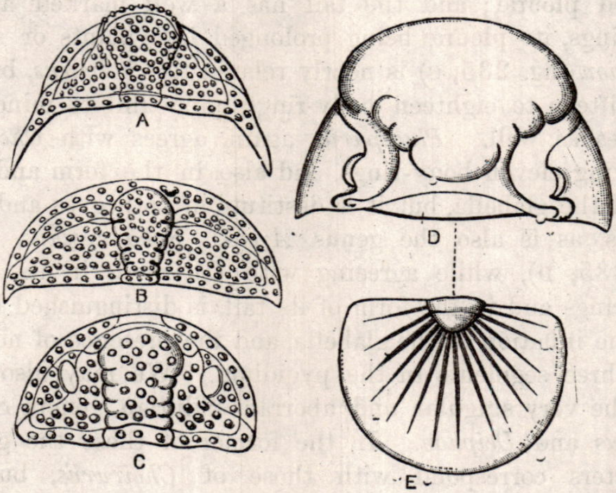

Fig. 236.—A, Head-shield of *Encrinurus punctatus*; B, Head-shield of *Cromus intercostatus*; C, Head-shield of *Cybele bellatula*; D, Head-shield of *Bronteus campanifer*; E, Tail of the same. Silurian. (After Barrande.)

In *Encrinurus* (fig. 236, A) the glabella is pyriform, strongly tuberculated, with the lateral furrows almost obsolete. The body-rings are eleven in number, and the axis of the pygidium is composed of extremely numerous rings. The genus is characteristic of the Middle Silurian or of the base of the Upper Silurian (Llandovery period). In *Cromus* (fig. 236, B) the glabella has four well-marked lateral grooves; the eyes are small and ovoid; and the pygidium looks like a continuation of the thorax, its axis being composed of from twelve to twenty-eight segments. Lastly, *Cybele* (not *Zethus*) has four lateral grooves to the glabella (fig. 236, C), the eyes being clavate, the body-rings twelve in number, and the ends of the pleuræ being spinose, and directed conspicuously backwards in the region of the tail.

17. DINDYMENIDÆ (= ZETHIDÆ of Barrande).—In *Dindymene*, of the Silurian formation, the only genus which can be certainly referred to this family, the head-shield is semicircular, with a tumid glabella, destitute of lateral grooves. There are no eyes, nor facial sutures; and the cheeks are tumid, as in *Trinucleus*. There are ten body-rings, and the tail is large and distinctly segmented. Barrande calls this family *Zethidæ*, and includes in it the genus *Zethus*. This genus, however, is in a dubious position, the forms included under this name by British palæontologists being referable to *Cybele*, and therefore belonging to the *Encrinuridæ*, while the same appears to be the case with one of the two species —*Cybele* (*Zethus*) *bellatula*—quoted and figured by Barrande.

18. BRONTEIDÆ.—This family includes the single genus *Bronteus* (or *Brontes*) of the Silurian and Devonian. In this well-marked type (fig. 236, D and E) the most striking feature is the caudal shield, which is as large as, or larger than, the head, always more or less fan-shaped, the axis being rudimentary, and the "limb" greatly developed. The head-shield is trilobed, the glabella being dilated in front, the facial sutures being separate, and the eyes annular. There are ten thoracic segments, and the pleuræ are ridged.

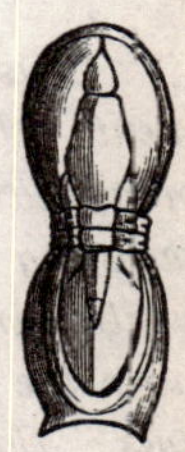

Fig. 237.—*Agnostus princeps*. Upper Cambrian. (After Salter.)

19. AGNOSTIDÆ.—This, the last, family of the Trilobites includes a number of small forms of this order, which range from the Upper Cambrian to near the summit of the Lower Silurian. The cephalic and caudal shields in this group are nearly equal, and are very similar to one another in form and markings. The thoracic segments are reduced to only two

in number, with grooved pleuræ; and both eyes and facial sutures are totally wanting. The type-genus of the family is *Agnostus* itself (fig. 237), which is represented by numerous forms in the Upper Cambrian and Lower Silurian. *Microdiscus* of the Upper Cambrian (sometimes placed in the *Trinucleidæ*) agrees with *Agnostus* in its want of facial sutures and eyes, but it has four body-rings, and the axis of the tail is segmented. Lastly, the *Shumardia* of the Lower Silurian is like *Agnostus*, but has both the axis and lateral lobes of the pygidium distinctly segmented.

ORDER MEROSTOMATA.

Crustaceans, often of large size, in which the mouth is furnished with mandibles and maxillæ, the terminations of which become walking or swimming feet, or organs of prehension (figs. 238, 239).

The order *Merostomata* comprises the two sub-orders of the *Xiphosura* and *Eurypteridæ*. The former appears to have commenced its existence in the Upper Silurian period, and is represented at the present day by the *Limuli* or King-crabs. The latter is wholly extinct, and is exclusively Palæozoic, none of its members being known out of the Silurian, Devonian, and Carboniferous formations.

SUB-ORDER I.—EURYPTERIDA.

"*Crustacea with numerous, free, thoracico-abdominal segments, the first and second (?) of which bear one or more broad lamellar appendages upon their ventral surface, the remaining segments being devoid of appendages; anterior rings united into a carapace, bearing a pair of larval eyes (ocelli) near the centre, and a pair of large, marginal, or sub-central eyes; the mouth furnished with a broad post-oral plate, or metastoma, and five pairs of movable appendages, the posterior of which form great swimming-feet; the telson, or terminal segment, extremely variable in form; the integument characteristically sculptured.*"—(Henry Woodward.)

In the typical *Eurypterids*, such as *Pterygotus* (fig. 239) and

Eurypterus, the anterior portion of the body is covered by a buckler or carapace, which bears a pair of minute larval eyes, and a pair of large compound eyes placed marginally or subcentrally. On the under surface of the carapace are five pairs of appendages. The first pair of these is usually re-

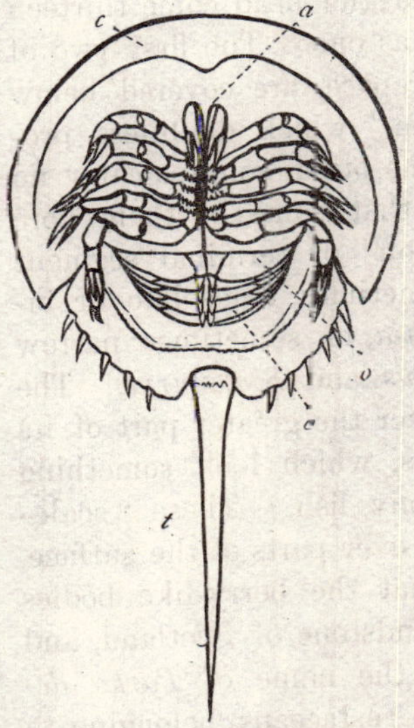

Fig. 238.—Xiphosura. *Limulus polyphemus*, viewed from below. *c*, The cephalic shield carrying the sessile eyes upon its upper surface; *o*, "Operculum," covering the reproductive organs; *b*, Branchial plates; *a*, First pair of antennæ (antennules) ending in chelæ. Below these is the aperture of the mouth, surrounded by the spiny bases of the remaining five pairs of appendages, which are regarded by Woodward as being respectively, from before backwards, the great antennæ, the mandibles, the first maxillæ, the second maxillæ, and a pair of maxillipedes. All have their extremities chelate.

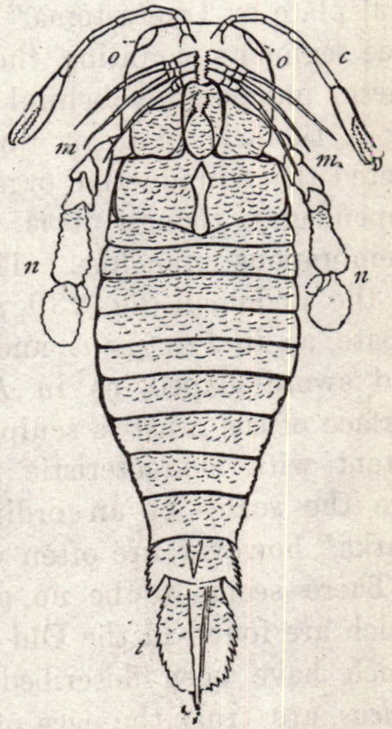

Fig. 239.—Eurypterida. *Pterygotus Anglicus*, restored (after H. Woodward). *c*, *c*, Chelate antennæ; *o*, *o*, Eyes, situated at the anterior margin of the carapace; *m*, *m*, The mandibles, and the first and second maxillæ; *n*, *n*, The maxillipedes—the basal margins of these are serrated, and are drawn as if seen through the metastoma or post-oral plate, which serves as a lower lip. Immediately behind this is seen the operculum or thoracic plate, which covers the two anterior thoracic somites. Behind this are five thoracic and five abdominal somites; and lastly, there is the telson (*t*).

garded as representing the antennæ. The appendages of this pair are mostly chelate, or converted into nipping-claws, but they are sometimes simple, and they sometimes are spinous towards their bases, and officiate as masticatory organs (*Eurypterus* and *Slimonia*). The next three pairs of appendages are simply-pointed spinous organs ("pedipalps"), but the last pair is sometimes converted into rowing-organs (*Stylonurus*). The last pair of appendages constitute two greatly-developed swimming-feet, the bases of which are furnished with spines, and form powerful jaws. The bases of these jaw-feet are covered by a greatly-developed post-oral plate or "metastoma." Behind the head come thirteen free segments, counting the telson as one. The first two of these, immediately behind the carapace, are covered below by a thoracic plate or "operculum," which doubtless protected the reproductive organs. The other somites carry no appendages, though it is certain that some of them bore membranous branchiæ. The "telson" or terminal segment of the abdomen (fig. 239, *t*) is sometimes lanceolate or bilobate, as in *Pterygotus* and *Slimonia*, or sometimes narrow and sword-shaped, as in *Eurypterus* and *Stylonurus*. The surface of the crust is sculptured over the greater part of its extent with characteristic markings, which look something like the scales of an ordinary Bony fish. These "scale-marks," however, are often wanting over parts of the surface.

There seems to be no doubt that the berry-like bodies which are found in the Old Red Sandstone of Scotland, and which have been described under the name of *Parka decipiens*, are truly the eggs of large Crustaceans belonging to the Eurypterids.

Besides such typical Eurypterids as *Pterygotus, Eurypterus*, and *Slimonia*, we have to refer to this order a number of forms which may be regarded as in many respects transitional between the *Eurypterida* and the *Xiphosura*. The most important of these connecting links is *Hemiaspis* (fig. 240, A), of the Upper Silurian, in which the head-shield has a general resemblance to that of *Limulus*, though the eyes have not been determined. The thorax consists of six segments, which are not amalgamated, but agree with those of the Eurypterids in being free, and which resemble those of the Trilobites in being markedly trilobed. There are three *free* segments in the abdomen, and these are followed by a long spine-like telson. Closely allied to the above, especially in the distinct division of the body into a cephalic, a thoracic, and an abdominal region, the presence of free thoracic segments, and the possession of a long tail-spine, are the genera *Pseudoniscus* and *Exapinurus* of the Upper Silurian (fig. 240, B and C.)

The *Eurypterids* range from the Upper Silurian, where they attain their maximum, through the Devonian, into the Carboniferous rocks, where they appear to die out. Traces, however, of these large Crustaceans are by no means wanting in the Lower Silurian, though not as yet fully

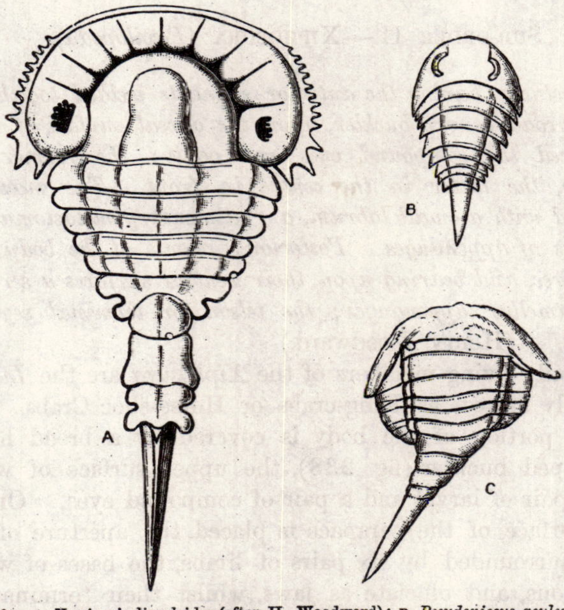

Fig. 240.—A, *Hemiaspis limuloides* (after H. Woodward); B, *Pseudoniscus aculeatus* (after Nieszkowski); C, *Exapinurus Schrenkii* (after Nieszkowski). All from the Upper Silurian.

described. Of the typical genera, *Pterygotus* extends from the Upper Silurian to the Upper Devonian, and species of this genus seem to have attained a gigantic size. (*Pterygotus Anglicus* is calculated to have reached a length of six feet.) *Slimonia* is Upper Silurian, and *Stylonurus* is both Upper Silurian and Devonian. *Eurypterus* is not known in the Silurian, but is represented by many species in the Devonian, and extends into the Carboniferous rocks. *Hemi-*

aspis, with only nine segments and the telson behind the carapace, is exclusively Upper Silurian. Lastly, *Pseudoniscus*, with the same number of free segments, is found in the passage-beds between the Upper Silurian and Devonian. In conclusion, it is interesting to note that these ancient Crustaceans present many larval features, resembling the larvæ of the *Decapoda*, especially in the fact that the hinder portion of the body is composed of free segments, which carry no appendages.

SUB-ORDER II.—XIPHOSURA (*Pœcilopoda*).

"*Crustacea having the anterior segments welded together to form a broad convex buckler, upon the dorsal surface of which are placed the compound eyes and ocelli; the former subcentrally, the latter in the centre in front. The mouth is furnished with a small labrum, a rudimentary metastoma, and six pairs of appendages. Posterior segments of the body more or less free, and bearing upon their ventral surfaces a series of broad lamellar appendages; the telson, or terminal segment, ensiform.*"—(Henry Woodward.)

The only living members of the *Xiphosura* are the *Limuli*, commonly known as King-crabs or Horse-shoe Crabs. The anterior portion of the body is covered by a broad horse-shoe-shaped buckler (fig. 238), the upper surface of which bears a pair of larval and a pair of compound eyes. On the lower surface of the carapace is placed the aperture of the mouth, surrounded by six pairs of limbs, the bases of which are spinous, and officiate as jaws, whilst their terminations are converted into chelæ or nipping-claws. The first pair of appendages is placed in front of the mouth, and represents the antennæ, so that the antennæ of the King-crabs are chelate. Behind the cephalic buckler comes a second shield, composed by six amalgamated segments, below which are carried the reproductive organs and branchiæ, the former protected by a thoracic plate or "operculum," the latter borne by five pairs of lamellar appendages. Lastly, articulated to the posterior margin of the abdominal shield, is a long sword-like spine or "telson" (fig. 238, *t*).

The *Xiphosura* seem to have commenced existence in the Upper Silurian period, where they are represented by the *Neolimulus falcatus* of Dr Henry Woodward. In this remarkable genus (fig. 241) the head-shield has a general resemblance to that of the recent *Limulus*, and there are traces of a divisional line crossing the head and apparently corresponding with the "facial suture" of the Trilobites. Compound eyes and ocelli seem to be present, and there are six *free* thoracic segments, followed, probably, by three *free* abdominal rings, of which only two have been preserved. A long spiniform telson may be conjecturally added to complete this ancient Limuloid Crustacean. In the Devonian

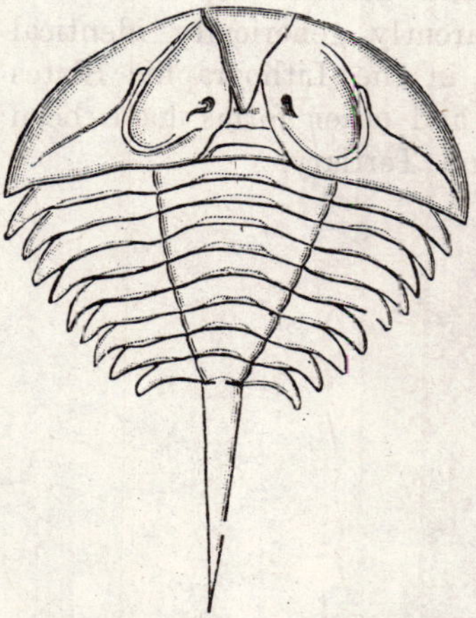

Fig. 241.—*Neolimulus falcatus*, enlarged about three times. Upper Silurian. (After Henry Woodward.)

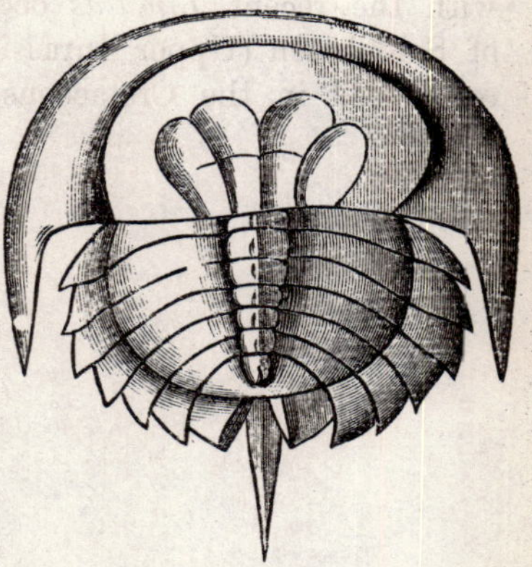

Fig. 242.—*Prestwichia rotundata*. Coal-measures.

no certain traces of Xiphosurans have been yet detected, but several types occur in the Carboniferous. The most important of these is the genus *Prestwichia* (fig. 242), which has the general form of *Neolimulus*, but in which the thoracic and abdominal segments are not marked off from one another, and are all anchylosed. The genus *Euproöps*, from the Coal-measures of North America, is hardly separable from the preceding, but the eyes are situated on the anterior edge of the cephalic buckler. Another well-known Carboniferous genus is *Belinurus*, which agrees with the preceding in having five thoracic and three abdominal segments, together with a long tail-spine, but in which the thoracic rings are free and movable, while those of the abdomen are anchylosed with one another. According to the researches of Dr Henry Woodward, it would also appear to be established that the singular Crustacean fossils which have been grouped together under the common title of *Cyclus* (or *Halicyne*) should be regarded as really larval forms of the Carboniferous Limuloids *Prestwichia* and *Belinurus*. Leaving the Carboniferous period, we find that Limuloid Crustaceans are not unknown in the Permian and Triassic deposits, while several species of a type apparently generically identical with the recent *Limulus* occur in the Lithographic Slates of Solenhofen (Upper Jurassic), and other forms have been recognised in the Cretaceous and Tertiary.

CHAPTER XIX.

CRUSTACEA (*Concluded*).

MALACOSTRACA.

THE Malacostracous Crustaceans are distinguished by the possession of a definite number of body-segments, seven somites generally going to make up the thorax, and an equal number entering into the composition of the abdomen (counting the telson as a somite). The *Malacostraca* are divided into two primary sections, termed respectively *Edriophthalmata* and *Podophthalmata*, according as the eyes are sessile or are supported upon eye-stalks.

DIVISION A. EDRIOPHTHALMATA.—The division of the Sessile-eyed Crustaceans comprises those *Malacostraca* in which the eyes are not supported upon stalks or peduncles, and there is mostly no carapace. The eyes are sometimes compound, sometimes simple, and are placed on the sides of the head. The head is almost always distinct from the body; and there are typically seven pairs of feet in the adult. (Hence the name of *Tetradecapoda* applied to this division by Agassiz.) The *Edriophthalmata* include the orders *Læmodipoda*, *Amphipoda*, and *Isopoda*, of which the two latter are alone known in a fossil condition.

ORDER AMPHIPODA.

Small Crustaceans in which the respiratory organs have the form of membranous vesicles attached to the bases of the thoracic limbs. Abdomen well developed, and composed of seven segments. Seven pairs of thoracic limbs, directed partly backwards and partly forwards.

The most familiar recent forms of the *Amphipoda* are the "fresh-water Shrimps" (*Gammarus*), the Sand-hoppers (*Talitrus*), and the Shore-hoppers (*Orchestia*). The oldest representative of the order is a doubtful form, which has been described by Mr Woodward from the Upper Silurian rocks under the name of *Necrogammarus*. The Carboniferous genus *Gampsonyx* has been referred here, but is perhaps more properly placed amongst the *Stomapoda*, or among the *Isopoda*. From the Permian rocks, however, Mr Spence Bate has described an Amphipod to which he has given the name of *Prosoponiscus problematicus* (fig. 243), and which he believes to be very closely related to the living *Phædra antiqua*.

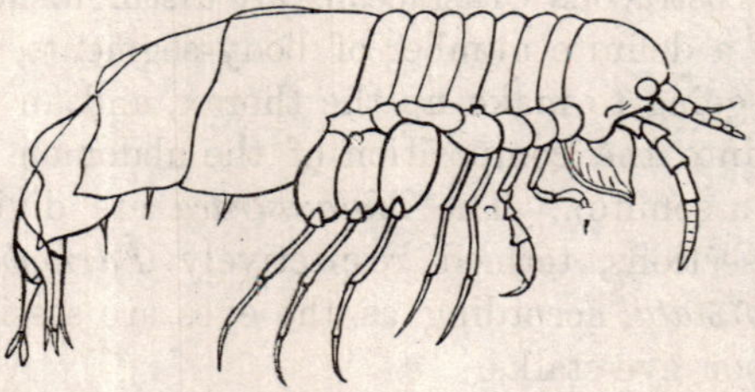

Fig. 243.—*Prosoponiscus problematicus*, viewed from one side, and partially restored. From the Magnesian Limestone (Permian) of Durham. (After Spence Bate.)

Order Isopoda.

Crustaceans in which the head is distinct from the segment bearing the first pair of feet. The eyes are compound and sessile. There are usually seven pairs of thoracic appendages, borne upon seven movable segments. The animal sometimes has the power of rolling into a ball. The abdominal segments are coalescent, and form a broad caudal shield, beneath which the branchiæ are carried.

Of the living Isopods, some (*Bopyridæ*) are parasitic in their adult condition upon other Crustaceans. Others, such as the common Wood-lice (*Oniscus*), live habitually upon the land. Others, again, are littoral in their habits, or frequent the sea.

The oldest known Isopod is a large form which has been described by Dr Henry Woodward from the Devonian rocks under the name of *Præarcturus.* It is believed to resemble the living *Arcturus Baffinsii.* From the Carboniferous rocks we have various forms which may be Isopodous, but the true affinities of which are rendered doubtful by the fact that the condition of the eyes has not been accurately determined. The most important of these is the genus *Acanthotelson,* of the Coal-measures, in which there are seven thoracic and seven abdominal segments, with apparently six pairs of thoracic legs, and with a pointed telson, flanked by two caudal stylets. Another Carboniferous genus, of dubious position, is *Gampsonyx,* which combines characters of the *Amphipoda, Isopoda,* and *Stomapoda,* but which is stated by Dr Anton Fric to have sessile eyes, and which, therefore, would seem to be referable to one or other of the two first-named of these groups. In the Upper Oolites (Purbeck beds) occurs the *Archæoniscus Brodiei* (fig. 244), often in large numbers. In the Chalk occurs the genus *Palæga,* which ranges to the Miocene Tertiary. Lastly, several forms, some of which are of very uncertain affinities, have been described from the Tertiary rocks.

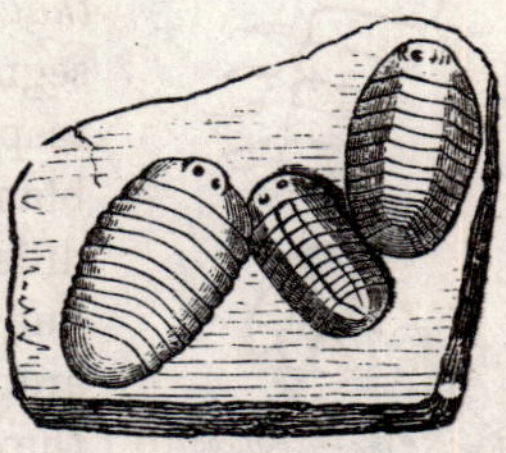

Fig. 244.—*Archæoniscus Brodiei,* a fossil Isopod from the Upper Oolites.

Division B. Podophthalmata.—The members of this division are Malacostracous Crustaceans, in which the eyes are compound, and are supported upon movable stalks or peduncles, and the anterior portion of the body (cephalothorax) is protected by a carapace. In this division are included the two orders of the *Stomapoda* and *Decapoda.*

Order Stomapoda.

Stalk-eyed Crustaceans in which there are generally from six to eight pairs of legs, and the branchiæ are not enclosed in a cavity beneath the thorax, but are either suspended beneath the abdomen, or, more rarely, attached to the thoracic legs.

Of the living Stomapods the best-known forms are the Locust-shrimps (*Squilla*), the Glass-shrimps (*Erichthys*), and the Opossum-shrimps (*Mysis*).

As regards fossil *Stomapoda*, considerable uncertainty still exists, owing to the imperfect condition of preservation of many extinct Crustaceans. The most ancient type which has been usually referred here is the *Palæocaris typus* (fig. 245), of the Coal-measures of North America. This form is clearly related to *Gampsonyx*, and if it shall be shown to possess sessile eyes, it must be removed to the *Isopoda* or *Amphipoda*. Here, however, we shall regard it as an early and "comprehensive" type of the *Podophthalmata*, characterised by the persistent segmentation of the thorax, but in other respects presenting considerable resemblance to the Macrurous Decapods. It differs from *Gampsonyx* in various points, but especially in the fact that the legs are undivided, whereas in the latter genus they are bifid. The *Pygocephalus* of the Coal-measures appears also to be a Stomapod. In deposits later than the Carboniferous, Stomapodous Crustaceans have been occasionally detected, but the genus *Squilla* itself does not appear to be represented in rocks older than the Eocene Tertiary.

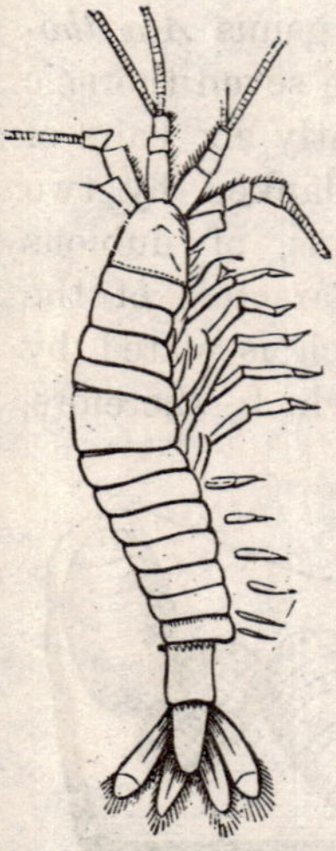

Fig. 245. — *Palæocaris typus*, slightly enlarged. From the Coal-measures of North America. (After Meek and Worthen.)

ORDER DECAPODA.

Crustaceans with five pairs of ambulatory legs, of which the first pair is modified to form nipping-claws, some of the other pairs behind this being often chelate as well. There is a large cephalothoracic carapace, and the branchiæ are contained in cavities at the sides of the thorax.

The order *Decapoda* includes the highest forms of the entire class of the Crustaceans, such as the Lobsters, Hermit-crabs, and Crabs, and it is divided into the following three tribes :—

TRIBE I. MACRURA.—The members of this tribe, such as the Lobsters, Shrimps, Prawns, and Cray-fish, have a long and well-developed abdomen, the posterior extremity of which forms a powerful natatory organ or caudal fin. This is constituted by a greatly-expanded pair of natatory appendages ("swimmerets") borne upon the last segment but one of the abdomen, between which is placed the last segment or "telson."

The Macrurous Decapods are unquestionably of a lower type than the Brachyurous Decapods or Crabs; so that it is no matter of surprise to find that the former, so far as is known, have enjoyed a considerably higher antiquity than the latter. The *Brachyura*, with two exceptions, are not known

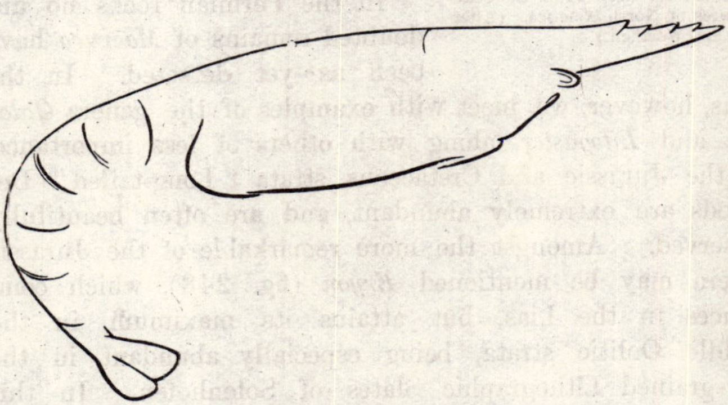

Fig. 246.—*Anthrapalæmon Salteri*. Carboniferous. (After Salter.)

in deposits older than the Oolites. The *Macrura*, on the other hand, were in existence before the close of the Palæozoic period. In the Carboniferous formation we have several forms of Macrurous Crustaceans belonging to the genus *Anthrapalæmon* (or *Anthracopalæmon*), of which a species is figured here (fig. 246). In both Europe and North America forms referable to this ancient type have been detected, and one of the most remarkable of these is the *Anthrapalæmon gracilis* of the Coal-measures of Illinois (fig. 247). In this species, as in all belonging to the genus, there is a well-developed carapace, which is furnished with a beak or "rostrum" in front, and, typically, with serrated lateral margins. There are five pairs of thoracic legs; the abdomen is composed of free segments; and there is a caudal fin, formed by the telson along with the last pair of "swimmerets." The genus *Anthrapalæmon* is not allied, as its name would seem to imply, to the true Prawns (*Palæmon*), but seems to be rather related to the living *Galathea*.

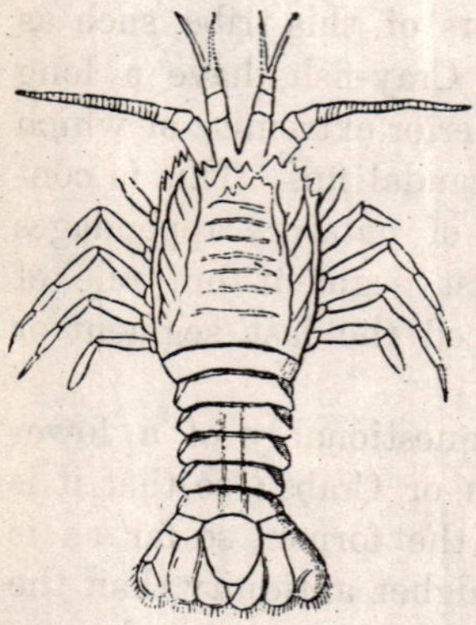

Fig. 247.—*Anthrapalæmon gracilis*, of the natural size. From the Coal-measures of North America. (After Meek and Worthen.)

In the Permian rocks no undoubted remains of *Macrura* have been as yet detected. In the Trias, however, we meet with examples of the genera *Galathea* and *Litogaster*, along with others of less importance. In the Jurassic and Cretaceous strata "Long-tailed" Decapods are extremely abundant, and are often beautifully preserved. Amongst the more remarkable of the Jurassic genera may be mentioned *Eryon* (fig. 248), which commences in the Lias, but attains its maximum in the Middle Oolitic strata, being especially abundant in the fine-grained Lithographic Slates of Solenhofen. In this singular genus the carapace is large and broad, and nearly quadrate in figure, whilst the antennæ are very small. The genus is nearly allied to the living *Willemoësia*. Another singular genus from the Solenhofen Slates is *Mecochirus* (or *Megachirus*), in which the first pair of legs is enormously elongated, but not terminated by chelæ. Von Seebach has also shown that the same deposits contain the remains of "Phyllosomæ," which may be regarded as the larval forms of such Palinuroid genera as *Eryon*. In the Cretaceous rocks are numerous Macrourans, belonging to the genera *Meyeria*, *Enoploclytia*, *Hoploparia*, &c. In many parts of the Tertiary series, especially in the London Clay (Eocene), are numerous remains of *Macrura*, some of which have been referred, with more or less doubt, to such living genera as *Astacus* and *Palinurus*.

TRIBE II. ANOMURA.—The Anomurous Decapods are distinguished by the condition of the abdomen, which is neither so well developed as in the *Macrura*, nor so rudimentary as in the *Brachyura*. The abdomen does not take any part in locomotion, and does not terminate posteriorly in a caudal fin. The penultimate segment of the abdomen, however, is mostly furnished with more or less well-developed appendages. The best-known living *Anomura* are the Hermit-crabs or Soldier-crabs (*Paguridæ*), the Crab-lobsters (*Porcellanæ*), and the Sponge-crabs (*Dromia*).

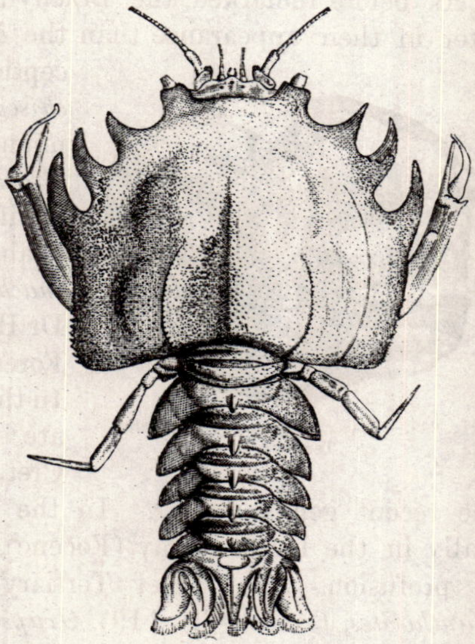

Fig. 248.—*Eryon arctiformis*. Middle Oolites. (Solenhofen Slates.)

The *Anomura* are of small importance as fossils. They commence in the Secondary period, a few forms having been described from the Oolites, and a greater number from the Cretaceous rocks. In the Tertiary period Anomurous Crustaceans are not uncommon; and the genus *Pagurus* itself appears to be represented in the Red Crag (Pliocene). The *Dromilites* of the London Clay is supposed to be related to the living *Dromia*.

TRIBE III. BRACHYURA.—The "Short-tailed" Decapods or Crabs are distinguished by having a rudimentary abdomen, which is tucked up beneath the cephalothorax. The carapace is usually very large, and the extremity of the abdomen is not provided with any appendages. Most of the Crabs are littoral in their habits, and have legs formed for walking. Others are adapted for swimming, and the Land-crabs habitually live inland.

As before remarked, the Brachyurous Decapods are much later in their appearance than the *Macrura*. With the exception the minute *Hemitrochiscus paradoxus* of the Permian, and the still more ancient *Brachypyge carbonis* of the Carboniferous, the oldest known Crab, at present, is the *Palæinachus longipes*, described by Dr Henry Woodward from the Forest Marble (Lower Oolites). In the Cretaceous series Crabs are tolerably abundant, one Cretaceous form belonging to the recent genus *Cancer*. In the Tertiary rocks, and especially in the London Clay (Eocene), remains of Crabs occur in profusion. The chief Tertiary genera are *Xanthopsis*, *Xantholites*, *Cancer* (fig. 249), *Grapsus*, and *Ebalia*.

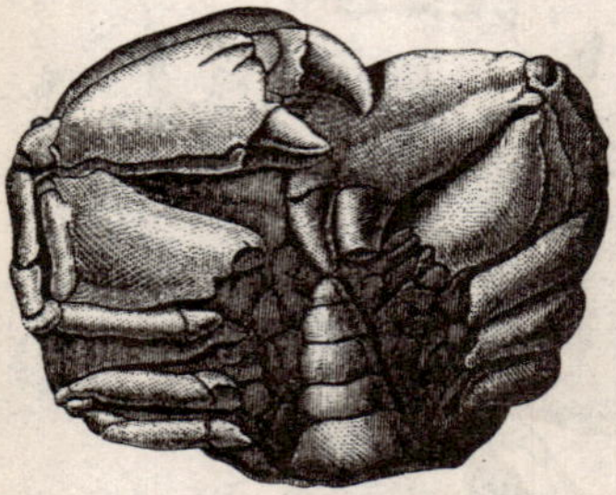

Fig. 249.—*Cancer* (*Carpilius*) *macrochelus*. Tertiary.

LITERATURE.

General.

1. "Glieder-füssler : Arthropoda," in 'Bronn's Klassen und Ordnungen des Thier-Reichs.' Gerstaecker. (Not yet completed.)
2. Article "Crustacea," in 'Encyclopædia Britannica,' 9th ed. vol. vi. Henry Woodward. 1877.
3. Article "Crustacea." 'Todd's Encyclopædia of Anatomy and Physiology.' Milne-Edwards. 1835.
4. "Histoire Naturelle des Crustacès." Milne-Edwards. 1834-1840.
5. "Histoire Naturelle des Crustacès fossiles." Brongniart and Desmarest. 1822.
6. "Catalogue of the British Fossil Crustacea." Henry Woodward. 1877.

Cirripedia.

7. "Monograph of the Cirripedia." Darwin. 'Ray Society.' "Lepadidæ," 1851. "Balanidæ," 1854.
8. "Fossil Lepadidæ of Britain." Darwin. 'Palæontographical Society.' 1851.

9. "Fossil Balanidæ of Britain." Darwin. 'Palæontographical Society.' 1854.
10. "On a New Genus of Cirripedia." Henry Woodward. 'Quart. Journ. Geol. Soc.,' vol. xxi. 1865. (*Turrilepas.*)
11. "Système Silurien de la Bohême." Barrande. Vol. i., Supplement. 1872. (*Plumulites* and *Anatifopsis.*)
12. "On Zoocapsa dolichorhamphia, a Sessile Cirripede from the Lias of Lyme Regis." H. G. Seeley. 'Ann. and Mag. Nat. Hist.,' ser. 4, vol. v. 1870.

OSTRACODA.

13. "Monograph of the Entomostraca of the Cretaceous Formation of England." Rupert Jones. 'Palæontographical Society.' 1849.
14. "Monograph of the Tertiary Entomostraca of England." Rupert Jones. 'Palæontographical Society.' 1856.
15. "Palæozoic Bivalved Entomostraca." Rupert Jones. 'Ann. and Mag. Nat. Hist.' 1855-1869.
16. "Monograph of the Recent British Ostracoda." G. S. Brady. 'Trans. Linn. Soc.,' vol. xxvi. 1868.
17. "Oversigt af Norges Marine Ostracoder." G. O. Sars. 1865. (Recent forms.)
18. "Carboniferous Entomostraca of Britain." Rupert Jones, J. W. Kirkby, and G. S. Brady. Part i., the Cypridinadæ. 'Palæontographical Society.' 1874.
19. "Post-Tertiary Entomostraca of Scotland." G. S. Brady, H. W. Crosskey, and D. Robertson. 'Palæontographical Society. 1874.
20. "Monograph of the Ostracoda of the Antwerp Crag." G. S. Brady. 'Trans. Zool. Soc.,' vol. x. 1878.
21. "Système Silurien de la Bohême." Barrande. Vol. i., Supplement. 1872.

PHYLLOPODA.

22. "Fossil Estheriæ." Rupert Jones. 'Palæontographical Society. 1862.
23. "Peltocaris, a New Genus of Crustacea." Salter. 'Quart. Journ. Geol. Soc.,' vol. xix. p. 87. 1863.
24. "Hymenocaris." Salter. 'Mem. Geol. Survey,' vol. iii., Appendix. 1866.
25. "Crustacean Teeth of the Carboniferous and Silurian." Henry Woodward. 'Geol. Mag.,' vol. ii. 1865.
26 "Discinocaris." Henry Woodward. 'Quart. Journ. Geol. Soc.,' vol. xxii. 1866.
27. "Aspidocaris." Reuss. 'Proc. Imp. Geol. Inst., Vienna.' 1867.
28. "Sixth Report on Fossil Crustacea." Henry Woodward. 'Rep. Brit. Ass.' 1872. (*Aptychopsis.*)

29. "Système Silurien de la Bohême." Barrande. Vol. i., Supplement. 1872. (*Cryptocaris* and *Pterocaris.*)
30. "Dithyrocaris." Henry Woodward and R. Etheridge, jun. 'Geol. Mag.,' vol. x. 1873.

TRILOBITA.

31. "Monograph of the British Trilobites." Salter. 'Palæontographical Society.' 1863-67.
32. "Système Silurien de la Bohême." Barrande. Vol. i., 1852, and supplement, 1872.
33. "Organisation of Trilobites." Burmeister. 'Ray Society.' 1846.
34. "British Organic Remains." Salter. 'Mem. Geol. Survey,' decade ii. 1849. Also 'Mem. Geol. Survey,' vol. iii. 1866.
35. "Catalogue of British Fossil Crustacea." Henry Woodward. 1877.

(Among other works dealing more or less extensively with the Trilobites, the student may consult the "Catalogue of Cambrian and Silurian Fossils in the Woodwardian Museum," by Salter (1873); the "Catalogue and Chart of Fossil Crustacea," by Salter and H. Woodward (1865); the "Palæozoic Fossils in the Woodwardian Museum," by M'Coy (1851-55); the "Geological Report of Londonderry," &c., by Portlock (1843); the "Palæontologia Suecica," by Angelin (1854); the "Untersuchungen über Trilobiten," by Beyrich (1846); the "Prodrom einer Monographie der Böhmischen Trilobiten," by Hawle and Corda (1847); the "Palæontology of New York," by James Hall (vols. i. and ii., 1847 and 1852); the "Palæozoic Fossils of Canada," by Billings (vol. i., 1865), &c.)

MEROSTOMATA.

36. "Monograph of the Genus Pterygotus." Huxley and Salter. 'Mem. Geol. Survey,' monograph i. (with atlas). 1859.
37. "Monograph of the British Fossil Crustacea belonging to the Order *Merostomata*." Henry Woodward. 'Palæontographical Society.' 1866-78.
38. "British Palæozoic Crustacea belonging to the Order *Merostomata*." Henry Woodward. 'Geol. Magazine,' vol. ix. 1872. (*Hemiaspis.*)
39. "On a New King-Crab from the Upper Silurian." Henry Woodward. *Ibid.*, vol. v. 1868. (*Neolimulus.*)
40. "Contributions to British Fossil Crustacea." Henry Woodward. 'Geol. Magazine,' vol. vii. 1870. (*Cyclus.*)
41. "Exapinurus and Pseudoniscus." Nieszkowski. 'Archiv für Naturkunde Liv. Ehst. und Kurlands,' ser. i. vol. ii. 1858.
42. "Eurypterus." James Hall. 'Pal. N. York,' vol. iii. 1859.
43. "Belinurus." Baily. 'Ann. and Mag. Nat. Hist,' ser. iii. vol. xi. 1863.

44. "On the Relation of the Xiphosura to the Eurypterida." Henry Woodward. 'Quart. Journ. Geol. Soc.' 1872.

45. "Catalogue of the British Fossil Crustacea." Henry Woodward. 1877.

MALACOSTRACA.

(AMPHIPODA, ISOPODA, STOMAPODA, DECAPODA.)

46. "Histoire naturelle des Crustacés fossiles." Brongniart and Desmarest. 1822

47. "Histoire naturelle des Crustacés." Milne-Edwards. 1834-40.

48. "Monograph of the Malacostracous Crustacea of Great Britain." Bell. 'Palæontographical Society.' 1857 and 1862.

49. "Fossilen Krebse von Bayern." Münster. 'Beiträge zur Petrefaktenkunde,' vol. ii. 1839.

50. "Neue Gattungen fossiler Krebse." Von Meyer. 1840.

51. "On a Fossil Crustacean found in the Magnesian Limestone of Durham," &c. Spence Bate. 'Quart. Journ. Geol. Soc.,' vol. xv. 1859. (*Prosoponiscus.*)

52. "Palæozoic Macrura." Salter. 'Quart. Journ. Geol. Soc.,' vols. xvii. and xix. 1861 and 1863. (*Anthrapalæmon.*)

53. "Palæontology of Illinois." Meek and Worthen. Vols. ii. and iii. 1866 and 1868. (*Palæocaris, Gampsonyx, Acanthotelson,* and *Anthrapalæmon.*)

54. "Archæoniscus Brodiei." Westwood, in Brodie's 'Insects of the Secondary Rocks.' 1845.

55. "Contributions to British Fossil Crustacea." Henry Woodward. 'Geol. Magazine,' vol. vii. 1870. (*Palæga.*)

56. "Palæozoic Macrura." Huxley. 'Quart. Journ. Geol. Soc.,' vols. xiii. and xviii. 1857 and 1862. (*Pygocephalus.*)

CHAPTER XX.

ARACHNIDA, MYRIAPODA, AND INSECTA.

CLASS ARACHNIDA.

THE *Arachnida* are *Articulate animals, in which the respiratory organs, when present, are in the form of pulmonary chambers or sacs, or of ramified tubes ("tracheæ") formed by an involution of the integument and fitted for breathing air directly; or both these organs are combined. In no case are the breathing-organs in the form of gills. There are four pairs of locomotive limbs, and there are no limbs attached to the segments of the abdomen. There is only one pair of antennæ, and these are not present as antennæ, but are converted into jaws or pincers. The head is amalgamated with the thorax to form a cephalothorax, the eyes are sessile, and the integuments are more or less chitinous.*

The *Arachnida* are mainly distinguished from the *Crustacea*[1] by the absence of gills, and the general presence of organs adapted for breathing air directly. They are distinguished from the Insects by the possession of four pairs of legs, by never possessing wings, and by having simple eyes, whilst the head is amalgamated with the thorax. From the Myriapods they are distinguished by the fact that the legs of the latter are never less than nine pairs in number, whilst the segments of the thorax are distinct from one another and

[1] Van Beneden would refer the Trilobites, King-crabs, and Eurypterids to the *Arachnida*, but such a radical change must be supported by overwhelming evidence before it can be accepted.

from the head, and the segments of the abdomen carry legs. As is the case with all the air-breathing Articulates, the remains of *Arachnida*, though of considerable theoretical interest, are of very rare occurrence as fossils. They will therefore be very briefly noticed here. Of the groups of the *Arachnida*, the Mites (*Acarida*), the Harvest-spiders (*Phalangidæ*), the Book-scorpions (*Pseudoscorpionidæ*), the Scorpions (*Pedipalpi*), and the true Spiders (*Araneida*), have all been detected in a fossil condition. The three first groups require little consideration here, being almost unknown except as occurring in amber, which is a fossil resin of late Tertiary age. The Scorpions and Spiders both appear to have come into existence in the Carboniferous period, and some of the forms which then existed do not appear to have been strikingly different from living types.

ORDER PEDIPALPI.—The typical members of this order are the Scorpions (*Scorpionidæ*), in which *the abdomen is distinctly segmented, and not separated from the thorax by any marked constriction.* The respiratory organs are in the form of pul-

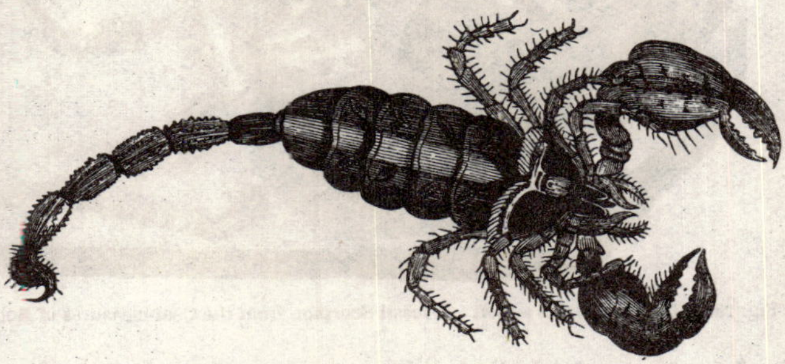

Fig. 250.—A recent Scorpion (reduced). The great nipping-claws of the Scorpion are not legs, but are a development of organs belonging to the mouth.

monary sacs opening on the under surface of the abdomen by distinct apertures or "stigmata." The jaws (maxillæ) carry an enormously-developed pair of nipping-claws (fig. 250), and the antennæ are also converted into chelæ. The head carries six, eight, or twelve simple eyes, and the last joint of the abdomen (telson), terminates in a hooked claw, perforated for the transmission of the duct of a poison-gland.

As regards their distribution in time, the Scorpions commence in the Carboniferous period, where they are represented by the genera *Eoscorpius* and *Cyclophthalmus.* The most celebrated fossil Scorpion is the *Cyclophthalmus senior* (fig. 251) of the Bohemian Coal-measures. This remarkable form resembles the living *Androctonus* in having twelve eyes, but these are disposed in a circle, whereas in the latter there are six eyes on each side of the head. The *Eoscorpius carbonarius,* of the Coal-measures of Illinois, is also a true Scorpion, and is most nearly related to the recent genus or sub-genus

Fig. 251.—*Cyclophthalmus senior.* A fossil Scorpion from the Coal-measures of Bohemia.

Buthus, from which it does not seem to be separated by any characters of great importance.

Besides the typical Scorpions, the order *Pedipalpi* likewise contains various forms like the living *Phrynus* and *Thelyphonus,* in which the maxillary palpi are not chelate; and we appear to have ancient representatives of these in the *Eophrynus* and *Arthrolycosa* of the Carboniferous rocks. In the former of these (fig. 252, A) the dorsal surface of the abdomen is tuberculated, and the ventral surface exhibits six pairs of stigmata. The posterior border of the abdomen carries four short spines, and the long and slender maxillary palpi do not seem to have been chelate. This singular form was described by Dr Henry Woodward from the Coal-measures of Dudley, and is believed by him to be most nearly allied to the recent *Phrynus*. *Arthrolycosa*, from the Coal-measures of Illinois, seems to have chelate maxillary palpi, while the abdomen is distinctly segmented, and consists of

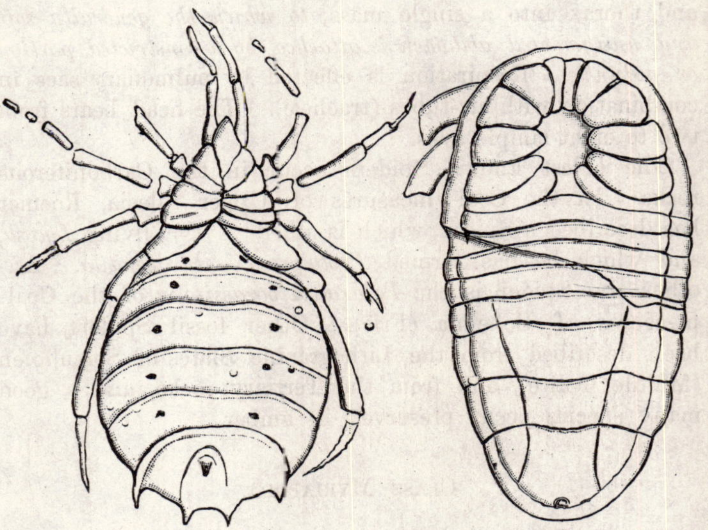

Fig. 252.—A, *Eophrynus Prestvicii*, viewed ventrally, and somewhat enlarged—Carboniferous (after H. Woodward); B, *Architarbus subovalis*, enlarged four times, and viewed from below—Carboniferous (after H. Woodward).

seven somites. It is apparently allied to *Thelyphonus*, though showing resemblances to the *Phalangidæ*.

Another comprehensive type, with alliances to both *Phrynidæ* and *Phalangidæ*, is the *Architarbus* of the Carboniferous rocks of North America and Britain. In the *A. subovalis* of the latter region, the anterior portion of the body is composed of amalgamated somites (fig. 252, B), carrying four pairs of legs and a pair of palpi, the form of which is unknown. Behind come four narrow somites, followed in turn by three very wide segments, forming together nearly half the length of the body. Another Carboniferous genus is *Mazonia*, which has been founded for the reception of an Arachnide of uncertain affinities, but in some respects intermediate between the *Scorpionidæ* and the *Pseudoscorpionidæ*, from the Coal-measures of Illinois. Lastly, the *Microlabis* of the Coal is believed to be most nearly allied to the Pseudoscorpions.

ORDER ARANEIDA.—This order includes the true Spiders, which are characterised by the amalgamation of the head and thorax into a single mass, *to which the generally soft and unsegmented abdomen is attached by a constricted portion or peduncle.* Respiration is effected by pulmonary sacs in combination with air-tubes (tracheæ). The head bears from two to eight simple eyes.

The oldest known Spiders occur in the Carboniferous rocks. In the Coal-measures of Upper Silesia, Roemer has described a Spider, which is allied to the living *Lycosa*, and which he has termed *Protolycosa anthracophila*. Another true Spider is the *Palaranea borassifolia* of the Coal-measures of Bohemia (Fric). Other fossil Spiders have been described from the Lithographic Slates of Solenhofen (Middle Oolite), and from the Tertiary rocks, and a good many species occur preserved in amber.

CLASS MYRIAPODA.

The Myriapods are *Articulate animals in which the head is distinct, and the remainder of the body is divided into nearly similar segments. There is no marked boundary-line between the thorax and abdomen, and the segments of the latter carry locomotive limbs. There is one pair of jointed antennæ, and the number of legs is always more than eight pairs. Respiration is effected by air-tubes (tracheæ).*

The living Myriapods are divided into the four orders *Chilopoda*, *Chilognatha*, *Pauropoda*, and *Onychophora*. In the *Chilopoda* are the Centipedes, characterised by their masticatory mouth and carnivorous habits, by the possession of legs in single pairs (usually from fifteen to forty pairs), and by having antennæ of from fourteen to forty or more joints. In the *Chilognatha* (fig. 253) are the Millipedes and Galleyworms, characterised by their vegetarian habits, by having the segments of the body so amalgamated that the legs appear to be arranged in double pairs, and by having antennæ of six or seven joints. In the *Pauropoda* is the single genus *Pauropus*, characterised by having only nine pairs of legs, and the antennæ bifid, with three long multi-articulate appendages. Lastly, in the *Onychophora* we have only the single and very anomalous genus *Peripatus*, comprising worm-like Myriapods, with annulated bodies and hooked or bristled feet. No member of the *Chilopoda*, *Pauropoda*, or *Onychophora* is known with certainty to occur in the fossil condition, but various extinct types of the vegetable-eating *Chilognatha* have been preserved to us, beginning in the Carboniferous period.

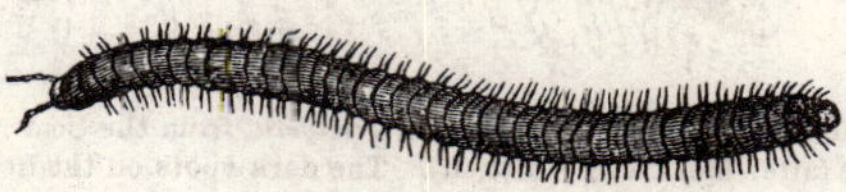

Fig. 253.—Millipede (*Iulus*).

The oldest-known Myriapods occur in the Coal-measures, the two best-known genera being *Xylobius* and *Archiulus*. These genera belong to the order *Chilognatha*, and comprise, therefore, vegetable-feeders. In *Xylobius* (fig. 254) the segments are divided by cross sutures into numerous fragments, in a manner wholly unknown amongst recent forms. Several species of this genus are known, of which the one figured above derives its specific name from the fact that it is found in the hollow trunks of *Sigillaria*. It must have possessed, like the living Galleyworms, the

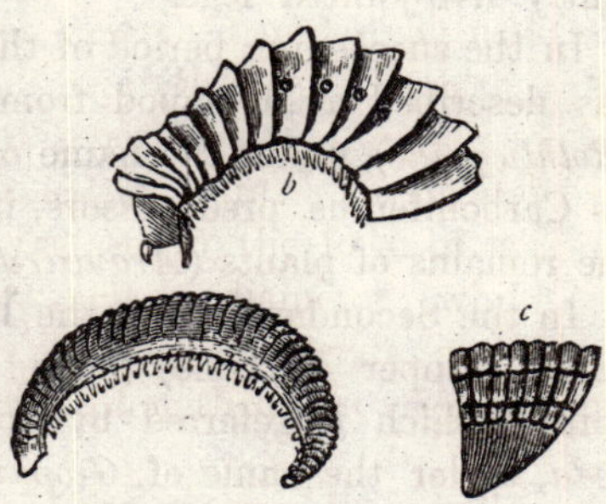

Fig. 254.—*Xylobius Sigillariæ*, a Carboniferous Myriapod (after Dawson). *a*, Natural size; *b*, Anterior portion, enlarged; *c*, Posterior portion, enlarged.

power of rolling itself up into a ball (Dawson). In the allied genus *Archiulus*, the segments are not broken up into sections, as they are in *Xylobius*. The characters of both these genera are so peculiar that they have been placed in a separate family under the name of *Archiulidæ*. Other Myriapods have been discovered in the Carboniferous rocks of North America and Britain, and have been referred to the genus *Euphoberia* (fig. 255). The true place of this genus is somewhat uncertain, owing to its possession of several abnormal features, though there can be little hesitation in referring it to the *Chilognatha*. Thus the dorsal surface shows a series

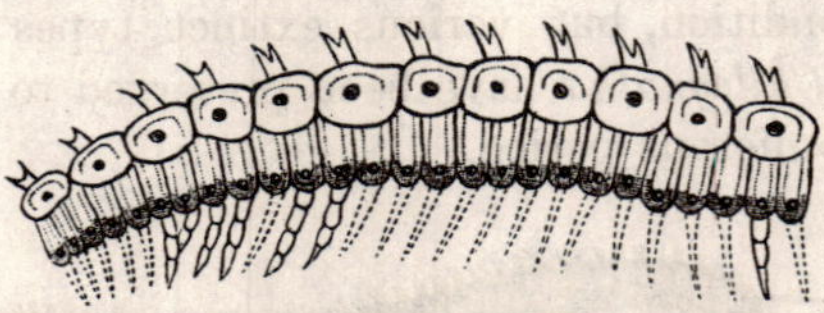

Fig. 255.—Portion of the body of *Euphoberia armigera*, from the Coal-measures of Illinois, of the natural size (after Meek and Worthen). The dark spots on the dorsal shields are pits left by the breaking off of the dorsal spines.

of large shields, which are armed with projecting spines, each of these shields corresponding with *two* ventral segments. Each of the latter bears a pair of slender, apparently five-jointed legs.

In the succeeding period of the Permian, Professor Geinitz has described a Myriapod from the base of this formation (*Rothliegende*), under the name of *Palæojulus Dyadicus*. Like its Carboniferous predecessors, it occurs in association with the remains of plants (*Araucarites*).

In the Secondary rocks, the Lithographic Slates of Solenhofen (Upper Jurassic) have yielded the remains of an animal which is referred by Count Münster to the *Myriapoda*, under the name of *Geophilus proavus*. Other Myriapods, lastly, have been described from Tertiary strata and from amber.

Class Insecta.

The Insects are *Articulate Animals, in which the head, thorax, and abdomen are distinct from one another. The thorax consists of three segments, each of which carries a pair of legs. Mostly there are two pairs of wings borne by the two hinder segments of the thorax. The abdomen never carries locomotive limbs, but the last abdominal segments may carry reproductive or sensory appendages. A single pair of jointed antennæ is present, and the eyes are generally compound. Respiration is effected by air-tubes (tracheæ).*

The number of Insects known from Tertiary deposits is already extremely large, while they are comparatively scarce in the Secondary period, and are among the very rarest of fossils from the Palæozoic formations. Moreover, the remains of this class of animals almost always occur in a fragmentary condition, and under any circumstances they cannot be satisfactorily deciphered except by practised workers in the department of Entomology. For these reasons, among others, it will be sufficient to give here a brief outline of the geological history of the class.

As regards the general distribution of the *Insecta* in time, the oldest-known forms are from the Devonian rocks of North America. Here occur the remains of several insects which belong to the order of the Neuropterous Insects (or to the *Pseudoneuroptera*). Amongst the most remarkable of these is the *Platephemera antiqua* of Mr Scudder (fig. 256). This species must have attained a large size—five inches in expanse of wing—and it is regarded by Mr Scudder as being referable to the *Ephemeridæ* (the May-flies). This eminent authority, however, regards it as a "synthetic type"—that is to say, as a form combining peculiarities of structure which are now only found in different groups. Other genera belonging to the *Neuroptera* have been described from the Devonian rocks of North America, under the names *Homothetus*, *Lithentomum*, and *Xenoneura*, the last having attached to its wing the remains of a "stridulating organ," similar to that possessed by recent grasshoppers.

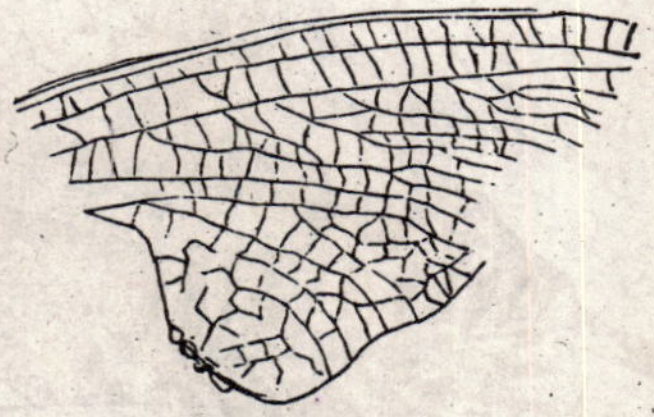

Fig. 256.—Wing of *Platephemera antiqua* (after Dawson). Devonian.

In the Carboniferous period, the remains of Insects, though still far from common, have been discovered in greater abundance, as might have been anticipated from the comparative extent of our knowledge of the land-plants of this formation. The order *Neuroptera* is now represented by a number of remarkable forms, of which one of the most remarkable has been referred to the *Ephemeridæ*, under the name of *Haplophlebium Barnesii* (fig. 257). This insect must have attained a size much larger than that of any recent Ephemerids, measuring fully seven inches in expanse of wing.

Other Carboniferous *Neuroptera* have been described under the generic titles of *Dictyoneura*, *Miamia*, *Ephemerites*, *Hemeristia*, and *Paolia*. The order of the *Orthoptera* makes its first appearance here, so far as known, and is represented by various species of *Blattina*, allied to the

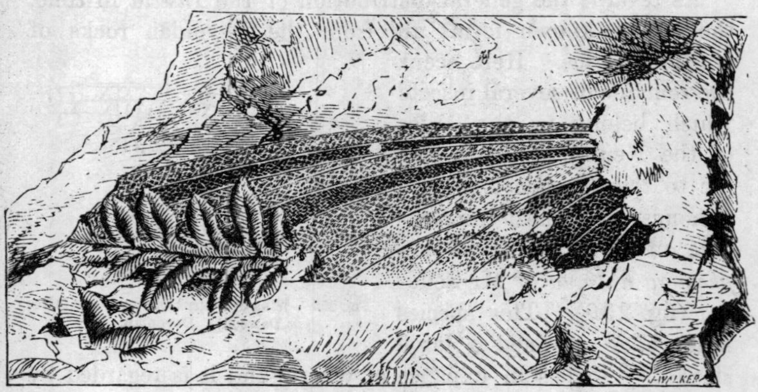

Fig. 257.—*Haplophlebium Barnesii* (after Dawson). From the Carboniferous rocks of Canada. *a*, Profile of base of wing.

living Cockroaches (*Blatta*). One of the most remarkable of these Carboniferous Cockroaches is the *Archimulacris Acadicus* (fig. 258) of the Canadian Coal-measures, which presents peculiarities unknown in any of its living allies. The *Lithomantis carbonarius* of the English Coal-measures appears to be an ancient representative of the *Mantidæ* (H. Woodward); and two Carboniferous species have been referred to the recent genus *Gryllacris* (A. H. Swinton). One of the latter—viz., *Gryllacris* (*Corydalis*) *Brongniarti*—exhibits traces of a stridulating organ or "file," similar to that of the males of the living Crickets. The order of the *Coleoptera* (Beetles) is sparingly represented by forms of uncertain affinities (*Curculioides* and *Troxites*). Lastly, the great order of the *Lepidoptera* (Moths and Butterflies) is dubiously represented by a wing from the Coal-measures of Belgium, which has been regarded as belonging to a moth of the family *Saturniidæ*, and has been described under the name of *Breyeria Borinensis* (Preudhomme de Borre).[1] No representatives of the *Hemiptera*, *Diptera*, or *Hymenoptera* have yet been detected in the Carboniferous rocks.

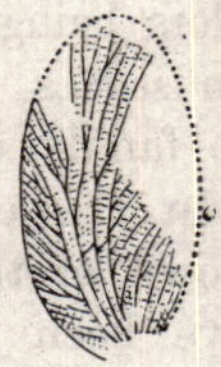

Fig. 258.—*Archimulacris Acadicus* (after Dawson). From the Carboniferous rocks of Canada.

In the Permian deposits few remains of insects have been yet discovered. The *Eugereon* of Dr Dohrn, however, is a remarkable Neuropterous insect from the Lower Permian of Birkenfeld.

Coming to the Secondary period, we are as yet very partially acquainted with the characters of the Insect-fauna of the Trias; but an abundant insect-life is already known to have existed in the Jurassic period. Even in deposits as old as the Lias, we find Beetles of such well-known existing types as the *Curculionidæ*, *Elateridæ*, and *Melolonthidæ*; while the *Orthoptera* are represented by Grasshoppers and Cockroaches, and the *Neuroptera* by Dragon-flies and May-flies. In the higher portion of the Jurassic series, in addition to various types of the *Coleoptera*, *Orthoptera*, and *Neuroptera*, we appear to have a true representative of the Butterflies. This is the *Palæontina Oolitica* of the Stonesfield Slate (Lower Oolites), which is regarded by Mr Butler as intermediate between the existing groups of the *Nymphalinæ* and *Satyrinæ*, and as allied to certain tropical American genera (*Brassolis*). In the Cretaceous deposits, as might have been expected from their being so largely

[1] According to the high authority of Mr R. M'Lachlan, *Breyeria* is one of the *Pseudoneuroptera*, and belongs to the *Ephemeridæ*.

marine in origin, insects are rarely found. We meet, however, with remains of *Orthoptera, Hemiptera, Neuroptera,* and *Lepidoptera;* the most remarkable form of the last of these orders being the *Neorinopsis sepulta* of the Aix-la-Chapelle Sandstones (Upper Cretaceous ?), a Butterfly belonging to the family of the *Satyridæ.*

In the early part of the Tertiary period little is yet known of the Insects, but a vast number of fossil forms have been preserved to us in deposits of Miocene and Pliocene age. It would be quite hopeless to attempt to give any detailed account of these here, but we meet during these periods with representatives of all the existing orders of Insects, with the exception of the *Aptera,* many of the forms being closely allied to types now in existence. Even the fragile and delicate Butterflies are fairly represented; one of the most celebrated examples being the *Mylothrites* (*Vanessa*) *Pluto* of the Lower Miocene Marlstone of Radaboj in Croatia, which has been variously referred to the *Nymphalinæ* and the *Pierinæ.* Amber, a resin of Post-Tertiary date, has, lastly, yielded a great number of insects, among which are the only undoubted examples of any order of the Apterous Insects (*Thysanura*).

LITERATURE.

ARACHNIDA.

1. "Cyclophthalmus." Corda. 'Böhmische Verhandlungen,' 1835. See also Buckland, 'Geology and Mineralogy' (Bridgewater Treatise), 1836.
2. "Eoscorpius." Meek and Worthen. 'Geology of Illinois,' vol. iii., 1868. See also Henry Woodward in 'Quart. Journ. Geol. Soc.,' vol. xxxii., 1876.
3. "Eophrynus." Henry Woodward. 'Geol. Magazine,' vol. viii. 1871.
4. "Architarbus." Scudder. 'Geology of Illinois,' vol. iii., 1868. Also Henry Woodward, in 'Geological Magazine,' vol. ix., 1872.
5. "Arthrolycosa." Harger. 'Amer. Journ. Sci. and Arts.' 1874.
6. "Protolycosa." Ferd. Roemer. 'Leonhard und Geinitz Neues Jahrbuch.' 1866.

MYRIAPODA.

7. "Xylobius." Dawson. 'Quart. Journ. Geol. Soc.,' vol. xvi. 1859. Also in 2d ed. of 'Acadian Geology,' with note by Mr Scudder (pp. 385 and 495), 1868. See also H. Woodward, in 'Trans. Geol. Soc. Glasgow,' 1866.
8. "Euphoberia." Meek and Worthen. 'Geology of Illinois,' vol. iii., 1868. See also H. Woodward, 'Geol. Magazine,' vol. viii., 1871.
9. "Palæojulus Dyadicus." Geinitz. 'Sitz. Nat. Ges. Isis.' 1872.

INSECTA.

10. "Haplophlebium and Archimulacris." Scudder, in Dawson's 'Acadian Geology,' 2d ed. 1868.
11. "Miamia," Scudder. 'Geology of Illinois,' vol. iii. 1868.
12. "Hemeristia and Miamia." Dana. 'Amer. Journ. Sci. and Arts, 1864; and Scudder, ibid., 1865.
13. "Die fossilen Insecten der Kohlenformation von Saarbrücken." Von Goldenberg. 'Palæontographica.' 1854.
14. "Eugereon Böckingi, eine neue Insectenform, aus dem Todtliegende." Dohrn. 'Palæontographica.' 1866.
15. "On Fossil Orthoptera claiming Affinity with Gryllacris." Swinton. 'Geol. Magazine.' 1874.
16. "Palæontina Oolitica." Butler. 'Geol. Magazine.' 1874.
17. "Fossil Butterflies." Scudder. (Published for the American Assoc. for the Advancement of Science.) 1875.
18. "Primæval World of Switzerland." Heer. (Trans. by Heywood.) 1876.
19. "Insect-Fauna of the Tertiary Formations of Œningen and Radaboj." Heer. 'Denkschr. der Schweitz. Nat. Ges.,' 1847, 1850, 1853, and 1867. See also 'Naturkundige Verhandelingen van der Hollandsche Maatschappij der Wetensch.,' &c. Haarlem. 1862.
20. "Insects of the Secondary Rocks." Brodie. 1845.

CHAPTER XXI.

SUB-KINGDOM MOLLUSCA.

POLYZOA.

SUB-KINGDOM MOLLUSCA.—The *Mollusca* comprise the animals ordinarily known as Shell-fish, from their commonly possessing an exoskeleton or shell. The Molluscs are *soft-bodied and destitute of any evident segmentation. Commonly the integument secretes a hard calcareous or horny envelope, but this may be absent. The alimentary canal is always present, and never communicates with the body-cavity. The nervous system consists typically of three pairs of ganglia, disposed in a characteristically scattered manner; but in the lower forms a single ganglion alone is present. A heart may or may not be present, and there may or may not be distinct respiratory organs.*

As a matter of course, it is only with the *shell* of the *Mollusca* that the palæontologist has to deal, and those forms which are destitute of this structure are wholly unknown in the fossil condition. The special characters of the shell will be treated of in speaking of the separate classes. In the meanwhile it is sufficient to draw attention to some general considerations. In the Sea-mosses and Sea-mats (*Polyzoa*), the animal is compound, and the hard structures secreted by the colony would not come under the common designation of a "shell." In these cases the investment of the colony would rather be termed a "polypidom," and when of a horny nature, it does indeed show a very close resemblance to the "polypary" of the Sertularian Zoophytes. In the Ascidian Molluscs or Sea-squirts (*Tunicata*), the animal is simply enclosed in a leathery or cartilaginous case, in which calcareous matter is very rarely developed. Hence we need feel no surprise that the Tunicaries are not known in the fossil state. The Lamp-shells and their allies (*Brachiopoda*) possess a *bivalve* shell consisting of two pieces or "valves," which are more or less highly calcareous. Coming to the higher *Mollusca*, the true Bivalve Shell-fish (*Lamellibranchiata*), as their common name implies, have also a bivalve shell; but this is distinguished from the shell of the Brachiopods by sufficiently good characters. No Lamellibranch is destitute of a shell, and the remains of this class occur more or less abundantly in all deposits except the most ancient. The ordinary Univalve Shell-fish (*Gasteropoda*), as indicated by their common name, have usually a shell composed of a single piece or "valve." In many Gasteropods, however, there is either no shell at all, when the animal is said to be "naked" (as in the Sea-slugs), or the shell is quite rudimentary, and is concealed within the mantle (as in the ordinary slugs). In other Gasteropods, again (viz., in the Chitons), the shell is "multivalve," consisting of eight pieces or valves placed one behind the other. Most, however, of the "multivalve" shells of older writers are really referable to the *Cirripedia*. In the minute Oceanic Molluscs, or "Winged Snails," which form the class *Pteropoda*, the animal is sometimes naked, but is more usually protected by a symmetrical glassy shell, which is always univalve. In the class of the *Cephalopoda*, finally, great diversity exists in the character of the skeleton. All the ordinary Cuttle-fishes have an *internal* skeleton, embedded in the mantle, and not visible externally. This internal skeleton may be calcareous or horny, and it may be of a very complicated nature; but it merely serves to support the soft parts of the animal, and it does not form an external case in which the animal lives. In one Cuttle-fish only (viz., the Argonaut or Paper Nautilus) is there an external shell, but the nature of this is quite peculiar, and it cannot be compared with the shell of any of the ordinary

Molluscs. In another group of the *Cephalopoda*, represented at the present day by the Pearly Nautilus, there is a well-developed external shell, which is always composed of a single piece, and is always chambered, the animal living in the last and largest chamber of the shell.

In composition the shell of the higher *Mollusca* consists of carbonate of lime—usually having the atomic arrangement of calcite—with a small proportion of animal matter. In the *Pholadidæ*, however, the calcareous matter exists in the allotropic condition of arragonite, which is very much harder than calcite. As regards their texture, three principal varieties of shells may be distinguished—viz., the "porcellanous," the "nacreous," and the "fibrous." In the nacreous or pearly shells, as seen in "mother-of-pearl," the shell has a peculiar lustre, due to the minute undulations of the edges of alternate layers of carbonate of lime and membrane. The "fibrous" shells are composed of successive layers of prismatic cells. The "porcellanous" shell has a more complicated structure, and is composed of three layers or strata, each of which is made up of very numerous plates, "like cards placed on edge." The direction in which these vertical plates are placed is sometimes transverse in the central layer, and lengthwise in the two others; or longitudinal in the middle, and transverse in the outer and inner strata.

From their so commonly possessing hard structures, whether external or internal, no fossils are more abundant or important than Molluscs. As regards the general distribution of the *Mollusca* in time, the sub-kingdom commences its existence in the Cambrian period, in which the classes of the *Polyzoa*, *Brachiopoda*, *Lamellibranchiata*, *Pteropoda*, *Gasteropoda*, and *Cephalopoda* are more or less certainly represented.[1] Speaking generally, the chief representatives of the *Mollusca* in Palæozoic time are the chambered Cephalopods (*Tetrabranchiata*) and the *Brachiopoda*; in Mesozoic

[1] The *Brachiopoda* and *Pteropoda* have now been detected in the Lower Cambrian. The Gasteropods and Lamellibranchs are found in the Upper Cambrian. The *Cephalopoda* do not make their appearance till near the summit of the formation, and are then poorly represented (so far as known, by *Orthoceras* only). Lastly, the *Polyzoa* are of doubtful occurrence even in the Upper Cambrian, unless we place the genus *Dictyonema* in this class.

time, the Cuttle-fishes (Dibranchiate Cephalopods), the chambered Cephalopods, and the *Polyzoa*; in Kainozoic time, the Lamellibranchs and Gasteropods. The *Polyzoa* are comparatively poorly represented in Palæozoic rocks, and attain their maximum (as fossils) towards the close of the Mesozoic period. The *Brachiopods* are vastly more abundant in Palæozoic deposits than in Mesozoic, and have gradually declined to the present day. The *Lamellibranchiata* seem to have been gradually increasing in importance since their first appearance in the Upper Cambrian seas, and they have attained their maximum at the present day. The *Gasteropods*, upon the whole, like the Bivalves, seem to have reached their culminating point in recent seas; whilst the *Pteropods* seem to have been as abundant in Silurian seas as they are at present. The history of the *Cephalopoda* is a remarkable one. The *Tetrabranchiate* forms, with chambered shells, attained their maximum in the earlier portion of the Silurian period, as regards their simpler types; but the more complex types of the group swarmed in the seas of the Secondary period, and finally disappeared at the close of this epoch. This group at the present day is represented solely by the Pearly Nautilus. The *Dibranchiate Cephalopoda*, on the other hand, represented at the present day by the Cuttle-fishes, did not make their appearance till the commencement of the Secondary period, and seem to have reached their maximum in existing seas.

The sub-kingdom *Mollusca* is divided into two great divisions, termed respectively the *Molluscoida* and the *Mollusca Proper*. The division *Molluscoida* comprises the three classes of the *Polyzoa*, *Tunicata*, and *Brachiopoda*, characterised by having *a nervous system consisting of a single ganglion or principal pair of ganglia, whilst there is either no distinct organ of the circulation or an imperfect heart*. In the division of the *Mollusca Proper* are comprised the classes of the *Lamellibranchiata* (Bivalves), *Gasteropoda* (Univalves), *Pteropoda*, and *Cephalopoda*. All these classes are distinguished by having *a nervous system composed of three principal pairs of ganglia; whilst there is a well-developed heart, consisting of at least two chambers.*

Class Polyzoa or Bpyozoa.

Animal composite, forming colonies, all the members of which are produced by budding from a primitive being (zoöid). Each member of the colony (zoöid) is enclosed in a double-walled sac, the outer coat of which is mostly hardened by horny or calcareous matter. There is no heart, and the mouth is surrounded by a circle or crescent of hollow ciliated tentacles. The colonies are all but invariably fixed to some foreign object, and are in many cases plant-like in form.

All the *Polyzoa* live in an associated form in colonies or "polyzoaria," which are sometimes foliaceous, sometimes branched (fig. 259) and plant-like, sometimes encrusting, and very rarely are free. Each "polyzoarium" consists of

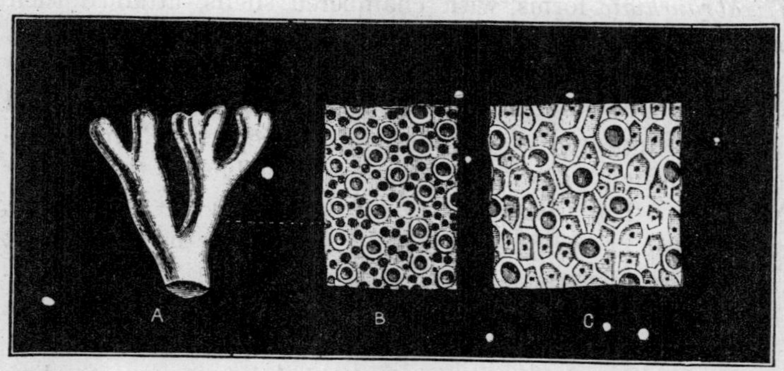

Fig. 259.—A, Fragment of an undescribed living species of *Heteropora*, from New Zealand, of the natural size; and B, Surface of the same, enlarged (original); C, Surface of a branch of *Heteropora subreticulata*, from the Tertiary, enlarged (after Reuss).

an assemblage of distinct but similar zoöids arising by continuous gemmation from a single primordial individual. The colonies thus produced are in very many respects, closely similar to those of many of the Hydroid Polypes, with which, indeed, the *Polyzoa* were for a long time classed. The "polyzoarium," or "cœnœcium," however, of a *Polyzoön* differs from the polypidom of a composite Hydroid in the *general* fact that the separate cells of the former do not communicate with one another otherwise than by the continuity of the external integument; whereas the zoöids of the latter are united by an organic connecting medium, or "cœnosarc," from which they take their origin. On this point Mr Busk observes:—

"It has been before said that the *Polyzoa* are always associated into compound growths, made up of a congeries of individuals, which, though distinct, yet retain some degree of intercommunication, comparable in kind perhaps, though not in degree, to what obtains in many of the compound Ascidians. That this community exists is proved by the otherwise inexplicable circumstance that the polyzoaria in many instances present elements common to the whole growth, and not belonging specially to any individual. The chief bond of connection would appear to reside partly in the continuity of the external integument, and partly also, in all probability, in a slow interchange of the vital fluid with which the cavities of the cells are charged."

In one sub-order of the *Polyzoa* (*Ctenostomata*), the polyzoarium consists of a series of cells arising from a common tube, but this exception does not affect the value of the above *general* distinction between the *Polyzoa* and the *Hydroida*.

A second point of difference is found in the invariably corneous (or chitinous) texture of the polypidoms of the *Hydroida*, whereas those of the *Polyzoa* may be corneous or fleshy, but are in the majority of instances more or less highly charged with carbonate of lime.

As before remarked, the colonies of the *Polyzoa* are produced by a process of continuous budding from a primitive being or zoöid. The budding takes place according to a determinate law, differing in different forms, and the resulting colony varies in shape according to the method of budding in each species. All the zoöids of the colony are termed "polypides," and the entire colony consists simply of an aggregation of precisely similar polypides, which may be simply united by their external integuments or, more rarely, spring from a common tube. It is only with the outer investment of the colony that the palæontologist has to deal; but it may be well briefly to describe the structure of a typical polypide.

The polypide of a *Polyzoön* (fig. 260, 2) consists essentially of a double-walled sac, filled with fluid, and perforated by an aperture where the mouth of the polypide is situated. In the majority of cases the outer wall of the sac (termed the "ectocyst") is of a horny consistence, or may be more or less highly calcareous. It forms a little chamber, which is technically called the "cell." At one point, varying in its position, the cell is furnished with an aperture or "mouth" (fig. 260, 1), whence the polypide can protrude its tentaculate head. The inner wall of the sac (termed the "endocyst") is invariably flexible and membranous, and the space included within it is filled with fluid, in which floats the alimentary canal. The commencement of the alimentary canal is surrounded by a series of hollow ciliated tentacles,

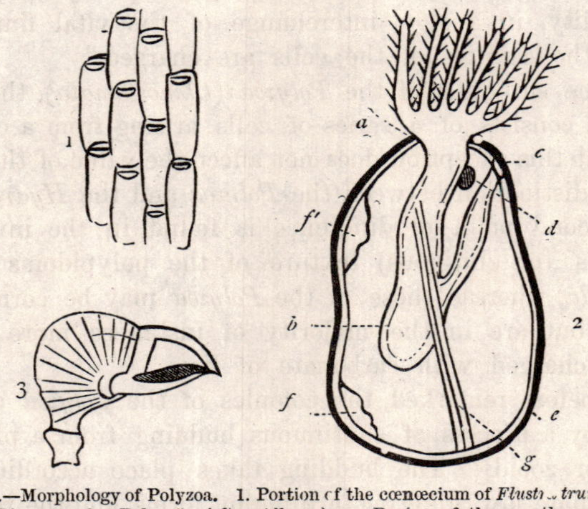

Fig. 260.—Morphology of Polyzoa. 1. Portion of the cœnœcium of *Flustra truncata*, magnified. 2. Diagram of a Polyzoön (after Allman): *a*, Region of the mouth surrounded by tentacles; *b*, Alimentary canal; *c*, Anus; *d*, Nervous ganglion; *e*, Investing sac (ectocyst); *f*, Testis; *f′*, Ovary; *g*, Retractor muscle. 3. Bird's-head process, or "avicularium," of a Polyzoön.

which are mostly arranged in a circle in the marine *Polyzoa*, but are disposed in the shape of a horse-shoe in most of the fresh-water forms. The digestive canal passes through the body-cavity, without opening into it, and terminates in a distinct anus placed near the mouth. The only other organs possessed by the polypide are a nervous ganglion, and the organs of reproduction, each zoöid being hermaphrodite.

In order, then, to arrive at a clear conception of the structure of a *Polyzoön*, we have simply to imagine that such a polypide as above described should have the power of repeating itself by gemmation, "thus producing one or more precisely similar systems, holding a definite position relatively to one another, while all continue organically united."

The only element of the *Polyzoa* with which the palæontologist is concerned is the external investment of the colony—the "cœnœcium" or "polyzoarium." This is formed by the combined ectocysts of the various polypides, and it varies greatly both in form and actual composition. In form, it may be plant-like, rooted at one point, and rising into foliaceous expansions or arborescent growths; or it may spread over some foreign object as a continuous crust. In consistence, it may be fleshy, horny, sub-calcareous, or completely calcareous; the simply fleshy forms, as a matter of course, never occurring in a fossil condition.

The shape of the "cell," formed by the ectocyst or outer wall of each polypide, varies considerably, and important distinctions may be drawn from this character alone. In one large group of the *Polyzoa*—the *Cheilostomata*—the mouth of the cell is never quite terminal in position, but is always placed upon the front of the cell, generally close to one end (fig. 261); whilst the diameter of the mouth is less than the diameter of the cell. In most of these forms, also, the mouth of the cell is provided with a movable lid or shutter, by which it can be closed when the animal is retracted within it. In another great group—the *Cyclostomata*—the cells are tubular in form, and the mouth is terminal in position, whilst its diameter usually equals that of the cell. In these forms, also, there is no special apparatus for the closure of the mouth of the cell.

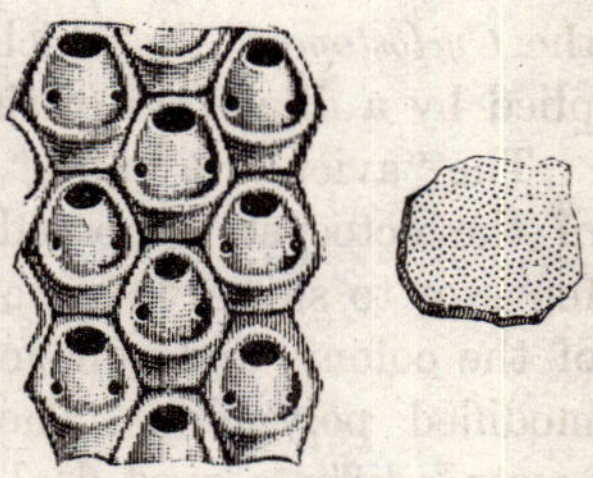

Fig. 261.—*Membranipora Oceani*, showing the sub-terminal mouths of the cells. Upper Cretaceous.

The surface of the cell may be "either smooth and entire,

spinous or granulous; perforated with minute pores, or cribriform with larger openings; reticulate or ribbed, &c.,—all of which conditions, with certain precautions, afford excellent diagnostic characters" (Busk). The margins of the mouth of the cell, also, may be "simple or thickened, unarmed or beset with erect 'marginal spines,' which again may be either rigid or articulated at the base, simple or branched."

There still remain three structures which are present in many forms, and especially in the *Cheilostomata*, which require some notice. The structures in question are known as the "ovicell," the "avicularia," and the "vibracula."

The "ovicell" is a structure especially characteristic of the Cheilostomatous *Polyzoa;* but its presence is not universal, and when present it may be inconspicuous. Its general form is that of "a more or less rounded eminence situated above or behind the cell. . . . The cavity of the organ is continuous with the perivisceral space, through a passage situated at the upper and back part of the cell, and through which it would appear the ova are conveyed as into a sort of marsupial pouch. This organ is wanting in the *Cyclostomata*, in which its functions are apparently supplied by a dilatation of the body of the cell itself." (Busk.)

The "avicularia" and "vibracula" are peculiar appendages of the ectocyst, supposed to be weapons of offence and defence, or to subserve some unknown function in the economy of the colony, and believed by Prof. Huxley to be peculiarly modified polypides. The avicularia, or "bird's-head processes," differ a good deal in shape, but consist essentially of a "movable mandible and a cup furnished with a horny beak, with which the point of the mandible is capable of being brought into apposition" (Busk).—In shape they are often closely similar to the head of a bird (fig. 260, 3), and they perform a peculiar snapping movement, which is continued long after the apparent death of the colony. In many respects the avicularia are comparable with the "pedicellariæ" of the Sea-urchins and Star-fishes. In the "vibracula," the place of the mandible of the avicularium is taken by a bristle or seta, which is capable of extensive movement

The following table exhibits the leading groups of the *Polyzoa* :—

TABLE OF THE DIVISIONS OF THE POLYZOA.

ORDER I.—PHYLACTOLÆMATA.

Tentacles arranged in the shape of a horse-shoe or crescent. Mouth furnished with a valve-like organ or "epistome."

Sub-order 1. *Lophopea* (fresh-water).—Arms of the tentacular disc ("lophophore") free or obsolete; consistence, horny or sub-calcareous.

Sub-order 2. *Pedicellinea* (marine).—Arms of the tentacular disc united at their extremities; consistence, soft and fleshy.

Sub-order 3. *Rhabdopleurea* (marine).—Cœnœcium branched, adherent, membranous, with a chitinous rod on its adherent side. Tentacular disc horse-shoe-shaped. No epistome (?).

ORDER II.—GYMNOLÆMATA.

Tentacles arranged in the form of a more or less complete circle. No valve-like organ, or "epistome," arching over the mouth.

Sub-order 4. *Paludicellea* (fresh-water).—Polypide completely retractile; evagination of tentacular sheath imperfect; consistence, horny or sub-calcareous.

Sub-order 5. *Cheilostomata* (marine).—Polypide completely retractile; evagination perfect; orifice of cell sub-terminal, of less diameter than the cell, and usually closed with a movable lip or shutter, sometimes by a contractile sphincter; cells not tubular; consistence, calcareous, horny, or fleshy.

Sub-order 6. *Cyclostomata* (marine).—Cell tubular; orifice terminal, of the same diameter as the cell, without any movable apparatus for its closure; consistence, calcareous.

Sub-order 7. *Ctenostomata* (marine).—Orifice of the cell terminal, furnished with a usually setose fringe for its closure; cells distinct, arising from a common tube; consistence, horny or carnose.

Of the above sub-orders of the *Polyzoa*, only the marine groups of the *Cheilostomata* and *Cyclostomata* are known to occur in the fossil condition; their preservation being due to their marine habits and their general possession of a calcareous or sub-calcareous cœnœcium. The general facts as to the distribution of the *Polyzoa* in past time have been already alluded to. The *Oldhamia* of the Cambrian rocks and the Graptolites have been referred to the *Polyzoa;* but the former is probably a plant, and the latter almost certainly belong to the *Hydrozoa.* The genus *Dictyonema* of the Upper Cambrian, Silurian, and Devonian (fig. 52), is, however, very possibly really a Polyzoön, and has, indeed, been compared by Sir Wyville Thomson with the living Polyzoan genus *Naresia.* Leaving these out of account, the *Polyzoa* seem to commence in the Lower Silurian, and are well represented in the Upper Silurian, Devonian, Carboniferous, and Permian rocks, but especially in the Carboniferous. None of the Palæozoic genera extend into the Secondary period. In the Secondary period *Polyzoa* are very abundant, and they attain their maximum of development in the Cretaceous period, the Chalk having yielded over two hundred species belonging to this class. In the Tertiary period, also, *Polyzoa* are abundant; the Coralline Crag (Pliocene) deriving its name from the great profusion of its Polyzoan remains.

The Palæozoic *Polyzoa* are both very numerous and very peculiar in their characters; but they may be briefly con-

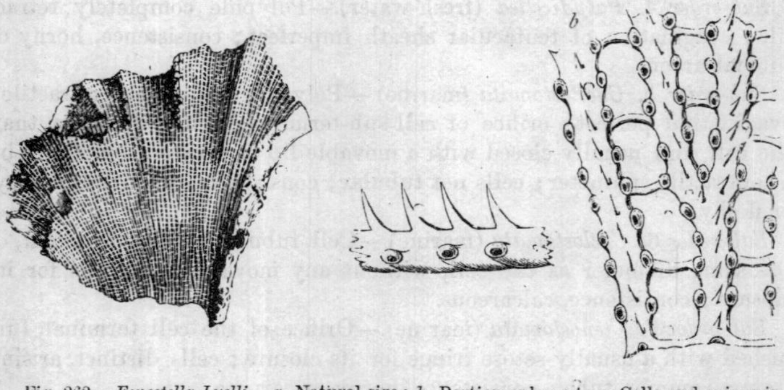

Fig. 262.—*Fenestella Lyelli.* *a*, Natural size; *b*, Portion enlarged; *c*, Cells and spines in profile. From the Carboniferous rocks of Canada. (After Dawson.)

sidered here under the head of a limited number of leading types. Foremost among these is the genus *Fenestella*, the type-form of the great group of the *Fenestellidæ* or "Lace-corals," ranging in geological time from the Lower Silurian to the Permian, where it seems, so far as known, to become extinct. In this genus, the cœnœcium (fig. 262) forms a funnel-shaped or fan-shaped expansion, the base of which is attached to some foreign object The cœnœcium is composed of a number of nearly parallel stems, united to one another by numerous cross-bars or "dissepiments," enclosing small interspaces. The outer surface of the branches is minutely porous or longitudinally striated. The inner surface of the branches exhibits a central ridge or keel, separating the mouths of two rows of cells. Sometimes there is an additional row of cells on the mesial keel, and the dissepiments are usually destitute of cells. The entire cœnœcium is calcareous. In the nearly allied genus *Phyllopora* or *Retepora*,[1] the cœnœcium is also a fan-shaped expansion, and is also of a calcareous consistence. In place, however, of transverse dissepiments, the branches of the cœnœcium unite with one another in such a manner as to form ovate interspaces or "fenestrules." The outer surface of the cœnœcium is non-celluliferous and minutely striated. The inner surface bears two or more rows of small cells. The genus ranges from the Lower Silurian to the Permian. *Actinostoma*, of the Carboniferous, is like *Fenestella* in having a fenestrated frond, poriferous on one side only, but the branches are not keeled, the cell-mouths are furnished with radiating teeth, and the cells have a second smaller aperture close to their ends. *Polypora* (fig. 264, *a*), abundant in the Devonian and Carboniferous, resembles *Fenestella* in general form; but the branches are not keeled, and are pierced by from three to five rows of cell-openings, while the "dissepiments" are solid and non-poriferous. *Synocladia*, of the Carboniferous and Permian, has the funnel-shaped and reticulated frond of *Fenestella* and *Polypora*, but the dissepiments are directed obliquely upwards, bearing two rows of pores each, while the branches carry from three to five rows of pores separated by keels. *Carinopora*, again, of the Devonian, has the external features of *Retepora*; but the branches are furnished internally with immense keels, bordered on each side by a single row of pores. Lastly, in *Ptilopora* (fig. 263) are forms essentially similar to *Fenestella*, but having a feather-like arrangement, consisting of a central stem giving off lateral

[1] The name *Retepora* has become so firmly attached to this genus, that it appears difficult to dislodge it now. The Palæozoic types so named are, however, altogether without relationship with the more modern true *Reteporœ*, and it would therefore be best to adopt for them Prof. King's name of *Phyllopora*.

branches, which are connected by dissepiments, leaving oval fenestrules.

Among the more abnormal forms of the *Fenestellidæ*, *Hemitrypa* comprises Devonian and Carboniferous forms, in which the cœnœcium is funnel-shaped and reticulate, as in *Fenestella*; but the branches have high ridges externally, and the crests of these are united by cross-bars or dissepiments, which are so thickened laterally as to leave only small round pores. The pores open into galleries between the branches, and these open internally by a second set of pores, the mouths of

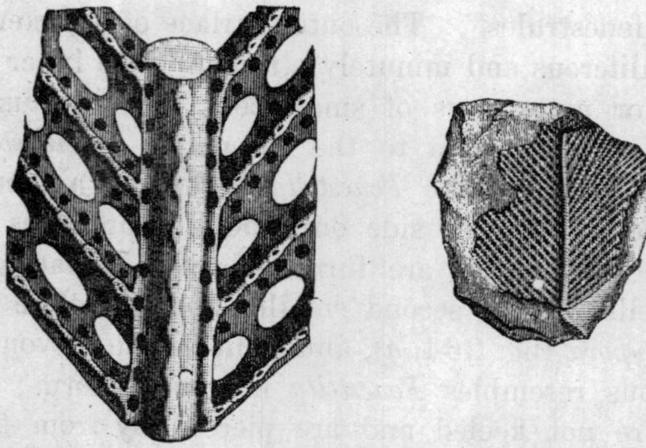

Fig. 263.—*Ptilopora pluma*; the right-hand figure of the natural size, the left-hand figure enlarged. Carboniferous.

the cells themselves opening into the galleries just mentioned. Another remarkable genus is *Archimedipora* (fig. 264, *c*), which occurs abundantly in parts of the Carboniferous series of North America. In this genus the cœnœcium is wound in an oblique spiral round a solid central screw-like axis. The fan-shaped expansions of the frond are simply reticulate externally, but the branches carry a double row of cell-pores internally. In the genus *Glauconome* (*Acanthocladia*, King), again, the cœnœcium is branched in an irregular manner, and the lateral off-sets (fig. 264, *b*) are not united by dissepiments. The one side is simply longitudinally striated, but the other side exhibits a double series of alternately placed cell-mouths, other cells being placed on the small side-shoots of the frond. The genus is Carboniferous and Devonian.

Lastly, in the *Lyropora* of the Carboniferous, the cœnœcium resembles that of *Fenestella* in general features, but the net-like frond is bounded by solid lateral supports, which spring, like the two branches of the letter U, from a small base of attachment.

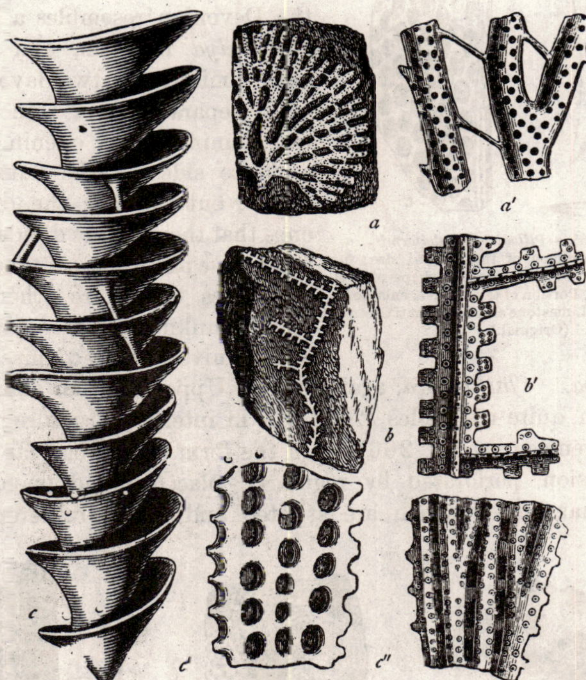

Fig. 264.—Carboniferous *Polyzoa*. *a*, Fragment of *Polypora dendroides*, of the natural size—Ireland; *a'*, Small portion of the same, enlarged to show the cells; *b*, *Glauconome pulcherrima*, a fragment, of the natural size—Ireland; *b'*, Portion of the same, enlarged; *c*, The central screw-like axis of *Archimedipora Wortheni*, of the natural size—Carboniferous, America; *c'*, Portion of the exterior of the frond of the same, enlarged; *c''*, Portion of the interior of the frond of the same, showing the mouths of the cells, enlarged. (After M'Coy and Hall.)

Another great group of Palæozoic *Polyzoa* is typified by the widely-distributed genus *Ptilodictya*, which ranges from the Lower Silurian to the Carboniferous. In this genus the cœnœcium is flattened, foliaceous, or more commonly dichotomously branched. The cellules are placed obliquely upon both sides of the thin flat frond, being separated by a delicate, striated, calcareous membrane (fig. 265), and opening on the surface by round or oval mouths, while the margins of the frond are striated longitudinally and destitute of cell-mouths. The *Heterodictya* of the Devonian resembles a large *Ptilodictya* in form, and consists similarly of two layers of cells, separated by a thin central lamina, and opening on opposite sides of the flattened frond; but there is the difference that the cells are divided by transverse partitions or "tabulæ," structures which are otherwise almost unknown amongst the forms universally recognised as *Polyzoa*. *Clathropora*, again, of the Upper Silurian and Devonian, quite resembles *Ptilodictya* in internal structure; but the cœnœcium (fig. 266) is in the form of a broad fan-like expansion, perforated by regularly-placed round openings, the margins of which are striated and non-celluliferous.

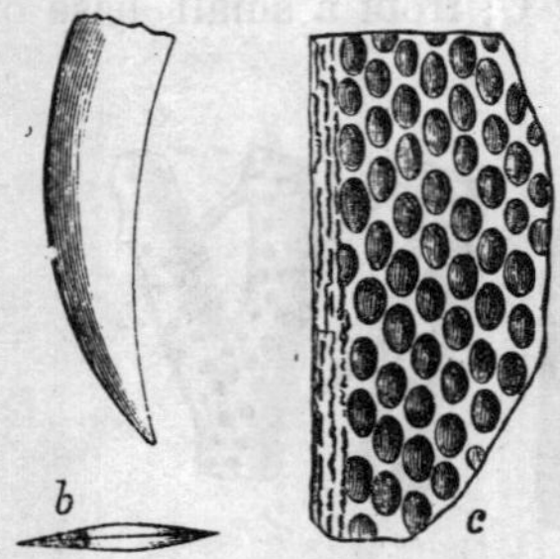

Fig. 265.—*Ptilodictya falciformis.* *a*, Small specimen of the natural size; *b*, Cross-section, showing the shape of the frond; *c*, Portion of the surface, enlarged. Trenton Limestone and Cincinnati Group, America. (Original.)

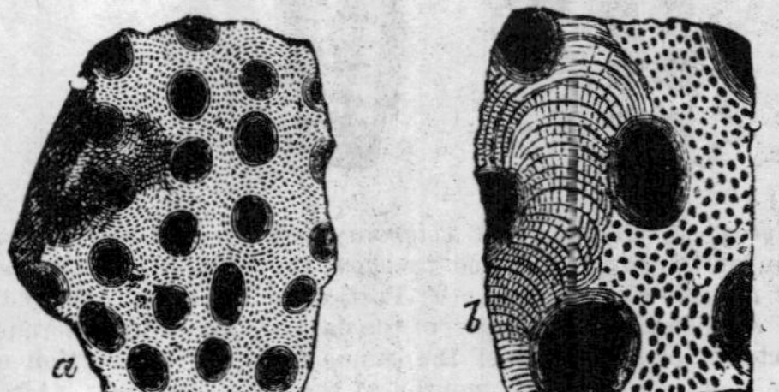

Fig. 266.—Fragment of *Clathropora intertexta*, of the natural size and enlarged. Devonian, Canada. (Original.)

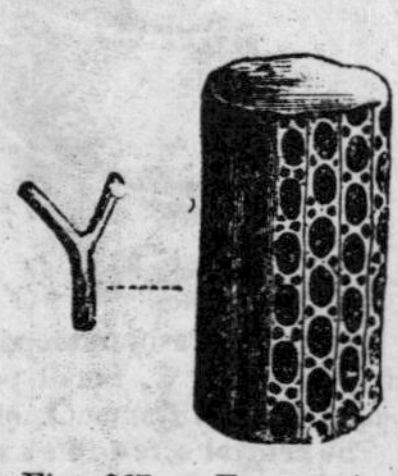

Fig. 267.—Fragment of *Ceriopora Hamiltonensis*, of the natural size and enlarged. Devonian, Canada. (Original.)

As the representatives of another great group of Palæozoic *Polyzoa*, we may take the Silurian, Devonian, and Carboniferous forms which have been described under the names of *Ceriopora*, *Rhabdomeson*, *Trematopora*, and *Helopora*. All these have the form of slender rounded stems, usually more or less branched, and consisting of small tubes radiating from a central line, and opening at all points of the surface, diminutive tubules sometimes occupying the interspaces between the larger cells. The minute structure of most of the forms generally placed under *Ceriopora*, as well as of those named *Trematopora* and *Helopora*, is still very little known; but the *Rhabdomeson* of the Carboniferous has been shown by Professor Young and Mr John Young to have the curious feature that the cells spring from a central hollow axis. It is possible that all of these, and perhaps some other Palæozoic forms, may be ultimately shown to be allied to the true *Cerioporidæ*; but their true affinities and relations are at present more or less uncertain.

We may here make a few remarks upon the position of a number of common and widely distributed Palæozoic fossils, belonging to such genera as *Chætetes*, *Monticulipora*, *Callopora*, *Fistulipora*, &c. These are now very generally regarded as being really *Polyzoa*, and *some* of the forms which have been described under some of the above generic titles are, doubtless, really of this nature. Moreover, there is a strong resemblance between certain of the above types and some undoubted Tertiary and Recent *Polyzoa* (*e.g.*, *Heteropora*). On the other hand, all the forms which can be allowed to stand finally under any of the above heads, have the feature that the tubes are crossed by transverse calcareous partitions or "*tabulæ*," these structures being quite unknown among recent Polyzoans. They have, also, strong and numerous points of agreement with various forms (such as *Alveolites*, *Favosites*, *Tetradium*, &c.) which cannot, at present, be removed from the Corals without violence. It has therefore seemed best, in the meanwhile at any rate, to consider the forms in question as being *Actinozoa*, and as not certainly referable to the class now under consideration.

A fourth group of Palæozoic *Polyzoa* includes a number of forms which form thin crusts, growing parasitically upon shells, corals, and other foreign bodies, and which may be taken, in a general way, as ancient representatives of the existing family of the *Membraniporidæ*. The Silurian *Pales-*

chætra, which generally forms thin crusts, composed of polygonal cells, may be taken as the type of this group, and appears to be allied to the recent *Flustræ*. Some of the forms which have been referred to *Monticulipora* will doubtless find their true place in this Polyzoan genus.

Still another group of the Palæozoic *Polyzoa* is typified by the *Hippothoa* of the Lower Silurian (fig. 268, C), a genus still in existence. In this form the cœnœcium is parasitic and decumbent, and consists of rows of pyriform cells, each with a single opening on its front face, lateral branches similarly composed being given off from the sides of certain of the cells.

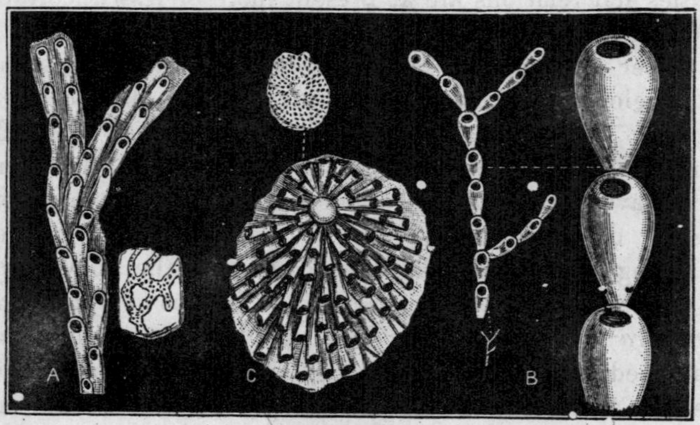

Fig. 268.—A, Portion of the cœnœcium of *Alecto auloporoides*, of the natural size and enlarged—Lower Silurian; B, Portion of the cœnœcium of *Ceramopora Huronensis*, of the natural size and enlarged—Devonian; C, Portion of the cœnœcium of *Hippothoa inflata*, of the natural size and enlarged—Lower Silurian. (Original.)

Lastly, we have a group of Palæozoic *Polyzoa*, typified by the genus *Alecto* (fig. 268, A), and belonging to the Cyclostomatous division of the class, and to the family *Tubuliporidæ*. In this form the cœnœcium is parasitic on foreign bodies, and the cells are tubular, with round mouths, and disposed either in single series or in irregular transverse rows.

In the *Ceramopora* of the Silurian and Devonian formations (fig. 268, B) we have another Cyclostomatous genus, with relations to both *Alecto* and *Diastopora*. The cœnœcium in this form has the character of thin, generally circular, though sometimes quite irregular crusts, growing upon foreign bodies, and often superimposed in successive layers one above the other; the cells being easily recognised by their arched, subtriangular, or crescentic mouths.

The Secondary and Tertiary *Polyzoa* are so numerous and varied, and from their small size are so difficult of study, that it will be sufficient here to give an extremely brief and general review of the leading groups. Taking the *Cheilostomatous Polyzoa* first, we find that only a certain number of the known families have been as yet detected in the fossil condition. The first of these is the family *Cellulariadæ*, in which the cœnœcium is branched and erect, and the branches are linear and composed of cells arranged in the same plane. As the type of this family we may take the genus *Scrupocellaria* (fig. 269, E), of the Tertiary and Recent periods. In the family of the *Salicornariadæ* the cœnœcium is erect and dichotomously divided, the branches being cylindrical, and the cells disposed round an imaginary axis. In *Salicornaria* (fig. 269, C), of the Tertiary and Recent seas, the surface is divided into rhomboidal or hexagonal spaces, representing the front walls of the cells, and irregularly-disposed avicularia are present. In the family of the *Hippothoidæ* the polyzoary is creeping and attached to foreign bodies, the cells being pyriform, distant, and arranged in linear series. The principal genus is *Hippothoa* (fig. 268, C), which is mostly Tertiary and Recent, but which, as we have seen, is represented by a well-marked form as early as the Lower Silurian. More important than any of the preceding is the family of the *Membraniporidæ*, in which there is an encrusting calcareous or corneo-calcareous polyzoary, composed of horizontal and contiguous cells. In the comprehensive genus *Membranipora* (fig. 261) the front wall of the cell is to a larger or smaller extent membranous, so that in all but actually fresh specimens it appears to be to a corresponding degree deficient and open, while the cells are surrounded by an elevated border. Numerous species of the genus are found in the Cretaceous, Tertiary, and Recent periods; and some ancient types (such as the previously-noticed *Paleschara* of the Silurian) appear to be nearly related to this. An equally comprehensive genus of this family is *Lepralia* (fig. 269, A), in which the cœnœcium usually forms circumscribed crusts

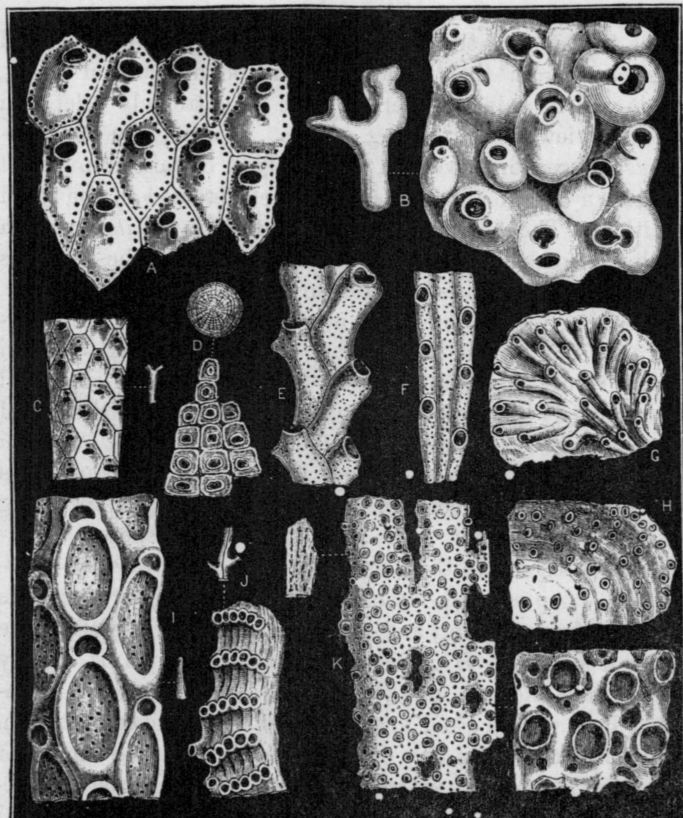

Fig. 269.—A, A few cells of *Lepralia violacea* (Pliocene and Recent), magnified; B, *Cellepora coronopus*, of the natural size, and a portion of the surface enlarged (Pliocene); C, A small piece of *Salicornaria crassa*, of the natural size and enlarged (Pliocene); D, *Lunulites quadrata*, of the natural size, and a small portion of the upper surface enlarged (Tertiary); E, A fragment of *Scrupocellaria elliptica*, viewed from behind, enlarged (Tertiary); F, A small piece of *Crisia denticulata* (Tertiary), enlarged; G, A fragment of *Tubulipora flabellaris* (Tertiary), enlarged; H, A fragment of *Diastopora simplex*, enlarged (Tertiary); I, A piece of *Vincularia Haidingeri*, natural size and enlarged (Tertiary): J, *Idmonea fenestrata*, natural size and enlarged (Tertiary); K, *Hornera reteporacea*, natural size and enlarged (Tertiary). (After Busk and Reuss.)

growing upon foreign bodies, and the front wall of the cells (though sometimes more or less minutely perforated) is completely calcified. This genus begins at least as early as the Cretaceous period, and is represented by numerous Tertiary and living types. In the family of the *Celleporidæ*, again, we have forms in which the polyzoary is irregular, often massive in its mode of growth, composed of cells which are more or less vertical to its own plane or axis, and which are heaped up or irregularly superimposed upon one another. *Cellepora* itself (fig. 269, B) is principally Tertiary and Recent. Another important family is that of the *Escharidæ*, in which the cœnœcium is erect and rigid, with quincuncially-arranged cells disposed in a single plane on one or both sides of the frond. In *Eschara* (fig. 270) the polyzoary may be broad and leaf-like, or branched and nearly linear, and the cells are placed back to back, amalgamated with each other by their walls, and opening on both surfaces of the frond. The genus is abundantly represented from the Jurassic to the present day. *Biflustra* (Cretaceous to Recent) resembles *Eschara* in the composition of the colony out of two layers of cells placed back to back, but the cells are incomplete in front, and possess a raised border, while they are readily separable from each other. A third genus of this family is *Retepora*[1], in which the cœnœcium is more or less branched, often reticulate, funnel-shaped, or fan-like, the cells opening only on one (the upper) surface. The genus is Tertiary and Recent.

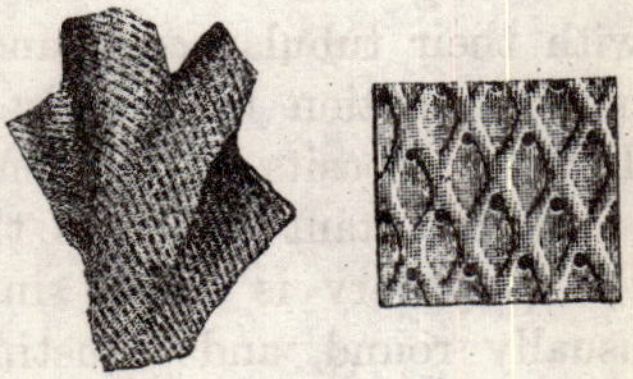

Fig. 270.—*Eschara Ranvilliana*. Jurassic.

Still another interesting group is the family of the *Selenariadæ*, with the curious allied types of *Selenaria*, *Cupularia*, and *Lunulites*, ranging from the Cretaceous to the present day. In these singular forms (fig. 269, D) the polyzoary is *free*, and consists of a plano-convex or concavo-convex disc,

[1] As previously remarked, this genus has nothing to do with the forms from the Palæozoic rocks generally known by the name *Retepora*. These latter have no claim to this generic title, and should stand under the genus *Phyllopora* of King.

composed of only one layer of cells, the mouths of which open on the convex surface. Lastly, we have the family of the *Vinculariadæ*, typified by *Vincularia* itself (fig. 269, I), in which the polyzoary is erect, rigid, and branched, the cells disposed alternately round an imaginary axis, and having a raised border in front. The undoubted species of the genus are Cretaceous, Tertiary, and Recent; but it seems probable that some of the Palæozoic *Polyzoa* may ultimately be referred to this family, if not to this genus.

Passing on next to the series of the *Cyclostomatous Polyzoa*, with their tubular cells and terminal cell-mouths, we need merely mention the family of the *Crisiadæ*, represented in Tertiary deposits by the genus *Crisia* (fig. 269, F). A much more important family is that of the *Idmoneidæ*, in which the polyzoary is erect, simple, or branched, the branches usually round, and sometimes anastomosing with one another. In the genus *Hornera* (fig. 269, K) the cells open on one side only of the branches, the mouths being placed in somewhat rhomboidal spaces marked out by wavy anastomosing ridges. The species of this type are Cretaceous, Tertiary, and Recent, but some Palæozoic forms may possibly find a place here. In *Idmonea* itself (fig. 269, J), on the

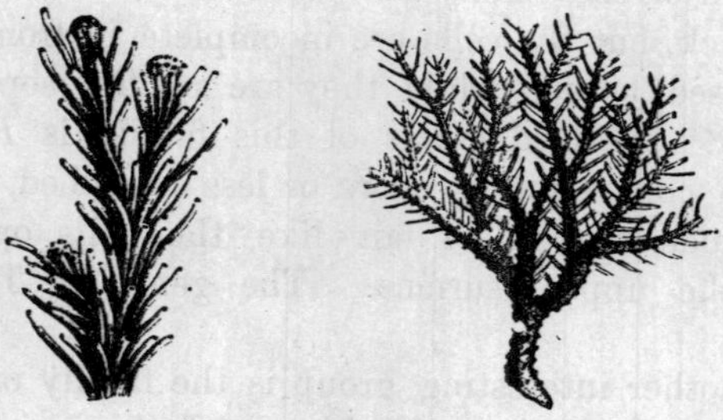

Fig. 271.—*Pustulopora* (*Entalophora*) *cellarioides*, natural size and enlarged. Jurassic.

other hand, the cells are disposed in transverse or oblique rows on each side of the front faces of the branches, which are divided mesially by an angulation or longitudinal keel. The genus ranges from the Cretaceous to the Tertiary. In the third genus, *Pustulopora* (*Entalophora*), the tubular cells open on all sides of the branches (fig. 271). The genus extends from the Cretaceous to the Recent period. The family of the *Tubuliporidæ* includes a number of well-known forms, in which the tubular cells of the polyzoary are more or less largely free and disconnected. In the genus *Tubulipora* (fig. 269, G) the tubes are free for a great part of their length, and the colony is attached more or less extensively by its base, the cells radiating from an excentric point. The genus seems to extend from the Cretaceous to the present day. *Alecto* (fig. 268, A) is very like the preceding, but the cœnœcium is creeping and irregularly branched, and the cells are only free close to their mouths, and do not ascend as in *Tubulipora*. The best-known forms of this genus are Tertiary and Recent, but examples which appear to be clearly referable to it are found in deposits as old as the Lower Silurian. *Mesenteripora* (*Bidiastopora*), again, resembles *Alecto*, but the cœnœcium is foliaceous, and the cells are in two layers, separated by a calcareous membrane and opening on both surfaces.

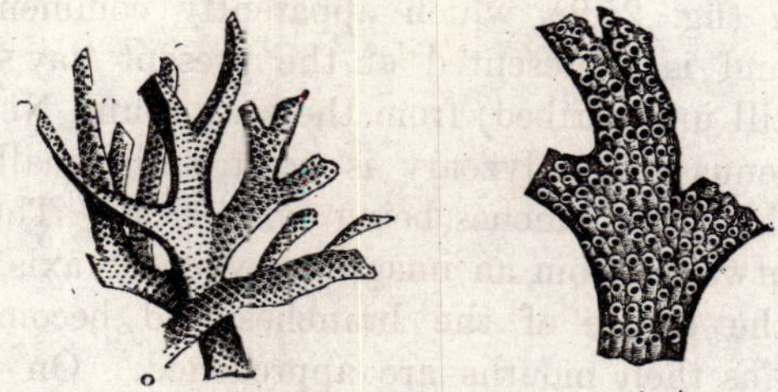

Fig 272.—*Mesenteripora* (*Bidiastopora*) *cervicornis*, natural size and enlarged Jurassic.

Closely allied to the *Tubuliporidæ* is the great family of the *Diastoporidæ*, chiefly distinguishable from the preceding by the fact that the tubular cells are not free even at their terminal portions. In this group the genus *Tubulipora* of the preceding is represented by *Diastopora* (fig. 269, H), in which the encrusting cœnœcium is discoidal, and more or less excentric in its mode of growth. The *Ceramopora* of the Silurian and Devonian (fig. 268, B), as before noted, is related to *Diastopora* in many respects. *Patinella* (ranging from the Cretaceous to the Recent period) has a discoidal polyzoary, with the tubes concentrically opening on the upper surface. *Berenicea* has the cœnœcium composed of a series of superimposed layers. *Discoporella*, of the Tertiary, is like *Patinella*, but there is an interstitial cancellated tissue between the tubes; and *Defrancia* (Jurassic to Recent) is chiefly distinguished from both of these by the fact that the cells are disposed in distinct elevated ridges radiating from the centre of the disc.

In the family of the *Ceriporidæ*, again, are forms with a solid, simple, or branched cœnœcium, composed of crowded contiguous cells. In *Fungella* (Cretaceous and Tertiary), the colony is stalked, and the cells open on its upper surface. *Ceriopora* itself is an ill-defined genus, the limits and range of which are not known, forms of very diverse affinities having been included by palæontologists under this name. It is possible, however, that in the reconstitution of this genus, certain of the Palæozoic *Polyzoa* may be found capable of inclusion in it, as a more than merely provisional arrangement. A better-defined genus of this family is *Heteropora* (fig. 259), which apparently commences in the Jurassic, and is represented at the present day by a single species (still undescribed) from the seas round New Zealand. In this genus the polyzoary is erect, and usually more or less branched, the branches being cylindrical. The cell-tubes radiate outwards from an imaginary vertical axis, being vertical in the centre of the branches and becoming nearly horizontal as their mouths are approached. On the surface are seen two sets of differently-sized openings, of which the largest are the cell-mouths, and are much fewer in number than the others. The smaller and more numerous openings, on the other hand, are generally termed "ostioles," on the belief that the small canals into which they lead are not of the nature of true "cells," but are possibly water-canals. These so-called "interstitial canals" may, however, possibly have lodged a peculiar set of zoöids, the colony being a dimorphic one. Be this as it may, the mouths of these canals ("ostioles") are often closed at the surface by a calcareous lid, which is usually perforated in the centre. In the course of growth a number of these hymen-like diaphragms may be formed in each of the interstitial canals, and these tubes may thus come to be crossed by a succession of perforated transverse partitions, which look like the "tabulæ" of the 'Tabulate Corals." No such partitions, however, are ever found in the larger tubes, which are regarded as the true "cells." *Heteroporella* agrees with *Heteropora* in essentials, but forms crusts upon foreign bodies.

We have dwelt at some length on the structure of *Heteropora*, as it has an important bearing upon the question as to the true affinities and systematic position of certain of the so-called "Tabulate" corals (such as *Chætetes*, *Fistulipora*, &c.) The general resemblance between specimens of *Heteropora* and certain species of *Chætetes* is undoubtedly very striking; and it seems quite probable that *some* of the forms which have been described under the generic names of *Chætetes*, *Monticulipora*, *Callopora*, &c., may really prove to be related to *Heteropora*, or to *Heteroporella*. The typical forms of the above-mentioned Palæozoic genera invariably, however, possess complete transverse partitions or "tabulæ" crossing both the larger and smaller tubes (when both sets of tubes are present). In spite, therefore, of the general resemblance between them, we are at present hardly justified in asserting that these types have any actual relationship with *Heteropora*.

The last family of the Cyclostomatous *Polyzoa* which needs

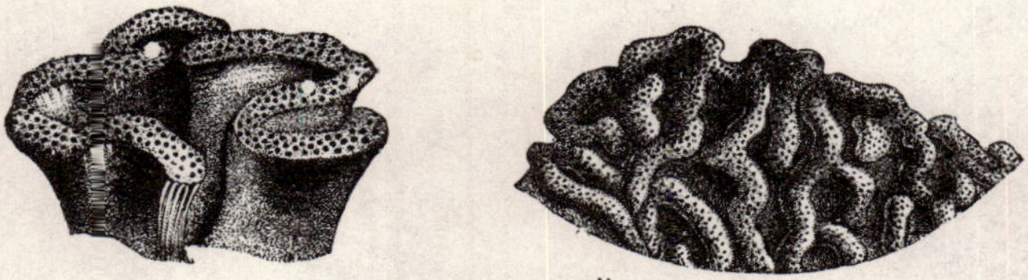

Fig. 273.—*Fascicularia* (*Meandropora*) *cerebriformis*. Tertiary.

to be mentioned is that of the *Theonoidæ*, comprising massive, globose, or irregular forms, in which the cell-tubes are contiguous and crowded together. The two principal genera are *Alveolaria* and *Fascicularia*, both found in the so-called "White Crag" or "Coralline Crag" (Pliocene), a famous depot of fossil *Polyzoa.* In the latter (fig. 273) the cœnœcium is globose and often of large size, the cell-tubes being arranged in distinct bundles, which radiate from the centre of the base in all directions, or may be disposed in vertical convoluted laminæ.

CHAPTER XXII.

BRACHIOPODA.

THE *Brachiopoda* are defined as *Molluscous animals in which the body is protected by a bivalve shell, which is lined by expansions of the integument or "mantle." The mouth is furnished with long, spirally-coiled, cirriferous processes or "arms." The animal is never composite.* (Fig 274.)

The *Brachiopoda* are essentially very similar in structure to the *Polyzoa*, from which they are distinguished by the fact that they are never composite, and by the possession of a bivalve, calcareous, or sub-calcareous shell. They are commonly known as "Lamp-shells," and are all inhabitants of the sea. All the living forms, save the aberrant *Lingula pyramidata* (fig. 274, A), are fixed to some solid object in their adult condition; but there is good reason to believe that some of the fossil forms were unattached and free in their fully-grown condition. From the presence of a bivalve shell, the Brachiopods have often been placed near the true bivalve Molluscs (the *Lamellibranchiata*); but their organisation is very much inferior, and there are also sufficient differences in the shell to justify their separation.

The two valves of the shell in any Brachiopod are articulated together by an apparatus of teeth and sockets, or are kept in apposition by muscular action alone. As regards the contained animal, the position of the valves is anterior and posterior, so that they are properly termed the "ventral" and "dorsal" valves. One of the valves is always slightly, sometimes greatly, larger than the other, so that the shell is said to be "inequivalve" (fig. 275). On the other hand, a line drawn vertically from the beak of the shell to its

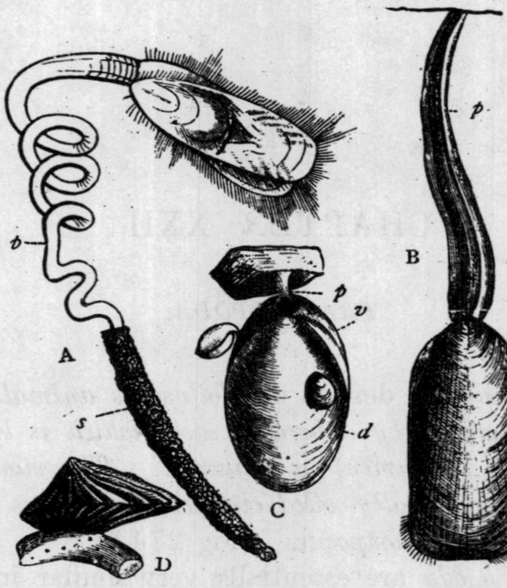

Fig. 274.—Morphology of *Brachiopoda*. A, *Lingula pyramidata* (after Morse): *p*, Peduncle; *s*, Sand-tube, encasing base of peduncle. B, *Lingula anatina* (after Cuvier): *p*, The peduncle. C, *Waldheimia cranium*, with adherent young, attached to a stone (after Davidson): *p*, Peduncle; *v*, Ventral valve; *d*, Dorsal valve. D, *Crania Ignabergensis*, attached by its ventral valve to a piece of coral (Chalk).

base (in fig. 275, B, from *c* to *f*) would divide it into two equal halves, so that the shell is said to be "equilateral." In the true bivalve Shell-fish (*Lamellibranchiata*), on the

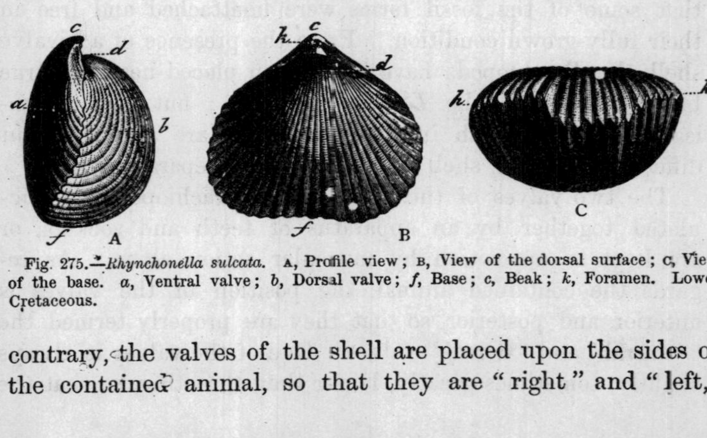

Fig. 275.—*Rhynchonella sulcata*. A, Profile view; B, View of the dorsal surface; C, View of the base. *a*, Ventral valve; *b*, Dorsal valve; *f*, Base; *c*, Beak; *k*, Foramen. Lower Cretaceous.

contrary, the valves of the shell are placed upon the sides of the contained animal, so that they are "right" and "left," instead of being dorsal and ventral. Further, the two valves are usually of the same size ("equivalve"), and a line drawn from the beak to the base would almost always divide the shell into unequal halves; so that the shell is "inequilateral."

Ordinarily the ventral valve of the shell of the Brachiopods is the larger of the two, and it is generally furnished with a prominent curved "beak." Very commonly the beak is perforated by a larger or smaller aperture, which is termed the "foramen" (fig. 275, B), and which serves for the transmission of a muscular peduncle or stem by which the shell is attached to some foreign object. In some cases, however (as in *Lingula*, fig. 274, B), the peduncle simply passes between the apices of the valves, and there is no foramen; whilst in others (as in *Crania*, fig. 274, D), the shell is merely attached by the substance of the ventral valve. The dorsal valve, which is also usually the smaller, is always free, and is never perforated by a foramen. Further, as already remarked, there is reason to believe that some fossil forms were free and unattached in their adult condition.

In most living Brachiopods the valves are articulated to one another by two teeth which are developed upon the ventral valve, and fit into corresponding sockets in the dorsal valve. Behind the dental sockets of the dorsal valve there is usually a prominent process ("cardinal process"), to which are attached two "cardinal muscles." These are inserted on each side of the centre of the ventral valve, and serve to open the shell. The valves of the shell are closed by proper "adductor muscles" (usually four in number), which also pass between the valves; and in those in which hinge-teeth are wanting, it is by these muscles that the valves are kept together. These muscles leave "scars" or impressions at their points of insertion and origin: and the number and form of these scars afford important diagnostic characters to the palæontologist.

Very commonly, the beaks of the dorsal and ventral valves are separated from one another by a narrower or wider space, which is termed the "hinge-area" (fig. 276). In some genera, as in *Spirifera*, the area is very conspicuous; in other cases it is very narrow, or even does not exist. In front of the foramen of the ventral valve, and very often forming part of its circumference, there is commonly a triangular plate, which may be composed of one or two pieces; and which is termed the "deltidium" (fig. 276, *m*). In other cases this structure is altogether wanting.

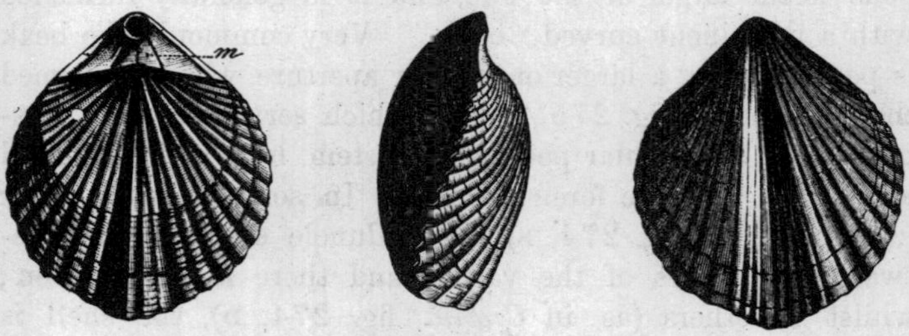

Fig. 276.—*Terebratella Astieriana*—Cretaceous. *e*, Hinge-area; *m*, Deltidium.

In intimate structure, the shell of most of the *Brachiopoda* (fig. 277) consists "of flattened prisms, of considerable length, arranged parallel to one another with great regularity, and at a very acute angle—usually only about 10° or 12°—with the surfaces of the shell."—(Carpenter.) In most cases, also, the shell is perforated by a series of minute canals, which pass from one surface of the shell to the other, in a more or less vertical direction, usually widening as they approach the external surface. These canals give the shell a "punctated" structure, and in the living animal they contain cæcal tubuli, or prolongations, from the mantle, which are considered by Huxley as analogous to the vascular processes by which in many Ascidians the muscular tunic, or "mantle," is attached to the outer tunic, or "test." In some of the *Brachiopoda* (as in the *Rhynchonellidæ*) the shell is "impunctate," or is devoid of this singular canal system.

Fig. 277.—Minute structure of the shell of *Terebratula*, showing the flattened prisms of the shell, and the canals.

The inner surface of the valves of the shell is lined by expansions of the integument which secrete the shell, and are called the "lobes" of the "pallium," or "mantle." The digestive organs and muscles occupy a small space near the beak of the shell, which is partitioned off by a membranous septum, which is perforated by the aperture of the mouth. The remainder of the cavity of the shell is almost filled by two long oral processes, which are termed the "arms," and from which the name of the class has been derived (fig. 278, D). These organs are lateral prolongations of the margins

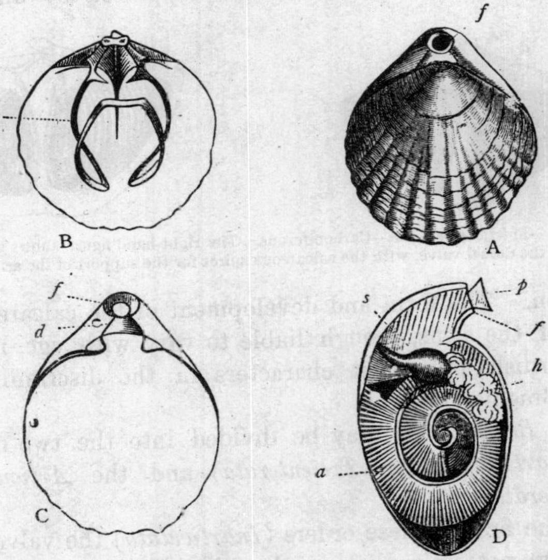

Fig. 278.—*Terebratula* (*Waldheimia*) *flavescens*. A, The shell viewed from behind, showing the dorsal valve, and the perforated summit of the ventral valve above it; B, Inner view of the dorsal valve, showing the shelly loop (*l*) which supports the spiral arms; C, Inner view of the ventral valve, showing the foramen or aperture (*f*) in the beak, through which the muscular stalk of attachment passes; D, Longitudinal and vertical section of the animal, showing the spiral arms (*a*), the stomach (*s*), and the liver (*h*). At *f* is the opening in the beak, with the stalk of attachment (*p*) passing through it. (After Davidson and Owen.) Some details have been omitted in figs. B, C, and D, for the sake of clearness.

of the mouth, usually of great length, closely coiled up, and fringed on one side with lateral processes, or "cirri." In many Brachiopods the arms are supported upon a more or less complicated internal calcareous framework or skeleton, which is sometimes called the "carriage-spring apparatus."

In some forms, as in the *Terebratulidæ* (fig. 278, B), the internal skeleton which supports the arms is a short shelly loop, of a very simple character. In these cases it is only at their bases that the arms are supported, and they are therefore more or less movable. In other cases, as in the *Spiriferidæ* (fig. 279), the arms must have been immovable, as they are supported by two thin spirally-rolled lamellæ, which form two calcareous spires in the interior of the dorsal valve. In some cases, the whorls of these spires are in turn furnished with minute calcareous spines, showing that the cirri of the arms were also supported by an internal

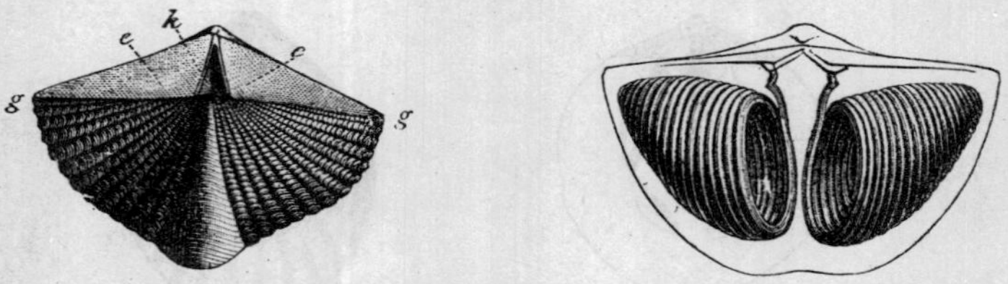

Fig. 279.—*Spirifera hysterica*—Carboniferous. The right-hand figure shows the interior of the dorsal valve, with the calcareous spires for the support of the arms.

skeleton. The form and development of the calcareous supports of the arms, though liable to vary with age, nevertheless furnish important characters in the discrimination of fossil Brachiopods.

The *Brachiopoda* may be divided into the two orders of the *Inarticulata* (or *Tretenterata*) and the *Articulata* (or *Clistenterata*).

In the first of these orders (*Inarticulata*) the valves of the shell are not united along the hinge-line, the mantle-lobes are completely free, and the intestine terminates in a distinct anus. In this division are included the three families of the *Craniadæ, Discinidæ,* and *Lingulidæ*—all very ancient, and all represented at the present day by living forms—together with the Silurian family of the *Trimerellidæ.*

In the second order (*Articulata*) the valves of the shell are united by teeth along the hinge-line, the lobes of the mantle are not completely free, and the intestine ends blindly. In this division are included the living families of the *Terebratulidæ, Rhynchonellidæ,* and the *Thecidiidæ,* and the extinct families of the *Spiriferidæ, Pentameridæ, Strophomenidæ,* and *Productidæ.* In the first two of these families the arms are supported upon a shelly loop, of variable shape and size; whilst in some of the extinct *Rhynchonellidæ*, and in the *Spiriferidæ*, the arms were supported by large spirally-coiled calcareous lamellæ.

The Brachiopods are of such importance as fossils that the characters of the more important groups, with their geological range and leading genera, will here be given, along with figures of the commoner types. As regards the general distribution of the class in time, the *Brachiopoda* are found from the Cambrian rocks up to the present day, and present us with an example of a group which appears to be slowly dying out. Nearly four thousand extinct species have been described, and the class appears to have attained its maximum in the Silurian epoch, which is, for this reason, sometimes called the "Age of Brachiopods." Numerous genera and species are found also in both the Devonian and Carboniferous formations. In the Secondary rocks *Brachiopoda* are still abundant, though less so than in the Palæozoic period. In the Tertiary epoch a still further diminution takes place, and at the present day we are not acquainted with more than a hundred living forms. Of the families of *Brachiopoda*, the *Productidæ*, *Strophomenidæ*, and *Spiriferidæ* are the more important extinct types. Of the genera, the most persistent is the genus *Lingula*, which commences in the Cambrian rocks, and has maintained its place up to the present day, though it appears to be gradually dying out.

According to Woodward: "The hingeless genera attained their maximum in the Palæozoic age, and only three now survive (*Lingula*, *Discina*, *Crania*)—the representatives of as many distinct families. Of the genera with articulated valves, those provided with spiral arms appeared first, and attained their maximum while the *Terebratulidæ* were still few in number. The subdivision with calcareous spires disappeared with the Liassic period, whereas the genus *Rhynchonella* still exists. Lastly, the typical group, *Terebratulidæ*, attained its maximum in the Chalk period, and is scarcely yet on the decline."

Of the families of the *Brachiopoda*, the *Productidæ* and *Strophomenidæ* are exclusively Palæozoic. The *Spiriferidæ* are mainly Palæozoic, but extend into the Lias, where they finally disappear. The *Lingulidæ* commence in the Cambrian period, and have survived to the present day. The *Rhynchonellidæ, Craniadæ,* and *Discinidæ* commence in the Silurian period, and are represented by living forms in existing seas The *Thecidiidæ* extend from the Trias to the present day; and the *Terebratulidæ* appear to commence in the Upper Silurian, and are well represented by living forms. In the following are given the leading characters and more important forms of the families of the *Brachiopoda*:—

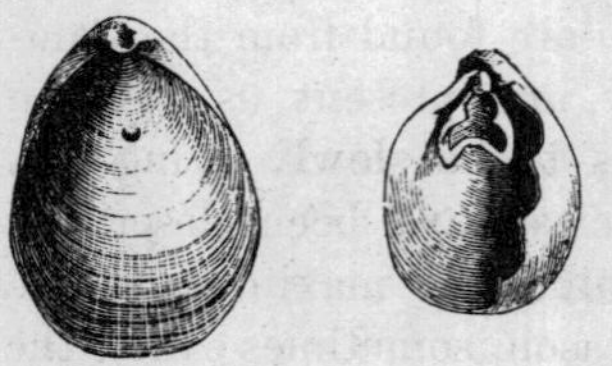

Fig. 280.—*Terebratula sacculus*—Carboniferous. The right-hand figure shows the interior of the dorsal valve with the loop. (After Dawson.)

FAM. I. TEREBRATULIDÆ.—Shell minutely punctate: ventral valve with a prominent beak, perforated by a foramen for the emission of a muscular peduncle, whereby the animal is fixed to some submarine object. Foramen partially surrounded by a deltidium of one or two pieces. Arms entirely or partially supported by calcified processes, usually in the form of a loop, and always fixed to the dorsal valve (fig. 280).

In the genus *Terebratula* (fig. 281) itself, and in *Terebratulina,* the loop supporting the arms is very short, the former commencing in the Devonian period, the latter in the Oolitic,

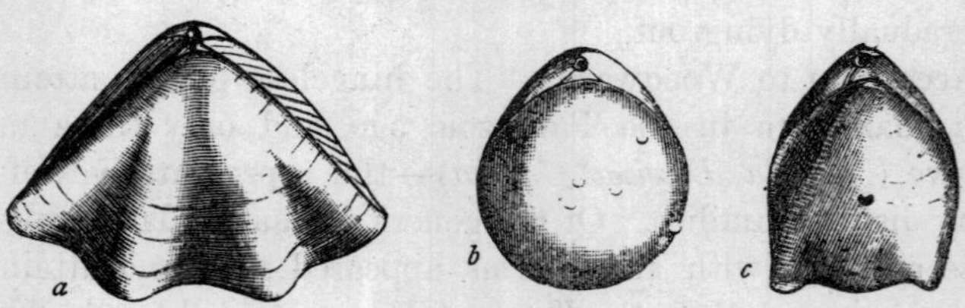

Fig. 281.—*a, Terebratula quadrifida*—Lias; *b, Terebratula sphæroidalis*—Inferior Oolite; *c, Terebratula digona*—Bradford Clay, Forest Marble, and Great Oolite (Jurassic). (After Davidson.)

and both being represented by living forms. In the genus *Waldheimia* (fig. 278) there is a very long loop, which is bent backwards, and the same is the case with *Terebratella* (fig. 276). The former appears to commence in the Trias, the latter in the Cretaceous rocks, and both have survived to the present day. *Terebrirostra* has a greatly elongated beak to the ventral valve, and appears to be confined to the Cretaceous period. Another Cretaceous form is the *Trigonosemus* of König, which is closely allied to *Terebratella*, especially in the fact that the brachial loop has a secondary attachment to a mesial septum, but which has an incurved beak, a large cardinal process, and an extensive hinge-area. *Rensselæria*, again, of the Devonian, is essentially a large ovoid *Terebratula*. The genus *Argiope*, ranging from the Cretaceous to the present day, is distinguished from *Terebratula* by the form of its brachial loop, which is attached to a single septum, or to three nearly parallel septa, directed inwards from near the border of the dorsal valve. In the Cretaceous *Magas* an even more highly developed septum is present, almost dividing the cavity of shell into two halves; while in the *Megerlia* of the same formation the brachial loop has an exceptionally complex character.

Lastly, we may place in the *Terebratulidæ* the Devonian genus *Centronella*, and the seemingly related *Leptocœlia* of the Upper Silurian and Devonian. Forming a section of the *Terebratulidæ*, or sometimes regarded as a separate family, are the two or three species which make up the genus *Stringocephalus*. These (fig. 282) are all Devonian, and are characterised by the possession of a long loop, and a widely-punctated shell. The beak of the ventral valve is very prominent, and is pierced by a foramen, which is large in the young, and small in the adult shell; and the ventral valve has a well-developed mesial septum.

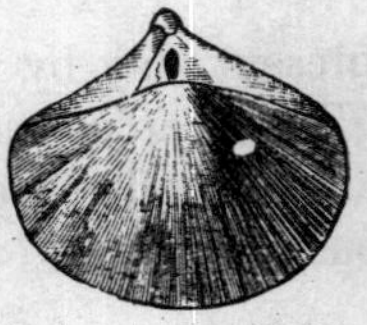

Fig. 282. — *Stringocephalus Burtini*—Devonian—reduced in size.

FAM. II. THECIDIIDÆ.—Shell fixed to the sea-bottom by the substance of the beak of the larger or ventral valve; structure punctated. Oral processes (arms) united in the form of a bridge over the visceral cavity; cirrated arms folded upon themselves, and supported by a calcareous loop. The shell is thickened, and has a granulated border (fig. 283).

The members of this family are all attached to some foreign body by a portion of the beak of the ventral valve, which, in the adult state, has either no foramen, or an exceedingly small one. The ventral valve has a well-marked hinge-area and an indistinct triangular deltidium. All the known species belong to the single genus *Thecidium*, represented at the

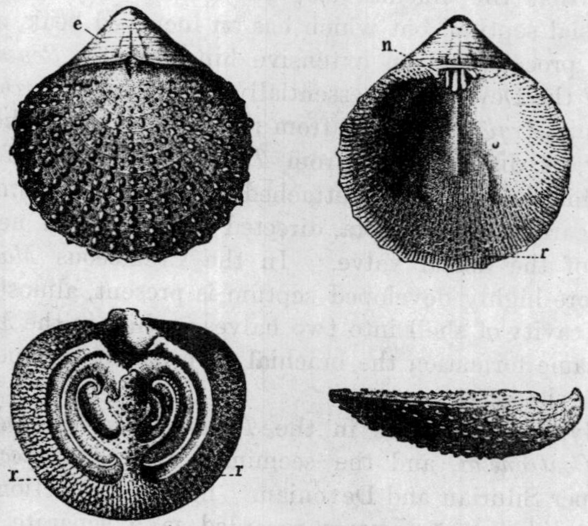

Fig. 283.—*Thecidium papillatum.* *e*, Hinge-area; *n*, Hinge-teeth of ventral valve; *r*, *r*, Granulated border of the interior of the dorsal valve.

present day by a single living species. In time, the genus *Thecidium* seems to have commenced in the Upper Trias, and is well represented in parts of the Jurassic and Cretaceous Series.

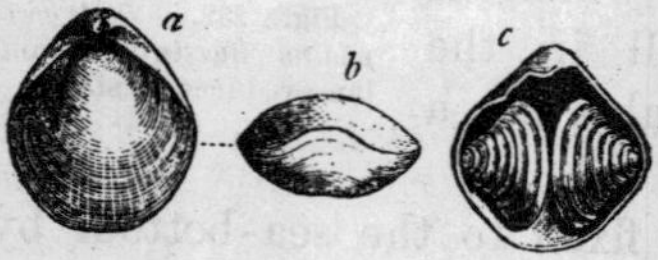

Fig. 284.—*Athyris subtilita*—Lower Carboniferous. The right-hand figure shows the interior of the dorsal valve, with the spiral supports for the arms. (After Dawson.)

FAM. III. SPIRIFERIDÆ.—Animal free when adult, or rarely attached by a muscular peduncle. Shell punctated or unpunctated. Arms greatly developed, and entirely supported upon a thin, shelly, spirally-rolled lamella (figs. 279 and 284).

The family of the *Spiriferidæ* is pre-eminently Palæozoic, but several forms extend into the older Secondary rocks. No member of the family, however, has yet been found in rocks younger than the Lias. Of the genera of the family, the most important is the genus *Spirifera,* or *Spirifer* (fig. 286), in which the valves are articulated by teeth and sockets, and the shell is not punctated. The hinge-line is long and straight, and the well-marked hinge-area is divided across in each valve by a triangular fissure, which in the ventral valve

Fig. 285.—*Spirifera sculptilis.* Devonian.

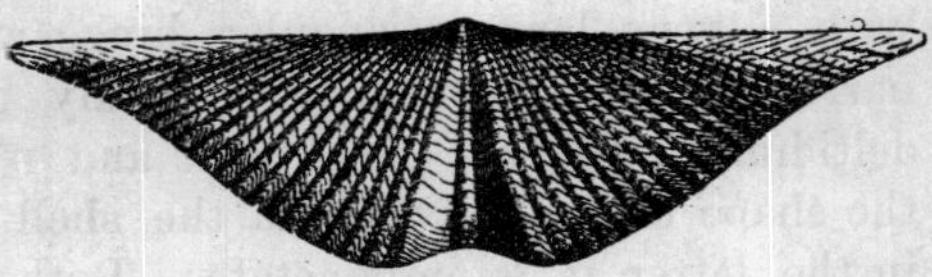

Fig. 286.—*Spirifera mucronata.* Devonian.

is closed more or less completely by a pseudo-deltidium, and in the dorsal valve is occupied by the cardinal process. The true Spirifers are mainly Upper Silurian, Devonian, and Carboniferous, and the forms of the second of these formations often have the shell winged, or drawn out at the lateral angles (fig. 286). In the Permian rocks a few species of

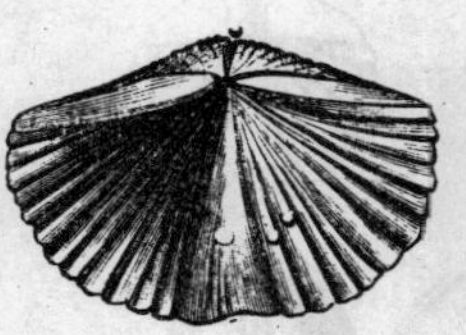

Fig. 287. — *Spirifera Niagarensis.* Upper Silurian, America.

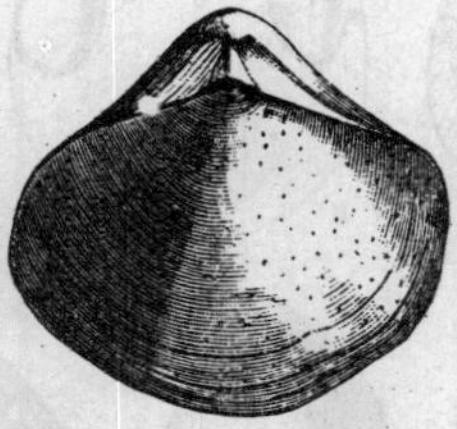

Fig. 288.—*Spiriferina rostrata.* Lias.

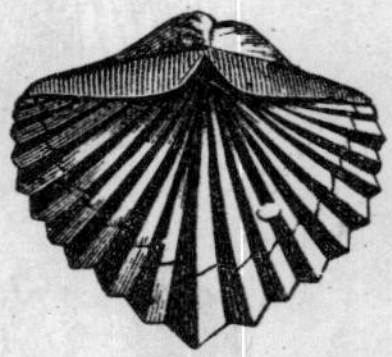

Fig. 289.—*Spirifera trigonalis.* Carboniferous Limestone.

the genus are found. If it be admitted that the possession of a non-punctated shell is an essential characteristic of the genus *Spirifera,* then no forms of this type have as yet been detected in deposits younger than the Palæozoic. We meet, however, in the Lias (Jurassic) with several forms (fig. 288) which agree in other respects with *Spirifera* proper, but which possess a punctated shell. These forms, along with others from the Carboniferous and Devonian, are now placed in the genus *Spiriferina,* which is separable from *Spirifera* principally by the character just mentioned, but also by the possession of a mesial septum. The *Suessia* of the Lias, and the *Syringothyris* of the Carboniferous, are also types very closely allied to *Spirifera.*

In the genus *Cyrtia* the shell resembles *Spirifera* in most respects; but the valves are very unequal, the dorsal valve is approximately flat, and the ventral valve is pyramidal, with a very large triangular hinge-area and a long and narrow foramen, which is partially closed by a pseudo-deltidium. *Cyrtina* (fig. 290, C and D) resembles *Cyrtia* in the shape of the valves, but the shell is punctate, whereas in the latter it is impunctate. Both genera are Palæozoic, and range from the Upper Silurian to the Carboniferous.

More important than either of the preceding is the genus *Athyris* (including under this head the *Spirigera* of D'Orbigny), which ranges from the Silurian to the Permian. The shell in this genus (fig. 290, E) is convex, with unequal

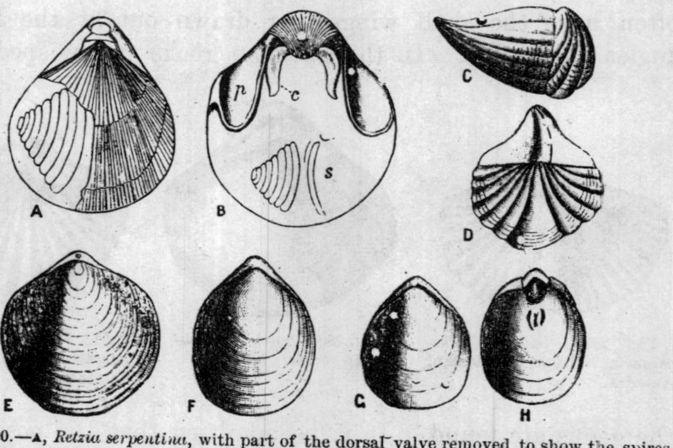

Fig. 290.—A, *Retzia serpentina*, with part of the dorsal valve removed to show the spires; B, *Uncites gryphus*, with the spires (*s*), from the Devonian; C, Side view of *Cyrtina heteroclita*—Devonian; D, The same viewed from the dorsal aspect; E, *Athyris concentrica*—Devonian; F, *Merista lævis*—Upper Silurian; G, *Meristella angustifrons*—Upper Silurian, enlarged; H, Cast of the same. (After Davidson and Hall.)

valves, the beak of the ventral valve being incurved, and either perforated by a small round foramen, or having the foramen concealed or closed in the adult state. (The *name Athyris,* like that of *Atrypa,* is a zoological misnomer, since in both genera the beak of the ventral valve is really perforated, in the young state at any rate. Some authorities, however, retain the name of *Spirigera* for those forms in which the foramen remains throughout life, and employ that of *Athyris* for those in which this aperture becomes closed in the adult condition, the latter having the additional distinctive character that the interior of the dorsal valve is partially divided by a longitudinal septum.) The spiral supports for the arms in *Athyris* are largely developed, and their pointed extremities are directed towards the lateral angles of the shell (fig. 284, *c*). *Merista* (fig. 290, F), of the Silurian and Devonian, is like *Athyris* in general character, but there is a longitudinal septum in the ventral valve, which is supported by strongly-arched transverse plates, together forming what

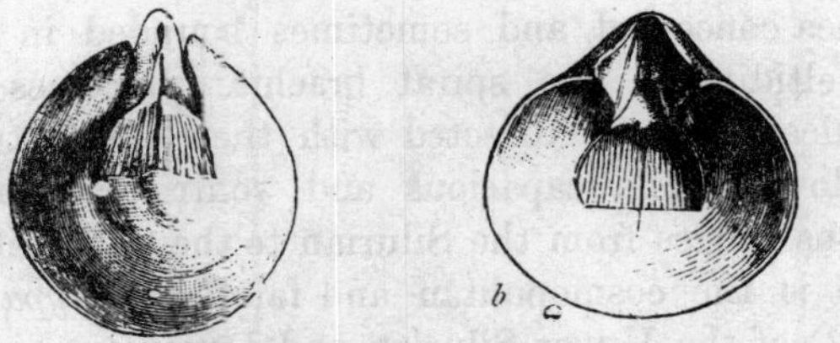

Fig. 291.—*a*, Cast of the interior of the ventral valve of *Meristella nasuta*, from the Devonian (original); *b*, Interior of the ventral valve of the same (after Billings).

is known as the "shoe-lifter process." *Meristella* (fig. 290, G and H, and fig. 291) closely resembles the preceding, but the septum and supporting arched plates are wanting in the ventral valve. The genus is Silurian and Devonian, and a well-known and familiar species is the *Meristella tumida* of the Upper Silurian.

The genus *Retzia* (fig. 290, A), with a geological distribution from the Silurian to the Carboniferous, has the spiral brachial supports of the *Spiriferidæ*, but the foramen is large, and the shell resembles that of *Terebratula*. The shell-structure is punctate, and there is a small hinge-area. *Uncites* (fig. 290, B), of the Devonian, resembles the preceding, but the beak of the ventral valve is strongly curved, the foramen disappears early, there is no hinge-area, and the shell-structure is impunctate.

The only other genus of the *Spiriferidæ* which needs mention here is *Atrypa*, which may be regarded as in many respects intermediate between the present family and the *Rhynchonellidæ*. Though named in accordance with the belief that the beak of the ventral valve was imperforate, a small foramen is really present in this genus (fig. 292),

Fig. 292.—*Atrypa reticularis*. Upper Silurian and Devonian of Europe and America. (After Billings.)

sometimes concealed, and sometimes bounded in front by a small deltidium. The spiral brachial processes are large and conical, and are directed with their apices turned into the hollow of the capacious and ventricose dorsal valve. The genus ranges from the Silurian to the Carboniferous, and its type is the cosmopolitan and familiar *Atrypa reticularis* (fig. 292) of the Upper Silurian and Devonian.

Fig. 293.—*Koninckina Leonhardi*, showing the spiral supports for the arms. Trias.

FAM. IV. KONINCKINIDÆ.—Animal unknown. Shell free; valves unarticulated (?). Oral arms supported by two lamellæ spirally coiled (fig. 293).

The only genus of this family is *Koninckina*, represented by the single species *K. Leonhardi* of the Trias of St Cassian. The shell resembles *Producta* in being eared, and the dorsal valve is concave, and follows the curve of the ventral valve. It differs from *Producta* in having the arms supported upon spiral processes.

FAM. V. RHYNCHONELLIDÆ.—Animal free, or attached by a muscular peduncle issuing from an aperture situated under the extremity of the beak of the ventral valve. Arms spirally rolled, flexible, and supported only at their origin by a pair of short, curved, shelly processes. Shell-structure fibrous and impunctate.

Fig. 294.—*Rhynchonella capax;* dorsal, profile, and ventral views. Lower Silurian.

The *Rhynchonellidæ* range from the Lower Silurian to the present day, in the person of the genus *Rhynchonella* itself; but the remaining genera of the family are exclusively Palæozoic. The genus *Rhynchonella* (fig. 294) has the valves more or less convex, smooth or plaited, united by teeth and sockets. The shell is trigonal, generally with a mesial fold and sinus (fig. 275), and having the beak of the ventral valve acute, incurved, or prominent. The foramen is situated under the beak, open to view, or concealed, and entirely or partially completed by a deltidium. The species of the genus *Rhynchonella* are very abundant in both Palæozoic and Mesozoic deposits, and two species are known at the present day.

In *Leiorhynchus* (fig. 295) are various Devonian Brachiopods, very closely allied to *Rhynchonella* proper, but having

Fig. 295.—*Leiorhynchus Huronensis*, viewed dorsally (*a*), ventrally (*b*), and in profile (*c*). Devonian. (Original.)

the plications of the shell obsolete on the lateral angles, while well marked on the mesial fold and sinus. The beak of the ventral valve is pierced by a foramen, in at any rate the early stages of growth, and there is a well-defined septum in the dorsal valve. *Triplesia* (Silurian) has a triangular pedicle-notch beneath the beak of the ventral valve, and *Eichwaldia* (also Silurian) has the peduncle transmitted through a foramen in the ventral umbo. *Porambonites*, again, of the Lower Silurian—possibly the type of a distinct family—though its shell-structure is really impunctate and fibrous, is readily recognised by the fact that the surface is ornamented with minute close-set circular pits. Lastly, the genus *Camarophoria* (fig. 296) may be regarded as a link between *Rhynchonella* and *Pentamerus*, since it possesses the general form of the first, together with the converging dental plates of the latter. The genus is found in the Carboniferous and Permian deposits.

Fig. 296. — *Camarophoria globulina*. Permian. (After King.)

All the preceding forms are naturally associated with one another by their structural characters; but there is another great group of Brachiopods usually placed in the *Rhynchonellidæ*, and agreeing with this family in many points, of which the genus *Pentamerus* is the type, and which presents certain distinctive features of its own. In *Pentamerus* (fig. 297) the shell is ovate, the valves articulated by teeth and sockets, the

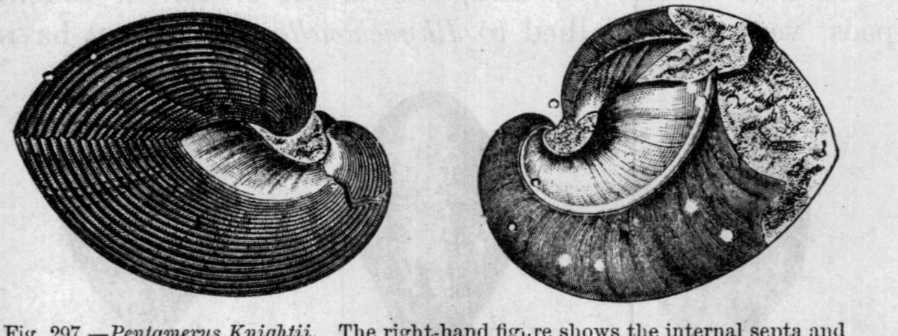

Fig. 297.—*Pentamerus Knightii*. The right-hand figure shows the internal septa and dental plates of the shell. Upper Silurian.

surface generally ribbed or striated, but sometimes smooth. The beaks are incurved, that of the ventral valve concealing a triangular fissure. Inside the ventral valve "two contiguous vertical septa coalesce into one median plate, extending from the beak to a greater or less distance; and then diverge and form the dental plates, enclosing a triangular chamber of much smaller dimensions than the lateral ones" (Davidson). The small central chamber must have been occupied by the digestive organs, and the spiral arms must have filled the great lateral spaces. In the interior of the smaller or dorsal valve are two longitudinal septa, which often form a chamber corresponding to and apposed to the median chamber in the ventral valve. The *Pentameri* range from the Lower Silurian to the Carboniferous inclusive; but they are especially characteristic of that portion of the Si-

Fig. 298.—Large Specimen of *Pentamerus oblongus*. Middle Silurian. (Original.)

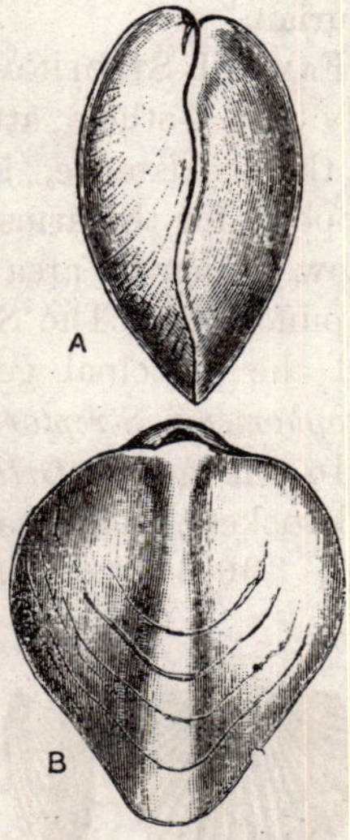

Fig. 299.—*Stricklandinia Davidsoni*, viewed sideways and dorsally. Middle Silurian. (After Billings.)

lurian series known as the Llandovery formation or Middle Silurian. They often occur in the greatest profusion, and the species have in many cases an enormous geographical range.

In *Pentamerus* (as typified by such forms as *P. Knightii, P. galeatus, P. pseudogaleatus, P. oblongus*, fig. 298, &c.) the shell is more or less globose, the ventral valve is much the largest, and the mesial septum in the same valve is very long. In other allied forms which have been termed *Strick-*

landinia (fig. 299), the two valves are not very disproportionately unequal, the shell is often more or less depressed, and there is but a short mesial septum in the ventral valve supporting a V-shaped chamber beneath the beak, whereas in the dorsal valve there are only two short socket-plates. *Pentamerella*, *Amphigenia*, *Gypidula*, and *Anastrophia* are other generic or sub-generic titles which have been proposed for forms more or less closely allied to *Pentamerus* itself. The first and second of these occur in the Devonian, but all the above forms are characteristic of the Middle and Upper Silurian.

FAM. VI. STROPHOMENIDÆ.—Animal unknown; some probably free—others attached, during the whole or a portion of their existence, by a muscular peduncle. No calcified supports for the arms. Shell with a straight hinge-line and a low triangular area in each valve. Shell-structure fibrous or punctated. The *Strophomenidæ* are exclusively Palæozoic, and the principal genera are *Orthis*, *Orthisina*, *Davidsonia*, *Strophomena*, *Streptorhynchus*, and *Leptæna*.

In the genus *Orthis* the valves are articulated by teeth and sockets, and usually are more or less transversely oblong (figs. 300-303). There is a straight hinge-line, generally

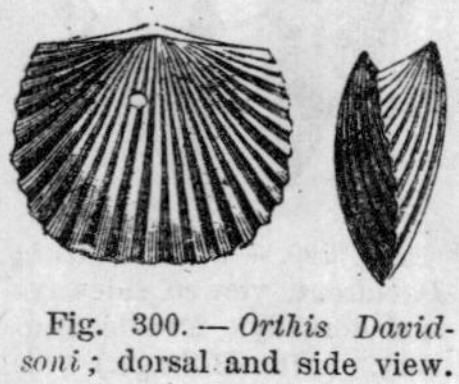

Fig. 300. — *Orthis Davidsoni*; dorsal and side view. Silurian.

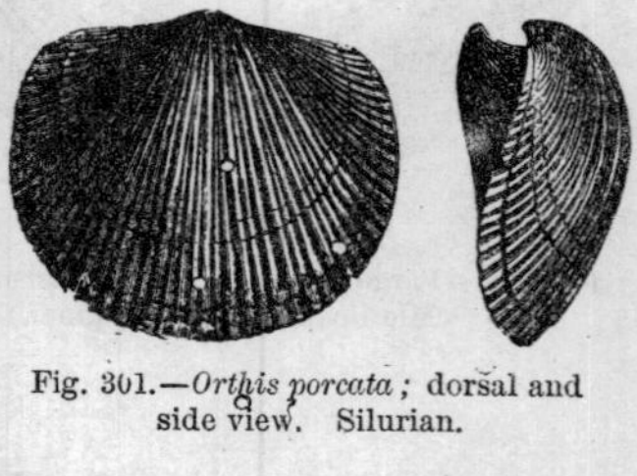

Fig. 301.—*Orthis porcata*; dorsal and side view. Silurian.

shorter than the width of the shell. Each valve has a hinge-area, notched in its centre by a triangular fissure through which the fibres of the peduncle were transmitted. The shell is often more or less flattened or depressed, and the surface may be smooth, but is more commonly ornamented with striæ, or furnished with well-marked longitudinal ribs. The species of the genus *Orthis* begin in the Cambrian, and abound in the Silurian, Devonian, and Carboniferous periods, especially in the first of these; but the genus is not known to have survived the Carboniferous period. This genus is

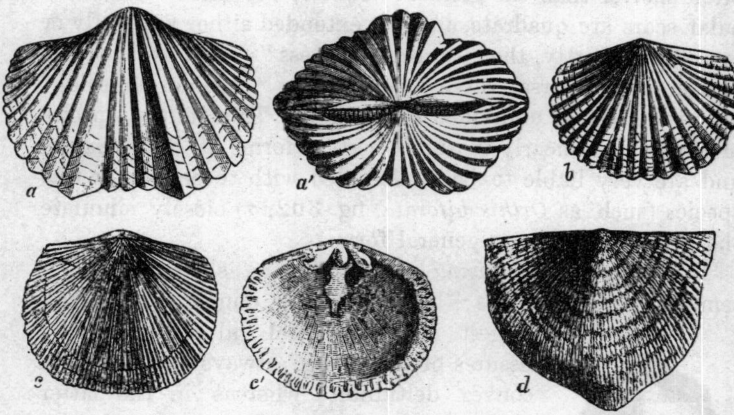

Fig. 302.—Lower Silurian Brachiopods. *a* and *a'*, *Orthis biforata*—Llandeilo-Caradoc, Britain and America; *b*, *Orthis flabellulum*—Caradoc, Britain; *c*, *Orthis subquadrata*—Cincinnati Group, America; *c'*, Interior of the dorsal valve of the same; *d*, *Strophomena deltoidea*—Llandeilo-Caradoc, Britain and America. (After Meek, Hall, and Salter.)

one of the most important and widely distributed groups of Palæozoic Brachiopods, and the species belonging to it may

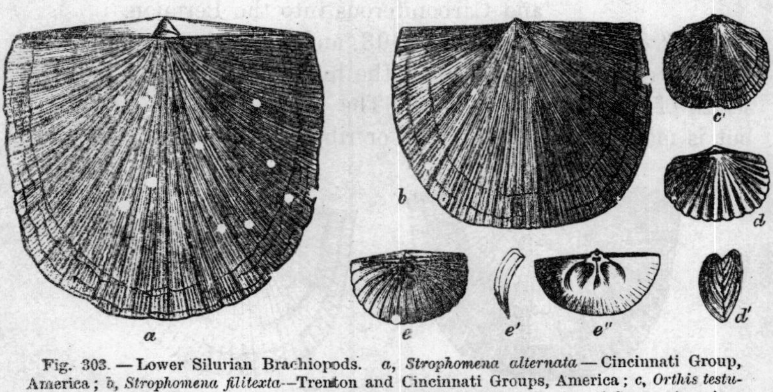

Fig. 303.—Lower Silurian Brachiopods. *a*, *Strophomena alternata*—Cincinnati Group, America; *b*, *Strophomena filitexta*—Trenton and Cincinnati Groups, America; *c*, *Orthis testudinaria*—Caradoc, Europe and America; *d*, *d'*, *Orthis plicatella*—Cincinnati Group, America; *e*, *e'*, *e''*, *Leptæna sericea*—Llandeilo-Caradoc, Europe and America. (After Meek and Hall, and original.)

generally be recognised from the closely related *Strophomenæ* by the fact that the shell is seldom flat, one valve being usually much more convex than the other, and its general form being compact and not extended; while the hinge-line is often shorter than the greatest width of the shell. The muscular scars are quadrate, and not extended either vertically or laterally. Lastly, the cardinal process"—that is, the projection of the dorsal valve to which the "cardinal muscles" are attached—is undivided and linear. Though the species of *Orthis* most nearly resemble certain forms of *Strophomena*, and are very liable to be confounded with these, one or two species (such as *Orthis biforata*, fig. 302, *a*) closely simulate the genus *Spirifera* in general form.

In the genus *Orthisina* (fig. 304) the shell nearly resembles that of *Orthis;* but there is a double hinge-area, largest in the ventral valve, the central fissures of which are always covered by a convex deltidium; whereas in the latter genus they are open. In some species (as in *O. Verneuili*) the deltidium is perforated by a foramen under the beak of the ventral valve. The typical species of *Orthisina* are Silurian; but the genus is stated by Mr Davidson to range through the Devonian and Carboniferous into the Permian.

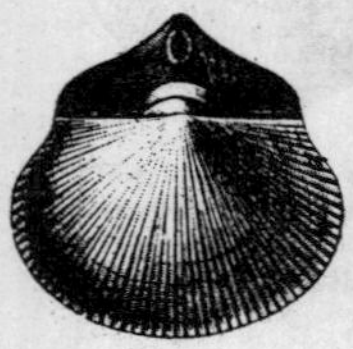

Fig. 304. — *Orthisina Verneuili*. Lower Silurian.

In *Strophomena* (figs. 302, 303, and 305) the shell is depressed, generally semicircular, the hinge-line as long as the width of the shell, or longer. The surface may be smooth, but is most commonly striated or ribbed. There is a double

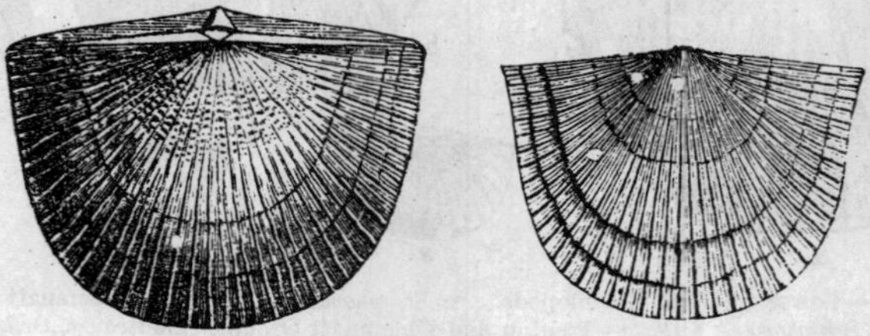

Fig. 305.—*Strophomena antiquata*. Silurian.

hinge-area, which is largest in the ventral valve. Each hinge-area has a median notch, which, in the ventral valve, is partially covered by a deltidium. The ventral valve may be convex or concave, and the dorsal valve follows the curvature of the ventral valve. The species of the genus *Strophomena* are very abundant in the Silurian, Devonian, and Carboniferous formations, often attaining a large size; but (except under a peculiar sub-generic type) they do not seem to have survived the close of the last-named period.

Speaking generally, the species of *Strophomena* may be distinguished from those of *Orthis* and *Leptæna*—both of which they much resemble occasionally—by attention to the following points: The shell is *usually* flat and semi-oval, its length and breadth being about equal, and the hinge-line always equalling and often exceeding in length the transverse diameter of the shell. Rarely, the shell is bent and transversely extended, as in *Leptæna*. The cardinal process is large and bifid; and the muscular impressions are quadrate and laterally expanded.

It may be appropriately added in this connection that the name *Strophomena* is employed above in the wide general sense in which it has been commonly used by most palæontologists. Certain forms of the genus, however, with twisted beaks to the ventral valve, and often a high false area, are commonly separated to form a distinct section (*Hemipronites* of Pander, or *Streptorhynchus* of King); and the name of *Strophodonta* has been given to others in which the hinge-line is crenulated, and there is no fissure in the hinge-area of the ventral valve.

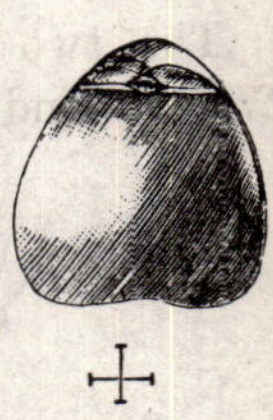

Fig. 306. — *Leptæna Liassica*. Jurassic. The small cross below the figure indicates the real size of the shell.

In the genus *Leptæna* are forms smaller than the *Strophomenæ*, but resembling them in many respects. The shell is more or less completely semicircular (figs. 306, 307), with a double hinge-area, notched in the centre, the fissure in the ventral valve being partly covered by a deltidium. The valves articulate by teeth and sockets, and the surface is generally striated. The *Leptænæ* extend from the Silurian rocks to the summit of the Lias, but are not known in any younger deposits. Speaking generally, *Leptænæ* can be separated from *Orthis* and *Strophomena* by the involute, transversely elongated shell, the valves being strongly bent, so that one is always very concave, and the other correspondingly convex. Moreover, the muscular impressions are elongated, and the lateral hinge-teeth are confounded in the dorsal or concave valve with the "cardinal process."

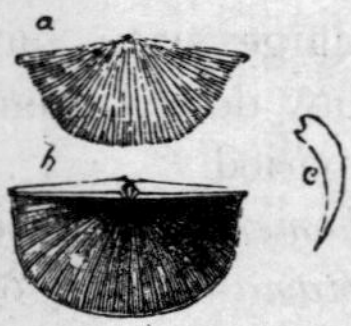

Fig. 307.—*Leptæna sericea*: *a*, Ventral valve; *b*, Dorsal valve; *c*, Section of the shell. Silurian.

Finally, the genus *Davidsonia* includes certain singular Brachiopods, ranging from the Devonian to the Trias, in which the shell is thick and solid, and attached to foreign bodies by the substance of the ventral valve. The ventral valve has a wide area, with a triangular fissure covered by a convex deltidium; and though there are no calcified brachial supports, the position of the arms is indicated by two spirally-grooved elevations in the interior of the valve.

FAM. VII. PRODUCTIDÆ.—Animal unknown. Shell entirely free, or attached to submarine objects by the substance of the beak or by means of spines, valves either regularly articulated, or kept in place by muscular action alone. No calcified supports for the oral processes. The *Productidæ* are exclusively Palæozoic, and are especially characteristic of the Devonian, Carboniferous, and Permian deposits.

The two most important genera of the *Productidæ* are *Chonetes* and *Producta*. In the genus *Chonetes* (fig. 308) the

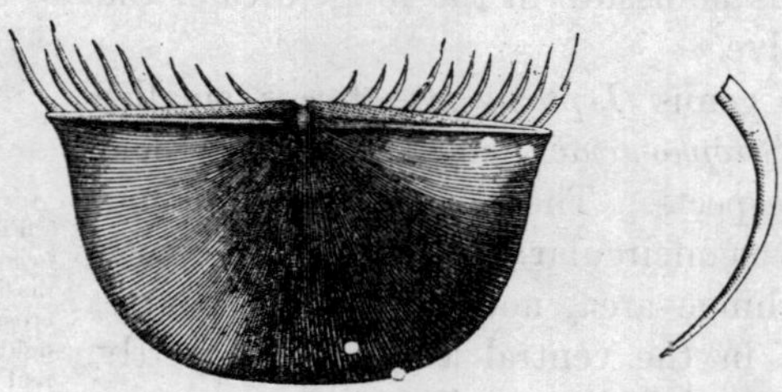

Fig. 308.—*Chonetes Dalmaniana*. Carboniferous.

shell is concavo-convex, transversely oblong, with a straight hinge-line. The hinge-line is as wide as the shell, or the shell is eared. The ventral valve is convex, the dorsal concave, and both have a distinct hinge-area, with a central fissure, closed in the ventral valve by a pseudo-deltidium. The upper edge of the hinge-area of the ventral valve is furnished with a row of delicate, tubular spines. The species of *Chonetes* are distributed in the Silurian, Devonian, and Carboniferous periods, but they are most abundant in the last of these.

In the genus *Producta* (figs. 309, 310) the valves are not articulated by any apparatus of teeth and sockets, and the shell would appear to have been generally free in the adult

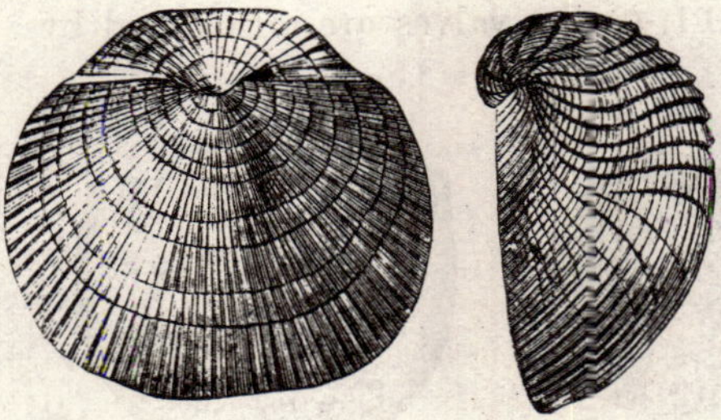

Fig. 309.—Dorsal and profile views of *Producta semireticulata*. Carboniferous.

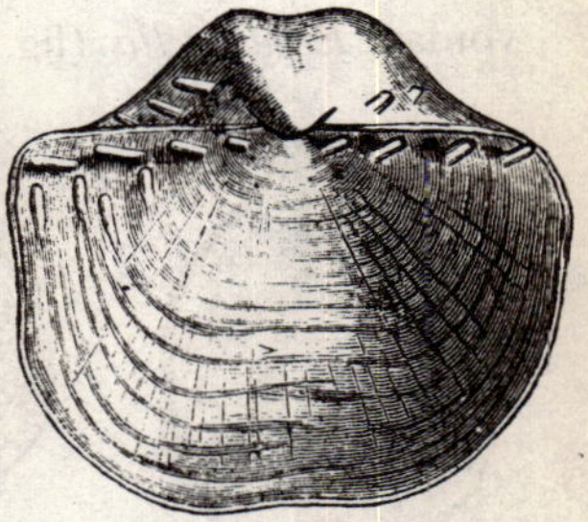

Fig. 310.—Dorsal aspect of *Producta horrida*. Permian. (After King.)

condition. In some cases, however, where the shell is furnished with long spines, we may believe that these structures served to moor the animal in the soft ooze of the sea-bottom. Mr R. Etheridge, jun., has also shown that in one species of the genus, at any rate, the shell was firmly fixed to one spot by the twisting of some of the spines of the ventral valve round the column of a Crinoid. The shell is generally transversely elongated, and is auriculate, or furnished with ear-like expansions. There is a straight hinge-line, usually shorter than the width of the shell. The hinge-area is rudimentary or wanting. The ventral valve is convex, the dorsal concave, following the curve of the former. The surface is ribbed or striated, and the ribs carry a greater or less number of longer or shorter tubular spines, which are especially abundant upon the auricular expansions. The species of *Producta* range from the Devonian to the Permian, but the genus is essentially and especially characteristic of the Carboniferous period.

The genus, or sub-genus, *Strophalosia* (fig. 311, B) ranges from the Devonian to the Permian, and is distinguished from *Producta* chiefly in the fact that the valves are not edentulous, but are articulated by teeth and sockets, while each valve has a distinct hinge-area, and the ventral valve has a foramen covered with a deltidium. *Aulosteges*, again (fig. 311, A), from the Permian, has no teeth or dental sockets —in this respect agreeing with *Producta*—but the ventral valve has a wide hinge-area, pierced by a foramen, which is covered by a convex pseudo-deltidium. Lastly, in the Devonian *Productella* (fig. 311, C) the valves are articulated by

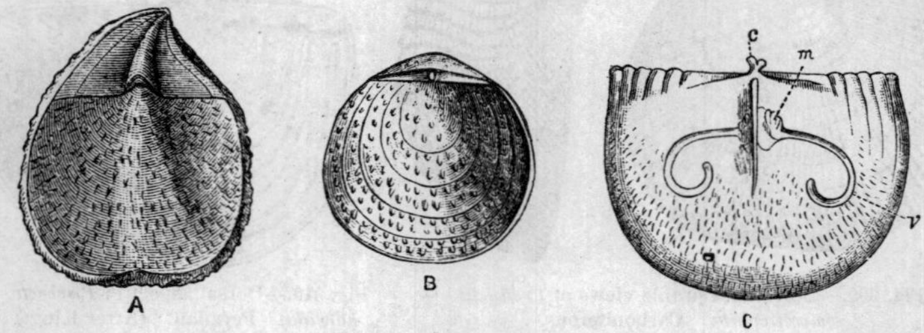

Fig. 311.—A, *Aulosteges Wangenheimii*—Permian, showing the hinge-area and deltidium; B, *Strophalosia Goldfussi*, viewed dorsally—Permian; C, *Productella obtusta*—Devonian—interior of the dorsal valve, showing the cardinal process (*c*), the muscular scars (*m*), and the reniform vascular impressions (*v*). (After Davidson and Hall.)

teeth and sockets, and a hinge-area is present in both valves, but the latter is narrow and linear, and the ventral valve is extremely convex and gibbous. All the three groups just mentioned are probably to be regarded as really nothing more than sub-genera of *Producta* itself.

FAM. VIII. CRANIADÆ.—Animal fixed to submarine objects by the substance of the ventral valve. Arms fleshy and spirally coiled. No hinge or articulating processes; upper or dorsal valve limpet-shaped (fig. 312). The family includes the single genus *Crania*, which commenced in the Silurian period, and has continued to exist without interruption to the present day. The shell may be smooth, or striated with radiating ribs, sometimes with spines or foliaceous expansions. The ventral valve is the shallowest, and the shell is usually attached by a portion of its substance.

The dorsal valve is more or less conical, and the valves are simply kept in apposition by muscular action.

FAM. IX. DISCINIDÆ.—Animal attached by means of a muscular peduncle passing through the ventral or lower

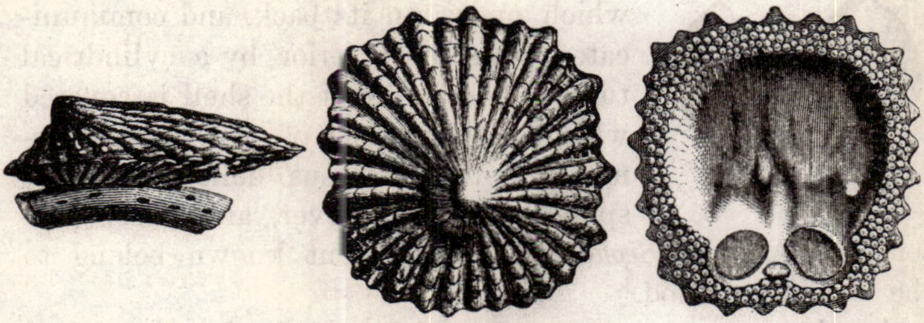

Fig. 312.—*Crania Ignabergensis.* Cretaceous.

valve, by means of a slit in its hinder portion or a circular foramen excavated in its substance. Arms fleshy. Valves unarticulated. The *Discinidæ* range from the Silurian period to the present day. The four most important genera are *Discina*, *Trematis*, *Siphonotreta*, and *Acrotreta*.

Fig. 313.—*Discina Circe.* Silurian.

In *Discina* (figs. 313, 314) the shell is generally circular or orbicular in shape, and is horny in texture. The upper valve is limpet-shaped, smooth or concentrically striated; the ventral valve is flat or partly convex, perforated by a longitudinal slit, which is placed in the middle of an oval depressed disc. The valves are not articulated to one another, but are kept together by muscular action alone. The species of the genus *Discina* (including *Orbiculoidea*) range from the Silurian rocks to the present day, seven species being now living.

Fig. 314.—*Discina Pelopea.* Silurian.

In the genus *Trematis* (fig. 316, A) both valves are more or less convex, and the general shape of the shell is more or less oval or sub-orbicular. The ventral valve is furnished with a marginal slit for the passage of a peduncle of attachment. The surface is radiately striated, and ornamented with small pits. This genus seems to be exclusively Silurian.

In the genus *Siphonotreta* (fig. 315) the shell is oval, inequivalve, with unarticulated valves. The beak of the ventral valve is perforated by a foramen which opens on its back, and communicates with the interior by a cylindrical tube. The surface of the shell is covered with concentric lines of growth, and furnished with numerous delicate tubular spines, which, however, are rarely preserved. All the *Siphonotretæ* at present known belong to the Silurian period.

Fig. 315.—*Siphonotreta verrucosa.* Silurian.

In the genus *Acrotreta* (fig. 316, E-G), lastly, the dorsal valve is nearly flat, and the ventral is conical and patelli-

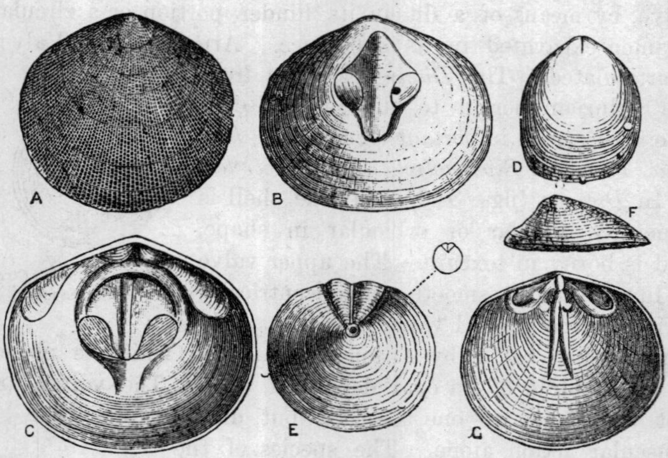

Fig. 316.—A, *Trematis filosa*—Lower Silurian; B, Cast of *Obolus Davidsoni*—Upper Silurian; C, Interior of the dorsal valve of the same; D, *Lingulella Davisii*—Upper Cambrian; E, Ventral valve of *Acrotreta Nicholsoni*, of the natural size and enlarged—Lower Silurian; F, Side view of the ventral valve of the same, enlarged; G, Interior of the dorsal valve of the same enlarged. (A is after Billings, and the other figures after Davidson.)

form, the surface being concentrically striated. The beak of the ventral valve is perforated by a round foramen, from which a pedicle-groove extends to the posterior margin, and is bounded laterally on either side by a small false area. The genus is wholly Lower Silurian.

FAM. X. LINGULIDÆ.—Animal fixed by a muscular peduncle passing out between the beaks of the valves. Arms fleshy, not supported by calcified processes. Shell unarticulated, sub-equivalve, of a horny texture, and partially phosphatic in composition. The *Lingulidæ* range, in the person of *Lingula* itself, from the Cambrian period to the present day.

In the genus *Lingula* (fig. 317) the shell is oblong, compressed, the dorsal valve little smaller than the ventral. The shell is oval, rounded, or satchel-shaped, tapering more or less towards the beaks. The surface is concentrically striated with lines of growth. The genus commences to be represented in the Cambrian rocks, and has continued without interruption, and with no perceptible change, to the present day.

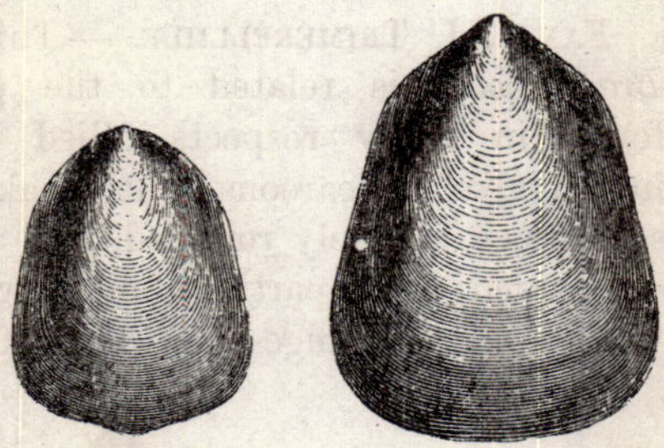

Fig. 317.—*Lingula Eva.* Lower Silurian. Dorsal and ventral valves. (After Billings.)

The genus or sub-genus *Lingulella* (fig. 316, D) differs from *Lingula* proper in the fact that there is a distinct groove for the passage of the peduncle. This type comprises the oldest of British Brachiopods, *L. ferruginea* being a small form which occurs low down in the Lower Cambrian. Another species—viz., *L. Davisii*—is extremely abundant in the so-called "Lingula Flags" (Upper Cambrian).

In *Obolus* (fig. 316, B) the valves are orbicular, subequal, smooth, the ventral valve having a longitudinal furrow for the passage of fibres of attachment, which pass out between the beaks. The valves are unarticulated, and are maintained in apposition by muscular action. All the known species of *Obolus* are confined to the Silurian period, and are especially characteristic of the Lower Silurian period. The genus *Obolella*, of the Cambrian and Lower Silurian, is nearly allied to *Obolus*, but the arrangement of the muscular impressions is different. Lastly, *Kutorgina* (fig.

318), of the Upper Cambrian, resembles *Obolella* in many respects, but possesses a straight hinge-line.

Fig. 318.—*Kutorgina cingulata.* (Billings.) Upper Cambrian.

FAM. XI. TRIMERELLIDÆ. — This, the last family of the *Brachiopoda*, is related to the preceding, and comprises forms in many respects allied to *Obolus*. The shell is inequivalve, calcareous, the beaks usually prominent, or sometimes obtusely rounded, and either massive and solid, or divided by a partition into two chambers. There is a well-developed hinge-area (fig. 320), and a wide deltidium,

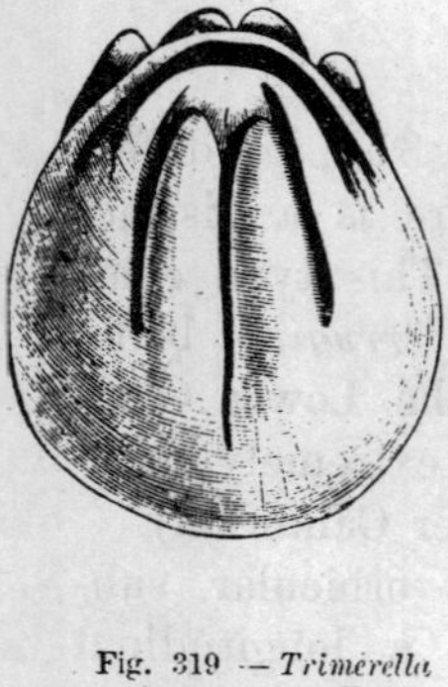

Fig. 319 — *Trimerella acuminata*—cast. Upper Silurian. (After Davidson and King.)

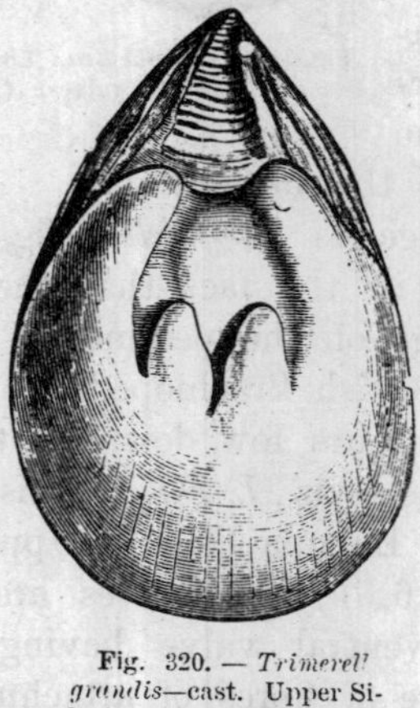

Fig. 320. — *Trimerella grandis*—cast. Upper Silurian. (After Davidson and King.)

Fig. 321. — *Monomerella prisca*—cast. Upper Silurian. (After Davidson and King.)

bounded by two ridges, the inner ends of which serve as teeth, though true teeth are not present. Each valve is furnished with muscular platforms, which, in the typical species, are elevated and doubly vaulted. The principal, or only, genera of the family are *Trimerella*, *Dinobolus*, and

Monomerella, all of which are confined to the Lower and Upper Silurian rocks.

The genera of this family are especially characteristic of parts of the Upper Silurian, and especially of the so-called "Guelph Formation" of North America. The differences between them are founded mainly upon internal characters, and as specimens usually occur in the state of casts of the interior, their study is attended with unusual difficulties.

CHAPTER XXIII.

LAMELLIBRANCHIATA.

THE *Lamellibranchiata*, *Pelecypoda*, or Bivalve Shell-fish, are distinguished by the fact that *the head is not distinct, and the mouth is destitute of any apparatus of teeth. The body is more or less completely protected by a bivalve shell, composed of two, usually symmetrical, pieces or valves. There are generally two leaf-like lamellar gills upon each side of the body.*

The Lamellibranchs include all the ordinary Bivalve Shell-fish, such as Oysters, Cockles, Mussels, and the like, and they are all either marine or inhabitants of fresh water.

Though they agree with the *Brachiopoda* in possessing a shell which is composed of two pieces or valves, there are, nevertheless, many points in which the shell of a Lamellibranch is distinguishable from that of a Brachiopod, irrespective of the great difference in the structure of the animal in each. The shell in the *Brachiopoda*, as we have seen, is rarely or never quite equivalve, and always has its two sides equally developed (equilateral); whilst the valves are placed antero-posteriorly as regards the animal, one in front and one behind, so that they are "dorsal" and "ventral." In the *Lamellibranchiata*, on the other hand, the two valves are usually of nearly equal size (equivalve), and are more developed on one side than on the other (inequilateral); whilst their position as regards the animal is always *lateral*, so that they are properly termed "right" and "left" valves, instead of "ventral" and "dorsal."

It is to be remembered, however, that many of the Bivalves, such as the Oysters, habitually lie on one side, in which case the valves, though really right and left, are called "upper" and "lower." It is to be borne in mind also that the two valves, especially in the attached Bivalves, may be very unsymmetrical, one valve being much larger or deeper than the other. Lastly, there are some cases in which the shell becomes very nearly equilateral, the line drawn from the beaks to the base dividing the shell into two almost equal halves.

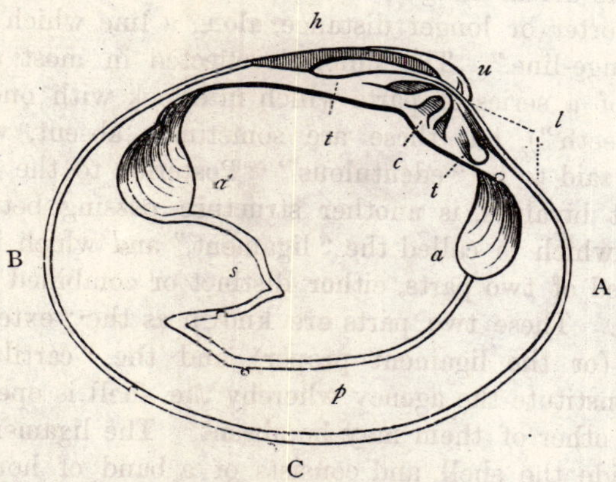

Fig. 322.—Left valve of *Cytherea chione* (after Woodward). A, Anterior margin; B, Posterior margin; C, Ventral margin or base. *u*, Umbo; *h*, Ligament; *l*, Lunule; *c*, Cardinal tooth; *t*, *t*, Lateral teeth; *a*, Anterior adductor; *a'* Posterior adductor; *p*, Pallial line; *s*, Pallial sinus caused by the retractor muscles of the siphons.

The following are the chief points to be noticed in connection with the shell of any Lamellibranch: Each valve of the shell may be regarded as essentially a hollow cone, the apex of which is turned more or less to one side; so that more of the shell is situated on one side of the apex than on the other. The apex of the valve is called the "umbo," or "beak" (fig. 322, *u*), and is almost always turned towards the mouth of the animal. Consequently the side of the shell towards which the umbones are turned is the "anterior" side, and it is usually the shortest half of the shell. In some Bivalves, however, the beaks are "reversed," and are turned towards the posterior side of the shell. The longer half of the shell, from which the umbones turn away, is called the "posterior" side, but in some cases this is equal to, or even shorter than, the anterior side. The side of the shell where the beaks are situated, and where the valves are united to one another, is called the "dorsal" side; and the opposite margin, along which the shell opens, is called the "ventral" side, or "base." The *length* of the shell is measured from its anterior to its posterior margin, and its *breadth* from the dorsal margin to the base.

At the dorsal margin the valves are united to one another for a shorter or longer distance, along a line which is called the "hinge-line." The union is effected in most shells by means of a series of parts which interlock with one another (the "teeth"), but these are sometimes absent, when the shell is said to be "edentulous." Posterior to the umbones, in most bivalves, is another structure passing between the valves, which is called the "ligament," and which is usually composed of two parts, either distinct or combined with one another. These two parts are known as the "external ligament" (or the ligament proper), and the "cartilage," and they constitute the agency whereby the shell is opened; but one or other of them may be absent. The ligament proper is outside the shell, and consists of a band of horny fibres, passing from one valve to the other just *behind* the beaks, in such a manner that it is put upon the stretch when the shell is closed. The cartilage, or internal ligament, is lodged between the hinge-lines of the two valves, generally in one or more "pits," or in special processes of the shell. It consists of elastic fibres placed perpendicularly between the surfaces by which it is contained, so that they are necessarily shortened and compressed when the valves are shut. To open the shell, therefore, it is simply necessary for the animal to relax the muscles which are provided for the closure of the valves, whereupon the elastic force of the ligament and cartilage is sufficient of itself to open the shell.

The hinge-line is mostly curved, but it may be quite straight. Generally the beaks are more or less contiguous, but they may be removed from one another to a greater or less distance, and in some anomalous forms they are not near one another at all. In the *Arcadæ* the two beaks are separated from one another by an oval or lozenge-shaped flat space or area. When teeth are present, they differ much in their form and arrangement. In some forms (fig. 322) the teeth are divisible into three sets—one group, of one or more teeth, placed immediately beneath the umbo, and known as the "cardinal teeth;" and two groups on either side of the preceding, termed the "lateral teeth." Sometimes there may be lateral teeth only; sometimes the cardinal teeth alone are present; and in some cases (*Arcadæ*) there is a row of similar and equal teeth.

In the interior of the shell of the Bivalves are found certain markings which are often of great importance to the palæontologist. The body is enclosed in an expansion of the dorsal integument, which constitutes the "mantle" or "pallium," whereby the shell is secreted. Towards its circumference the mantle is more or less completely united to the shell, leaving in its interior, when the soft parts are removed, a more or less distinctly impressed line, which is called the "pallial line" or "pallial impression" (fig. 323, *a*). In some of the Bivalves the two halves or "lobes" of the mantle are united at their margins, so that the animal is enveloped in an almost closed sac. In these cases it is necessary that there should be orifices in the mantle-sac by which water can be admitted to the gills, and can be expelled again from the body. The margins or lips of these orifices are usually drawn out or extended into longer or shorter muscular tubes, which are termed the "siphons," and which may be either separate, or may be united to one another along one side. The Bivalves which possess these siphons are said to be "siphonate," and there are two leading modifications in the arrangement of these tubes. In the Siphonate Bivalves which spend their existence buried in sand or mud, as well as in many other cases, the siphons are long, and can be partially or entirely retracted within the shell by means of special muscles, called the "retractor muscles of the siphons." In these cases, the pallial line does not run in an unbroken curve, but is deflected inwards posteriorly, so as to form an indentation or bay, which is termed the "pallial sinus" (fig. 323, 2). The presence, therefore, of *an indented pallial line shows that the animal possessed retractile siphons.* In other Bivalves the respiratory siphons are of small size, and are destitute of retractor muscles, so that they cannot be withdrawn within the shell. In these cases the "pallial line," or the impression caused by the attachment of the muscular border of the mantle, is unbroken in its curvature, and presents no indentation (fig. 323, 1). In another group of the Bivalves there are no respiratory siphons at all, and the mantle-lobes are free, and are not united to one another at their edges. In these cases also,

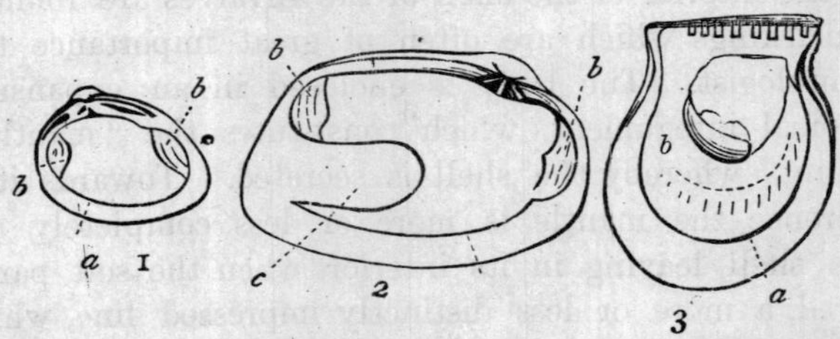

Fig 323.—Shells of Lamellibranchiata. 1, *Cyclas amnica*, a dimyary shell with an entire pallial line. 2, *Tapes pullastra*, a dimyary shell with an indented pallial line. 3, *Perna ephippium*, a monomyary shell (after Woodward). *a*, Pallial line; *b*, Muscular impressions left by the adductors; *c*, Siphonal impression.

the pallial line is unbroken or "simple." When, therefore, we find a Bivalve shell in which *the pallial line is not indented* by a sinus, we know that the animal which inhabited the shell either *possessed no siphons, or that if siphons were present, they were small and not retractile.*

In accordance with these considerations, the *Lamellibranchiata* are divided into two sections, according as respiratory siphons are present or absent, and according to their nature when they exist.

SECTION A. ASIPHONIDA. —Animal without respiratory siphons; mantle-lobes free; the pallial line simple and not indented (*Integro-pallialia*).

This section comprises the families *Ostreidæ, Aviculidæ, Mytilidæ, Arcadæ, Trigoniadæ,* and *Unionidæ.*

SECTION B. SIPHONIDA.—Animal with respiratory siphons; mantle-lobes more or less united.

Two subdivisions are comprised in this section. In the first *the siphons are short, and the pallial line is simple* (*Integropallialia*); as is seen in the families *Chamidæ*, *Hippuritidæ*, *Tridacnidæ*, *Cardiadæ*, *Lucinidæ*, *Cycladidæ*, and *Cyprinidæ*.

The second subdivision (*Sinu-pallialia*) is distinguished by the possession of *long respiratory siphons*, and *a sinuated pallial line*, and it comprises the families *Veneridæ*, *Mactridæ*, *Tellinidæ*, *Solenidæ*, *Myacidæ*, *Anatinidæ*, *Gastrochænidæ*, and *Pholadidæ*.

Besides the impressions left by the muscular border of the mantle, and by the retractor muscles of the siphons, when these are present, there are other impressions caused by the insertion into the shell of the muscles by which the valves are brought together—the "adductor muscles." The number of adductor muscles never exceeds two, but there may be only one; and in accordance with this distinction the Bivalves have been divided into the two groups of the *Dimyaria* and *Monomyaria*. These divisions, however, are not of very high value. In most Bivalves there are two adductor muscles passing between the inner surfaces of the valves, one being placed anteriorly, in front of the mouth, whilst the other is situated posteriorly, in the neighbourhood of the vent. In the Monomyary Bivalves it is the posterior adductor which remains, and the anterior adductor is absent. The adductors leave distinct "muscular impressions," or scars, in the interior of the shell, so that it is easy in any given specimen to determine where there was only one adductor, or whether two were present (see fig. 323).

The habits of the *Lamellibranchiata* are very various. Some, such as the Oyster (*Ostrea*), and the Scallop (*Pecten*), habitually lie on one side, the lower valve being the deepest, and the foot being wanting, or rudimentary. Others, such as the Mussel (*Mytilus*), and the *Pinna*, are attached to some foreign object by an apparatus of threads, which is called the "byssus," and is secreted by a special gland, and for the passage of which a special notch or aperture may exist. Others are fixed to some solid body by the substance of one of the valves. Many, such as the *Myas*, spend their existence sunk in the sand of the sea-shore or in the mud of estuaries. Others, as the *Pholades* and *Lithodomi*, bore holes in rock or wood, in which they live. Finally, many are permanently free and locomotive.

As regards the general *distribution in time* of the *Lamellibranchiata*, the class seems to have commenced in the Lower Silurian rocks, and to have steadily increased up to the present day, when it seems to have attained its maximum, both as regards numbers and as regards variety of type. The recent Bivalves are also superior in organisation to those which have preceded them. In the Palæozoic and earlier Secondary deposits the Bivalves belong mainly to the group of the *Asiphonida*, in which there are no respiratory siphons. In the later Secondary and Tertiary rocks, on the other hand, there is a predominance of *Siphonate* Bivalves, in which the mantle-lobes are united and there are respiratory siphons. Upon the whole, the *Lamellibranchiata* are sparingly represented in the Lower Silurian, more abundant in the Upper Silurian, reduced in numbers in the Devonian, very plentiful in the Carboniferous, scanty in the Permian and Trias, profusely represented in the Jurassic rocks, and very abundant in the Cretaceous and Tertiary periods (Lobley). In the Carboniferous rocks the family of the *Aviculidæ* is especially abundant. One very singular and aberrant family—viz., the *Hippuritidæ*—is exclusively confined to the Secondary period, and is not known to occur out of the limits of the Cretaceous formation. The *Veneridæ*, which are perhaps the most highly organised of the *Lamellibranchiata*, appear for the first time in the Oolitic rocks, and, increasing in the Tertiary period, have culminated in the Recent period. The remains of *Lamellibranchiata* are very abundant in many formations, and are of great palæontological importance. It will therefore be well to review the families[1] of the class briefly, giving the leading characters, more important genera, and geological distribution of each.

[1] In the following synoptical view of the *Lamellibranchiata*, the classification adopted in Woodward's admirable 'Manual of the Mollusca' has been mainly followed.

SECTION A.—ASIPHONIDA.

FAM. 1. OSTREIDÆ.—Shell inequivalve, slightly inequilateral, free or attached; hinge usually edentulous. Ligament internal. Lobes of the mantle entirely separated; the foot small and byssiferous, or wanting. A single adductor muscle.

In the typical Oysters, forming the genus *Ostrea* (figs. 324,

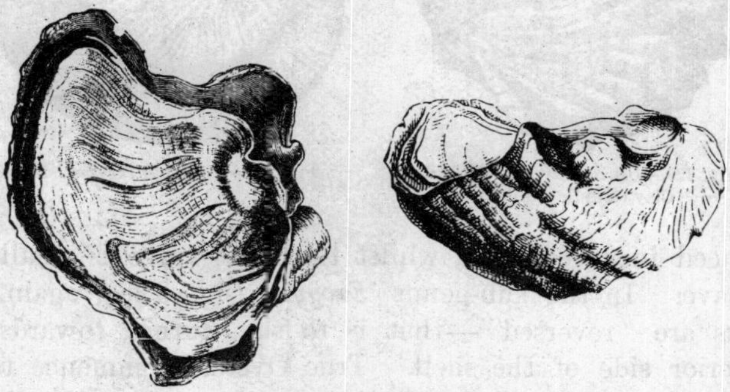

Fig. 324.—*Ostrea Couloni*. Lower Greensand.

325), the shell is irregular, and is attached by the left valve, which is also convex, and has a well-marked beak. The upper valve is generally flat or concave, and is the smallest of the two valves. The hinge is toothless, with a single

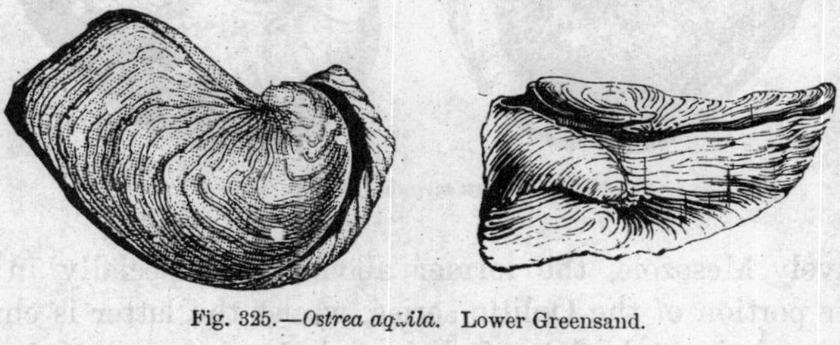

Fig. 325.—*Ostrea aquila*. Lower Greensand.

ligamental pit. Both valves may be more or less completely plain, and the upper one especially often is so. The lower valve, however, is commonly plaited, and both valves are sometimes thus ornamented, as in *Ostrea Marshii* of the Oolites (fig. 326).

In the sub-genus *Gryphæa* are included Oysters which were either quite free or very slightly attached. The left or lower valve (fig. 327) is much the largest, and has a very pro-

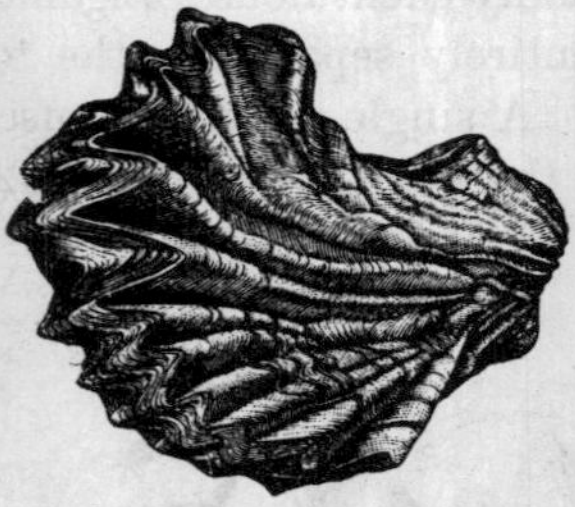

Fig. 326.—*Ostrea Marshii.* Oxford Clay (Middle Oolites).

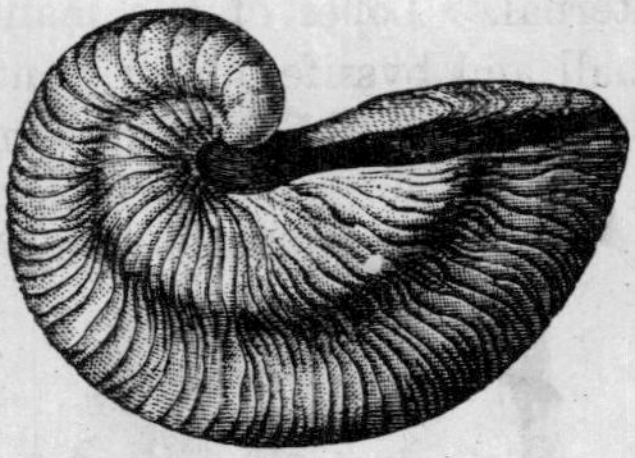

Fig. 327.—*Gryphæa incurva.* Lias.

nounced incurved beak, whilst the right valve is small and concave. In the sub-genus *Exogyra* (fig. 328), again, the beaks are "reversed"—that is to say, turned towards the posterior side of the shell. True Oysters commence to be represented in the Carboniferous seas, abound in the Secondary and Tertiary periods, and are very plentiful at the present day. The sub-genera *Gryphæa* and *Exogyra* are ex-

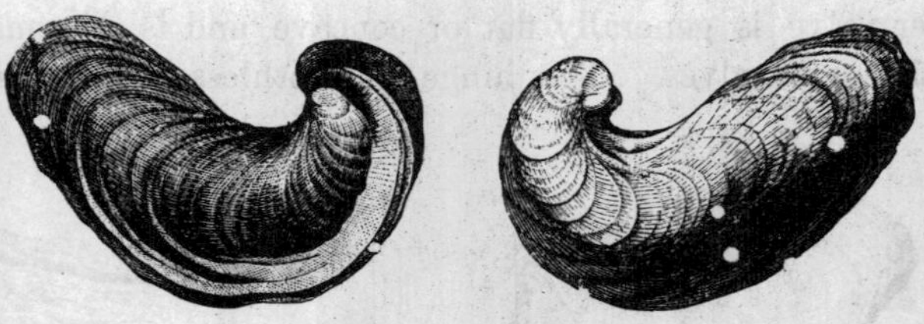

Fig. 328.—*Exogyra virgula.* Upper Oolites.

clusively Mesozoic, the former abounding especially in the lower portion of the Oolitic series, whilst the latter is chiefly characteristic of the later Oolitic and Cretaceous deposits.

Lastly, the genus *Pernostrea* has been founded for Jurassic Oysters, which differ from *Ostrea* proper in having the ligament contained in from four to eight transverse grooves or pits (as in *Perna*).

In a second section or sub-family (*Anomiadæ*) of the present family we have various forms, of which *Anomia* itself is the type. This genus comprises Ostreiform Bivalves, with a thin and translucent shell, which is fixed to some solid body by a plug which passes through a hole or notch in the right valve. The first appearance of *Anomia* seems to be in the Cretaceous rocks; but the *Limanomia* of the Devonian, the *Anomianella* of the Carboniferous, and the *Placunopsis* of the Jurassic, are earlier forms of the same type. The genus *Placuna*, dating from the Tertiary (?), comprises flattened and cake-like Oysters, allied in many respects to *Anomia*, but having the shell unattached.

A third group is that of the *Pectinidæ*, often ranked as a distinct family, and typified by the familiar genus *Pecten*, comprising the Scallops. In this genus (fig. 329) the shell

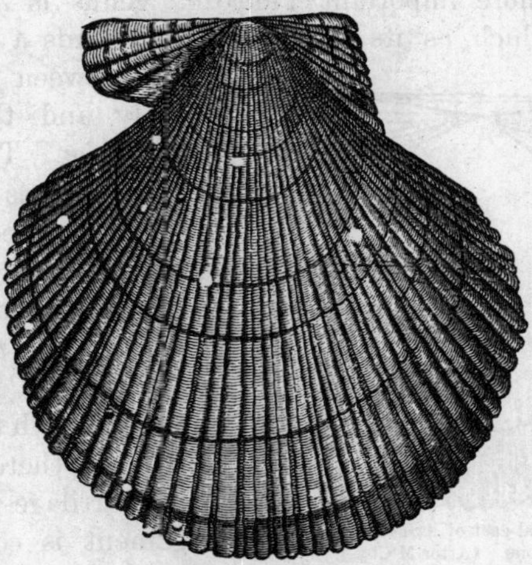

Fig. 329.—*Pecten Islandicus*, left valve. Post-Tertiary and Recent.

rests upon the right valve, and the beaks are furnished with ears. The anterior ears are usually the largest and most prominent, and the shell is generally furnished with ribs radiating from the umbos. There is a single, median, triangular cartilage-pit. The right valve is the deepest, and is notched below the anterior ear. Using the name *Pecten* in its modern and restricted signification, it is probable that all the Palæozoic shells so named are really referable to allied genera (*Aviculopecten* and *Pernopecten* more especially). The genus, however, is largely represented both in the Secondary and Tertiary deposits, and exists under numerous and varied forms at the present day. Closely allied to *Pecten* is the genus *Hinnites*, in which, however, the shell is attached, during the later stages of its life, to foreign bodies by the substance of the right valve. It still exists, and is represented by various Secondary and Tertiary species.

In the genus *Pernopecten*, of the Devonian and Carboniferous, we have a group of Scallops, with small, nearly equal, ears, the central triangular cartilage-pit being flanked by a row of smaller pits on each side (thus approaching *Perna*), and the surface being nearly or quite devoid of radiating ridges.

A much more important Palæozoic genus is *Aviculopecten* (fig. 330), which, as its name implies, affords a transitional link between the present family and that of the *Aviculidæ*. The shell in this genus has the general form and aspect of that of *Pecten* itself; but the anterior ear is flattened, and *smaller than the posterior one*. There is a byssal notch beneath the anterior ear; but there is no median cartilage-pit, and the ligament is confined to a narrow facet along the hinge-margin. The muscular impression and pallial line are as in *Pecten*. The species of *Aviculopecten* are distributed between the Devonian and Permian, but they are most characteristic of the Carboniferous rocks, in which they are extremely abundant. In the absence of a median cartilage-pit, and the lodgment of the ligament in a groove along the hinge-line, the genus approaches the *Aviculidæ*, but its shell is stated by Meek to have the corrugated and laminated structure of *Pecten*, and not the prismatic structure of the

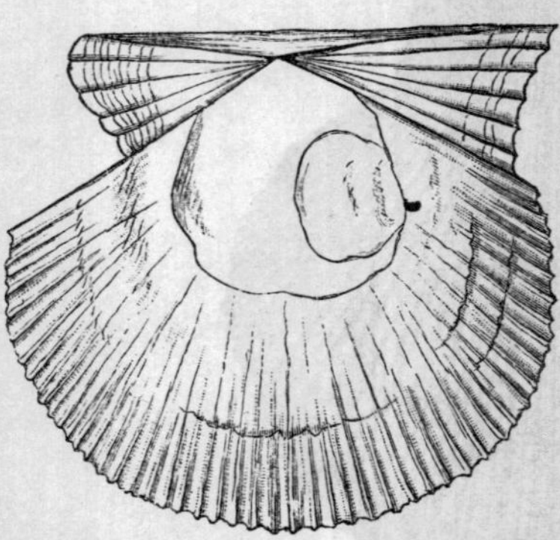

Fig. 330.—Internal cast of *Aviculopecten*. Carboniferous. (After M'Coy.)

former. The genera *Entolium*, of the Carboniferous, and *Streblopteria*, of the Carboniferous and Permian, comprise forms more or less closely related to the preceding.

A fourth group of the *Ostreidæ*, also often regarded as a distinct family (*Limadæ*), is that typified by the genus *Lima* or *Radula*. In this genus the shell is equivalve and free, and the beaks are separated from one another and eared (fig. 331). The surface is generally adorned with radiating ribs, and there is a median cartilage-pit, and a triangular hinge-area. The genus (including *Plagiostoma*) appears to occur in the Carboniferous, is abundantly represented in the Triassic, Jurassic, Cretaceous, and Tertiary rocks, and exists in diminished numbers at the present day. *Limæa*, ranging from the Trias to the Recent period, differs from *Lima* principally in having the hinge-line closely toothed; and the *Limoptera* of the Devonian may be an early representative of this group.

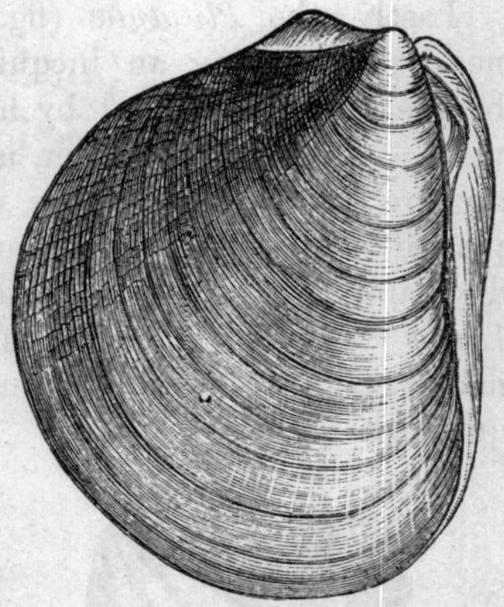

Fig. 331.—*Lima (Plagiostoma) gigantea.* Lias.

A fifth group of the *Ostreidæ*, sometimes raised to the rank of a distinct family (*Spondylidæ*), is that represented by the Thorny Oysters, forming the genus *Spondylus*. In this

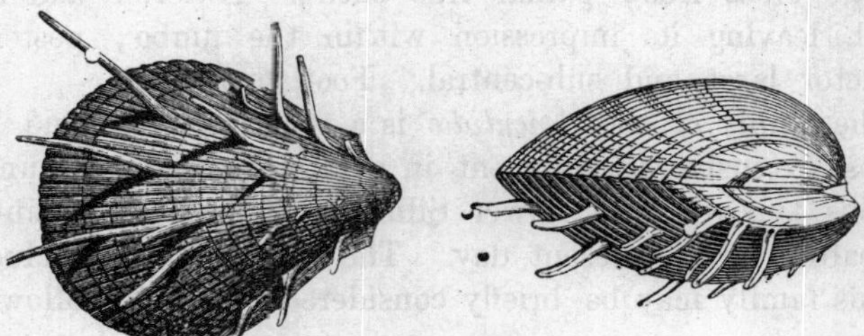

Fig. 332.—*Spondylus spinosus.* Chalk.

genus (fig. 332) the shell is inequivalve, and is fixed by the right valve to some foreign body. The beaks are apart and eared, and the shell is covered with spines, foliaceous expansions, or ribs radiating from the beak. The lower valve has a triangular hinge-area, and there are two teeth in each valve. The *Spondyli* seem to have commenced in the Jurassic period, are abundant in the Cretaceous, and have continued through the Tertiary period to the present day.

Lastly, the *Plicatulæ* (fig. 333) approach the *Spondyli* nearly, by having an inequivalve shell, which is attached by the right valve, and by having two hinge-teeth in each valve. The shell, however, is rarely eared, the hinge-area is obscure, and the valves are not spiny, though they may be plaited. The *Plicatulæ* extend from the Trias to the present day, and they abound to such an extent in parts of the Lower Greensand (Cretaceous), as to have given rise to the name of "Argile à Plicatules" applied to the beds in question.

Fig. 333.—*Plicatula placunea.* Lower Greensand.

FAM. 2. AVICULIDÆ.—Shell inequivalve, very oblique, attached by a byssus; hinge nearly or quite edentulous. Mantle-lobes free; pallial line entire. Anterior adductor small, leaving its impression within the umbo; posterior adductor large and sub-central. Foot small.

The family of the *Aviculidæ* is a very large one, and has a most extensive development in past time, beginning under various types in the Lower Silurian, and being continued thereafter to the present day. The various types included in this family may be briefly considered under the following groups:—

Firstly, we have the great group of shells of which *Avicula* itself (fig. 334) is the type. In this genus the shell is oblique and very inequivalve. The right valve has a notch under the anterior ear, and the hinge has one or two cardinal teeth, sometimes with an elongated posterior tooth. How far many of the Palæozoic shells which have been referred to *Avicula* can be regarded as truly referable to this genus is doubtful; but numerous species have been quoted from the Silurian upwards, and undoubted representatives of this type are abundant in the Secondary and Tertiary rocks. *Aucella*, of the Permian and Secondary deposits, resembles *Avicula*, but the anterior ear of the left valve is obsolete. In the genus *Monotis* (fig. 334 D), of the Permian and Triassic rocks, the shell is like that of *Avicula* in general form; but the anterior side is short and rounded, and the posterior is slightly eared. The curious genus *Posidonomya* fig. 334, F), ranging from the Silurian to the top of the Palæozoic rocks, comprises singular, thin, concentrically-

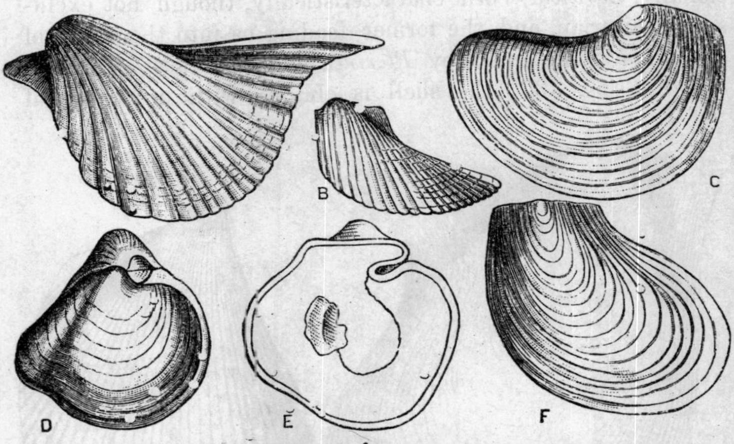

Fig. 334.—Types of *Aviculidæ*. A, *Avicula Cottaldina*—Cretaceous; B, *Avicula contorta*—Upper Trias; C, *Vulsella falcata*—Eocene; D, *Monotis speluncaria*—Permian; E, Interior of the shell of the same, showing the adductor impression and pallial line; F, *Posidonomya Bechei*—Devonian.

striated shells, with a straight hinge-line, and a general resemblance in shape to the valves of *Estheria*. The hinge is toothless, and no ears are developed. Closely allied to the preceding, and to one another, are the genera *Halobia* and *Daonella* (fig. 335), of the Triassic period, in both of which are comprised flat, equivalve, radiately-striated shells, slightly inequilateral, the hinge-line straight, without a ligamental pit or teeth, and with no ears. *Pseudomonotis* (*Eumicrotis*), of the Carboniferous and Permian, though in some respects allied to *Monotis*, has a very inequivalve shell, which is hardly at all oblique; while the Silurian *Amphicœlia* is more nearly allied to the forms immediately to be mentioned, the shell being nearly equivalve, eared, and having a triangular cartilage-pit beneath the beaks. We must also associate here the genera *Ambonychia*, and, provisionally, *Cardiola*, both characteristically, though not exclusively, Silurian, and the former leading us into the group of *Aviculidæ* represented by *Pterinea* and its allies. In *Ambonychia* (fig. 337) the shell is oblique, with nearly equal

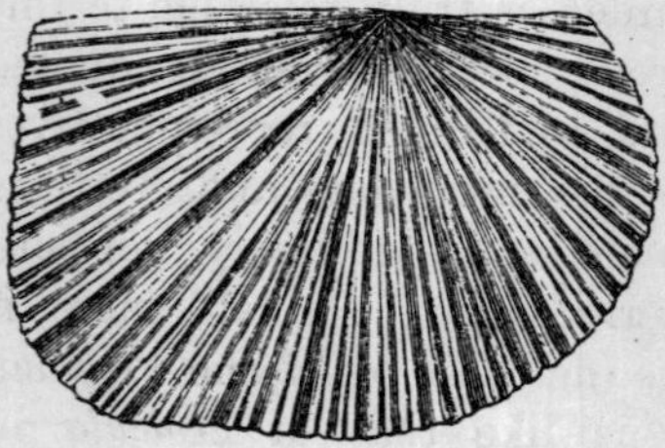

Fig. 335.—*Daonella* (*Halobia*) *Lommelli*. Trias.

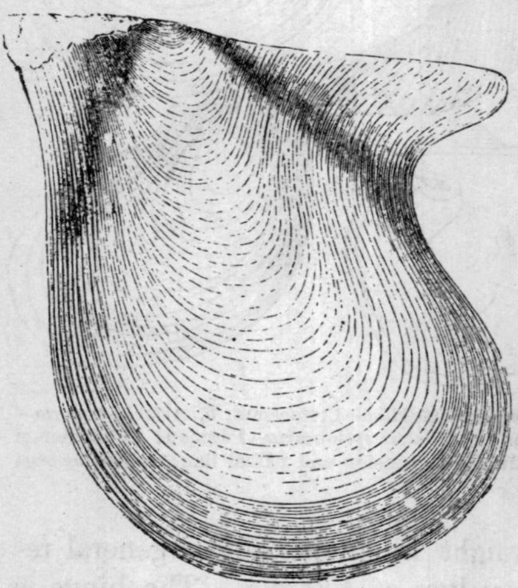

Fig. 336.—*Pterinea* (*Avicula*) *demissa*. Lower Silurian.

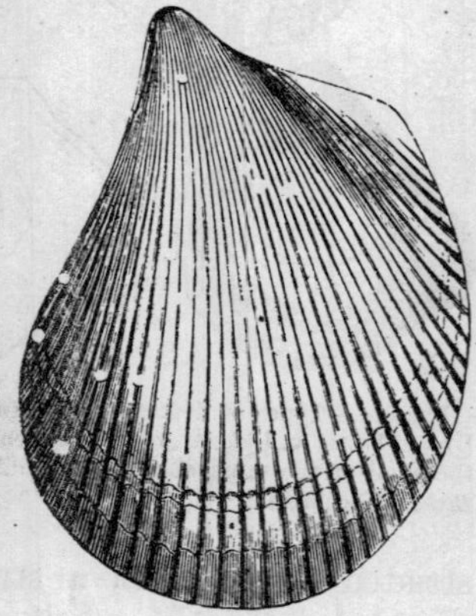

Fig. 337.—*Ambonychia radiata*. Lower Silurian.

convex valves, winged posteriorly, the anterior ears being nearly obsolete, and having the surface generally ornamented with prominent radiating ribs. The genus *Cardiola* (fig. 338, A and C) occupies an uncertain position, though the type-species—*C. interrupta*, of the Upper Silurian—would appear to be properly referable to the *Aviculidæ*. The genus, however, has been referred to the *Arcadæ*; and it is questionable how many of the forms so-called truly belong here. The shell in the type-form is oblique and equivalve, with prominent beaks, and radiating surface-ribs, both valves being convex, and the general appearance, as the name implies, reminding one of the Cockles (*Cardium*).

Closely related to the group we have been considering is another section of the *Aviculidæ*, of which *Pterinea* (figs. 336, 338, B) is the type. In this genus the shell resembles that of *Avicula* in being eared and oblique, but it is nearly equivalve, and the long straight hinge-line carries a few anterior radiating teeth along with elongated and oblique posterior teeth. The genus ranges from the Silurian to the Carboniferous.

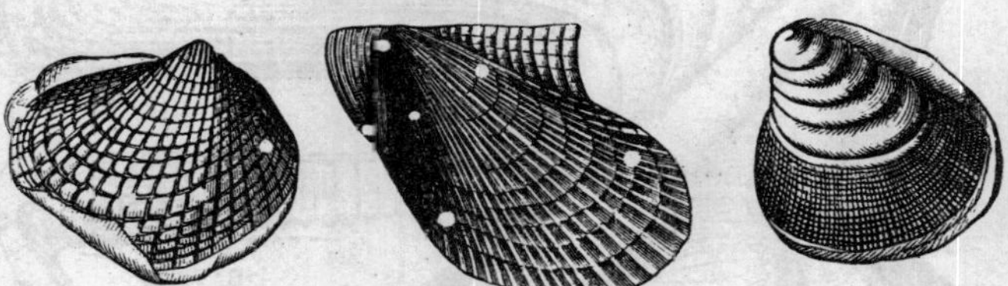

Fig. 338.—A, *Cardiola interrupta*; B, *Pterinea subfalcata*; C, *Cardiola fibrosa*. (After M'Coy and Salter.) Silurian.

Monopteria, separated from the preceding chiefly by its edentulous hinge, is Carboniferous, as is the thin-shelled *Pteronites* of the same formation; while the Lower Silurian *Megaptera* wants the posterior oblique teeth of *Pterinea*, and in other respects approaches *Ambonychia*. The Jurassic *Pteroperna*, again, is a link between *Pterinea* and the group represented by *Perna*. Lastly, the Carboniferous and Permian genus *Myalina* may be regarded as allied to *Pterinea* by the characters of its hinge, while in its general form and nearly terminal beaks it resembles the Mussels.

A third group, hardly separable from the first, is the small one represented by *Vulsella* and *Malleus*, together with some

Tertiary and Secondary forms of small importance, characterised by the common feature that the ligament is lodged in a single pit extending from the beak internally. *Vulsella* (fig. 334, C) is distinguished by its nearly equivalve, earless shell; while the "Hammer-oysters" (*Malleus*) have very long ears, and nearly central beaks. No fossil forms of the latter are known with certainty, but several species of the former have been detected in the Tertiary, and others have been described from the Secondary rocks.

A more important group of the *Aviculidæ* is that represented by *Perna* and its allies, all of which have a straight hinge-line, crossed by numerous transverse furrows for the lodgment of the ligament. In *Perna* (or *Melina*) itself (fig. 339) the shell varies in form, but there is generally a long posterior ear; there are numerous close-set cartilage-pits,

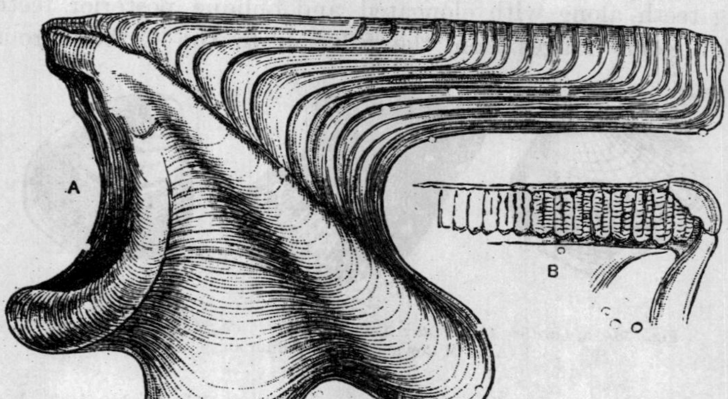

Fig. 339.—(A) *Perna Mulleti*, and a portion of its hinge-line (B). Lower Greensand.

and the right valve has a byssal notch. The genus seems to begin in the early part of the Secondary period, and is especially well represented in the Cretaceous and Tertiary deposits. A large and well-known species is the *Perna Mulleti* of the Lower Greensand. Allied to the preceding is the genus *Gervillia* (fig. 340, A), in which the shell is elongated and very oblique, with a broad and wing-like posterior ear. The hinge-line is furnished with numerous cartilage-pits (fig. 340, B), and the anterior ear is comparatively small. The species of the genus are principally Mesozoic, though a few Tertiary forms are known. *Bakewellia*, of the Permian, resembles *Gervillia* in general form, but the cartilage-pits are few in number, and the hinge is provided with anterior and posterior teeth. The Devonian *Actinodesma* is probably a still older relative of *Gervillia*. In this connection also must be placed the singular Triassic genera *Cassianella* and *Hoërnesia*, the former extending also into the Jurassic. In *Cassianella* (fig. 340, D and E) the

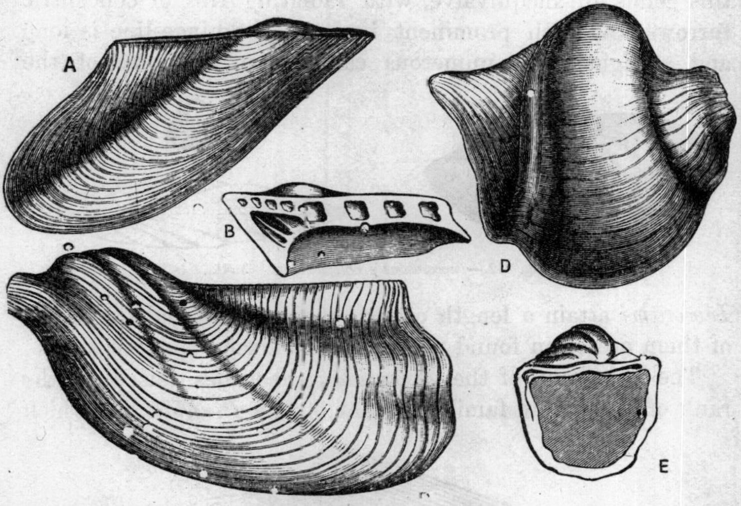

Fig. 340.—Types of *Aviculidæ*. A, *Gervillia Hartmanni*—Lias; B, Part of hinge of the same, enlarged, C, *Hoërnesia Joannis-Austriæ*, slightly enlarged—Trias; D and E, *Cassianella gryphæata*, of different ages—Trias. (After Münster and Laube.)

shell is thick and very inequivalve, the right valve being flat or concave, while the left is strongly convex. The hinge-line is straight, with a few small teeth, and there are short ears. In *Hoërnesia*, again (fig. 340, C), the left valve is similarly inflated, with a similarly incurved beak, but cartilage-pits are present, a single strong tooth exists in both valves, and the hinge-line is crenulated.

Nearly related to both *Gervillia* and *Perna* is the important and widely-distributed genus *Inoceramus* (figs. 341, 342), which is entirely confined to the Secondary period, and is mainly characteristic of the Cretaceous series. The shells of

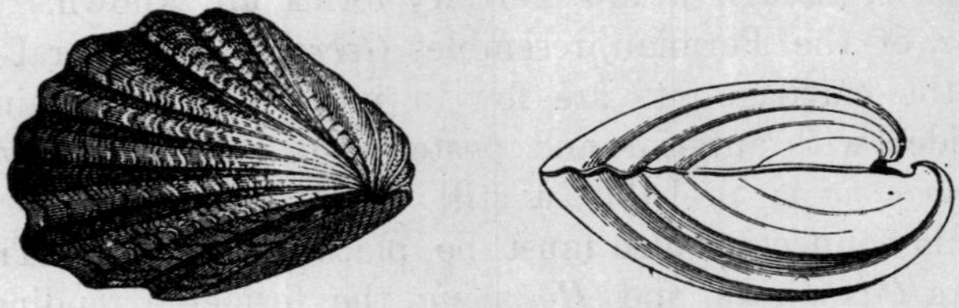

Fig. 341.—*Inoceramus sulcatus.* Gault (Cretaceous).

this genus are inequivalve, with radiating ribs or concentric furrows, and with prominent beaks. The hinge-line is long and straight, with numerous cartilage-pits. Some of the

Fig. 342.—*Inoceramus problematicus.* Chalk.

Inocerami attain a length of two or three feet, and fragments of them are often found perforated by boring sponges.

The last group of the *Aviculidæ*, sometimes raised to the rank of a separate family, is that of the *Pinnidæ*, by which

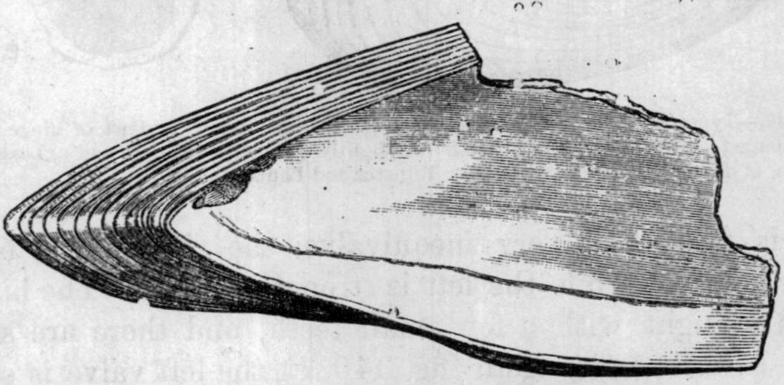

Fig. 343.—*Pinna* sp., interior of a broken right valve, showing the hinge. Carboniferous Limestone. (Original.)

we are naturally conducted to the true Mussels (*Mytilidæ*). In *Pinna* (fig. 343), the type-genus of this group, the shell is equivalve and wedge-shaped; the beaks are placed quite at the anterior end of the shell, and the posterior end is truncated and gaping. The hinge is toothless, and there is an elongated ligamental groove. The genus commences in the Devonian, is well represented in the Carboniferous as well as in later deposits, and exists under various specific forms at the present day. *Trichites*, of the Jurassic and Cretaceous, resembles the preceding, but has an inequivalve shell and twisted beaks; while the *Aviculopinna* of the Permian has the beaks not altogether terminal.

FAM. 3. MYTILIDÆ.—Shell equivalve; umbones *anterior;* hinge, typically, edentulous; anterior muscular impression small, posterior large. Shell attached by a byssus. Mantle-lobes united between the siphonal apertures. Foot cylindrical, grooved, and byssiferous.

In the genus *Mytilus* are the true Mussels, in which the shell is wedge-shaped, and the beaks terminal. Numerous fossil forms are known, commencing in the Permian. The *Modiolæ*, or "Horse-mussels," have the beaks anterior, blunt, not pointed, the hinge edentulous, and the shell oblong. More than one hundred fossil species have been described, commencing in the Lias, and extending to the present day. The Palæozoic *Modiolæ* are probably referable to different genera. The Date-shells (*Lithodomus*) form a sub-genus of *Modiola*, and are distinguished by their long, cylindrical, anteriorly-inflated shell, and by their habit of forming perforations in rocks, in which they live. They appear to date from the Carboniferous rocks, and are known to palæontologists by both their shells and their burrows. *Crenella*, of the Cretaceous, Tertiary, and Recent, is another ally of *Modiola*.

The genus *Dreissena* (including *Congeria*) comprises Mussel-shaped Bivalves, with terminal beaks, and a small byssal notch in the right valve, but differing from *Mytilus* in having keeled valves, and in the fact that the internal lining of the shell is not nacreous. Fossil species of this genus appear in the Tertiary rocks, in certain parts of which they are present in great abundance. The Carboniferous genus *Anthracoptera* may be regarded as probably related to *Dreissena*.

Of the Palæozoic Bivalves which have been more or less properly referred to the *Mytilidæ*, the following deserve mention. Firstly, we have the important Silurian genus *Modiolopsis* (fig. 344), in which the shell is equivalve, very inequilateral, the beaks anterior, and the surface smooth, or marked by fine concentric lines of growth. The shell is thin, and its posterior end is considerably broader than the

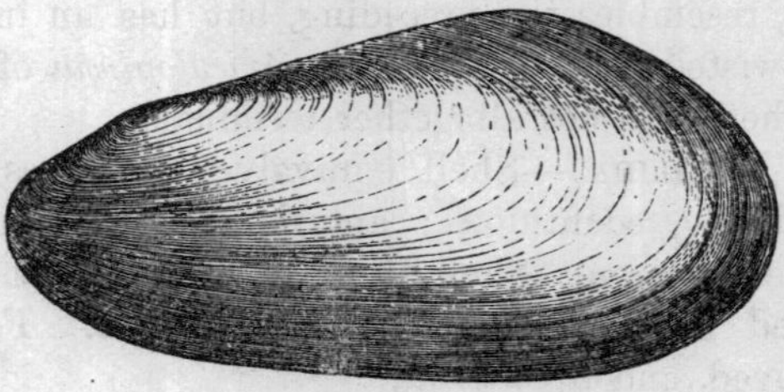

Fig. 344.—*Modiolopsis modiolaris.* Lower Silurian.

anterior. The hinge is edentulous, and there is a ligamental groove, which begins in front of the beak, and extends to the posterior extremity.

The genus *Orthonota* likewise comprises a number of Silurian Bivalves, and is also in a somewhat doubtful position. The shell (fig. 345) is elongated, equivalve, very inequilateral, having the beaks placed close to its anterior end. The shell is thin, and its margins are parallel. The hinge seems to be edentulous, and the hinge-line is typically long and straight. In one group, however, of the shells placed here (*Orthodesma*) the hinge-line is bent or contracted in front of the beaks, and straight behind (fig. 345).

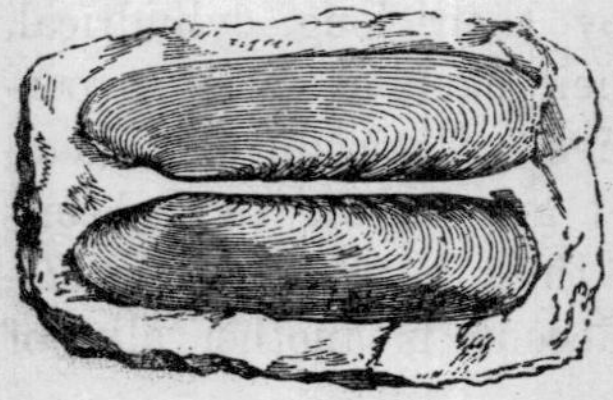

Fig. 345.—*Orthonota (Orthodesma) parallela.* Lower Silurian.

We may also, provisionally, place here the genera *Myoconcha* and *Hippopodium*, in accordance with the views of Stoliczka; though both these genera have a toothed hinge, and in other respects differ from the *Mytilidæ* and approach the genus *Cardita.* In *Myoconcha* the shell is mussel-shaped and thick, with nearly terminal beaks and an external ligament, and having an oblique tooth in the right valve. The Jurassic *Hippopodium* (fig. 346)—typified by the well-known

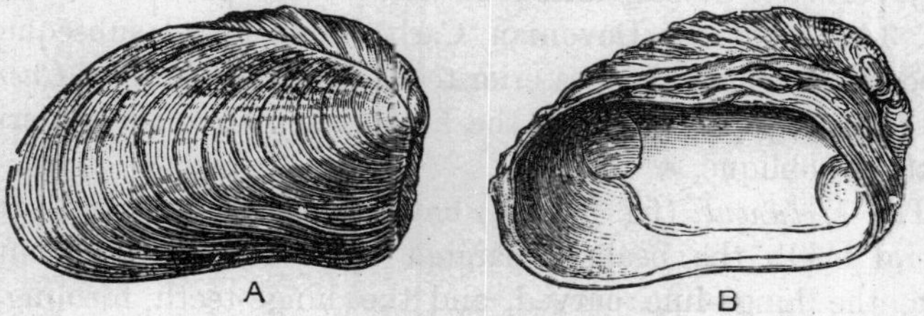

Fig. 346.—*Hippopodium ponderosum.* Lias.

H. ponderosum of the Lias—closely resembles the preceding, but the shell is more massive and ventricose, and there is a single thick oblique tooth in each valve.

FAM. 4. ARCADÆ. — Shell equivalve; hinge long, with many comb-like teeth; muscular impressions nearly equal; mantle-lobes separated; foot large, bent, and deeply grooved. This large and important family has existed under many and varied types from the Silurian upwards, and we may briefly consider some of the leading genera under the three groups of the *Arcadæ* proper, the *Nuculidæ*, and the *Nuculanidæ*, sometimes regarded as so many distinct families.

The type of the first of these groups is constituted by the Arks (*Arca*), which have a straight hinge-line, with remote beaks, separated from one another by an oval or lozenge-shaped ligamental area (fig. 347). The teeth are numerous and transverse, and the surface is generally strongly ribbed. Species of *Arca* have been described from the Lower Silurian rocks upwards. It is probable, however, that the older Palæozoic forms referred here really belong to other genera, especially to *Tellinomya* and *Cyrtodonta*.

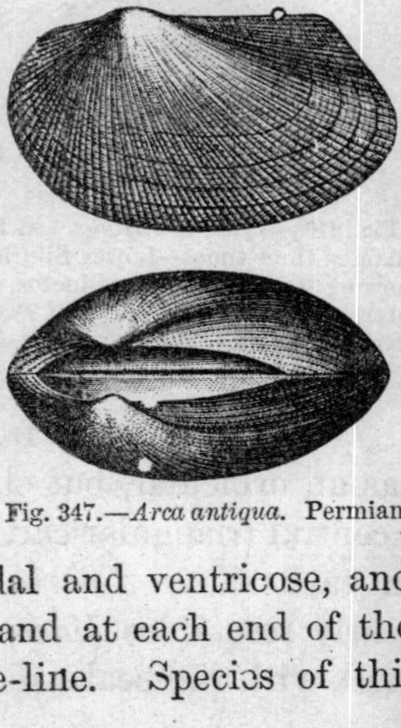

Fig. 347.—*Arca antiqua.* Permian.

In *Cucullæa* the shell is trapezoidal and ventricose, and the hinge-teeth are few and oblique, and at each end of the hinge become parallel with the hinge-line. Species of this genus have been described from the Lower Silurian upwards; but undoubted forms do not appear till we reach the Secondary period. Closely allied to this is the genus *Macrodon* (fig. 348, E) of the Devonian, Carboniferous, and subsequent deposits, in which the posterior teeth resemble those of *Cucullœa* in being parallel with the hinge-line, while the anterior teeth are oblique.

The *Pectunculi* (fig. 348, C) have a nearly round and equilateral shell, the beaks separated by a striated ligamental area, the hinge-line curved, and the hinge-teeth forming a semicircular row. *Pectunculus* is a comparatively modern genus, and does not seem to have come into existence before the Cretaceous period. Numerous species are known in the Tertiary rocks.

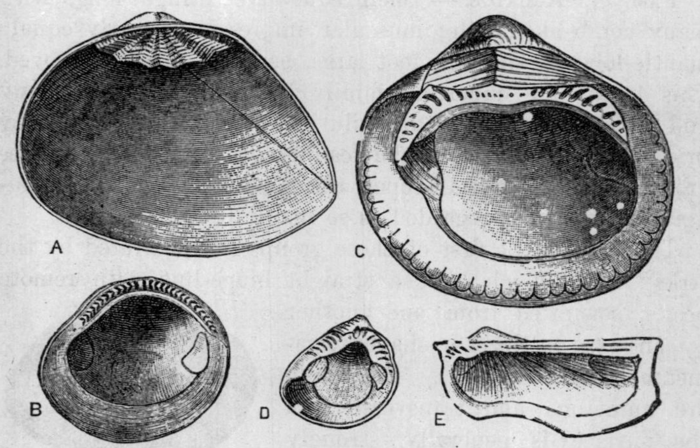

Fig. 348.—Types of *Arcadæ*. A, Interior of *Lyrodesma Cincinnatiensis*, showing the hinge, enlarged three times—Lower Silurian (after Hall); B, Interior of *Tellinomya pectunculoides*, showing the hinge and adductor scars, enlarged twice — Lower Silurian (after Hall); C, Interior of the right valve of *Pectunculus subpilosus* — Tertiary; D, Interior of valve of *Limopsis aurita*—Pliocene; E, Interior of *Macrodon Hirsonensis*—Jurassic.

Limopsis, ranging from the Jurassic to the Recent period, has an orbicular, but slightly oblique shell (fig. 348, D), with a central triangular cartilage-pit, and a row of transverse teeth on each side. *Isoarca*, of the Secondary rocks, again, more nearly resembles *Macrodon*, having an elongated inequilateral shell, with the beaks near the anterior end. The beaks, however, are inflated and incurved, the hinge-line is curved, and a row of transverse teeth exists on each side. *Mytilarca*, distributed from the Upper Silurian to the Carboniferous, possesses a mytiloid shell, with anterior beaks; a striated ligamental area being present, with lateral and cardinal teeth placed along a curved hinge-line. *Lyrodesma* (*Actinodonta*), of the Lower Silurian, is an extremely well-marked type, with an equivalve oblique shell, truncated behind, and with

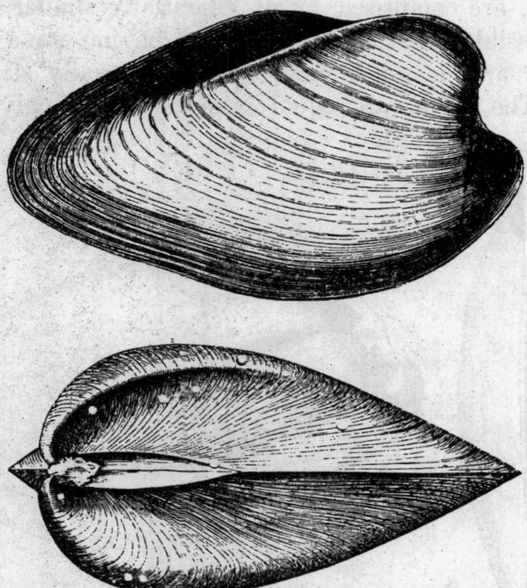

Fig. 349.—*Cyrtodonta Hindi.* (Billings.) Lower Silurian. *a*, Side view; *b*, Dorsal view.

a hinge of several transverse cardinal teeth radiating in a fan-shaped manner from the beak (fig. 348, A).

The shells of the genus *Cyrtodonta*[1] (*Palæarca* of Hall) are

[1] So much controversy has been carried on as to the Palæozoic shells known by the names of *Cyrtodonta*, *Palæarca*, *Megalomus*, *Megambonia*, and *Cypricardites*, that it would be impossible to discuss here which of the groups so named can be finally retained. So far as the author can judge, *Cyrtodonta* of Billings has priority over *Palæarca* of Hall, and the latter must be abandoned. *Megalomus* may be usefully retained as at least a sub-genus. *Megambonia*, of the Upper Silurian and Devonian, has affinities with the *Aviculidæ*, but may be really referable to *Cyrtodonta*. Lastly, *Cypricardites* appears to include a number of diverse types, and probably the best course would be to discard this name altogether.

ventricose and very inequilateral, the umbones being anteric (fig. 349). The hinge-area is undefined, and the surface generally smooth. There are a few (three) anterior cardinal teeth, and "two or three remote oblique posterior teeth parallel to the hinge-margin" (Salter). The species of *Cyrtodonta* appear to be exclusively confined to the Silurian and Devonian rocks. *Vanuxemia* comprises forms of the genus with nearly terminal beaks. In the genus *Megalomus*, again (fig. 350), are comprised forms essentially similar to *Cyrtodonta* (possibly identical with it), but having an excessively thickened and massive shell, which is usually strongly inflated. The valves are equal; the beaks strongly incurved

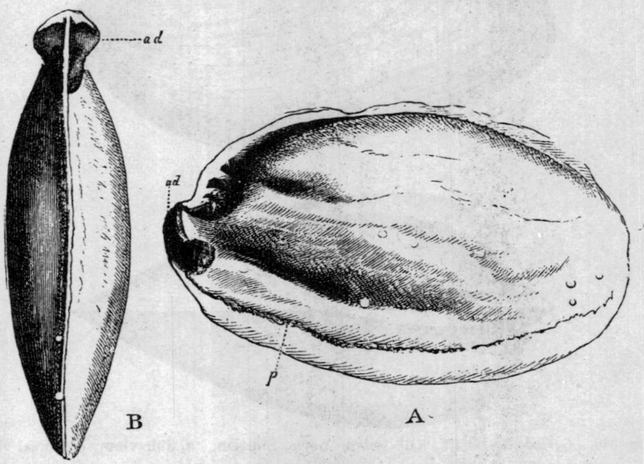

Fig. 350.—*Megalomus compressus*. A, Side view of the cast of the shell, natural size; B, The same viewed from above. *ad*, Cast of the adductor impression; *p*, Pallial line.

and placed anteriorly; the hinge-line furnished with three or four (?) strong transverse teeth; and the anterior muscular impression extremely deep. The forms of this genus appear to be confined to the Upper Silurian, and often attain a great size, but they usually occur in the state of casts only.

Coming next to the group of Arks included under the name of *Nuculidæ*, we find that *Nucula* itself (fig. 351) has a trigonal shell, the beaks of which are reversed, and turned towards the posterior side of the shell, which is also the shortest side. The hinge has numerous teeth on each side of a central internal cartilage-pit. The Palæozoic shells referred to *Nucula* probably belong to other genera. Many species, however, are known from the Secondary and Tertiary rocks. Allied to *Nucula* is the characteristic Silurian and Devonian

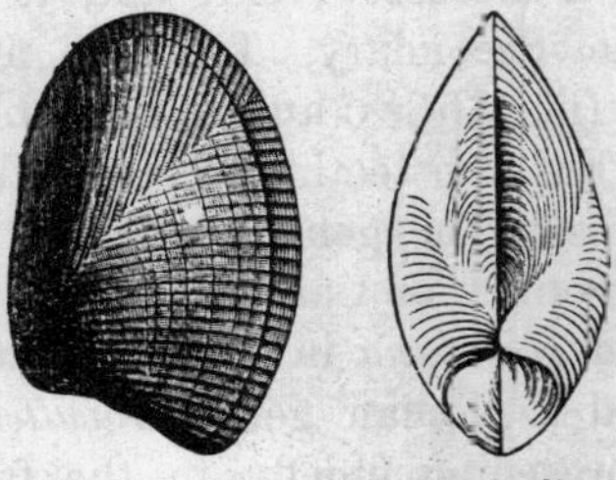

Fig. 351.—*Nucula bivirgata*. Gault.

Fig. 352.—*Tellinomya contracta*. Lower Silurian. *a*, Interior of right valve; *b*, Exterior of the same.

genus *Tellinomya* (= *Ctenodonta* of Salter). In this genus the posterior side is usually, though not always, the shortest, and there is an *external* ligament. The hinge is furnished with numerous small teeth (see fig. 348, B, and fig. 352).

Lastly, we have the group of shells included under the head of *Nuculanidæ*, which should probably form a family

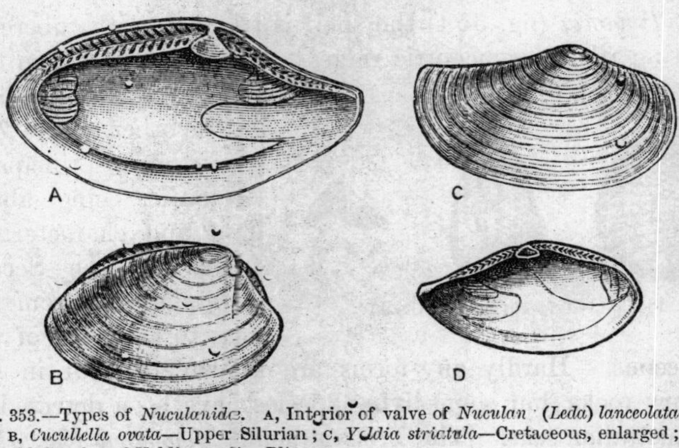

Fig. 353.—Types of *Nuculanidæ*. A, Interior of valve of *Nuculana* (*Leda*) *lanceolata*—Pliocene; B, *Cucullella ovata*—Upper Silurian; C, *Yoldia strictula*—Cretaceous, enlarged; D, Interior of left valve of *Yoldia myalis*—Pliocene.

by themselves, and which are distinguished from the preceding groups by having the shell more or less produced posteriorly, and by the fact that the pallial line is usually more or less sinuated (fig. 353, A). In *Nuculana*, or as it is more familiarly called, *Leda* (fig. 353, A), the shell resembles that of *Nucula*, but is rounded in front and produced behind, while the pallial line is slightly indented. The hinge has numerous small teeth on either side of a small central cartilage-pit. Various species of the genus have been described from the Palæozoic rocks, and it is abundantly represented in the Secondary, Tertiary, and Post-Tertiary deposits. *Yoldia* (fig. 353, C and D) resembles the preceding in most respects, but the teeth are comb-like, and there is a large pallial sinus. The genus seems to be represented as early as the Devonian or Carboniferous, but can hardly be distinguished from *Nuculana* in the fossil condition. Lastly, the Silurian and Devonian genus *Cucullella* (fig. 353, B) differs from the preceding chiefly in the fact that there is a long internal septum extending from below the beak towards the anterior muscular impression, while the pallial line is said to be entire.

FAM. 5. TRIGONIADÆ.—Shell equivalve, trigonal; hinge-teeth few, diverging; umbones directed posteriorly. Mantle open; foot long and bent. The most important genera of this family are *Trigonia*, *Myophoria*, and *Schizodus*.

In *Trigonia* (fig. 354) the shell is trigonal, with tubercles, radiating ribs, or concentric ridges. The hinge-teeth are two in one valve and three in the other. The *Trigoniæ* are principally Mesozoic, being abundant and characteristic fossils of the Secondary deposits from the Lias to the close of the Cretaceous. Hardly any forms are known to occur in the Tertiary rocks, but some living species have been detected in the Australian seas. The Triassic genus *Myophoria* (fig. 355) comprises sub-triangular shells, obliquely keeled, smooth, concentrically striated, or with a partial development of radiating ribs. The left valve has three, and the right two, cardinal teeth. Lastly, *Schizodus*, ranging from the Upper Silurian to the Permian, comprises trigonal shells in many

Fig. 354.—*Trigonia scabra*. Chalk.

respects closely allied to the preceding, but having the posterior side marked by an obscure oblique ridge, and not so markedly angular as in *Myophoria*.

We may also provisionally associate with the above the Palæozoic genera *Curtonotus* of the Devonian, *Anodontopsis* of the Silurian, and *Dolabra* of the Carboniferous, the structural characters of which are still but very imperfectly known.

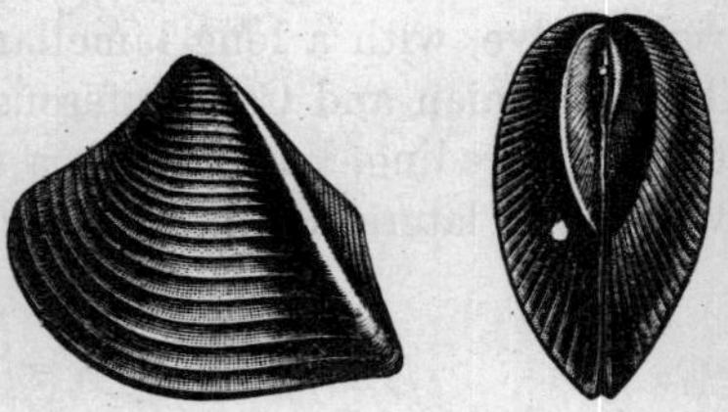

Fig. 355.—*Myophoria lineata*. Trias.

FAM. 6. UNIONIDÆ.—Shell usually equivalve, with a large external ligament. Anterior hinge-teeth thick and striated; posterior laminar or wanting. Mantle-lobes united between the siphonal apertures. Foot very large, compressed, byssiferous in the fry. All the members of the *Unionidæ* are inhabitants of fresh water, and they are therefore not known as fossils except in fluviatile and lacustrine deposits. The principal fossil genera of the family are *Unio*, *Anodon*, *Anthracosia*, and *Carbonicola*.

In the genus *Unio* (fig. 356) the shell is oval or elongated, somewhat resembling that of a mussel (hence the name of River-mussels commonly given to the *Unios*). The species of this genus appear to commence in the Lower Cretaceous rocks, and they are very abundant at the present day. The beaks of fossil specimens are often deeply eroded, as are those of living forms.

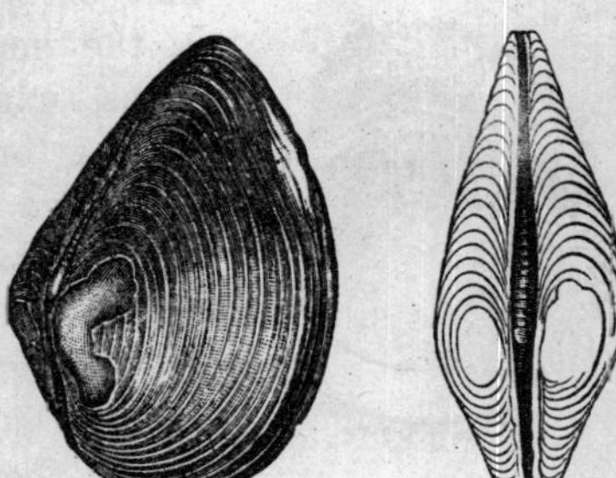

Fig 356.—*Unio Valdensis*. Wealden (Lower Cretaceous).

The genus *Anodon* or *Anodonta* (Swan-mussels) closely resembles *Unio*, but the shell is edentulous. The earliest undoubted fossil forms occur in the Lower Tertiaries

(Eocene). Some forms as old as the Upper Devonian have, however, been referred to this genus.

Carbonicola, of the Devonian and Carboniferous, comprises Unionoid Bivalves, with thick shells, an external ligament, and a concentrically-striated surface. The beaks are not eroded; and the hinge has a thick cardinal tooth in the right valve, with a long lamellar lateral tooth on each side. The Devonian and Carboniferous genus *Anthracosia* is allied to the preceding, but there is a single cardinal tooth in each valve, and lateral teeth are wanting.

SECTION B.—SIPHONIDA.

Subdivision I. Integropallialia.—Siphons short, pallial line simple.

FAM. 7. CHAMIDÆ.—Shell inequivalve, attached. Hinge-teeth 2-1 (two in one valve and one in the other). Impressions of the adductors large. Mantle closed; pedal and siphonal orifices small and nearly equal. Foot very small. The most important fossil forms of this family belong (if we exclude *Caprina* and its allies) to the genera *Chama*, *Diceras*, and *Requienia*.

In the genus *Chama* (fig. 357) the shell is attached usually by the beak of the left valve, but sometimes by that of the right. The upper valve is the smallest, and both bear foliaceous expansions. The free valve carries one tooth which articulates with two teeth in the attached valve. The *Chamas* do not appear as fossils till we reach the Cretaceous rocks, and they have continued to exist up to the present day.

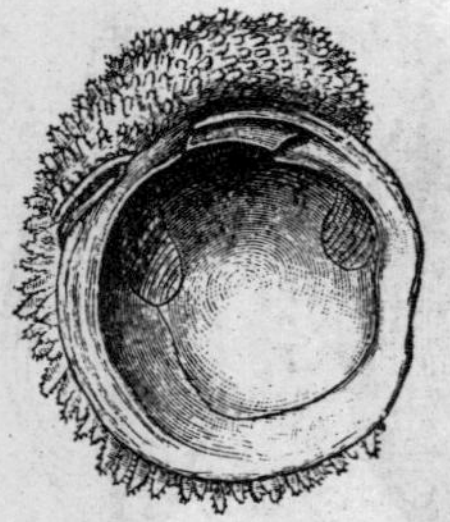

Fig. 357.—Interior of the attached valve of *Chama lamellosa*. Eocene Tertiary.

In the remarkable genus *Diceras* (fig. 358), the shell is "sub-equivalve, attached by either umbo; beaks very prominent, spiral, furrowed externally by ligamental grooves; hinge very thick, teeth 2-1, prominent; muscular impressions bounded by long spiral ridges, sometimes obsolete"

(Woodward). The species of *Diceras* are exclusively confined to the Middle Oolites. In this formation in the Alps they occur in such abundance as to give rise to the name of "Calcaire à Dicerates," applied to beds of the same age as the Coral Rag of Britain.

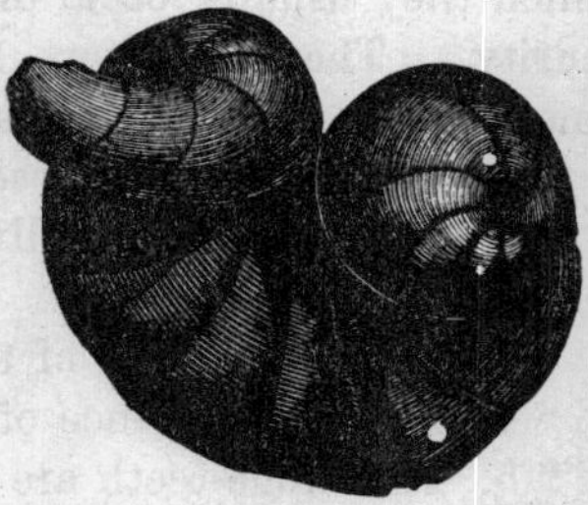

Fig. 358.—*Diceras arietina*. Middle Oolites.

The genus *Requienia* is exclusively confined to the Cretaceous period, and differs from *Diceras* chiefly in having a very inequivalve shell, always attached by its left valve. The attached valve is the largest, and is spiral, whilst the free valve is small and sub-spiral.

FAM. 8. HIPPURITIDÆ (*Rudistes* of Lamarck).—"Shell inequivalve, unsymmetrical, thick, attached by the *right* umbo; umbones frequently camerated; structure and sculpturing of the valves dissimilar; ligament internal; hinge-teeth 1-2; adductor impressions two, large, those of the left valve on prominent apophyses; pallial line simple, sub-marginal."—(Woodward.)

The *Hippuritidæ* are not only entirely extinct, but are exclusively confined to the Cretaceous rocks, whence more than one hundred species have been described. All the members of this family were attached, and lived in beds like oysters. The two valves of the shell are always altogether unlike in sculpturing, appearance, shape, and size; and the cast of the interior of the shell is often extremely different to the form of the shell itself. About a hundred species of the family are known, all of which are Cretaceous, occurring in Britain, Southern Europe, the West Indies, North America, Algeria, and Egypt. Species of this family occur in such numbers in certain compact marbles in the South of Europe, of the age of the Lower Chalk, as to have given origin to the name of "Hippurite Limestones" applied to these strata.

The *Hippuritidæ* have been especially studied by Dr S. P. Woodward, who makes the following remarks upon their structure and affinities: "They are the most problematical of all fossils; there are no recent shells which can be supposed to belong to the same family; and the condition in which they usually occur has involved them in greater obscurity. The characters which determine their position amongst the ordinary Bivalves are the following:—

"1. The shell is composed of two distinct layers.

"2. They are essentially unsymmetrical and right-and-left valved

"3. The sculpturing of the valves is dissimilar.

"4. There is evidence of a large internal ligament.

"5. The hinge-teeth are developed from the free valve.

"6. The muscular impressions are two only.

"7. There is a distinct pallial line.

"The outer layer of shell in the Hippurite and Radiolite consists of prismatic cellular structure; the prisms are perpendicular to the shell-laminæ, and subdivided often minutely. The cells appear to have been empty, like those of *Ostrea.* The inner layer which forms the hinge and lines the umbones, is sub-nacreous, and very rarely preserved. . . . The inner shell-layer is seldom compact, its laminæ are extremely thin, and separated by intervals like the water-chambers of *Spondylus.* . . . The chief peculiarity of the *Hippuritidæ* is the dissimilarity in the structure of the valves, but even this is deprived of much significance by its inconstancy. The free valve of *Hippurites* is perforated by radiating canals, which open round its inner margin, and communicate with the upper surface by numerous pores, as if to supply the interior with filtered water. . . . In the closely allied genus *Radiolites* there is no trace of such canals, nor in *Caprotina.*"

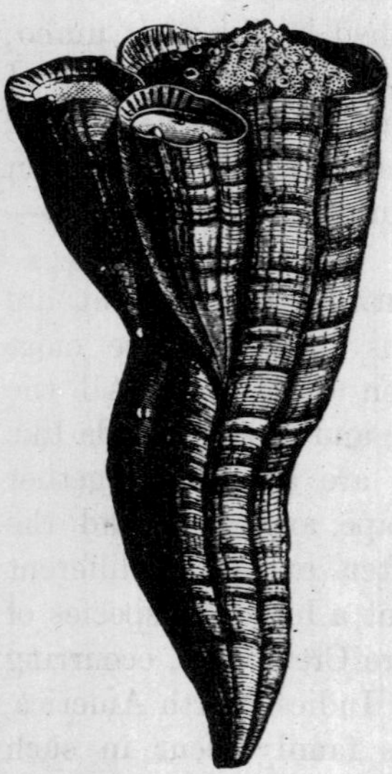

Fig. 359.—*Hippurites Toucasiana.* A large individual, with two smaller ones attached to it.

The shell of *Hippurites* (fig. 359) is inversely conical or cylindrical, and sometimes attains a length of a foot or more. The shell is attached by the larger conical valve, and is closed by a small depressed free valve, with a central umbo. In *Radiolites* the shell is inversely conical, bi-conical, or cylindrical, with dissimilar valves. The upper valve is sometimes flat, sometimes conical, and has a central umbo. In *Caprina* (often placed among the *Chamidæ*) the valves of the shell (fig. 360) are dissimilar, the fixed valve being conical, whilst the free valve is oblique, or is spirally rolled.

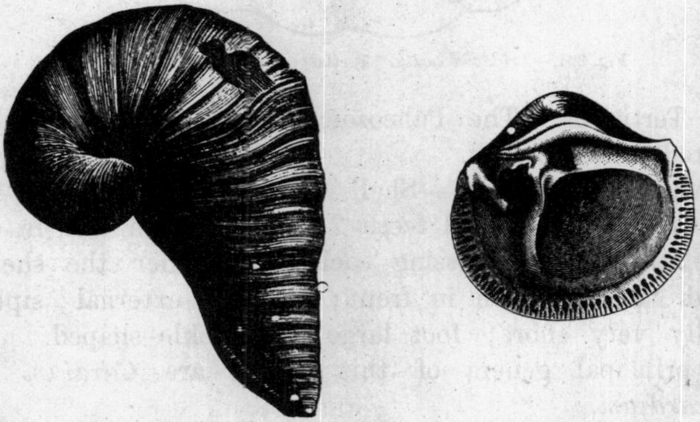

Fig. 360.—*Caprina Aguilloni*. The right-hand figure shows the interior of the left valve.

The free valve is thick, and is "perforated by one or more rows of flattened canals, radiating from the umbo, and opening all round the margin" (Woodward). The cavity of the free valve is sometimes chambered. Other genera allied to *Caprina* are *Caprinella*, *Caprinula*, and *Caprotina*.

FAM. 9. TRIDACNIDÆ.—Shell equivalve; ligament external; muscular impressions blended, sub-central. Animal attached by a byssus, or free. Mantle-lobes extensively united; pedal aperture typically large; siphonal orifices surrounded by a thickened pallial border. Foot finger-like and byssiferous. The shell is truncated in front, the surface ribbed, and the margins toothed. In *Tridacna* (fig. 361) itself the opening for the foot ("byssal sinus") is large, and placed just in front of the beaks; whereas in the closely allied *Hippopus* the shell is closed, and the pedal aperture is only indicated by small serrations of the margin of the valves. *Tridacna* is only known in a fossil condition by a few species from the

Fig. 361.—*Tridacna media.* Tertiary. Interior of right valve.

later Tertiaries. The Palæozoic *Eurydesma* has also been referred to this family.

FAM. 10. CARDIADÆ.—Shell equivalve, heart-shaped, with radiating ribs; cardinal teeth 2; lateral teeth 1-1 in each valve, the former crossing each other when the shell is closed. Mantle open in front; ligament external; siphons usually very short; foot large and sickle-shaped. The two principal genera of this family are *Cardium* and *Conocardium.*

In *Cardium* are comprised the true Cockles, in which the shell is ventricose, the beaks pronounced, and placed nearly in the centre of the dorsal margin, the margins crenated, and the pallial line more or less indented. It is doubtful if any true *Cardium* has been detected in the older Palæozoic rocks. With the Carboniferous, however, the genus begins to be represented, and it has continued up to the present day, attaining its maximum in existing seas. *Papyridea, Lævicardium,* and *Lithocardium* are Secondary to Recent types closely allied to *Cardium* proper. *Protocardium* (fig.

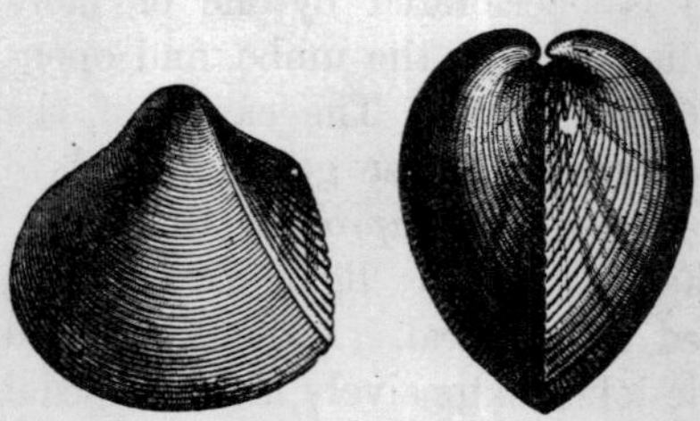

Fig. 362.—*Cardium (Protocardium) Hillanum.* Upper Greensand.

362) has the posterior slope of the shell radiately ribbed, while the rest of the shell is concentrically striated. *Hemicardium* has keeled valves, the shell appearing cordate as viewed from behind or in front. *Lunulicardium* comprises a number of Devonian Cockles, in which there is a deep excavated "lunule" beneath the beaks. *Lymnocardium* and *Adacna* include brackish-water and fresh-water Cockles, in which the cardinal teeth are small or obsolete; and the species of the former are common in some of the fluviatile and estuarine deposits of the Upper Tertiary period.

Lastly, we must place here the extraordinary genus *Conocardium* (= *Pleurorhynchus*), which ranges from the Silurian to the Carboniferous. The shell in this type (fig. 363) is

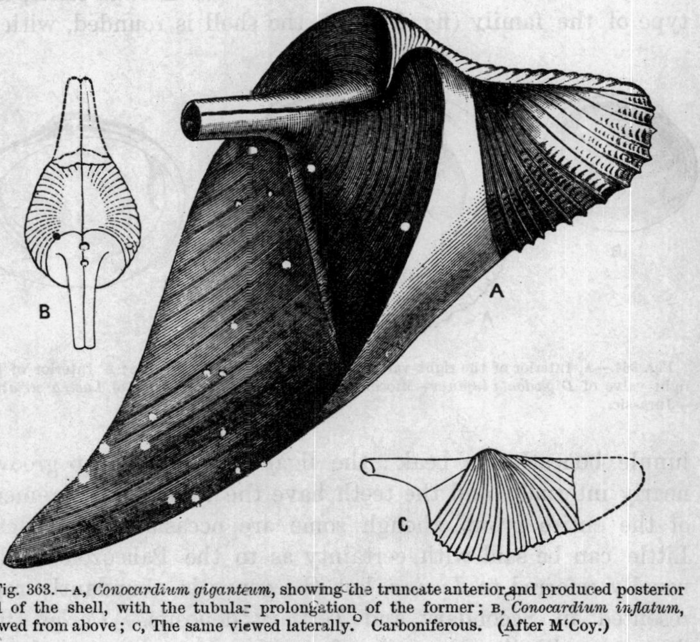

Fig. 363.—A, *Conocardium giganteum*, showing the truncate anterior and produced posterior end of the shell, with the tubular prolongation of the former; B, *Conocardium inflatum*, viewed from above; C, The same viewed laterally. Carboniferous. (After M'Coy.)

keeled, and very oblique, the *anterior* end of the shell being short and abruptly truncated, so as to appear as a cordate flattened area when the shell is viewed from the front. Just below the beaks the shell is produced anteriorly into a long cylindrical tubular projection or beak. *Posteriorly* the shell is elongated and contracted, the valves being widely deficient or gaping at the extremity. The hinge-line is long and straight; and two cardinal teeth, with a hinder lateral one, appear to be present. (It should be mentioned that this description is sometimes reversed, the truncated side being regarded as *posterior.*)

FAM. 11. LUCINIDÆ.—Shell orbicular, free; cardinal teeth 1 or 2; lateral teeth 1-1, or obsolete. Mantle-lobes open below, with one or two siphonal orifices behind; foot elongated, cylindrical, or strap-shaped. Taken as a whole, the family is principally Secondary, Tertiary, and Recent, its Palæozoic representatives being mostly imperfectly understood, and referred here with doubt. In *Lucina* itself, the type of the family (fig. 364, C), the shell is rounded, with a

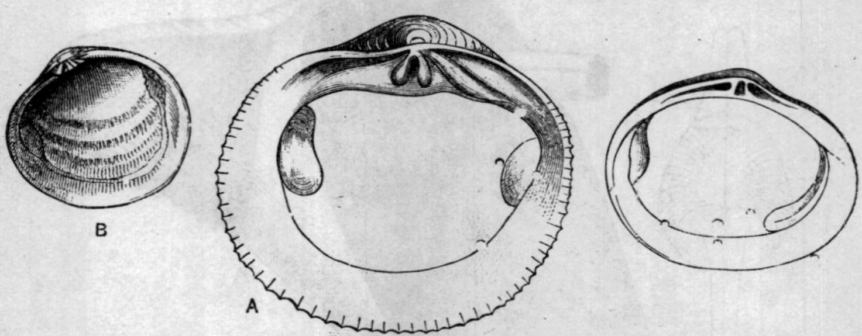

Fig. 364.—A, Interior of the right valve of *Corbis pectunculus*—Eocene; B, Interior of the right valve of *Diplodonta lupinus*—Miocene; C, Interior of the left valve of *Lucina striatula*—Jurassic.

lunule beneath the beak; the ligament is in a deep groove, nearly internal; and the teeth have the typical arrangement of the entire group, though some are occasionally obsolete. Little can be said with certainty as to the Palæozoic shells usually referred to *Lucina*, but the genus is abundantly represented in Secondary and Tertiary deposits. *Corbis* (fig. 364, A), with many species from the Jurassic onwards, is very like *Lucina*, but has the surface concentrically furrowed, with denticulate edges. *Diplodonta* (Cretaceous to Recent) has two cardinal teeth in each valve, the anterior in the right and the posterior in the left being bifid. *Kellia*, with numerous Tertiary species, has two cardinal teeth and an internal ligament. Lastly, the genera *Montacuta*, *Lepton*, and *Galeomma*, are all represented by fossil forms in the Pliocene Tertiary. It is probable that the genus *Axinus* (not the same as *Schizodus*), with various Tertiary species, should be referred here; and we may perhaps provisionally include in this family the singular genus *Solemya*, though its shell is elongated and cylindrical, gaping at both ends, and its hinge is edentulous. This genus has been variously placed, and is sometimes regarded as the type of a distinct family (Stoliczka). It begins as early as the Devonian, has several Carboniferous forms, and is represented by a few species at the present day.

FAM. 12. CYCLADIDÆ.—Shell sub-orbicular, closed; hinge with cardinal and lateral teeth; ligament external. Mantle open in front; a single siphon, or two more or less united. Foot large, tongue-shaped. The genera *Cyclas* and *Cyrena* are the two most important members of this family, and both are inhabitants of fresh water; though the latter not uncommonly frequents brackish water, and one species of the former has been described as marine.

In the *Cyclades* the shell is thin, and there are two hinge-teeth in one valve and one in the other. In *Cyclas* itself the shell is nearly equilateral, but in the sub-genus *Pisidium* it is inequilateral, with the anterior side the longest. In *Cyrena* (fig. 365) the shell is thick, and there are three hinge-teeth in each valve. Both *Cyclas* and *Cyrena* seem to have come into existence at the commencement of the Cretaceous period (Wealden), and they are abundantly distributed through the Tertiary rocks.

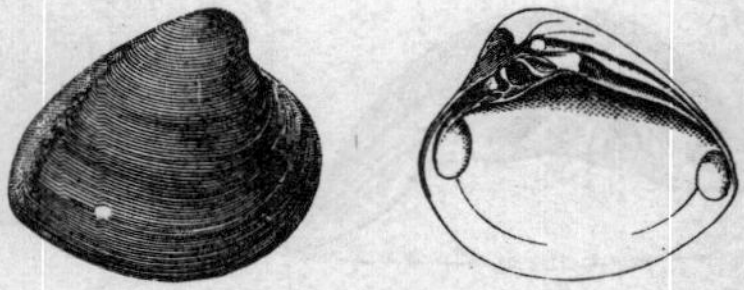

Fig. 365.—*Cyrena antiqua*. Eocene.

FAM. 13. CYPRINIDÆ.—Shell equivalve, closed; ligament external; cardinal teeth 1-3 in each valve, and usually a posterior tooth. Mantle-lobes united behind by a curtain pierced with two siphonal orifices. Foot thick and tongue-shaped. Taken as a whole, the *Cyprinidæ* have passed their acme, and have begun to decline in numbers and importance. The family is one of great palæontological interest, but only the more important forms belonging to it can be here considered.

The *Cyprinidæ* fall naturally into two great groups, the one represented by *Cyprina* and the other by *Astarte*, the latter being sometimes raised to the rank of a distinct family (*Astartidæ.*) In *Cyprina* (fig. 366, A and B) the shell is large, strong, and rounded, with a thick epidermis, a strong ligament, and two cardinal teeth in each valve, flanked by a single lateral tooth. Numerous fossil species are known, commencing in the Trias. *Cypricardia* (or *Trapezium*) has a trapezoidal shell, usually radiately striated, the hinge with three radiating cardinal teeth. The true *Cypricardiæ* are probably Secondary and Tertiary, several recent species being also known. We may, however, place in its vicinity the *Cypricardella* of the Carboniferous, the *Cypricardinia* of the Silurian, Devonian, and Carboniferous, and the *Goniophora* of the two first

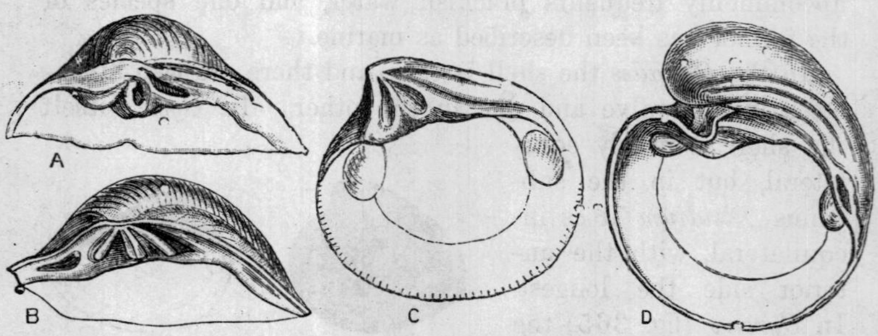

Fig. 366.—Types of *Cyprinidæ*. A, Hinge of *Cyprina tumida*—Pliocene; B, Hinge of *Cyprina Saussuri*—Cretaceous; C, Interior of *Astarte detrita*—Jurassic; D, Interior of *Isocardia crassa*—Pliocene. (After Pictet.)

of these formations. Lastly, *Isocardia* (fig. 366, D), including the recent "Heart-cockles," has a heart-shaped and inflated shell, and the beaks remote and sub-spiral. The *Isocardiæ* do not appear to have existed in the Palæozoic period, but commence in the Trias, are tolerably abundant in the Oolites and Cretaceous rocks, decline in numbers in the Tertiaries, and are represented by a few forms in existing seas.

In *Astarte* (fig. 366, c), the type of the second great group of the *Cyprinidæ*, the shell is thick and usually concentrically furrowed, with a well-marked "lunule," and two cardinal teeth in each valve, the front tooth of the right valve being large and thick. A very great number of fossil species are known, but the genus appears to be wholly Secondary, Tertiary, and Recent. In the neighbourhood of *Astarte*, however, we may place the Carboniferous genus *Astartella*.

Crassatella (fig. 367) comprises thick, solid, ventricose shells, attenuated posteriorly, and generally having a con-

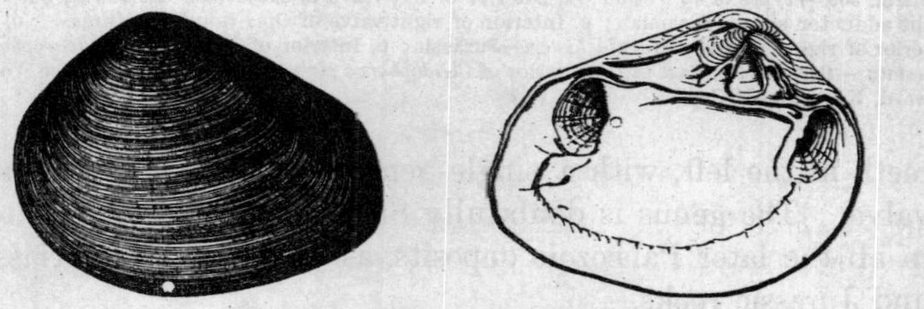

Fig. 367.—*Crassatella ponderosa.* Eocene Tertiary.

centrically-furrowed surface. Unlike the preceding genus, *Crassatella* has the ligament internal. There is a distinct "lunule," and the cardinal teeth are two in one valve and one in the other. The genus commences in the Cretaceous rocks, is abundant in the Tertiaries, and is well represented at the present day.

Pachyrisma is an extinct genus, in which the shell is also very thick and ponderous in its structure. It has large sub-spiral umbones, and is peculiar to the Great Oolite.

Megalodon (fig. 368, A) is likewise extinct, and includes massive shells, with sub-spiral beaks and an external ligament. The genus is doubtfully represented in the Silurian and Carboniferous rocks, and is characteristically Devonian.

The Secondary genus *Opis* (fig. 368, B) has a heart-shaped and keeled shell, with prominent incurved beaks, a distinct lunule, and a single cardinal tooth in each valve. Still another extinct genus is *Cardinia* (fig. 368, c), in which the shell is trigonal or ovate, posteriorly attenuated, not ventricose; an external ligament is present; there is a single cardinal tooth in the right valve, and two small

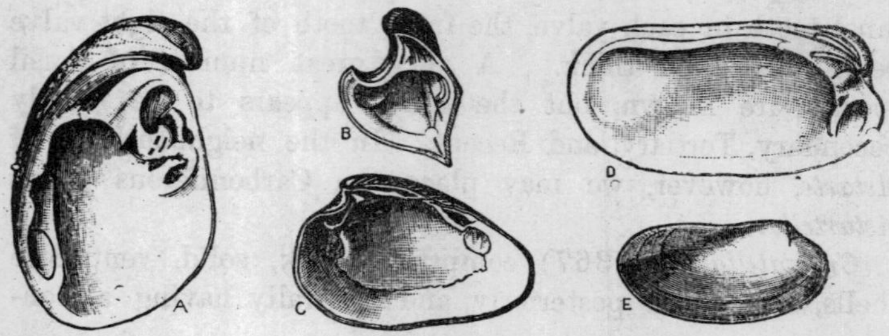

Fig. 368.—Types of *Cyprinidæ*. A, Interior of left valve of *Megalodon cucullatus*, showing the adductor scars—Devonian; B, Interior of right valve of *Opis lunulata*—Jurassic; C, Interior of right valve of *Cardinia Listeri*—Jurassic; D, Interior of left valve of *Pleurophorus costatus*—Permian; E, Cast of the interior of *Cleidophorus planulatus*—Silurian. (After Woodward, King, and M'Coy.)

teeth in the left, with a single remote lateral tooth in each valve. The genus is doubtfully Silurian, but is represented in all the later Palæozoic deposits, as well as in the Triassic and Jurassic rocks.

Another small group of *Cyprinidæ*, allied to the preceding, is represented by *Cardita* (fig. 369), including cockle-shaped shells, which have radiating ribs, an external ligament, and a toothed margin. The genus commences in the Trias, but

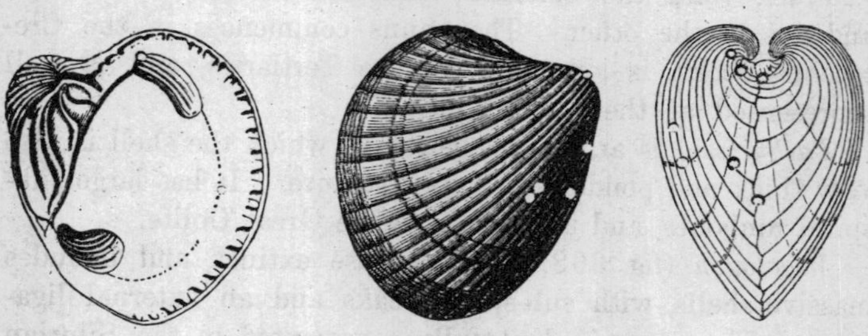

Fig. 369.—*Cardita planicosta.* Eocene Tertiary.

attains its maximum in the Tertiary period, about a hundred species having been enumerated from rocks of this age.

We may, finally, just mention here two or three Palæozoic genera, which would seem to be referable to this family. The first of these is the Lower Silurian *Matheria*, which has an external ligament, and has the beaks placed anteriorly. A second is the Permian *Pleurophorus* (fig. 368, D), in which the shell is oblong, with anterior beaks, two cardinal teeth in each valve, and a single elongated lateral tooth placed posteriorly, the anterior adductor impression being very deep. The third is the characteristically Silurian genus *Cleidophorus*, which has usually been regarded as identical with the preceding, though apparently upon insufficient grounds. The most peculiar feature in this genus is a vertical internal ridge, which commences in front of the beaks, and is continued downwards behind the anterior adductor, and which leaves a deep slit in the cast of the shell (fig. 368, E).

Subdivision II. Sinupallialia.—Respiratory siphons large; pallial line indented.

FAM. 14. VENERIDÆ.—Shell regular, sub-orbicular or oblong; ligament external; hinge with usually three diverging teeth in each valve. Animal usually free and locomotive;

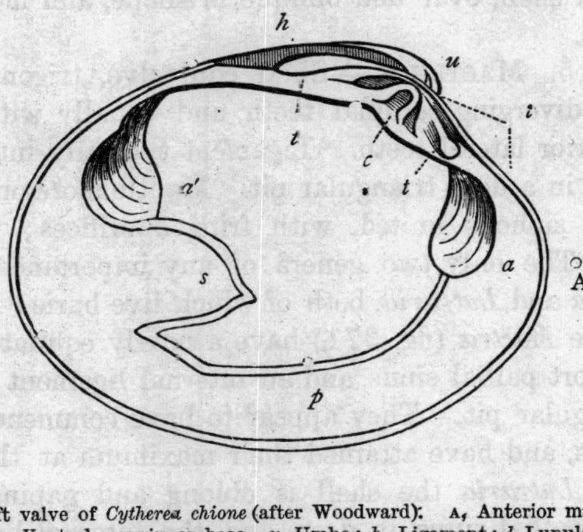

Fig. 370.—Left valve of *Cytherea chione* (after Woodward). A, Anterior margin; B, Posterior margin; C, Ventral margin or base. *u*, Umbo; *h*, Ligament; *l*, Lunule; *c*, Cardinal tooth; *t*, *t*, Lateral teeth; *a*, Anterior adductor; *a'*, Posterior adductor; *p*, Pallial line; *s*, Pallial sinus, caused by the retractor muscles of the siphons.

mantle with a rather large anterior opening; siphons unequal, more or less united. Foot tongue-shaped, compressed, sometimes grooved and byssiferous. The *Veneridæ* are the most highly organised of the Bivalves, and comprise some of the most beautiful examples of the class. They commence in the Oolitic rocks, are abundant in the Tertiaries, and have attained their maximum at the present day.

The *Veneridæ* have been divided into a number of closely allied genera and sub-genera, which, for the most part, are distinguishable only by minute characters. The most important fossil forms belong to the related types *Venus* and *Cytherea*, which begin in the Jurassic, are more abundant in the Cretaceous, and undergo a great development in the Tertiary rocks. *Artemis* (*Dosinia*) has an orbicular, compressed, and concentrically-striated shell, with a deep "lunule," and appears to be first clearly represented in the Cretaceous. *Tapes*, with various Tertiary species, has an oblong shell, with anterior beaks. *Venerupis*, also commencing in the Tertiary, is readily recognised by the fact that the shell is radiately ribbed, and at the same time furnished with concentric ridges, its general shape being oblong. Lastly, *Petricola*, ranging from the Cretaceous to the present day, has a thin shell, oval and oblique in shape, and more or less ventricose.

FAM. 15. MACTRIDÆ.—Shell equivalve, trigonal; hinge with two diverging cardinal teeth, and usually with anterior and posterior lateral teeth. Ligament typically internal, and contained in a deep triangular pit. Mantle more or less open in front; siphons united, with fringed orifices; foot compressed. The only two genera of any importance as fossils are *Mactra* and *Lutraria*, both of which live buried in sand or mud. The *Mactræ* (fig. 371) have a nearly equilateral shell, with a short pallial sinus, and an internal ligament contained in a triangular pit. They appear to have commenced in the Cretaceous, and have attained their maximum at the present day. In *Lutraria* the shell is oblong and gaping at both ends, the pallial sinus is deep and the internal ligament is supported by a prominent cartilage-plate. The genus is not certainly known in rocks earlier than the Cretaceous.

Gnathodon, ranging from the Cretaceous to the present day, has a thick ventricose shell, the beaks of which are often eroded; the lateral teeth are doubled in the right valve, and are transversely striated.

FAM. 16. TELLINIDÆ.—Shell free, usually equivalve and closed; cardinal teeth two at most, laterals 1-1, sometimes

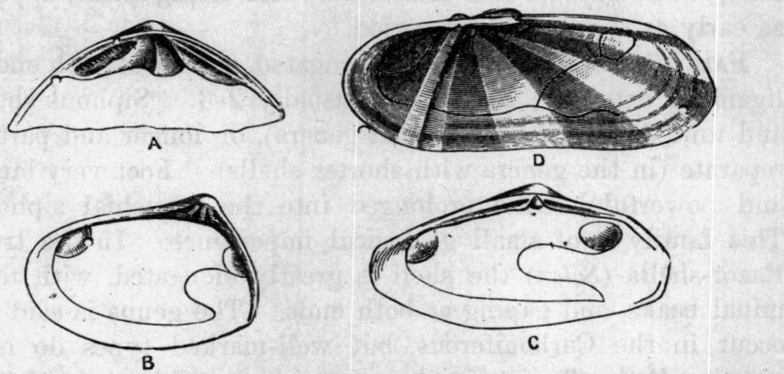

Fig. 371.—A, Hinge of *Mactra solida*—Recent; B, Interior of right valve of *Donax retusa*—Eocene; C, Interior of right valve of *Psammobia rudis*—Eocene; D, Interior of right valve of *Cultellus polita*—Recent.

wanting. Ligament on the shortest side of the shell, sometimes internal. Mantle widely open in front. Siphons long and slender; foot tongue-shaped, compressed. Pallial sinus very large. The chief fossil genera are *Tellina*, *Psammobia*, and *Donax*. In *Tellina* the shell is very slightly inequivalve (fig. 372) with a prominent external ligament. More than a hundred fossil species are known, dating from the Oolitic period; but the genus has attained its maximum at the present day. In *Psammobia* (fig. 371, C) the shell is oblong, compressed, and slightly gaping at both ends; whilst in *Donax* (fig. 371, B) the shell is wedge-shaped, the front rounded and produced, the posterior side short. Both genera commence in the Eocene Tertiary, and are represented by numerous species at the present day.

Fig. 372.—*Tellina proxima*, right valve. Post-Pliocene.

Among the remaining types of the *Tellinidæ*, the genera *Sowerbyia* and *Quenstedtia* are Jurassic, and we may perhaps place here the *Tancredia* of the same formation, which much resembles *Donax* in general form, but is said to possess an entire pallial line. *Sanguinolaria*, *Semele*, and *Scrobicularia* are represented both in the Tertiary rocks and in our own seas; while *Capsula* and *Mesodesma*, with living forms, appear as early as the Secondary period.

FAM. 17. SOLENIDÆ.—Shell elongated, gaping at both ends; ligament external; hinge-teeth usually 2-3. Siphons short and united (in the long-shelled genera), or longer and partly separate (in the genera with shorter shells). Foot very large and powerful. Gills prolonged into the branchial siphon. This family is of small geological importance. In the true Razor-shells (*Solen*) the shell is greatly elongated, with terminal beaks, and gaping at both ends. The genus is said to occur in the Carboniferous, but well-marked types do not appear till the Tertiary rocks are reached. *Solenopsis*, of the Carboniferous rocks, is believed to be allied to *Solen*, from which it differs in its prominent beaks and closed anterior end, but its true affinities are not certain. *Cultellus* (fig. 371, D) has a shorter shell than *Solen*, which gapes at both ends and has the beaks placed a little in front of the centre, and supported by an oblique rib. It commences in the Cretaceous, as does the genus *Solecurtus*, in which the beaks are also sub-central.

FAM. 18. MYACIDÆ.—Shell gaping posteriorly. Mantle almost entirely closed; siphons united, partly or wholly retractile. Foot very small. The more important genera of this family are *Mya*, *Corbula*, *Thetis*, *Panopæa* and *Saxicava*.

Fig. 373.—*Mya truncata*. Post-Pliocene and Recent.

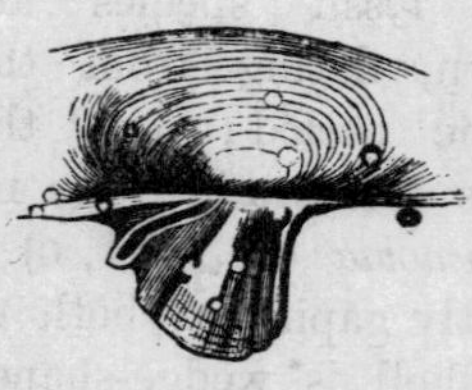

Fig. 374.—Portion of the hinge of *Mya arenaria*, showing the cartilage-process.

In the Gapers (*Mya*) the shell is oblong, inequivalve, and gaping at both ends. The left valve is the smallest, and it carries an internal ligament supported upon a prominent cartilage-process (figs. 373, 374). The Myas live buried vertically in sand or mud. They are not known to have existed before the period of the Middle Tertiary (Miocene), and almost all the fossil species are in existence at the present day.

In *Corbula* (fig. 376) the shell is inequivalve, the left valve the smallest, and with a prominent cartilage-process; but the shell is gibbous, and does not gape at its ends, whilst the pallial sinus is small. Numerous fossil species are known, commencing in the Lower Oolites.

Fig. 375.—*Saxicava rugosa*, left valve. Post-Pliocene and Recent.

Fig. 376.—*Corbula pisum*, viewed from the left and right sides. Eocene.

Neæra, commencing in the Jurassic period, is allied to *Corbula*, but the shell is nearly equivalve, the right valve being slightly the smallest, and the shell is produced and open posteriorly.

The genus *Thetis* is a small one, including thin, translucent, sub-orbicular shells, with an external ligament. A few species of the genus are known, commencing with the Lower Cretaceous rocks. *Poromya*, of the Eocene, appears to be related to *Thetis*.

Panopæa resembles *Mya* in having a thick oblong shell, gaping at each end; but the shell is equivalve, and the ligament is external. Very numerous fossil species of this genus are known, commencing in the Lower Oolites. *Glycimeris*, with Tertiary and Recent species, is nearly allied to *Panopæa*, the only living forms being characteristic of the Arctic seas.

Saxicava, as its name implies, includes shells which form burrows in rocks. The adult shell (fig. 375) is edentulous, equivalve, and oblong, gaping at the ends, and furnished with an external ligament. The genus seems to commence in the Eocene Tertiary, and has continued to the present day.

FAM. 19. ANATINIDÆ.—Shell often inequivalve, with an external ligament. Mantle-lobes more or less united. Siphons long, more or less united. Foot small. The family of the *Anatinidæ* has considerable palæontological importance, having a great development in Secondary times, and appearing to be also represented by numerous types in the Palæozoic formations, though the true systematic position of some of the old forms referred here must be regarded as uncertain.

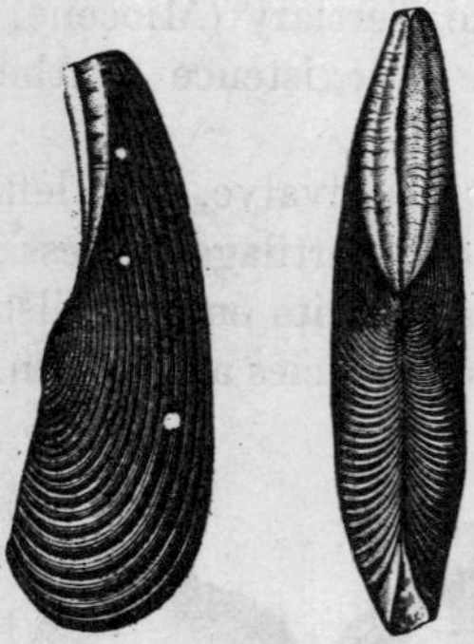

Fig. 377.—*Anatina spatulata*. Kimmeridge Clay (Upper Oolites).

Of the more ancient types of the family, the genus *Myacites* has a gaping ventricose shell, with the umbones directed anteriorly, and the ligament external. The species are known in the Palæozoic period, commencing in the Silurian; and they are represented in the earliest portion of the Secondary period; but they seem to have died out in the Chalk.

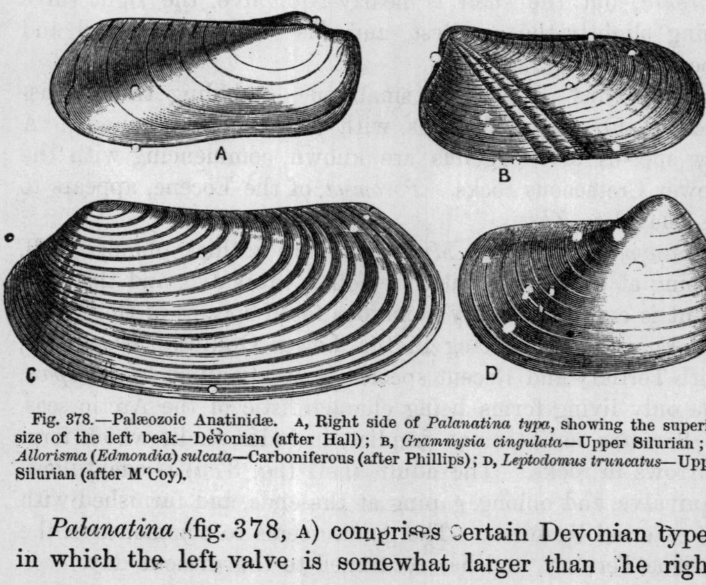

Fig. 378.—Palæozoic Anatinidæ. A, Right side of *Palanatina typa*, showing the superior size of the left beak—Devonian (after Hall); B, *Grammysia cingulata*—Upper Silurian; C, *Allorisma (Edmondia) sulcata*—Carboniferous (after Phillips); D, *Leptodomus truncatus*—Upper Silurian (after M'Coy).

Palanatina (fig. 378, A) comprises certain Devonian types, in which the left valve is somewhat larger than the right.

Grammysia (fig. 378, B), of the Upper Silurian, has an elongated shell, resembling the preceding in general form, but having a single or double fold extending from the beaks to the middle of the ventral margin. *Leptodomus* (fig. 378, D) and *Sanguinolites*, both characteristically Upper Silurian, but the latter extending to the Carboniferous, comprise thin, oblong or elongated, concentrically-striated shells, apparently referable to this family, but for the most part very imperfectly understood. The former is stated to possess a well-defined "lunule" beneath the beaks. *Allorisma* (fig. 378, C) and *Edmondia*, characteristic of the Carboniferous and Permian rocks more particularly, though not unknown in the Upper Silurian and Devonian, likewise comprise thin, equivalve, concentrically-striated shells, with anterior beaks. *Cardiomorpha*, ranging from the Silurian to the Carboniferous, as its name implies, possesses a shell with the general form of the Cockles, but smooth or concentrically-striated, the beaks being prominent, and the hinge edentulous. Lastly we may place here the Carboniferous genus *Anthracomya* (= *Naiadites*), in which there is a thin concentrically-striated oblong shell, with anterior beaks, and an external ligament.

In the Lantern-shells (*Anatina*) the shell is oblong, gaping posteriorly, and having the beaks directed towards the posterior side (fig. 377). The hinge of each valve carries a spoon-shaped cartilage-process. The *Anatinæ* are doubtfully represented in the Devonian, and still more dubiously in the Silurian rocks. They occur, however, abundantly in the Secondary rocks, and are present in smaller numbers in the Tertiaries. Nearly related to the preceding is the genus *Thracia*, which is first certainly known to make its appearance in the Jurassic rocks.

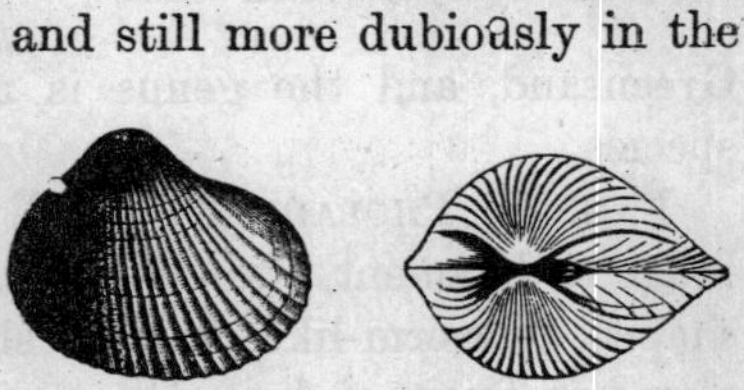

Fig. 379.—*Pholadomya æquivalvis*. Chalk.

The genus *Pholadomya* includes a large number of shells, which are equivalve, oblong, and gaping posteriorly (fig. 379). The shell is thin, ventricose, and adorned with radiating ribs on the sides. The ligament is external, and there is a large pallial sinus. The species of *Pholadomya* are very numerous in the Secondary rocks, where they attain their maximum. They are much reduced in number in the Tertiaries, and are barely represented at the present day.

FAM. 20. GASTROCHÆNIDÆ.—Shell equivalve, gaping, with thin edentulous valves, sometimes cemented to a calcareous tube. Mantle-margins thick in front, united, with a small pedal aperture. Siphons very long, united. Foot finger-shaped. The members of the *Gastrochænidæ* burrow in mud or stone, and the only two fossil genera are *Gastrochæna* and *Clavagella*, the existence of *Aspergillum* in a fossil state being doubtful.

In *Gastrochæna* the shell is wedge-shaped, gaping in front and closed behind. The fossil species commence in the Inferior Oolite, and the genus is represented at the present day. In *Clavagella* (fig. 380), the shell is oblong, one of the valves being free, whilst the other forms part of a more or less elongated calcareous tube, which is often divided by a longitudinal partition and terminates in tubular openings. The fossil *Clavagellæ* commence in the Upper Greensand, and the genus is represented by several living species.

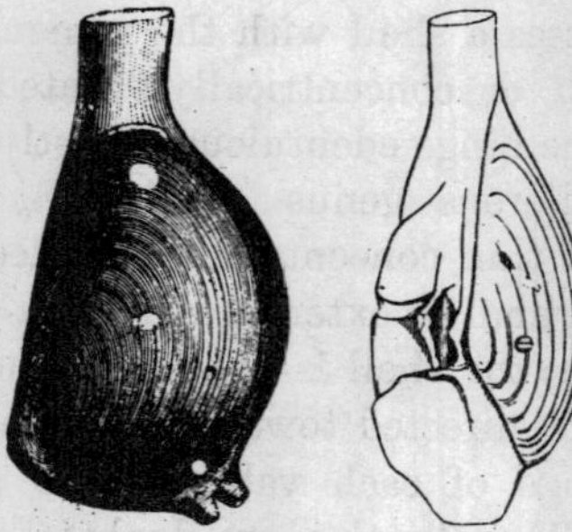

Fig. 380.—*Clavagella cretacea*. Chalk.

FAM. 21. PHOLADIDÆ.—Shell gaping at both ends, without hinge or ligament, often with accessory valves. Animal club-shaped or worm-like, with a short truncated foot. Mantle closed in front; siphons long, united to near their extremities.

In the genus *Pholas* the shell is cylindrical or oval, the valves are edentulous, and there is no ligament or a rudimentary one. The pallial sinus is very deep, and the dorsal margin of the shell is protected by accessory valves. The *Pholades* inhabit burrows which they form for themselves in clay, peat, or rock. Many fossil species of the genus are known, commencing in the Jurassic rocks. The living genus

Xylophaga is represented by the closely allied *Xylophagella* of the Cretaceous.

In the genus *Teredo* the shell is "globular, open in front and behind, lodged at the inner extremity of a burrow partly

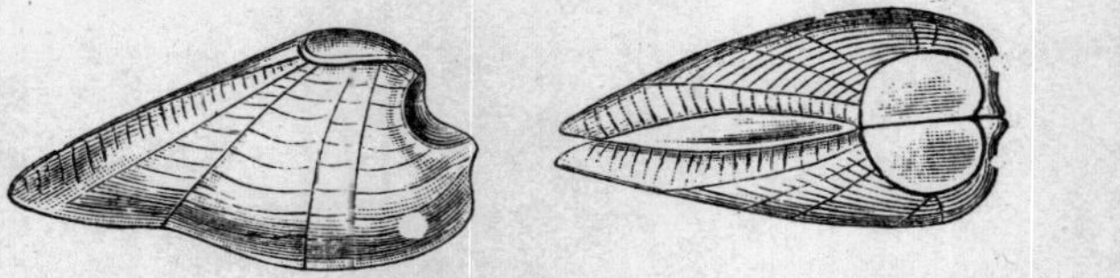

Fig. 2?1.—*Parapholas mersa*, viewed from one side and above. Cretaceous. (After Stoliczka.)

or entirely lined by shell; valves three-lobed, concentrically striated, and with one transverse furrow; hinge-margins reflected in front, marked by the anterior muscular impressions; umbonal cavity with a long curved muscular process" (Woodward). Species of *Teredo* occasionally reach a very large size, and they are known in the fossil state both by their shells and by their burrows in wood. The genus seems to have commenced in the Lias, and is well represented at the present day. The Devonian genus *Cimitaria* is probably referable to this family, in which case it is the oldest representative of the group.

END